Developing the Environment

Developing the Environment:

Problems and Management

C J BARROW

University of Wales Swansea
Swansea UK

Routledge
Taylor & Francis Group
LONDON AND NEW YORK

First published in 1995 by Pearson Education Limited

Published in 2014 by Routledge
2 Park Square, Milton Park, Abingdon, Oxfordshire OX14 4RN
711 Third Avenue, New York, NY 10017, USA

First issued in hardback 2017

Routledge is an imprint of the Taylor & Francis Group, an informa business

British Library Cataloguing in Publication Data
A catalogue for this title is available from the British Library.

ISBN 13: 978-1-138-16672-1 (hbk)
ISBN 13: 978-0-582-08700-2 (pbk)

Library of Congress Cataloging-in-Publication data
Barrow, Christopher J.
Developing the environment: problems and management / C.J. Barrow
p. cm.
Includes bibliographical references and index.
ISBN 0-470-23453-9
1. Economic development—Environmental aspects. 2. Environmental protection. I. Title
HD75.6.B36 1995
333.7—dc20 94-31121
CIP

Set in 9/11pt Times by 22

To Anne and Anna

Contents

Preface

Humanity has exploited the Earth's natural resources and modified the environment for thousands of years, but in the last two centuries human impact has increased hugely, in part because of population growth, in part because of technological changes and partly as a consequence of the way that development has been allowed to proceed. The last few decades have witnessed a growing awareness of, and concern for, environmental issues and since the mid-1980s these have been linked with calls for better approaches to development. In 1972, when the United Nations convened the Stockholm Conference on the Human Environment, environmental problems were widely disregarded and environmental protection was commonly seen either as a luxury or to conflict with development. Twenty years later there had been a sea-change, such that most of the representatives attending the 1992 Rio 'Earth Summit' accepted that environment and development were crucially interdependent and that associated problems deserved urgent attention from rich and poor nations.

Interest in environmental issues has generated a rapidly expanding literature ranging from specialist texts and journals to broader coverage, some too pessimistic, some too optimistic, some objective, but much subjective ill-informed advocacy or 'polarized perception'. There are repeated warnings that environmental problems are getting worse and pose a serious threat to human well-being or even survival, and there is a growing concern for the welfare of other organisms and the environment in general. Humanity is at a 'crossroads', about to face or already facing a crisis or crises. Some of these problems are due to natural causes, but many are wholly or in large part the result of human activity, which now often has supra-national (trans-boundary), in some cases global, impacts. Problems are often caused by more than one group of people or more than one nation and those causing difficulties have so far tended to escape the consequences.

Those dealing with environment and development problems have commonly mistaken symptoms for causes, and have concentrated on technical responses. For many of today's environmental problems, technical solutions are possible but the motivation to act and the funds to do so are lacking. Recently, there have been increasing efforts to understand causes and to address policy, political and practical aspects of development and environment.

Ultimately at the root of many environmental problems are unsound ethics and concepts of development and modernization. There is a need to promote new development ethics and to identify critical thresholds so that action can be taken in time to avoid, mitigate or respond to threats. It is also important to review opportunities to maintain or improve the well-being of people, other organisms and environment.

These are the issues that this book addresses. The aim is to present an overview of environmental problems, some of which we generate, and to review the responses that we could make to counter them. In pursuing this aim it is easy to fall into the trap either of excessive pessimism or of gross optimism; given the desirability that complacency be discouraged, the former is probably the lesser 'sin' and is the one I at times tend to succumb to.

C.J. Barrow
University of Wales Swansea
Swansea UK

15 December 1993

Acknowledgements

We are grateful to the following for permission to reproduce copyright material:

Christian Aid for figure 3.2b; Financial Times for figure 3.6 (Johnson and Taylor 1989); Food and Agriculture Organization of the United Nations for figure 1.1; IIASA for figure 5.2 (*IIASA Annual Report* 1992); Longman Group Ltd. for figures 6.1 (Barrow 1987) and 10.2 (Burke and O'Hare 1984, previously published by Oliver and Boyd); Methuen and Co. for figure 10.1 (Dickenson *et al. A geography of the third world.* 1983); Alessandra Muto and the Department of the Environment for figure 10.13 (Department of Environment's leaflet, *Helping the Earth Begins at Home*); New Scientist for figure 4.3 (*New Scientist* 1992); Swansea Museum for figure 10.4; The Ecologist for figure 9.4 (Lohmann 1991); Times Newspapers Ltd. for figures 3.2a (*Sunday Times* 1991) and 8.2a (*The Times* 1992); United Nations Environment Programme (UNEP) for figures 6.2 and 6.3 (*Ecoforum* 1993).

Whilst every effort has been made to trace the owners of copyright material, in a few cases this has proved impossible and we take this opportunity to offer our apologies to any copyright holders whose rights we may have unwittingly infringed.

CHAPTER 1 Development and environment

Attitudes towards the environment and humankind's ability or willingness to modify it are not fixed. At any point in time different groups, or individuals, may vary in their outlook on environment and development. Employees of a company earning the same salary are unlikely to have the same idea of what constitutes 'quality of life' or scenic beauty nor the same spending priorities. In the early 1980s I visited Tomé-Açu in Amazonian Brazil; this settlement illustrates how differently peoples with roughly similar resources at their disposal fare in managing their environment and development. Japanese settlers founded Tomé-Açu in the 1920s and have sustained agriculture on land in many cases abandoned by earlier Brazilian settlers as 'unfarmable'. It would seem that the Tomé-Açu settlers developed (and continue to evolve) appropriate resource use strategies that 'fit' the environment and the changing marketing conditions. Without supportive social institutions this evolution would probably have failed. Tomé-Açu is one example of how people can overcome environmental and other difficulties, if development is managed in the right way.

Development

Definitions of development reflect the current values of those making the definition. Thus, what was deemed development in the past may not be regarded so today. It has been suggested that development is the economic component of a wider process – modernization (Mabogunje 1980) – and that the former involves transformation from colonial-type dependence whereas the latter need not, indeed may involve increased dependence (Brookfield 1973: 8). Some would see development more as a learning process. A typical western definition of development would be 'an ambiguous term for a multidimensional process involving material, social and organizational change, accelerated economic growth, the reduction of absolute poverty and inequality'. The concept of development affects how we see and study the world. Escobar (1992) went further and argued that development is a mechanism used by richer nations for the management of less developed countries since the mid-1940s. Development has, until recently, been measured by insensitive statistics like gross national product (GNP) – total monetary value of goods and services produced by a nation in a year – or GNP divided by national population (per capita GNP). The UNDP formulated a human development index to try and provide a more sensitive multidimensional measure (UNDP 1991).

Modernization today is generally accepted to be a change toward economic, social and political systems that developed in western Europe and North America, particularly between the seventeenth and nineteenth centuries, and later around the Pacific Rim. Development and modernization at the national, regional or local level are often linked to external economic or social changes. In theory, a group of people could develop in a range of ways, but remain unmodernized or vice versa. Development is increasingly seen as requiring reduction of inter-group disparity (UNDP 1991: 31) or a 'social transformation' – the use of capital, technology and knowledge, to alter culture and society. Countries, regions or groups of people may have above or below 'average' development or modernization, and it has often been argued that richer countries should 'assist' others to develop; however, opponents feel that people must do it for themselves (Adams 1990: 72, 83).

People with different views from those held in Europe, America and the Pacific Rim since the 1930s may be reluctant to embrace western ways of development and modernization and may choose to follow different paths. The west has come to see development as an incremental process but it can cease at a given level and decline is possible.

It seems unlikely that development, as practised so far, is a process through which the world's poor will reach and sustain standards of living achieved in rich countries (M. Caldwell 1977: 98; Berger 1987: 116). Humankind probably has a limited time to set in motion development that will sustain indefinitely as many people as the Earth can support with a satisfactory 'quality of life', and as little environmental damage as possible. En route to that goal it will probably be necessary to support too large a population and to counter excessive environmental demands for perhaps several decades.

From the early nineteenth century 'underdevelopment' has been used, often in a pejorative way, to indicate backwardness or the failure of the development process (Dupâquier and Grebenik 1983: 247). Between the 1870s and 1930s supporters of Social Darwinism and environmental determinism (see chapter 4) saw underdevelopment to be rooted in racial or environmental disadvantage (Bernstein 1973: 13–30; Hettne 1990; Corbridge 1991). More recently, failure to develop has been attributed to a range of factors including: dependency (for trade, aid and technology); colonization; de-colonization; neo-colonialism; side-effects of the USA–USSR/China 'Cold War'; and, less often since the 1940s, to harsh, often tropical, conditions (Barrow 1987: 27; Toye 1987: 1–21).

The pursuit of development

An observation often made is that environmental problems frequently relate to development: this relationship is seldom simple. Through history human needs and wants have changed, and use of the environment usually alters it and limits the opportunities for future development (e.g. if land becomes salinized through irrigation it may be of less value than it was pre-development). Problems can appear in environments where there has been little attempt at development or modernization as well as those where there has. The process of development for one culture or locality may be easier than for others, even under apparently similar circumstances. Efforts by one group to develop may generate problems for others, perhaps coeval but different cultures, or those possibly quite distant in space or time, or for the globe as a whole. It is often claimed that countries or corporations that have developed did so mainly by exploiting others (the 'periphery').

Roughly 77 per cent of the world's total population (over 5400 million in 141 countries in 1992) subsisted on around 15 per cent of the world's income in the early 1990s; these are variously termed the underdeveloped countries, less developed countries, the South, the Third World, or developing countries. 'Third World' was put forward after 1945 as a title for countries not included in the capitalist (First World) and socialist (Second World – consisting of the centrally planned economies of the former USSR, the eastern bloc People's Republic of China [PRC] and Cuba) categories; a Fourth World consisting of very poor people, often women and refugees has also been identified. Following changes in the eastern bloc, the label Third World is of less value. Throughout this book the expression less developed countries (LDCs) is used.

The LDCs are not a homogeneous group: using per capita income and adult literacy rate data, UNCTAD subdivided them into 42 'least developed countries' (in effect the poorest), 86 non-oil-exporting 'developing nations' and 13 relatively rich oil-exporting 'OPEC countries' (Todaro 1989). In stark contrast are the 28 'developed nations' (abbreviated in this book to DCs), also referred to as the North, the First World or the more developed countries (MDCs). The seven major industrial DCs (Canada, France, Germany, Italy, Japan, UK and the USA) are often called the G-7 countries.

Since 1945 a few countries have become or are close to becoming DCs; some are the newly industrializing countries: South Korea, Taiwan, Singapore, Hong Kong and Hungary. The former Yugoslavia, India, Brazil and Romania have also made progress with industrialization and improvement of living standards; and others, like Malaysia (whose Prime Minister recently predicted that his country would be fully developed by AD 2020), have advanced rapidly.

Beginning with Adam Smith's *Wealth of Nations* (1776) (Smith 1975) development has mainly been perceived in terms of, and has been pursued by means of, economic growth, generally with neglect for social and environmental conditions. In the 1950s and 1960s it was widely held that nations passed, with or (less likely) without help, through a series of successive, predictable stages of economic growth one of which was a point of 'take-off' to a state of development (Rostow 1960). Others argue that development is achieved through the transfer of resources from poor to rich, and that this continues to cause increasing disparity between LDCs and DCs (Stretton 1976; M. Caldwell 1977). There has been a belief that economic growth in the DCs was via urban-industrial systems and capital-intensive technology and so LDCs were encouraged along such paths, in some cases unwisely.

Since 1945 development promotion has had its fashions, although economic growth has been an ongoing goal. During the First (UN) Development Decade (the 1960s) it became apparent that when selected recipients were aided, the benefits of growth might not appear or did not 'trickle down' to the poor as was hoped. So the aim in the Second Development Decade (the 1970s) was to widen the range of beneficiaries of economic growth: variations on the slogan 'redistribution with growth' were often used. By the 1980s underdevelopment was seen to stem from international or internal LDC institutional causes or power relations, particularly the terms of world trade. The Third Development Decade (the 1980s) therefore aimed to establish a New International Economic Order (an idea presented to the UN in 1974), to improve terms of world trade, and for better provision of basic needs (healthcare, education, clean water, adequate food, etc) for the poor. From 1980 funding agencies promoted 'structural adjustment' – changes designed to ensure LDCs had the 'correct' economic adjustment. In the Fourth Development Decade (the 1990s), economic growth is still seen to 'drive' development, and efforts are under way to change world trade, but there is also concern for democratic and participatory approaches to promote 'human development' ('adjustment with a human face'); for sustainable development and; environmental quality (UNDP 1991; World Bank 1992). In the 1990s the 'space age' seems to be giving way to an 'environment age'.

Environment

Environment can be defined as the sum total of the conditions within which organisms live; it is the result of interaction between non-living (abiotic) physical and chemical and where present living (biotic) parameters. Interest in the struggle of organisms, including people, with one another and their surroundings ('the survival of the fittest') was stimulated by publication of *The Origin of Species by Means of Natural Selection* (1859) by Charles Darwin. Ecologists, who study relationships between organisms and between organisms and environment, sometimes use 'natural environment' to indicate a situation where there has been little human interference and 'modified environment' where there has been significant modification or 'development' by people.

Changing attitudes towards environment and development

Many organisms alter the environment: the change may be slow or rapid, localized or global. Humans have the potential to recognize and to consciously respond to opportunities and to threats – natural or anthropogenic – perhaps to avoid or mitigate them. However, it is not clear whether humanity will successfully exploit that potential, or continue to destroy other organisms and perhaps the human species. Technology could offer humanity and other organisms escape from catastrophes which have in the past afflicted the Earth. Gould (1989: 323) noted: 'We are the offspring of history, and must establish our own paths in this most diverse and interesting of conceivable universes'. Recognition of problems and reaction to them depend on what individuals and communities think of themselves and how they relate to their specific environment.

At the roots of many of the world's environmental problems lie unsound concepts of development and modernization. Some, like Riddell (1981: 1), argue that DCs would be better described as 'overdeveloped', effectively in an unsustainable state of temporary development: consuming more than their fair share of resources and contributing more than their fair share of pollution. In the 1990s we have reached a crossroads where there is realization that a lot of what is done is maldevelopment (Amin 1990). Much of the damage results from 'Faustian bargains' – development activities which, like Marlowe's (1564–1593) Dr Faustus, sacrifice long-term well-being (in Faustus' case his soul) for short-term gains. Soussan (1992) felt that an understanding of human–environment interactions could be gained through examination of how the social relations of power relate to the control

Fig. *1.1* Roman olive press in a region of North Africa now devoid of treecover (*Source: Unasylva* (1985) 37 (150): 67 – original FAO).

and use of environment – the political economy approach.

The following passages review changing attitudes to environment and development to explore how present environment and development ethics evolved.

Prehistoric to medieval attitudes

That primitive people 'rooted in nature' cause little environmental damage is largely an 'Arcadian myth' (M. Smith, 1993). Prehistoric people, few in numbers, with access only to stone, bone, leather, wood and fire, still had considerable impact. A number of researchers recognize a Pleistocene 'overkill': the loss of large mammals from Africa, Eurasia, the Americas and Australia, attributed to human hunters (Martin 1967; Martin and Klein 1984; Simmons 1989: 34, 75). Fire was probably being used by hominids in Africa by 1.5 million BP; by 0.5 million BP, humans were probably causing vegetation changes with fire (Blaney 1982; Goudie 1989; Simmons 1989: 32) and in Asia by 18 000 BP non-sedentary agriculturalists appear to have begun to alter natural vegetation (Bunney 1990). By Neolithic times anthropogenic vegetation changes, notably forest clearance, were huge. Mining activity was underway in Eurasia by 10 000 BP, at first for flints, later for metal ores, millstones, building stone, lime and marl (Shepherd 1980). Peoples settling Madagascar, New Zealand, Polynesia, and other islands over the last 400 or 500 years, with little more technology than Europe or America's prehistoric peoples, quickly exterminated many animal and plant species.

As agriculture developed, more people became sedentarized and the size of settlements reached the point at which they generated environmental problems. The populations of ancient Sumaria, Egypt, classical Greece, and later Rome, were small by present standards; nevertheless they obtained grain, timber and other resources from wide areas. Rome set provincial governors precise quotas for grain, olive oil, etc, and considerable manipulation of the environment took place to meet these demands (Fig. 1.1). As early as 3700 BP Sumarian cities were probably abandoned because supporting irrigated cropland had degraded. Plato in *Critias* around 2400 BP mentioned serious deforestation and soil erosion in Attica (North Greece), and Pliny recorded poor husbandry and soil erosion by the first century AD in the Roman Empire (McCormick 1989; Mannion 1991: 104), other Roman sources and archaeology indicate centuries of environmentally damaging mining and smelting activity, for example by *c.* 450 BC

Artaxerxes I was trying to restrict cutting of Lebanese cedar (*Cedrus libani*) (Grove 1992).

Improvements to the plough in the seventh century AD enabled European farmers to clear new land, but with more soil disturbance than earlier scratch-ploughs and hand-tools had caused. In Europe, monastic orders supported large-scale expansion of farmlands. South Asia and China carried considerable populations by medieval times. The New World was spared the mouldboard-plough until the late sixteenth century, but supported large settlements: Cortés (1485–1547) and his troops found cities far larger than any they knew in Europe. By medieval times deforestation, peat cutting, mining and scorched earth warfare had damaged the environment of considerable parts of Europe, the Mediterranean fringes, the Middle East, China and the Americas. Demand for wood to build warships led to considerable deforestation in Europe, around the Mediterranean and, later, in the West Indies, Brazil, West Africa and the Far East. England, for example, had probably lost about 80 per cent of its forest cover by AD 1086 (by the early twentieth century the loss stood at about 96 per cent) (Young 1990). Human damage to the environment lessened when bubonic plague ('Black Death') struck Europe in the fourteenth century, decimating a population who may have been vulnerable to infections thanks to climatic cooling and failed harvests that partly reflected overuse of resources.

For many in the past and today, the struggle to survive has meant unwise exploitation of the environment. Exploiters may be aware of the damage they cause, some may not; Gandhi noted that people are driven by need or greed, but the latter has greater environmental impact. Some cultures have a concern for nature, but in practice, even where environmentally sympathetic beliefs (like China's Confucianism and Taoism) have been widespread, there has still been damage.

It has been asserted that the west has from Roman times had an anthropocentric Judaeo-Christian tradition, that this greatly affects how the environment is perceived, and is in large measure to blame for the present world environment and development predicament (White 1967). Having been derived from similar cultural roots, Marxism and Islam have been seen to be similarly anthropocentric (De Bardeleben 1985: 80). The expression anthropocentric interpreted as meaning holding the view that creation was for humankind to be a 'collaborative' finisher or master of – a subjugator of nature rather than steward (Toynbee 1972; Young 1990) (for a discussion of this 'dominion versus stewardship' perception see Cooper 1990 or Atkinson 1991: 129).

Western ways are often contrasted with those of aboriginal peoples held to be more sympathetic towards nature: the indians of the Americas, the aboriginies of Australia, the Maori of New Zealand and various tribes of Africa. To be more attuned to nature may be easier when populations are low, but even then some ruthlessly exploited resources before contact with westerners could corrupt them. Waddington (1977: 28) was sceptical that westerners alter the environment any more than other cultures: in his view all great civilizations have damaged nature, although he did concede that western Europe and North America have an industrial component to their development which leads to greater pollution.

Attempts to question the exploitative western attitude toward nature and to urge more humility and stewardship were rare before the late seventeenth century (Passmore 1974; Pepper 1984: 8). St Francis of Assisi probably argued for such changes in the late twelfth to early thirteenth centuries (Sorrell 1988: 35), and St Benedict of Nursia may have made similar arguments in the sixth century (Simmons 1989: 394) but such voices were rare and were suppressed. The Judaeo-Christian (more accurately Graeco-Christian) hold on western Europe discouraged utopian thought before the fourteenth century, and perhaps as late as the seventeenth.

The Renaissance, the Enlightenment, the Protestant Ethic and the modern era

By the late eleventh century European scholars were translating Arabic works of mathematics, philosophy and medicine. Within a century Europe had taken science from Islam and had added it to a Judaeo-Christian worldview which had also drawn heavily from classical sources (i.e. from ancient Latin or Greek writings), particularly Aristotle. Aristotelian tradition held that the universe could be understood through rational thought. But there was little attempt at rational thought before Galileo (1564–1642). The medieval view, much influenced by St Thomas Aquinas, was that the universe reflected Divine Perfection, and few saw the need or dared to question established truths, to undertake empirical observation or publish utopian ideas (one of the first changes was in 1516 when Thomas Moore published *Utopia*). Empirical inquiry became more common after Copernicus published his sun-centred cosmological views in 1543, encouraging science to compete in areas previously reserved for Church scholars (Young 1990: 78).

Dualistic (i.e. 'mechanistic view of nature': spirit and mind as opposed to matter deemed to be separate, so with study the world can be explained, but in so-doing setting humankind apart from nature), positivist (a philosophical system of recognizing positive facts and observable phenomena) and Cartesian philosophy (dualistic rationalism), which underpins modern systematic science, gained acceptance after publication of Descartes's (1663) *Discourse on Method.* René Descartes – (1596–1650), a French physicist, showed that three coordinates could locate a point anywhere in three-dimensional space and so related geometry to algebra. Descartes, Keplar, Galileo and other thinkers of the Renaissance (a revival of arts and letters in the fourteenth to sixteenth centuries, particularly in Italy under the influence of classical models: neo-Platonism, etc, and an increase in economic activity, development of a capitalist class and increased secular thought) prepared the way for early-eighteenth-century Newtonian rationalism, and the application of reductive science to understanding the environment (Capra 1982). Newtonian science stressed the separation of observer from nature; it generated knowledge but critics feel that this approach reduced spiritual and holistic perceptions (values and facts became separated). There are still cultures today who reject the western separation of religion and temporal matters (secularism), including some followers of Islam, certain Christian sects and many late-twentieth-century environmentalists.

Around 1600, with 'roots' in France, there began a 'modern', 'scientific revolution', an Enlightenment and widening world experience (partly as a consequence of rational study, partly of exploration), and lessening of superstition. Contact beyond Europe had begun well before the thirteenth century – Marco Polo embarked on his travels in 1275, Christopher Columbus discovered the New World in 1492 – and settlement began by the sixteenth century. Established social norms and attitudes, seldom questioned before the late seventeenth century, came under more frequent attack from intellectuals like Rousseau, Voltaire, Condorcet and Turgot, and then in 1789 was the iconoclasm of the French Revolution. Christendom's medieval worldview had been eclipsed by the scientific revolution.

In Capra's view (1982), since the late sixteenth century the world has come to be seen as a machine that could be rationalized and understood. Francis Bacon in 1597 had argued that, provided it was understood and its rules were obeyed, nature could be to some extent controlled (Bacon 1939: 117). What exactly constitutes 'modern', and whether the modern period ended in the 1960s or 1970s or continues, is debated. It has been held that the essence of modernism is reductionism – the *systematic* breaking down of problems into their simple constituents to facilitate study. Latour (1993) argued that modern also implies a separation of politics, science and technology. Redclift and Sage (1994: 19) suggested that the ideas of progress, reason and freedom underpin modernism, and an attitude that humanity were in charge of the environment.

Some lump all western philosophical traditions since as early as 7000 BP as 'modern', preceded by 'primitive' or 'classical' and followed by 'post-modern' sometime after 1960 (Smith 1993). The modern period has been characterized by the development of a capitalistic world market, mechanical skills and technology, dependance on hydrocarbon fuels (coal, oil and gas) individual enterprise and empirical 'compartmentalized' study (i.e. divisions between disciplines) (Cosgrove 1990). During this period economic growth became the driving force both at the level of individuals and at the national scale (O'Riordan 1976: 23; Chisholm 1982: 159: K. Thomas 1983). Tawney (1954) suggested that the Reformation (a sixteenth-century popular movement for reform of doctrines and practices of the Roman Catholic Church, ending in establishment of the Reformed or Protestant Churches) released individuals from some restrictions previously imposed by religion and social hierarchy, encouraging individual effort. Weber (1958) explored the possibility that Protestant, especially Calvinistic, ethics spread after the sixteenth century and helped to create the liberal capitalist economy of modern western Europe. Basically the Protestant ethic (or Puritan work ethic) allowed that an individual is responsible for his or her own salvation through good actions, with hard work, orderly behaviour and neatness seen as desirable traits (Hill 1964; Merton 1970).

The 'industrial revolution' is generally held to have started in AD 1752 in England when the iron-master Abraham Darby began smelting with coal. Long before that date there had been localized industrial activity in England and elsewhere in western Europe. The English population in 1752 was still only about 13 per cent urbanized (i.e. living in towns of more than 5000) but that soon changed: industry expanded and increasingly took place in urban factories rather than small workshops or scattered cottages, population grew and more people lived in cities. Technological advance led to greater longevity and together with rising city birth-rates meant population growth. Population growth, colonial settlement, plus improved means of transport, led to global trade in resources and manufactured goods which in turn helped to 'drive' the process of industrialization.

Misguided and outmoded nineteenth- and twentieth-century concepts about environment and development

Between the 1830s and the late 1940s many westerners believed that nature could and should be used, even 'tamed' (M. Park 1969). While engineers built the Suez and Panama Canals, great bridges and railways, Victorian and Edwardian scientists, missionaries and adventurers explored and catalogued nature. These scientific efforts provided baseline data, but as knowledge grew it became clear that the abiotic and biotic components of the environment were more complex, vulnerable and difficult to understand and manage than had been expected. A range of concepts were put forward to try and explain the humankind–environment relationship.

Concepts about humankind–environment put forward in Victorian and Edwardian times

Environmental determinism

Publication of Darwin's *The Origin of Species by Means of Natural Selection* (1859) encouraged some to apply concepts like 'natural selection' and 'survival of the fittest' to human development. From the 1870s environmental determinists argued that the human–nature relationship was such that physical factors (like climate) influence, even substantially control, behaviour and thus society and development. Race too was often seen as a determinant of development. Pepper (1984: 111–12) recognized 'crude' and 'scientific' environmental determinists. Crude environmental determinism, and associated concepts like comparative advantage, were expressed by intellectuals such as Richter, Kant, Ritter, Ratzel, Semple (1911) and List. Scientific environmental determinists, for example Ellsworth Huntington (1915), were a little more objective (Chisholm 1982: 35, 157; Simmons 1989: 3; Corbridge 1991). At its crudest, the professed logic would be: 'bracing climate + Caucasian race = development; hot climate + lazy native = underdevelopment'. This subjective linking of development to environmental conditions and race was used for political ends in the 1920s and 1930s, particularly in Germany (and may help to explain why many social scientists were unenthusiastic about environment until the 1980s).

There can be no doubt that human fortunes often reflect natural events, for example in western Europe archaeology, historical records and palaeoenvironmental evidence link volcanic eruptions, poor weather conditions and collapse of agricultural communities. However, the main criticism of environmental determinism is that it ignores that humans can make different choices under similar environmental conditions and often modify the environment.

Social Darwinism

Some, notably Herbert Spence (not Charles Darwin), developed the idea that humans were part of an evolutionary hierarchy, similar to that which Darwin had observed in the animal kingdom. Competition, self-interest and struggle, rather than cooperation and mutual aid, were seen as natural and justifiable ways to behave; the group best able to adapt to environment would become dominant (Pepper 1984: 134). By the 1920s eugenics was supported by many as a way to improve a particular human group's genetics and so better their long-term survival and achievements. Eugenicists encouraged breeding of desirable people and suppressed undesirables: the approach was embraced in Nazi Germany and even some states in the USA as late as the 1960s.

Environmental possibilism

A concept put forward by Vidal de la Blanche and later by Febvre (1924): that environment constrains human endeavour and sets limits, but that choices of course of action are possible within those limits and the same environmental opportunities may be used differently by the various cultures.

Disillusionment with development: early conservationists and early environmentalists

Some societies protect certain plants and animals for reasons of religion or local economy (e.g. baobab trees are conserved in most of Africa) and here and there rulers have established reserves, for example in parts of India before the fifteenth century. Through established laws and traditions many resources, including common land, resisted catastrophic degradation in Europe, although in the last few hundred years some have suffered. Islands were among the first regions beyond Europe to be settled by Europeans, and were vulnerable environments; by 1677 the forests and wildlife of Mauritius had been degraded and the timber had been cut on Madeira and in the Cape Verde Islands. By the 1760s there was legislation to try to protect forests on Tobago, Mauritius, St Helena, and (by the 1790s) on St Vincent (Grove 1992).

During the eighteenth and nineteenth centuries the industrial revolution, led, especially in Europe and North America, to swollen cities, damaged countryside, loss of commons and misery (the squalor and poverty of the period is vividly described by Ponting 1991). In the UK, Europe, Russia and America, a diverse group of artists, writers, poets, anarchists and others increasingly questioned the value of capitalism, agricultural modernization and industrial growth. Many were dubbed 'romantics', and saw nature as a source of inspiration or solace (Bate 1991 reviews romantic ecology). The writings of Baron Alexander von Humboldt (1769–1859), German geographer, explorer and naturalist, stimulated considerable academic interest in natural history and conservation. Amateur naturalists like Gilbert White (1720–93) popularized natural history and called for better understanding and treatment of nature (White's 1789 *A Natural History of Selborne* is reputedly the fourth most published English language text). Poets like Wordsworth (1770–1850), Blake (1752–1827) and Coleridge (1772–1834) drew inspiration from and glorified natural landscape as did artists like William Holman Hunt, Turner and utopian liberals such as William Morris (1891). Nineteenth-century romantic social reformers like Robert Owen (1771–1858), a cotton mill director who founded utopian colonies in the UK, Ireland and the USA in the 1820s and political theorists like the Russian Prince Pyotr Kropotkin (1842–1921) professed forms of 'utilitarian environmentalism', their concern being to improve humankind through better working and living conditions (Kropotkin 1974; C. Ward 1974). Kropotkin, an anarcho-communist, argued for small, decentralized communities, closer to nature and avoiding industrialization and the division of labour – something quite close to what many in the green movement would seek nowadays.

By the 1860s the UK's filthy and overcrowded urban conditions prompted utilitarians like Chadwick, and Chamberlain, to campaign for better public health measures, with the result that the UK Parliament passed the Alkali Act in 1863 to control air pollution and began to introduce municipal sewerage systems and a range of other pollution control and public health measures.

Away from the cities of western Europe and North America, in the lands supplying them with resources or being settled by their surplus population, damage to forests, wildlife and soil aroused concern by the early nineteenth century. Two groupings can be distinguished among those seeking to curb the degradation: 'romantics' concerned about the ethical, moral and inspirational losses and 'utilitarian conservationists'. In the second half of the nineteenth century the British sought assistance from German foresters to sustain timber production in Burma and India. In South Africa, other African colonies and India, utilitarian legislation was passed to try and reduce soil erosion, control game hunting and conserve areas of forest and outstanding natural beauty. By the turn of the century, reserves had been established in Kenya and South Africa.

To some it was clear that frontiers were closing or about to close and limitless resources and land to settle would soon be things of the past.

American middle-class interest in nature was aroused by John James Audubon's publication of *The Birds of America* (1827). The writings of Henry David Thoreau from the 1830s (Thoreau 1960) and of Ralph Waldo Emerson paralleled European 'romantics', whose writings were also influenced by German forestry and by European colonial conservation efforts. In 1864 George Perkins Marsh published an influential, if somewhat deterministic, text on environment and development, *Man and Nature*.

Awareness led to action: by the 1860s the US National Parks Service and the US Forest Service were established. Two groups were concerned for the US environment in the late nineteenth and early twentieth centuries: 'preservationists', like John Muir, who wished to maintain unspoilt wilderness areas, and 'conservationists', like Gifford Pinchot, who were prepared to see environmental protection combined with careful land use (McCormick 1989). Pinchot, chief of the US Forest Service between 1890 and 1908, was a major force in establishing parks and reserves (and probably coined the term 'conservation' in 1907, although the British already had conservancies in India). Muir, 'high priest of the Sierras' and 'father of the US environmental movement', founded the influential Sierra Club in California in 1892 (in 1969 this non-governmental organization [NGO] spawned

Table 1.1 Conservation and wildlife protection 1831–1934

Date established	
1831	*Mt Audubon Cemetery* (USA) probably the first conservation area to be established in the USA
1864	*Yosemite Conservation Area* (USA)
1864/5	*National Trust for Places of Historic Interest and Natural Beauty* (UK)
1865	*British Commons, Open Spaces and Footpaths Preservation Society* (UK) possibly the world's first environmental non-governmental organization (NGO)
1867	*East Riding Association for the Protection of Seabirds* (UK)
1868	*Convention on Protection of Wildlife* (Vienna) probably world's first international convention on wildlife protection, led in 1902 to a covenant signed by 12 nations
1872	*Yellowstone Park*, Wyoming (USA) world's first national park
1898	*Royal Society for the Protection of Birds* (UK) roots go back to 1893: many founders were women disgusted with the fashion trade in bird-feathers
1898	*Kruger Reserve* (South Africa) became Kruger National Park in the 1920s
1903	*Society for the Preservation of Wild Fauna of the Empire* (UK/Empire) hunters worried by the extinction of creatures like the quagga and bloubok in the 1860s ('penitent butchers')
1909	*Swiss League for the Protection of Nature* (Switzerland) similar bodies in Sweden and Germany soon afterwards
1911	*Convention on Fur Seal Hunting* (several sealing nations) controls to counter excessive hunting
1912	*Society for Promotion of Nature Reserves* (UK)
1913	*Act of Foundation of the Consultative Committee for International Protection of Nature* (Bern) probably the first international environmental protection body: 17 nations signed
1916	*Canadian–American Treaty for the Protection of Migrating Birds* (USA/Canada)
1922	*International Committee for Bird Protection*
1934	*International Office for Protection of Nature*

Note: A history of nature conservation in the UK has been produced by Evans (1991)
Source: Pepper 1984: 14; Eckersley 1988; Adams 1990: 16; L.K. Caldwell 1990: 31–54; Stone 1992: 50–73)

Friends of the Earth). Men like Pinchot and Powell (founder of the US Geological Survey) achieved a great deal in the first decade of the twentieth century: they and President Theodore Roosevelt must be credited with some considerable part in promoting environmental concern (Kuzmiak 1991).

Conservation and nature protection bodies began to spread in America, Europe and European colonies before the First World War (see Table 1.1). After 1917 divergence of development paths between Russia, and later other socialist economies, and the free-enterprise west made little difference: socialist countries also had serious environmental problems (Gerasimov *et al.* 1971; Smil 1984; De Bardeleben 1986), clearly followers of Marxism also dealt inadequately with environmental issues relating to development (Redclift 1984; for a discussion of Marxism and environment see Atkinson 1991: 30–50). The eastern bloc has, however, played an active part in international conservation and environmental protection activities, and the former USSR and the People's Republic of China (PRC) have established many national parks and reserves.

Environmentalism from the 1920s to 1960s

Not long after the 1929 New York stock-market 'crash' and the Great Depression that followed, there was serious drought in the US midwest and soil erosion in the Great Plains 'Dustbowl', especially between 1932 and 1938. Wind-blown dust from the Dustbowl was clearly visible in Chicago and Washington, DC, and large numbers of farming families were displaced. The folksinger Woody Guthrie and novelist John Steinbeck (author of *The Grapes of Wrath*, 1939) publicized the Dustbowl misery; although at first seen as subversives, they helped to provoke public and government concern. Soon the 'New Deal' policies of President Franklin D. Roosevelt sought to counter the Depression, social and environmental problems, not only by promoting integrated development of natural resources (Friedmann and Weaver 1979: 5) but also by forming in 1933 the US Soil Erosion Service and in 1935 its successor, the US Soil Conservation Service, designed to fight land degradation.

The lessons of the Dustbowl were soon forgotten. Between 1860 and 1973 overall costs of extraction and processing of most of the world's natural resources declined; few people were concerned that there might be limits to growth (O'Riordan 1976: 37). Teilhard de Chardin looked at the human–environment relationship in the late 1930s but the 1939–45 War hindered concern for the environment, accelerated the development of resources and led to the production of new threats like DDT and atomic weapons. During the first decade or so after 1945, efforts were focused on the economic and industrial reconstruction of Europe and Japan, and on raising agricultural production (to assist this, the UN Food and Agriculture Organization [FAO] was founded in 1945).

Publications on environment and development began to appear from the late 1940s (Osborn 1948; Leopold 1949; Dale and Carter 1954; W.L. Thomas 1956). For example, Vogt (1948) expressed views which in the 1970s would have been called neo-Malthusian and helped establish (in 1952) an influential US environmental body, Resources for the Future Inc. The writings of Aldo Leopold (1949) stimulated many of the USA's environmentalists in the 1960s and beyond.

In 1949 the UN held one of the first post-War environmental meetings, the Conservation Conference at Lake Success, New York State. During the early 1950s, UNESCO (founded in 1945) established the International Union for the Protection of Nature, which in 1956 changed its name to the International Union for Conservation of Nature and Natural Resources (IUCN). From 1966 the IUCN has regularly published *Red Data Books*, which list organisms at risk of extinction; as one of the world's most important environmental bodies, IUCN has promoted conservation and sustainable development (in 1990 it became the World Conservation Union).

By the 1950s a number of public figures associated with the media, government or conservation bodies worked to bring environmental and conservation issues to the attention of the public and politicians; particularly active in the UK were Sir Peter Scott, HRH Prince Philip and Sir David Attenborough.

Environmentalism in the 1960s and 1970s

Reacting to post-War publications on population growth and environmental problems, Sir L. Dudley Stamp tried to give an unbiased picture in *Our Developing World* (1960); however, the real growth of environmental interest took place eight to ten years later. Studies of newspapers and journals from the USA, UK and Japan indicate that environment articles increased in the 1960s and early 1970s (Sandbach 1980: 2–6). Popular 'landmark' texts from this period include: *Silent Spring* (Carson 1962); *Blueprint for Survival* (Goldsmith *et al.* 1972); *The Limits to Growth* (Meadows *et al.* 1972); *Small is Beautiful* (Schumacher 1973).

NGOs, notably the Sierra Club (USA), began to speak out on environmental issues in the late 1950s, and President Kennedy's appointment of Stewart L. Udall as Secretary of the Interior assisted the cause of environmentalism in the USA. By the early 1960s, there was some progress outside the USA, for example the signatories of the 1959 Antarctic Treaty agreed to set aside territorial claims, ban military activity and nuclear waste dumping, and declare Antarctica a zone of scientific research.

By the mid-1960s there had developed what has been variously called an environmental(ist) movement, environmentalism, ecology movement, environmental revolution or conservation movement. Environmentalist movement is probably the best term, although some argue that it did not get underway until 1972 (Pepper 1984: 13; Simmons 1989: 6; Atkinson 1991). The distinction has been drawn between the ecology movement, who stop at mere reforms of pollution control, conservation, etc, and environmentalists, who acknowledge the need for radical changes of social relations, development goals, ethics, etc (Young 1990: 133). The prominent activists included a number of ecologists, operating with what Rees (1985: 2) called 'messianic fervour'. Environmentalism was especially strong roughly between 1962 and 1970 (some argue 1967–74), after which media coverage and public interest declined until the mid-1980s.

Environmentalism has been described as 'a moral code or a set of mediating values – in effect a philosophy of human conduct' (O'Riordan 1976: viii) or 'a concern for environment elevated to a political pursuit' (McCormick 1989: ix). It is pointless seeking a precise definition as ideologies and objectives vary, although most environmentalists would nowadays profess the desirability of sustainable development (Cosgrove 1990: 345) (see Table 1.2).

The environmentalists got underway first in the USA, particularly California, where public interest law firms supported by grants or foundations acted on behalf of citizens or groups of citizens (previously action had to be undertaken by individuals) to protect the environment (e.g. the Environmental Defense Fund or the Natural Resources Defense Fund) (B. Harvey and Hallett 1977: 62). The emergence was probably encouraged by development of the media; increased leisure; reduction of poverty; growing popular awareness of environmental problems associated with development and the Cold War arms race; and information on the structure and function of the environment in-part generated by the International Geophysical Year (1957–8), the International Biological Programme (IBP) (1964–75) and the International Hydrological Decade (1965–74).

The US Civil Rights movement, 'hippies', the anti-Vietnam War movement, European anti-nuclear weapons protests, student unrest (especially in 1968) and the 1960s–70s 'pop culture' in general encouraged people to ask awkward questions about development. McCormick (1989: vii) suggested that environmentalism was 'Spawned by Victorian nature-lovers and philanthropists, nurtured by amateur naturalists and professional planners, and finally thrust onto the public policy agenda by a rebellious and idealistic new generation'.

Environmentalists in the USA and Europe were mainly middle-class liberals or democrats; by no means all were youthful or part of the 'pop culture' (McEvoy 1971; O'Riordan 1976; Morrison 1986) and there was no unified creed or strategy. Environmentalists, although active in publication, litigation and protest, were relatively non-political (in New Zealand and Germany politically active green movements were developing). The focus was often on over-population (Ehrlich 1962), conservation of wildlife and problems associated with technology (Farvar and Milton 1972), particularly pollution. Rachel Carson's *Silent Spring* (1962), one of the first 1960s environmental books, focused popular attention on pesticide pollution and the hubris of modern development. Many of the publications that followed between the mid-1960s and mid-1970s were rather dogmatic; warning of coming crisis, so that some authors (e.g. Paul Ehrlich, Barry Commoner, Garrett Hardin) became known as 'prophets of doom' or 'ecocatastrophists' (L. White 1967; Commoner 1972). Although criticized along with other ecocatastrophists for predicting doom that has yet to materialize, Commmoner played a key role in achieving the ban on atmospheric testing of nuclear weapons.

The IBP, and later the UNESCO Man and Biosphere Program (MAB) (begun 1971), and the space programme in the USA and former USSR, helped foster an awareness that global-scale problems were real and the Earth was finite. In 1965 US Ambassador to the UN, Adlai Stevenson, used the metaphor 'spaceship Earth' (the speech was drafted by a prominent environmentalist Barbara Ward); Boulding (1971) took up the concept – that the world was a vulnerable, effectively closed system – and 'spaceship Earth' became an 'ikon' of 1960s and 1970s environmentalists. Resource exploitation problems had become apparent by the 1960s for example a number of marine oil-spill accidents and over-exploitation of sea fisheries that led to the UK–Iceland 'cod wars' between 1952 and 1976.

Those who identified population growth as the primary cause of environment and development problems became known as neo-Malthusians (Ehrlich 1972; Ehrlich *et al.* 1970; Ehrlich and Ehrlich 1990). Some neo-Malthusians like Hardin and Paul and Anne Ehrlich even went so far as to discuss the possibility of triage – the withholding of assistance from LDCs with severe overpopulation and little chance of improvement, in order to concentrate available resources on recipients who might with help achieve control. Hardin (1968; 1972) published an essay on the fate of common property resources in the face of population growth. His 'the tragedy of the commons' argument was that people will tend to over-use commonly owned resources (pasture, ocean fisheries, etc), in all probability destroying them, because without overall agreement each user will seek to maximize short-term interests. However, Hardin's views have been widely attacked (for example by Harrison 1993) and are now generally dismissed: seldom is use of commons a 'free-for-all' and communities generally have some controls. The analogy of a full lifeboat adrift in a sea, full of would-be passengers who could cause a capsize if ruthless care was not taken, was developed by Hardin (1974a; 1974b), with those in the metaphorical boat practising 'lifeboat ethics' being the DCs and those in the sea the LDCs. These and other neo-Malthusian views have been much criticized as simplistic and invalid (for a critique see Boserüp 1990: 41). There were, however, some more moderate critics of western development and neglect of environment, like Maddox (1972) and B. Ward and Dubos (1972).

The World Bank, a major funding agency for LDC development, appointed its first Environmental Advisor and established an Office for Environmental Affairs in 1969. Another important development, called an 'environmental Magna Carta' by some, was the 1970 US National Environmental Policy Act (NEPA). NEPA helped the USA to pioneer a proactive approach to environment and development. Under the Act all major federal projects likely to 'significantly' affect the environment required a pre-project environmental impact assessment (EIA), the results of which were to be published as an environmental impact statement so that alternatives could be seen and short- and long-term problems could be identified. Under NEPA a Council on Environmental Quality was formed to draw up annual environmental quality reports and to advise the President on the state of the environment.

As well as the passing of NEPA, 1970 saw the establishment of the US Environmental Protection Agency (an independent body responsible for regulating, coordinating and enforcing US environmental laws and for reviewing the impact

Table 1.2 Forms of environmentalism

Forms of environmentalism are diverse, but can be classified using criteria like ideology, philosophical stance, attitude to society, etc. The resulting classification is not precise but provides a broad guide:

Form	Characteristics	Activists/date
Moderate environmentalists ('light-greens' and 'mid-greens')		
Early conservation/nature protection	• focus on non-human environment • area-specific	Most environmentalists before the 1960s
Modernism (dominant western paradigm)	• nature seen as resources • humans dominant over nature • main goal material economic growth • consumerism • centralized, urban-centred • high technology • nowadays mainly existentialist	Now widespread
Ecocentric/Biocentric Environmentalism	• sometimes non-scientific • humans seen to upset nature and therefore have to attune to it • humility and adaption to environment • bioethics: human obligation to protect nature • humans seen to evolve with nature • decentralization often advocated • value assigned to all living and non-living components of the environment	Nineteenth-century romantics/ Modern Gaianists/(biocentric) Deep Ecology
Economic Environmentalism	• utilitarianism • seeks optimal use of resources • more critical of western stance than 'light-greens' • seek revision of economic/political relationship with environment	'mid-greens'
Technocentric Environmentalism	• accept current human–environment relationship • scientific and rational • humans seen to be 'in charge' managerial approach to human/environment • given study and care humans can better use environment • appropriate technology • rational management • willing to accept centralization • technological optimism • may not accept there is an environmental crisis	'light-greens'/many feel they can control nature (cornucopians)
Ecological Scientific Rationalism	• first priority is seen as viable environment, second as economic growth • use of ecology/systems analysis	Ehrlich (1970) and other 1970s activists
Radical environmentalists ('deep-greens')		
Deep Ecology movement	• human-centred view is seen as 'shallow', non-anthropocentric is seen as 'deep' • seeks harmony with nature • secular/humanistic/mystic • not necessarily concerned about fate of humanity • humans = part of nature • nature has value in itself • subjective, radical • ignores social aspects of environmental problems? • extremists may resort to civil disobedience • tend to accept Malthusian views • Bookchin (1987) critical • seek to transform ethics/politics	A 'platform', not a single/simple dogma Inspired by writings of Naess (1973) Earth First! (movement in USA) Some anti-humanist/eco-anarchists/spiritual greens Some animal rights groups Back-to-landers Primitivism Neo-Luddites/Eco-fascists Radical ecofeminism

Table 1.2 (cont.)

Form	Characteristics	Activists/date
Social Ecology (ecosocialism)	• social problems = cause • often anti-militaristic • anti-paternalism/feminist • promotion of profit less important than social and environmental justice and well-being • humanistic • 'cutting edge' now in USA	European Greens Arne Naess (Norwegian ecophilosopher) (Naess 1973; Devall and Sessions 1985) Murray Bookchin (US social ecologist)
Shallow Ecology	• humans seen as separate from nature • environmental management judged by its usefulness to human interests • humans dominant over nature • environment seen as a resource • high-tech solutions • authoritarian • prepared to accept technological fixes	
Radical Ecology	• the 'cutting edge' of social ecology • seeks new consciousness of human responsibilities to nature and humanity	Merchant (1992)
Postmodern Environmentalism	• holistic worldview • humans seen as integral part of nature	Some 1960s 'hippies'?
Anti-Establishment Environmentalism		Ginger groups with specialist aims 'Hippies'
Ecofeminism	• blame traditional exclusion of women from institutions for environmental problems	Some green feminists
New Age Movement	• diverse intellectual, spiritual and cultural group, active in USA, UK, Germany • no well-defined creed • seek/expect global spiritual reawakening and see universe as at one with humans (monistic)	Chandler (1988)

Note:
Deep-greens seek drastic changes to human/environment relationship attitudes, outlook.
Mid-greens seek revision of economics/politics to improve human/environment relationship.
Light-greens more or less accept current relationship between humans and environment but may wish to change aspects of the 'western' approach to development.
Source: Naess 1973; 1988; 1989; O'Riordan 1976: 3; Sandbach 1980: 22; Cotgrove 1982; Devall and Sessions 1985; Pepper 1984: 26–7; Eckersley 1988; Foley 1988; Sylvan and Bennett 1988; Tokar 1988; Cosgrove 1990: 355; Simmons 1990; Young 1990; Lewis 1992

statements produced by NEPA), and the declaration of Earth Day (22 April 1970) which encouraged large numbers of Americans to demonstrate annually their sympathy for environmentalism.

Experts and decision-makers came together to discuss environment and development issues at a number of international conferences beginning in the late 1960s:

- 1968 UNESCO 'Biosphere Conference', Paris
- 1970 European Conservation Conference
- 1971 'Founex I' UN Conference on the Human Environment, Founex
- 1972 UN Conference on the Human Environment, Stockholm
- 1974 'Founex II' Conference, Cocoyoc

Founex I issued a statement that concern for the environment and development did not conflict, and that it was dangerous to ignore environmental issues. The 1972 Stockholm Conference is seen as the watershed, after which was gradually broken down the widespread view that environmental concern conflicted with development and was either a 'luxury' that poor nations could not afford, or was part of a conspiracy ('green imperialism') to hold back LDC development. In 1973 the establishment of the UN Environmental Program (UNEP), the UN's main environmental body, boosted environmental awareness (B. Ward and Dubos 1972).

Two early 1970s publications helped shake the west's complacency; the first, *Blueprint for Survival* (Goldsmith *et al.* 1972), was an ecocentric attempt to outline the global predicament, popularize the view that infinite growth cannot be sustained by finite resources, and suggest how to react (in 1899 Kropotkin had made similar observations and recommendations). The second, *The Limits to Growth* (Meadows *et al.* 1972), was a non-technical report by the Club of Rome, an informal international group concerned for the present and future predicament of humankind. In 1970 the Club of Rome had commissioned Dennis Meadows and colleagues, using systems dynamics approaches developed by Jay Forrester at the Massachusetts Institute of Technology (USA), to run a 'world-model' (see Box 1.1) intended to predict the likely behaviour of the global environment between AD 1900 and 2100.

The Limits to Growth was intended to promote concern and further research; it explored a range of possible future scenarios which depended on how population and other key development parameters were managed (McCormick 1989: 75). A second Club of Rome report was published by Mesarovic and Pestel (1975) and a heated 'futures debate' developed with some advocating slow or even 'zero (economic) growth', and others like Kahn *et al.* (1976) or Simon (1981) of the view that a free market might well overcome environmental difficulties associated with growth before limits were met, and that there was no need for 'slowed growth' (O'Riordan 1976: 52–65; Freeman and Jahoda 1978; Hughes 1980). Critics have termed such optimism about limits 'cornucopian' and supporters 'prophets of boom' (Cotgrove 1982). A sequel to *The Limits to Growth* appeared 20 years later – *Beyond the Limits* (Meadows *et al.* 1992) – with the message that the world has already overshot some of the limits flagged in 1972 and, if present trends continue, severe global problems are virtually certain within 50 years. However, Meadows *et al.* (1992) still feel that catastrophe can be avoided, provided development approaches are changed soon.

In an influential book, *Small is Beautiful*, Schumacher (1973) warned that the west's pursuit of profit and development had promoted giant organizations, increased specialization, economic inefficiency, environmental damage and inhuman working conditions. The remedies he offered included 'Buddhist economics', 'intermediate technology' (technology with a human face, using smaller working units, local labour and resources) and respect for renewable resources. Environmental concern in the 1970s proved to be something of a 'false start' (Newby 1990: 1), and the interest of many administrators and the public waned between roughly 1974 and 1987.

Warnings delivered by late 1960s to mid-1970s environmentalists appear, with hindsight, to have often been premature and dogmatic, ignoring important issues to concentrate on the threat of demographic growth. Regardless of their faults, these environmentalists helped promote interest: in 1972

Box 1.1 Frequently used development terms

Model a simplification of reality (sometimes only a caricature) which assists in understanding often complex systems or processes (models can describe, explain and predict). Once an understanding is reached, a model may be used to assist with monitoring and prediction. Models may be *mechanical* (analog), e.g. a model aircraft in a wind-tunnel or a model estuary or hydrological system; *mathematical* (nowadays such models are usually computerized), e.g. General Circulation Models (world models); or *word-models*, i.e. schematic representations. The accuracy of a model depends on how well it mimics reality, on how good the data input is, and on the skill of those interpreting results.

Scenario not a prediction, more an attempt to assess 'what would happen if . . . ?', i.e. the consequences of an assumed set of conditions (simulation is roughly synonymous).

Prediction a possibly imprecise assumption of trends, open to discussion and modification. It should not be confused with forecast.

Forecast interpolation forwards of the trends presently evident, to assess what will probably occur.

Projection a more precise form of forecast.

Network shows linkages between impacts, and so can be used to trace cause–effect relationships.

Systems diagram links components in a diagram on the basis of energy flows between them.

Source: B. Harvey and Hallett 1977; Munn 1979; and others

only 10 nations had environmental agencies, by 1976 there were 70 which had them.

The OPEC oil crisis, LDC debt, sustainable development, postmodern environmentalism, post-environmentalism and the 1992 UN Conference on Environment and Development, Rio

In 1973/74 the Organization of Petroleum Exporting Countries (OPEC) raised their oil prices, after decades of generally declining commodity and energy prices, causing widespread concern and difficulty (another marked oil price rise occurred in 1979–80; prices then fell in 1986 and since that time). Faced with rising fuel import bills and with less to spend as a consequence of servicing debts incurred through foreign borrowing, many LDCs increased their exploitation of natural resources and had little to spend on things like environmental management, family planning and social welfare.

Between 1974 and the early 1980s concern focused on the 'energy crisis', although attempts were made to review environmental threats and suggest strategies for mitigation and avoidance (for example Dasmann *et al.* 1973; B. Ward 1976; Council on Environmental Quality and Department of State 1980; 1981 – like *Limits to Growth* drawing on modelling). The most significant publications of this period were, however, the *World Conservation Strategy* (IUCN, UNEP and WWF 1980) and the Brandt Report (Independent Commission on International Development Issues 1980). The *World Conservation Strategy* promoted living resource conservation for 'sustainable development' (the first time that the latter phrase was widely publicized). The Brandt Report stressed that many world problems would be solved only if it was recognized that DCs and LDCs had a mutual interest: the solution of Third World problems was a question not only of charity but also of interdependence.

The World Commission on Environment and Development set out in 1984 to re-examine critical environment and development problems and to formulate realistic proposals for solutions. The commission's findings, the Brundtland Report (World Commission on Environment and Development 1987), said little new, but it provoked interest in more pragmatic approaches, highlighted the need for sustainable development and urged a 'marriage' of economics and ecology (publications on environmental economics had so far been limited, e.g. Cooper 1981). If we singled out 'landmark' texts for the 1980s and 1990s, one would be the Brundtland Report and another, *Blueprint for a Green Economy* (Pearce *et al.* 1989): these may be said to have initiated a new relationship between social science, natural science, economics and policy making.

By the late 1980s the World Bank had adjusted its policies to give greater support to environmental management (Warford and Partow 1989). Oil prices fell, and a 'green movement' had become active, particularly in Europe, where it embarked on policy advocacy (not just the environmentalist rhetoric widespread between the 1960s and mid-1980s). By 1988 environmental matters were back on the agendas of politicians and decision-makers and had a higher public profile than ever before (Young 1990). Also, between the 1970s and 1985, a number of environmentalist groups developed, members of which were willing to alter their own lifestyles and sometimes to encourage (or force) others to do so in order to try and halt environmental damage (Buttel 1978).

The concept of postmodernism has been discussed since the late 1970s by poets, novelists, artists, philosophers, geographers, environmentalists, mathematicians and scientists. While there is wide usage, the concept is confused (Funtowicz and Ravetz 1992). Many recognize a postmodern period, beginning during the early 1960s (Cosgrove 1990: 355), characterized by collapse of 'normality' and increasingly post-industrial activity and a holistic world view. Toffler (1980) foresaw the collapse of capitalism and socialism and the dawn of a high tech./information technology revolution – a 'third wave' following the industrial and agricultural 'revolutions' (the second and first waves respectively). Postmodernism might be a useful approach for environmentalism and others engaged in multidisciplinary study of problems (Capra 1982; Cheney 1989; Warford and Partow 1989; Kirkpatrick 1990). It has been called the 'cultural logic of late-capitalism'. It might offer a scheme for understanding cultural phenomena and there are signs that maths and fundamental physics are moving from 'Cartesian order' and the systematic/reductionist approach to understanding chaotic complexity, toward postmodern holism embracing 'chaos theory' or fractals (Peat 1988: 341; Lewin 1993) (D. Harvey 1989: 43 provided a comprehensive study of postmodernism). It would be misleading to imply that there are no opponents to the concept of postmodernism (for a critique see Berg 1993 and M. Smith 1993).

The postmodern concept may prove useful, given that it is increasingly difficult to maintain a separation between science and politics, etc. Although the concept of holism was used long ago by Smuts (1926), modern holism is difficult to define, but implies acceptance that 'the whole is greater than the sum of the parts' and that modern science unwisely tends toward excessive reductionism, empiricism and 'compartmentalization' (isolation of fields of study from each other). In short, postmodern researchers seek to understand the totality of problems, rather than components. Not everyone is happy with these trends: Atkinson (1991: 154) for example warns of risks involved in adopting a holistic approach.

Serious problems still confront environmentalists. One is that economics has only recently begun to seriously consider and allow for the finite nature and complexity of the environment and resources; one route might be the adoption of environmental economics, another might be through ecological economics (Cooper 1981; Pearce *et al.* 1989). Furthermore, the sciences are increasingly called upon to forecast environmental impacts and trends; with inadequate data and understanding of often complex processes they are pressed to give advice and policy recommendations (a problem familiar to social scientists). A quite widespread shift has taken place: from people who in the past would say they did not know, to those who now think they do, but who fail to check facts or who are driven by subjective attitudes. Advances in computing and monitoring offer some hope that these challenges will be better addressed.

Recent huge changes in eastern Europe and the former USSR, triggered by President Gorbachev's *perestroika* (restructuring) and *glasnost* (openness) and resulting in the end of the Cold War, might release money spent on armaments for environmental management, but they could also mean that aid is diverted from LDCs to the former Second World. Another development that may spell problems for those seeking to manage the environment is the 'culture of consumerism' which has taken hold in DCs and is spreading almost globally, even in the PRC (a wit remarked that dialectic materialism has been replaced by 'mail-order materialism'). It may therefore prove difficult to control irrational and wasteful consumption patterns.

Clearly DCs need to alter their practices, in particular address problems of resource wastage and pollution. The relevance of established economics and environmentalism has increasingly been called into question and there are now signs of a sea-change taking place – perhaps to a form of post-environmentalism (Pearce *et al.* 1989; Barde and Pearce 1990; Pearce and Turner 1990; Young 1990; Pearce 1991). *Blueprint for a Green Economy* (Pearce *et al.* 1989) suggested that environmentally benign behaviour could to a large extent be encouraged through free choice in a market in which the pricing system has been adjusted to account for environmental impacts.

The 1992 Earth Summit – the UN Conference on Environment and Development at Rio (UNCED) seems to mark a greater awareness and willingness to discuss environment and development issues. But it is also apparent that at least some of the LDCs will have to pursue paths toward development that are very different from those taken by DCs (Holmberg *et al.* 1993). Although to some extent separated from the government representatives at Rio, NGOs were clearly able to pressure politicians for change. It was also apparent at Rio that such issues as global climatic change, access to biotechnology, and control of deforestation can be and are used as political levers: support for an issue is won by conceding to what the supporter or potential supporter wants on another (a country might back global warming controls if it can be given access to biotechnology).

Some explanations invoked to explain why environmental problems occur

Natural disaster

Environmental damage is often blamed on nature or on the peasantry. Human activity may have made the environment more vulnerable, e.g. a drought may be more severe because of livestock overgrazing.

Neo-Malthusian

This claims that there is a direct link between population increase and environmental degradation, especially in marginal lands. It is criticized for being too simplistic by failing to look at population increase in its social and historical context. It fails to consider that roughly one-third of the world's people use most of the world's resources, produce much of the pollution and consume about six times as much energy as the poor majority. Boserüp (1965; 1981; 1990) suggested that population variation determines technological change which affects demographic change. This is opposite to Malthusian/neo-Malthusian views, in that population increase is seen as an independent variable that can determine agricultural development (Boserüp argued that rising population can lead to intensification of production and less land damage). Other critics like Simon (1981) have also turned neo-Malthusian theory on its head to argue population increase need not always be a threat. (For reviews of the population–environment relationship see Blaikie and Unwin 1988: 111–24 and Redclift and Sage 1994: 39–41.)

Poverty

People are too poor to avoid degrading the environment, some LDCs are too poor to protect or rehabilitate the environment.

Impacts of technology

Problems result from technological innovation *per se* or because the technology is misused or misapplied; appropriate technology is seen as a means to solve these problems.

Human greed

Various groups point to this as a cause. Supporters of the 'Gandhian' approach advocate reduction of per capita demands for consumer goods and discouragement of 'planned obsolescence'.

Economic perspectives

'Tragedy of the commons explanation' – the economics of production, particularly faulty property relationships and difficulty in managing commons resources, is blamed.

Externalities explanation

Argument that population increase leads to destruction of common resources as individuals act to maximize their benefits and in so doing harm society as a whole. Capital accumulation may cause accelerated exploitation of resources to the point of degradation, with profits invested elsewhere.

Dependency perspective

External factors affect resource use in a given country or region; these include technology transfer, inappropriate agriculture, trade and aid, access to expertise and information.

Terms of trade

The world market has been beyond the control of producer countries and in recent years produce prices have fallen. The same harvest buys less abroad so the producer is forced to produce more, resulting in environmental damage (Hayter 1989: 5).

Economics thinking

Economists have seen earth resources as limitless and concentrate on short-term gains, neglecting longer-term costs and difficult to measure (in economic terms) intangibles.

Neo-Marxist

Wealth of DCs has been achieved, at least in part, by transfer of resources from the world's poor. In so doing poor countries are impoverished and suffer environmental damage.

Ethical stance

Humans have seen themselves as above nature, in control, keen to exploit, but unwilling to act as stewards. In practice individuals, regardless of stance, will tend to be biased toward short-term gains which will benefit them.

Ignorance

Lack of knowledge or data results in problems (e.g. CFCs were seen as safe when first introduced).

Attitudes

People may be aware that their actions damage the environment and have the opportunity to alter activities, but are unwilling. Greed is a reason already listed; however, there may be social attitudes which prevent good management of the environment that do not involve greed. Warfare is a significant cause of environmental damage and may result from non-greed motives.

Cold War

Between the 1940s and 1990s attention and resources were diverted from environmental matters and resource use was subservient to strategic 'needs'.

Postmodern

Postmodern explanations reject the modern approach to development as mechanistic and reductionist (compartmentalized), which leads to problems. They argue for a paradigm shift towards holistic, less reductionist problem-solving, with little distinction made between humans and nature and a multidisciplinary approach to try and understand the complex whole.

References

Adams, W.M. (1990) *Green development: environment and sustainability in the third world.* London, Routledge
Amin, S. (1990) *Maldevelopment: anatomy of a global failure.* London, Zed
Atkinson, A. (1991) *Principles of political ecology.* London, Belhaven
Bacon, F. (1937) *Essays.* London, Dent (first published 1597)
Barde, J.-P. and Pearce, D. (1990) *Valuing the environment.* London, Earthscan
Barrow, C.J. (1987) *Water resources and agricultural development in the tropics.* Harlow, Longman
Bartelmus, P. (1986) *Environment and development.* London, Allen and Unwin
Bate, J. (1991) *Romantic ecology: Wordsworth and the environmental tradition.* London, Routledge
Berg, L.D. (1993) Between modernism and postmodernism. *Progress in Human Geography* 17(4): 490–507
Berger, P.L. (1987) *The capitalist revolution.* Aldershot, Wildwood House
Bernstein, H. (1973) *Underdevelopment and development: the third world today.* Harmondsworth, Penguin
Blaney, G. (1982) *The triumph of the nomads: a history of ancient Australia.* Melbourne, Macmillan
Bookchin, M. (1987) Social ecology versus deep ecology. A challenge for the ecology movement. *Green Perspectives* nos 4 and 5
Boserüp, E. (1965) *The conditions of agricultural growth: the economics of agrarian change under population pressure.* London, Allen and Unwin
Boserüp, E. (1981) *Population and technology.* Oxford, Basil Blackwell
Boserüp, E. (1990) *Economic and demographic relationships in development.* Baltimore, Md., Johns Hopkins University Press
Boulding, K.E. (1971) The economics of the coming spaceship earth. *Development Digest* IX(1): 12–15 First published 1966 In H. Jarrett (ed.) *Environmental quality in a growing economy.* Baltimore, Md., Johns Hopkins University Press
Brookfield, H.C. (1973) On one geography and a third world. *Institute of British Geographers Transactions* old series no. 58: 1–20
Bunney, S. (1990) Prehistoric farming caused 'devastating' soil erosion. *New Scientist* 125 (1705): 29
Buttel, F.H. (1978) Environmental sociology: a new paradigm? *American Sociologist* 13: 252–6
Caldwell, L.K. (1990) *International environmental policy: emergence and dimensions.* Durham, N.C., Duke University
Caldwell, M. (1977) *The wealth of some nations.* London, Zed
Capra, F. (1982) *The turning point: science, society and the rising culture.* London, Wildwood House
Carson, R. (1962) *Silent spring.* St Louis, Mis., Houghton Mifflin
Chandler, R. (1988) *Understanding the new age.* London, Word
Cheney, J. (1989) Postmodern environmental ethics: ethics as bioregional narrative. *Environmental Ethics* 11(2): 117–34
Chisholm, M. (1982) *Modern world development: a geographical perspective.* London, Hutchinson
Commoner, B. (1972) *The closing circle.* New York, Bantam
Cooper, C. (1981) *Economic evaluation and environment: a methodological discussion with particular reference to developing countries.* London, Hodder and Stoughton
Cooper, C. (1990) *Green Christianity: caring for the whole creation.* London, Hodder and Stoughton
Corbridge, S. (1991) Definitions of development. *Geography Review* 5(2): 15–18
Cosgrove, D. (1990) Environmental thought and action: premodern and post-modern. *Transactions of the Institute of British Geographers* (new series) 15(3): 344–58
Cotgrove, S. (1982) *Catastrophe or cornucopia? The environment, politics and the future.* Chichester, Wiley
Council on Environmental Quality and Department of State (1980) *The global 2000 report to the president: entering the twenty-first century.* Washington, DC, US Government Printing Office (Penguin edn, 1982)
Council on Environmental Quality and Department of State (1981) *Global future: a time to act.* Washington, DC, Council on Environmental Quality
Dale, T. and Carter, V.G. (1954) *Topsoil and civilisation.* Norman, University of Oklahoma Press
Darwin, C. (1859) *The origin of species by means of natural selection: or the preservation of favoured races in the struggle of life.* London, John Murray (Penguin edn, 1968)
Dasmann, R.F., Milton, J.P. and Freeman, P.H. (1973) Ecological principles for economic development. London, Wiley
De Bardeleben, J. (1986) *The environment and Marxism–Leninism: the Soviet and East German experience.* Boulder, Colo., Westview
Devall, B. and Sessions, G. (1985) *Deep ecology: living as if nature mattered.* Salt Lake City, Ut., Peregrine Smith
Dudley Stamp, Sir L. (1960) *Our developing world.* London Faber
Dupâquier, J. and Grebenik, E. (eds) (1983) *Malthus past and present.* London, Academic Press
Eckersley, R. (1988) The road to ecotopia? Socialism versus environmentalism. *The Ecologist* 18(4/5): 142–8
Ehrlich, P.R. (1972) *The population bomb.* New York, Ballentine
Ehrlich, P.R. and Ehrlich, A.H. (1990) *The population explosion.* New York, Simon and Schuster
Ehrlich, P.R., Ehrlich, A.H. and Holdren, J.P. (1970) *Ecoscience: population, resources, environment.* San Francisco, Calif., Freeman
Escobar, A. (1992) Reflections on development: grassroots approaches and alternative policies in the third world. *Futures* 24: 411–36
Evans, D. (1991) *A history of nature conservation in Britain.* London, Routledge
Farvar, M.T. and Milton, J.P. (eds) (1972) *The careless technology.* New York, Garden City Press
Febvre, L. (1924) *A geographical introduction to history.* London, Routledge and Kegan Paul
Foley, G. (1988) Deep ecology and subjectivity. *The Ecologist* 18(4/5): 120–3
Fox, M. (1983) *Original blessing.* Santa Fe, N.M., Bear
Freeman, C. and Jahoda, M. (eds) (1978) *World futures: the great debate.* Oxford, Martin Robertson

Friedmann, J. and Weaver, C. (1979) *Territory and function: the evolution of regional planning*. London, Edward Arnold
Funtowicz, S.O. and Ravetz, J.R. (1992) The good, the true and the post-modern. *Futures* 24: 963–76
Gerasimov, I.P., Armand, D.L. and Yefron, K.M. (eds) (1971) *Natural resources in the Soviet Union: their use and renewal* (translated from 1963 Russian edition by J.I. Romanowski). San Francisco, Calif., Freeman
Goldsmith, E. (1988) The way: an ecological world view. *The Ecologist* 18(4/5): 160–7
Goldsmith, E. (1990) Evolution, neo-Darwinism and the paradigm of science. *The Ecologist* 20(2): 67–73
Goldsmith, E., Allan, R., Allaby, M., Davol, J. and Lawrence, S. (1972) *Blueprint for survival*. Harmondsworth, Penguin (also published in *The Ecologist* 2(1): 1–43)
Goudie, A. (1989) The changing human impact. In L. Friday and R. Luskey (eds) *The fragile environment: the Darwin College lectures*. Cambridge, Cambridge University Press.
Gould, S.J. (1989) *Wonderful life: the Burgess Shale and the nature of history*. Harmondsworth, Penguin
Grove, R.H. (1992) The origins of western environmentalism. *Scientific American* 267(1): 22–7
Gupta, A. (1988) *Ecology and development*. London, Routledge
Hardin, G. (1968) The tragedy of the commons. *Science* 162(3859): 1243–8
Hardin, G. (1972) The survival of nations and civilization. *Science* 172: 129
Hardin, G. (1974a) Lifeboat ethics: the case against helping the poor. *Psychology Today* 8: 38–43, 123–6
Hardin, G. (1974b) *The ethics of a lifeboat*. Washington, DC, American Association for the Advancement of Science
Harrison, P. (1993) *The third revolution: environment, population and a sustainable world*. Harmondsworth, Penguin
Harvey, B. and Hallett, J.D. (1977) *Environment and society: an introductory analysis*. London, Macmillan
Harvey, D. (1989) *The condition of postmodernity: an enquiry into the origins of cultural change*. Oxford, Basil Blackwell
Hayter, T. (1989) *Exploited earth: Britain's aid and the environment*. London, Earthscan
Hettne, B. (1990) *Development theory and the three worlds*. Harlow, Longman
Hill, J.C. (1964) Puritanism, capitalism and the scientific revolution. *Past and Present* 29: 88–97
Holmberg, J., Thomson, K. and Timberlake, L. (1993) *Facing the future: beyond the Earth Summit* London, Earthscan
Hughes, B.B. (1980) *World modeling: the Mesarovic–Pestel world model in the context of its contemporaries*. Lexington, Ky., Lexington Books
Huntington, E. (1915) *Civilisation and climate*, New Haven, Conn., Yale University Press
Independent Commission on International Development Issues (1980) *North–South: a programme for survival*. London, Pan
IUCN, UNEP and WWF (1980) *World conservation strategy: living resources for sustainable development*. Gland, Switzerland, International Union for Conservation of Nature and Natural Resources
Johnston, R. (1989) *Environmental problems: native, economy and state*. London, Belhaven
Kahn, H., Brown, W. and Martel, L. (eds) (1976) *The next 200 years*. London, Abacus
Kirkpatrick, D. (1990) Environmentalism: the new crusade. *Fortune* 12 February 1990: 24–30
Kropotkin, P. (1974) *Fields, factories and workshops* (original 1899 in Russian; English edn ed. C. Ward). London, Unwin
Kuzmiak, D.T. (1991) A history of the American environmental movement. *Geographical Journal* 157(3): 265–78
Latour, B. (1993) *We have never been modern* (translated from the French by C. Porter). Hemel Hempstead, Harvester-Wheatsheaf
Leopold, A. (1949) *A Sand County almanac*. London, Oxford University Press
Lewin, R. (1993) *Complexity: life at the edge of chaos*. London, Dent
Lewis, M. (1992) *Green delusions: an environmentalist critique of radical environmentalism*. Durham, N.C., Duke University Press
Mabogunje, A.L. (1980) *The development process: a spatial perspective*. London, Hutchinson
Maddox, J. (1972) *The doomsday syndrome*. London, Maddox Educational
Mannion, A.M. (1991) *Global environmental change: a natural and cultural environmental history*. Harlow, Longman
Marsh, G.P. (1864) *Man and nature: or physical geography as modified by human action*. New York, Charles Scribner
Martin, P.S. (1967) Prehistoric overkill. In P.S. Martin and H.E. Wright (eds) *Pleistocene extinctions*. New Haven, Conn., Yale University Press
Martin, P.S. and Klein, R.G. (eds) (1984) *Quaternary extinctions*. Tucson, University of Arizona Press
McCormick, J. (1989) *Reclaiming paradise: the global environmental movement*. Bloomington, Ind., Indiana University Press
McEvoy, J. III (1971) A comment: conservation an upper-middle class social movement. *Journal of Leisure Research* 3: 127–8
Meadows, D.H., Meadows, D.L. and Randers, J. (1992) *Beyond the limits: global collapse or a sustainable future*. London, Earthscan
Meadows, D.H., Meadows, D.L., Randers, J. and Behrens, W.W. III (1972) *The limits to growth (a report for the Club of Rome's project on the predicament of mankind)*. New York, Universal
Merton, R.K. (1970) *Science, technology and society in seventeenth century England*. New York, Howard Fertig
Mesarovic, M. and Pestel, E. (1975) *Mankind at the turning point (the second report to the Club of Rome)*. London, Hutchinson (published 1974 in USA)
Miller, G.T. jnr (1990) *Living in the environment: an introduction to environmental science* (6th edn). Belmont, Calif., Wadsworth
Morris, W. (1891) *News from nowhere*. London, Reeves and Turner
Morrison, D.E. (1986) How and why environmental conciousness has trickled down. In A. Schnaiberg, N. Watts and K. Zimmerman (eds) *Distributional conflicts in environmental resource policy*. Aldershot, Gower

Munn, R.E. (1979) *Environmental impact assessment* (SCOPE 5), 2nd edn. Chichester, Wiley
Naess, A. (1973) The shallow and the deep, long range ecology movement, a summary. *Inquiry* 16: 95–100
Naess, A. (1988) Deep ecology and ultimate premises. *The Ecologist* 18(4/5): 128–32
Naess, A. (1989) *Ecology, community and lifestyles: outline of an ecosophy.* Cambridge, Cambridge University Press
Newby, H. (1990) *Environmental change and the social sciences.* Paper to 1990 Annual Meeting of the British Association for the Advancement of Science (mimeo), 22 pp.
Nicholson, E.M. (1970) *The environmental revolution.* London, Hodder and Stoughton
O'Riordan, T. (1976) *Environmentalism.* London, Pion
Osborn, F. (1948) *Our plundered planet.* Boston, Mass., Little Brown
Osborn, F. (1953) *Limits of the earth.* Boston, Mass., Little Brown
Park, C. (1980) *Ecology and environmental management.* Folkstone, Dawson
Park, M. (1969) *Travelling in Africa.* London, Dent (originally published 1811)
Passmore, J. (1974) *Man's responsibility for nature.* London, Duckworth
Pearce, D.W. (1991) *Blueprint 2: greening the world economy.* London, Earthscan
Pearce, D.W. and Turner, R.K. (1990) *Economics of natural resources and the environment.* Hemel Hempstead, Harvester-Wheatsheaf
Pearce, D.W., Markandya, A. and Barbier, E.B. (1989) *Blueprint for a green economy.* London, Earthscan
Peat, D. (1988) *Superstrings and the search for the theory of everything.* London, Cardinal-Sphere Books
Pepper, D. (1984) *The roots of modern environmentalism.* London, Croom Helm
Pinchot, G. (1910) *The fight for conservation.* Seattle, University of Washington Press
Ponting, C. (1991) *A green history of the world.* London, Sinclair-Stevenson
Progress Publishers (1977) *Current problems, society and environment: a Soviet view* (translated from Russian by J. Williams). Moscow, Progress
Redclift, M. (1984) *Development and the environmental crisis.* London, Methuen
Redclift, M. and Sage, C. (1994) *Strategies for sustainable development: local level agendas for the southern hemisphere.* Chichester, Wiley
Rees, J. (1985) *Natural resources: allocation, economics and policy.* London, Methuen
Riddell, R. (1981) *Ecodevelopment: economics, ecology and development.* Farnborough, Gower
Rostow, W.W. (1960) *The stages of economic growth: a non-communist manifesto.* Cambridge, Cambridge University Press
Sandbach, F. (1980) *Environment, ideology and policy.* Oxford, Basil Blackwell
Schumacher, E.F. (1973) *Small is beautiful: a study of economics as if people mattered.* London, Bland and Briggs
Scolimowski, H. (1988) Eco-philosophy and deep ecology. *The Ecologist* 18(4/5): 124–7
Semple, E.C. (1911) *Influences of geographic environment.* New York, Henry Holt
Shepherd, R. (1980) *Prehistoric mining and allied industries.* London, Academic Press
Simmons, I.G. (1989) *Changing the face of the earth: culture, environment, history.* Oxford, Basil Blackwell
Simmons, I.G. (1990) Ingredients of a green geography. *Geography* 75(2): 98–105
Simon, J.L. (1981) *The ultimate resource.* Princeton, N.J., Princeton University Press
Simonis, U.E. (1990) *Beyond growth: elements of sustainable development.* Bonn, Edition Sigma
Simpson, E.S. (1987) *The developing world: an introduction.* Harlow, Longman
Smil, V. (1984) *The bad earth: environmental degradation in China.* London, Zed
Smith, A. (1975) *An enquiry into the nature and causes of the wealth of nations* (original 1776). London, Dent
Smith, M. (1993) Cheney and the myth of postmodernism. *Environmental Ethics* 15(1): 3–18
Smuts, J.C. (1926) *Holism and evolution* (3rd edn 1936). London, Macmillan.
Sorrell, R.D. (1988) *St Francis of Assisi and nature: tradition and innovation in western Christian attitudes towards the environment.* New York, Oxford University Press
Soussan, J.G. (1992) Sustainable development. In A. Mannion and S.R. Boulby (eds) *Environmental issues in the 1990s.* Chichester, Wiley
Stone, R.D. (1992) *The nature of development: a report from the rural tropics on the quest for sustainable economic growth.* New York, A.A. Knopf
Stretton, H. (1976) *Capitalism, socialism and the environment.* Cambridge, Cambridge University Press
Sylvan, R. and Bennett, D. (1988) Taoism and deep ecology. *The Ecologist* 18(4/5): 148–58
Tawney, R.H. (1954) *Religion and the base of capitalism.* New York, Mentor (original edn 1926)
Teilhard de Chardin, P. (1964) *The future of man.* London, Collins
Teilhard de Chardin, P. (1965) *The phenomenon of man.* New York, Harper and Row (first published in 1938 in French)
Thomas, K. (1983) *Man and the natural world: changing attitudes in England 1500–1800.* Harmondsworth, Penguin
Thomas, W.L. (ed.) (1956) *Man's role in changing the face of the earth.* Chicago, University of Chicago Press
Thomas-Hope, E.M. and Hodgkiss, A.G. (1983) *A geography of the third world.* London, Methuen
Thoreau, H.D. (1960) *Walden, or life in the woods* (1854 edn, Boston, Mass., Houghton Mifflin – originally published in 1843). New York, New American Library
Todaro, M.P. (1989) *Economic development in the third world* (4th edn). Harlow, Longman (5th edn 1994)
Toffler, A. (1980) *The third wave.* London, Collins
Tokar, B. (1988) Social ecology, Deep Ecology and the future of green political thought. *The Ecologist* 18(4/5): 132–42
Toye, J. (1987) *Dilemmas of development: reflections on the counter-revolution in development theory and policy.* Oxford, Basil Blackwell
Toynbee, A. (1972) The religious background to the present environmental crisis. *International Journal of Environmental Studies* 3: 141–6
Tudge, C. (1977) *The famine business.* Harmondsworth, Penguin

UNDP (1991) *Human development report 1991*. Oxford, Oxford University Press
Vogt, W. (1948) *Road to survival*. New York, William Sloane
Waddington, C.H. (1977) *Tools for thought*. London, Jonathan Cape
Ward, B. (1976) *The home of man*. Harmondsworth, Penguin
Ward, B. and Dubos, R.E. (1972) *Only one earth: the care and maintenance of a small planet*. Harmondsworth, Penguin
Ward, C. (ed.) (1974) *Peter Kropotkin: fields, factories and workshops tomorrow*. London, Allen and Unwin
Warford, J. and Partow, Z. (1989) Evolution of the World Bank's environmental policy. *Finance & Development* December 1989: 5–9
Weber, M. (1958) *The protestant ethic and the spirit of capitalism* (English translation by T. Parsons, originally published in 1904 and 1905 as 2 vols). New York, Charles Scribner
White, G. (1789) *A natural history of Selborne* (1977 edn). Harmondsworth, Penguin
White, L. jnr (1967) The historical roots of our environmental crisis. *Science* 155(3767): 1203–7
World Bank (1992) *World development report: development and the environment*. Oxford, Oxford University Press
World Commission on Environment and Development (1987) *Our common future* (Bruntland Report). Oxford, Oxford University Press
Young, J. (1990) *Post environmentalism*. London, Belhaven

Further reading

Adams, W.M. (1990) *Green development: environment and sustainability in the third world*. London, Routledge [Introduction to environment and development, sustainable development and the background of environmentalism]
Bartelmus, P. (1986) *Environment and development*. London, Allen and Unwin [Concise introduction to environment and development]
Crump, A. (1991) *Dictionary of environment and development: people, places, ideas and organizations*. London, Earthscan [Brief definitions and information]
Pearce, D., Markandya, A. and Barbier, E.B. (1989) *Blueprint for a green economy*. London, Earthscan [The Pearce Report – proposals for reaching and financing a stable environment]
Pepper, D. (1984) *The roots of modern environmentalism*. London, Croom Helm [Introduction to environmental thought and environmentalism]
Ward, B. and Dubos, R.E. (1972) *Only one earth: the care and maintenance of a small planet*. Harmondsworth, Penguin [An unofficial report commissioned to help prepare delegates and the public for the 1972 Stockholm Conference – a 1970s 'classic']
Young, J. (1990) *Post environmentalism*. London, Belhaven [Stimulating guide to environmentalism]

CHAPTER 2 Structure, function and management of the environment

Environment, ecology, systems theory and ecosystems

Environment may be defined as the external conditions and influences affecting the life and development of organisms. Less commonly used nowadays, biogeocoenosis is a broadly synonymous term, meaning life and Earth functioning together. Living organisms and non-living elements of the environment interact in complex ways. The study of these interactions, ecology was founded as an academic subject (oecology) in 1869 by Ernst Haeckel and by 1914 the *Journal of Ecology* had been established. In 1927 Charles Elton described ecology as 'scientific natural history'; modern definitions would include: the study of the structure and function of nature; the study of interactions between organisms (biotic) and non-living (abiotic) environment; the science of the relations of organisms to their total environment, and the interrelationships of organisms inter-specifically and between themselves within a species (Fraser–Darling 1963; Odum 1975: 1; Park 1980: 33).

Humans either adapt to or seek to modify their environment to achieve security and well-being. In making modifications (development) people often create a 'human environment' (Treshow 1976: xi), often a degraded, rather than improved, version of the original conditions. People's behaviour and culture are partly a consequence of physical surroundings and partly human genetics (just how much of each is debated). The expression 'human ecology' was probably first used to describe the study of people and their environment in 1910, but interest in the field has grown mainly since the 1970s (Sargent 1974; Richerson and McEvoy 1976). Sociocultural mechanisms mediate between people and the environment, and human ecology (some prefer the expression cultural ecology) examines these mechanisms and the conditions that they help create (Steward 1955). Political ecologists seek to build foundations for sustainable relations between society and the environment (Atkinson 1991; Blaikie 1985).

Since the early 1970s ecology has also come to mean a viewpoint, a holistic approach, rather than a single discipline or study (O'Riordan 1976), and many countries now have an 'ecology party'.

The global complex of living and dead organisms forms a relatively thin layer, mainly between approximately 60 metres below to 6100 metres above sea-level, known as the biosphere. The term 'ecosphere' is used to signify the biosphere interacting with the non-living environment, biological activity being capable of affecting physical conditions even at the global scale. The ecosphere is subject to various climates; Fig. 2.1 gives a generalized picture of the present-day situation. It should be noted, however, that the pattern of climate can change, so a world map of climate for say 20 000 years ago would be very different from that of today, as might be that for AD 2100. Climate could be affected by a range of factors including:

1 Variation in incoming solar energy due to fluctuation of the Sun's output or possibly dust in space.
2 Variation in the Earth's orbit around or change in its rotation about its axis.
3 Variation in the composition of the atmosphere or in quantity of dust, gases or water vapour present (biological activity may alter atmospheric composition).
4 Altered distribution of continents, changes in oceanic currents or of sea-level that may expose or submerge continental shelves.
5 Formation and removal of topographic barriers.

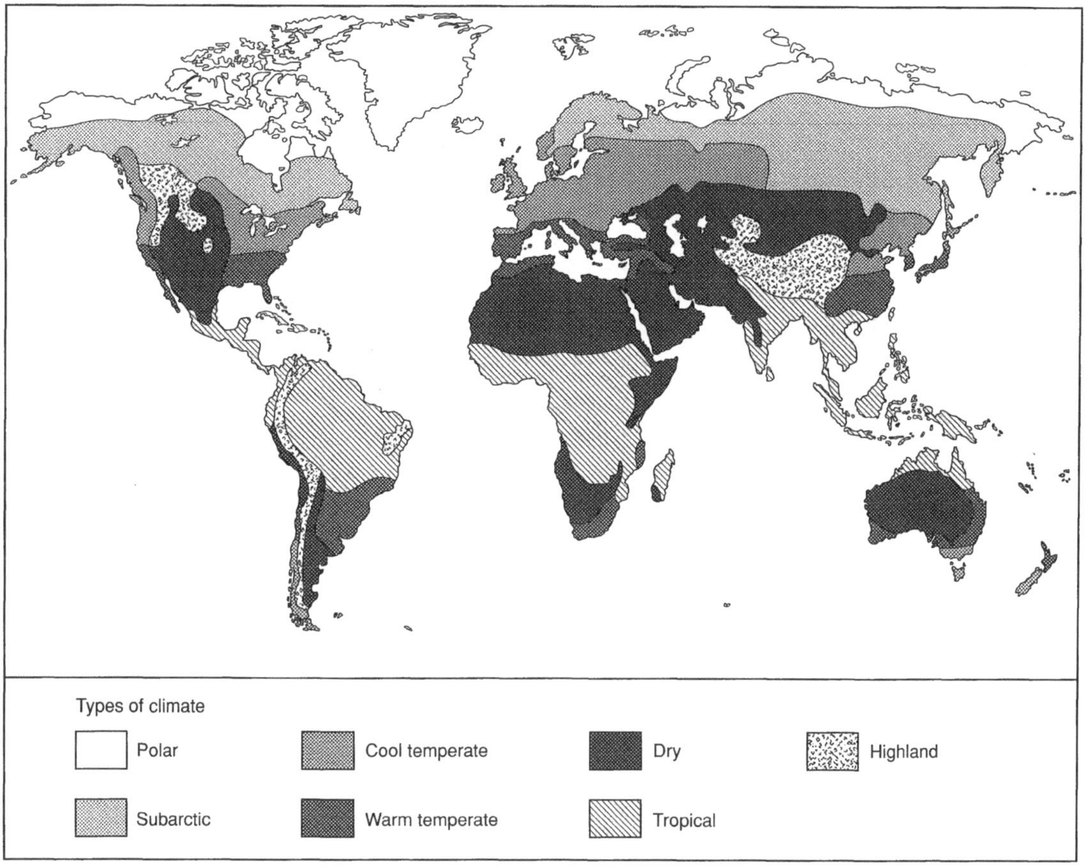

Fig. 2.1 Generalized map of world's major climates (*Source*: author from various sources).

The biosphere is composed of many interacting ecosystems (ecological systems), the boundaries between which are seldom sharp and often take the form of transition zones known as ecotones (while organisms from either of the adjoining zones may be present, it is also possible for some to be restricted to the ecotone only and not the adjoining zones). Large land ecosystems or biomes (synonymous with biotic areas) can be recognized, such as tundra biome, desert biome, and grassland biome. Biome is a concept that seeks to extend the ideas of community among vegetation and animal populations to cover the patterns of life within both (Watts 1971: 186).

The ecosystem, the basic functional unit of ecology (Park 1980: 107), was discussed and defined by Sir Arthur Tansley (1935); it refers to an assemblage of organisms living and interacting in association under certain environmental conditions. Organisms occupy, and events and energy flows take place within an ecosystem, such that change in one variable may affect one or more, perhaps all other variables. According to Miller (1991: 112) an ecosystem has six major features: interdependence, diversity, resilience, adaptability, unpredictability and limits.

The concept of ecosystem may be applied to cities or agriculture (urban ecosystems and agroecosystems respectively), although these are not actually true, discrete units in terms of energy flows, function and so on. No two ecosystems are *exactly* the same; however, one may recognize general rules and similarities useful to environmental managers.

Some environmental and ecological definitions

Biodiversity (biological diversity) the variety and variability among living organisms and the ecological complexes in which they occur (biodiversity can be distinguished at various levels, from the chemical structure of cells up to whole ecosystems).

Biome major regions in which distinctive plant and animal groups usually live in harmony, and are also well adapted to the environment.

Biosphere complete global complex of living and dead organisms at or near to the Earth's surface. An artificial enclosed environment with a crew of eight designed to study ecosystem function and biogeochemical cycles was established in the USA in 1991, named 'Biosphere-2' (to emphasize the separation from the Earth's biosphere). This complex of hermetically sealed greenhouses maintained a breathable atmosphere and provided almost enough food for its crew for two years (although there had to be imports of oxygen, arousing some scepticism about the research) (Allen 1991).

Biota collective, plant, animal, fungal and microbial life characterizing a given region.

Ecology study of the structure and function of nature. The study of relationships between organisms of different species with each other and with biological, physical and chemical components of the environment in which they live.

Autecology study of how individual organisms function and relate to environment.

Synecology study of how communities composed of different organisms function and relate to environment.

Ecosystem (ecological system) a functional system that includes organisms together with their physical environment.

Ecotone transitional zone between ecosystems.

Environment external conditions and influences affecting life and the development of organisms.

Evolution process of change in the characteristics of organisms, by which descendants come to differ from their ancestors.

Gene the basic unit of inheritance found in chromosomes in the nucleus of living cells, genes control the genetic identity of each individual. A small section of DNA which contains information that can be passed from one generation to another.

Gene pool the collection of genes in an interbreeding population.

Genetic diversity the variety of genes within a particular species, variety, or breed.

Genetic resources useful characteristics of organisms that are transmitted genetically.

Habitat place where an organism or group of organisms live.

Species (abbreviated to sp. or – plural – spp.) taxonomic category of closely related, morphologically similar, individual organisms.

Trophic level organisms whose food is obtained from plants by the same number of steps are said to be at the same trophic level ('nourishment level').

Systems theory

In the 1920s systems theory was developed by ecologists as a means of assessing the laws governing the life of organisms. By the late 1940s, systems diagrams were being constructed to show energy flows between components of ecosystems and now it has been adopted by many social scientists and business managers as a framework for study and as a means of prediction (as discussed earlier, a systems approach was used by the Club of Rome to try to model global limits).

Physical scientists tend to try to treat social and physical processes as interacting parts of an ecosystem. Social scientists tend to consider humans as acting within an environmental, political and economic structure which contains their actions in the short term but which can be changed or modified in time (Mannion and Bowlby 1992: 4).

According to systems theory, changes in one component of a system will promote changes in other, possibly all, components – as is the case for ecosystems. As subsystems may interact in different ways the ecosystem approach is essentially holistic. Given time, natural, undisturbed ecosystems theoretically reach a state of dynamic equilibrium (steady state). Regulatory mechanisms of checks and balances (positive and negative feedbacks) counter changes within and outside the ecosystem (e.g. climatic changes) tending to maintain the steady state. Increasingly humans have upset the regulating mechanisms – checks and balances – of natural ecosystems.

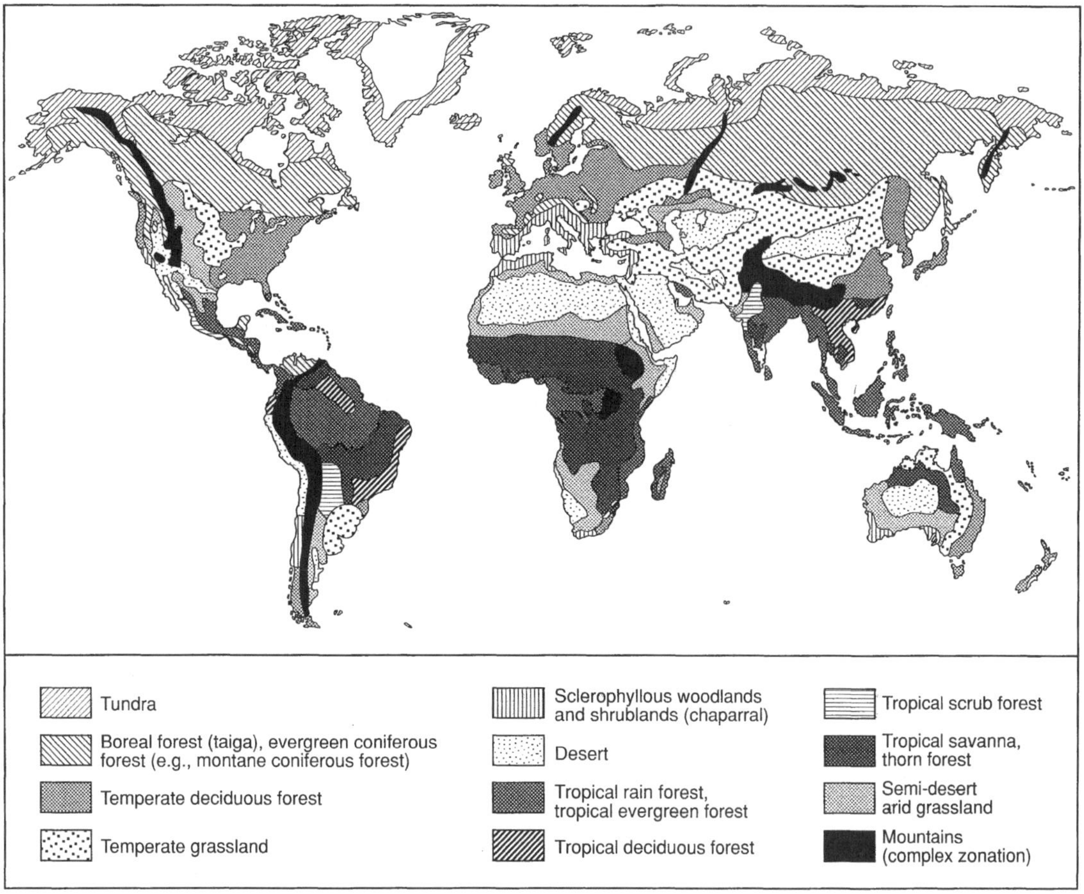

Fig. 2.2 Generalized map of world's major biomes ⟨*Source*: author from various sources, including Lean *et al.* 1990: 12–13; Barrow 1991: 12 (Fig. 1.2)⟩.

Ecosystems can be subdivided, according to local physical conditions, into habitats (places where an organism or group of organisms live) populated by characteristic mixes of plants and animals (e.g. a pond ecosystem may have a gravel bottom habitat and a mud bottom habitat). Environments can be divided more coarsely into biomes – a prevailing regional climax vegetation and its associated animal life (Park 1980: 150): in effect regional scale ecosystems (see Fig. 2.2). Biomes not only reflect climate, but also are likely to be shaped by incidence of fire, drainage, soil characteristics, type of grazing, etc. Biomes and habitats may be subdivided into communities, which consist of several populations of different species that live together in a particular place and which interact.

In a stable ecosystem each species will have found a position, primarily in relation to its functional needs: food, shelter, etc. This position, or niche, is where a given organism can operate most effectively. Some organisms have very specialized demands and so occupy very restricted niches (e.g. the water-filled hollow of a particular bromeliad plant, itself with a restricted niche), others can exist in a very wide niche. Niche demands are not always simple; in some situations a species may be using only a portion of its potential niche; it is also possible that alteration of a single environmental parameter may suddenly open, restrict or deny a niche for an organism, competition for the niche with other organisms is one such parameter.

The expression ecological diversity refers to the range of biological communities that interact with each other in a given environment. Biological diversity (biodiversity) refers to species diversity plus genetic diversity within those species. Loss of biological diversity occurs when species extinctions exceed the rate of species creation – which it is suggested in nature takes between 2000 and 100 000 years. Extinction is a natural process, sometimes sudden, perhaps catastrophic, otherwise an ongoing, gradual process (humans have accelerated the rate of extinction: loss of biodiversity is one of today's most serious problems).

Ecosystem function

Biogeochemical and biogeophysical cycles

Within the biosphere cyclic processes move and renew supplies of energy, water, chemical elements, sediment and air. These biogeochemical and biogeophysical cycles affect physical features and lifeforms and some cycles are affected by lifeforms. In general, barring occasional catastrophic events, biogeochemical and biogeophysical cycles were in a state of dynamic stability before human activity started to upset them. Upset to one process may affect others, problems could then become magnified and perhaps uncontrollable (a runaway positive feedback).

Biogeochemical cycles largely control the movement of nutrients between the non-living environment and living organisms (for an introduction to biogeochemical cycles see Park 1980: 54). All chemical elements are subject to some degree of biogeophysical or biogeochemical cycling, but there are cycles which are crucial for the nutrition of organisms or for the maintenance of atmospheric gas mix and global temperature within acceptable limits; these include water, oxygen, carbon dioxide, nitrogen, phosphorus and sulphur (there are over 30 known biogeochemical cycles). Some biogeochemical cycles involve gases and have a turnover of as little as a few days; some involve sediments, with a cycle time so slow (perhaps millions of years) that the material is practically non-renewable as far as humans are concerned.

Biogeochemical cycles are not fully understood, for example there is still much to learn about the crucially important cycling of carbon dioxide. Incoming solar energy 'drives' most biogeochemical cycling, the climate and weather. Very few ecosystems are not ultimately dependent on sunlight; exceptions include deep ocean hydrothermal vent ('smoker') communities mainly sited along the Earth's mid-oceanic ridges supported by hot water flows, typically 350° to 400°C and rich in sulphides that nourish bacteria which higher organisms consume (Cann and Walker, 1993). Biogeochemical and biogeophysical cycles could be classified as (1) natural, (2) upset by humans and (3) recycling (managed by humans and sustainable) (Chadwick and Goodman 1975: 4). Many of the first group have already been converted to the second and the threat of this grows; conversion to the third class is an important, challenging development goal.

Regulatory mechanisms

Organisms in an ecosystem can be grouped by function according to their trophic level (the level at which they gain nourishment). Each successive trophic level's organisms depend upon those of the next lowest for their energy requirements (food). The first trophic level, primary producers or (autotrophs), in virtually every case convert solar radiation (sunlight) to chemical energy (the exceptions include 'smoker' communities), the cliché 'all flesh is grass' is largely correct, in that the first trophic level is almost always photosynthetic plants. At level two are consumers of level one organisms, the primary consumers (carnivores, parasites or scavengers/ decomposers); at level three are consumers of level two organisms, the secondary consumers. Seldom are there more than four or five trophic levels including the first, because organisms expend energy living, moving, perhaps generating body heat, and when consumed by those of the next level have incurred losses. Transfer of energy from one trophic level to the next is unlikely to be better than 10 per cent efficient. Within an ecosystem there is likely to be a pyramid shaped pattern of trophic levels for organisms that consume others, with greater mass and number of organisms at lower levels (Fig. 2.3a).

Given the losses in energy transfer between trophic levels it is possible to feed more people if they eat at a low rather than high trophic level: put crudely, a diet of grain would support more people than would be possible if that grain were used to fatten animals for consumption as meat, eggs or milk (Fig. 2.3b). One ecologist calculated that only about one part in 100 000 of solar energy makes it through to a carnivore.

The energy cost of various food production systems

To remain healthy humans need appropriate amounts of carbohydrate, vitamins, minerals protein and

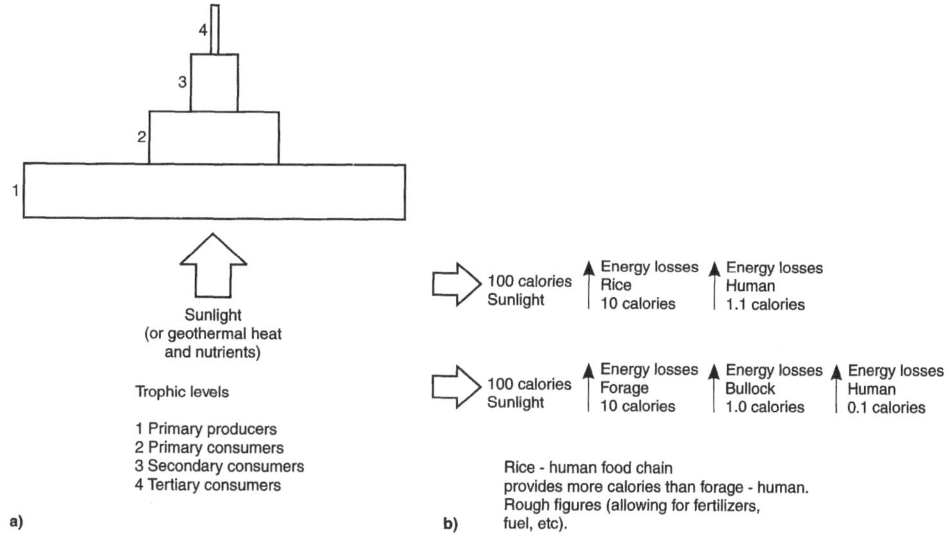

Fig. 2.3 Energy flows in ecosystems. a) Energy pyramid. b) Hypothetical food chains. *Note*: Biomass pyramids can also be drawn and often resemble a) However, for oceans these may show less bulk at level 1 than 2, 3, 4 because phytoplankton fix energy better than land plants, multiply fast and are moved by currents to consumers.

various fats, determined by the local environmental conditions, activity they follow and individual metabolism. Rational agriculture production should produce the most and best nutritional quality food with the minimum energy expenditure and environmental damage on a sustained basis (subsidiary aims might include minimization of cruelty to any livestock; ensuring the system is resilient against climatic changes, disease or other perturbations; avoiding inclusion of toxic or other harmful compounds in the food).

Food production systems vary in their energy inputs (sunlight, labour, fuel, fertilizer, etc) and energy output (food). In DCs the production system is likely to include considerable (hydrocarbon) energy inputs for transport, storage, processing and packaging, in LDCs much less (and most of that manual labour). In the USA a typical commercial agricultural system might use 10 energy units (much of which is from non-renewable fossil fuel) to put 1.0 unit of (food) energy on the consumer's table. In an LDC 1.0 units of energy (mainly human labour) might provide 10 units of (food) energy (Miller 1990: 265–6). Obviously these are gross generalizations because there is considerable range of energy input to output ratio within DC and LDC agriculture, for example between shifting cultivation and sedentary smallholder production (also LDC consumers probably have to expend energy processing unprepared foods).

The DCs tend to over-consume protein and sugars (to the detriment of health of many people) and obtain this from livestock or milk products, yet adequate protein can be had from vegetable sources (e.g. soya, groundnut, cereals or potatoes, etc). In terms of protein yield per unit area (as well as energy inefficiency and risk of waste products pollution) livestock are clearly a poor foodsource, as the following list of produce indicates: note that these are very rough figures and yields of a particular crop vary a lot (Tudge 1977: 41–4).

Produce	*kg protein yield per unit area*
potatoes	150
cereals	140
soya	100
broiler chickens	35
beef/sheep	10

The sum total of biomass (organism mass, expressed as live weight, dry weight, ash-free dry weight or carbon weight) produced at each trophic level at a given point in time is termed the standing crop. This needs to be treated with caution: if taken at the end of an optimum growing period, it indicates potential; if taken during a drought, cool season, period of agricultural neglect or insect damage, it does not. Ecologists commonly measure the energy produced at an ecosystem's first trophic level, minus the estimated respiration losses, which gives net primary productivity in units of:

$g\ m^{-2}\ d^{-1}$ (biomass/area/day) or

$g\ m^{-2}\ y^{-1}$ (biomass/area/year)

Net primary productivity gives a measure of the total amount of usable organic material produced per unit time at the first trophic level. Studies so far show that most cultivated ecosystems, i.e. human efforts to maximize food and commodity production, are well below the net primary production of several natural ecosystems. There is thus, in theory, very great potential for the improvement of agriculture.

Within even the most simple ecosystems there are complex relationships among organisms and between organisms and environment. Seldom is there a simple food chain: more likely are intertwined chains forming a food-web – complex pathways along which energy (food) and perhaps pollutants are passed. There is a tendency for pesticides, radioactive isotopes, heavy metals and other pollutants to become concentrated in organisms feeding at higher trophic levels: apparently 'harmless' background contamination could, through this 'biological magnification' (bio-accumulation), prove harmful to humans and other animals (a threat made public by Carson 1962).

Ecosystems adjust to perturbation through regulatory mechanisms. When the relationship between input and output to the system is inverse (e.g. increased sunlight causes more cloud which reduces the impact of that sunlight on the surface), it is termed a negative feedback. The opposite is a positive feedback, whereby an effect is magnified. There is always a risk that a feedback could result in an uncontrollable runaway reaction. Human impact upon the environment is particularly dangerous if it causes perturbations leading to runaway reactions affecting critical biogeochemical or biogeophysical cycles and one role of environmental management is to watch for such threats.

Ecological stability and succession

It is widely held that, given long enough, a steady state will be reached by an ecosystem with a web of relationships that allow it to adjust to serious localized or moderate widespread disturbances (Park 1980: 99). The ecosystem should remain in such a steady state unless a critical parameter alters sufficiently; if change then occurs it is termed ecological succession or biotic development. Over a very long period of time organisms may evolve to an evolutionary maturity; over a shorter period a successional maturity may be reached before such evolution can occur (Johnson and Steere 1974: 8).

The concept of ecological succession, pioneered by Clements (1916) is complex and still debated. Organisms, by occupying an environment, may modify it to assist others – a birch wood may act as a 'nursery' for a pine forest, which ultimately kills and replaces the birch – thus birch is a successional stage en route to a pine stage. These transitional vegetation stages leading to a mature climax community are known as seres. Each vegetational stage or sere will have a characteristic assemblage of macrofauna and micro-organisms.

Two types of successions are recognized: primary succession and secondary succession. Primary succession is the sequential development of biotic communities from a bare lifeless area (the site of a fire, volcanic-ash, newly de-glaciated land, etc). Secondary succession is the sequential development of biotic communities from an area where the environment has been altered but has not had all life destroyed (cut forest, abandoned farmland, land that has suffered a flood, or been lightly burnt, etc). Where succession is taking place from a bare area the first stage is known as the pioneer stage, although in practice the expression may be applied to growth taking place in areas that do have some life – such as regrowth after logging (natural forests may be assumed to maintain maturity, rather than becoming senile and degenerating, through 'patch and gap' dynamics – clearings caused by storms, etc, allow regeneration). Pioneer communities have a high proportion of plants and animals that are hardy, have catholic niche demands, and disperse well (weeds with wind-carried seeds, insects which can fly, etc). Mature, climax communities tend to have more species diversity, tend to recycle dead matter better (in effect manage nutrient cycling well) and are believed to be more stable. Many communities do not reach maturity before being disturbed by natural forces or humans; increasingly people are disturbing successions moving toward stability or stable mature climax communities.

Systems approach to environmental study and management

As already discussed, ecologists commonly use an ecosystems approach when seeking to understand and monitor a given situation, and it is a useful scale at which to work (Bartelmus 1986: 1; a critique of the ecosystems approach may be found in Pepper 1984: 107–10). There is also a growing interest in understanding and managing global environmental systems. Adoption of a systems approach means that less attempt is made to break down the environment into its components for study, but rather to take a holistic view to see how components work together.

A system is a set of linked components (the linkages may not be direct); the next section suggests classifications of types of environmental systems. An ecosystem approach is not without its problems: it can be difficult to recognize boundaries; measurement of what goes in and comes out can be difficult; establishing whether an ecosystem is genuinely natural, rather than modified, can be difficult; and it is possible for organisms to migrate into or out of a particular ecosystem. Nevertheless, it is often possible to get some idea of an ecosystem's energy and material distribution, even if a detailed and completely accurate picture is impractical. Without understanding all of the complex interactions one could model the behaviour of an ecosystem, although the more complex the ecosystem, the more difficult this becomes. There is also a chance that some of the processes operating work at random and therefore cannot be modelled satisfactorily with currently available approaches.

Classifications of environmental systems

By function

Isolated systems

Boundaries are closed to import and export of material and energy.

Closed systems

Boundaries prevent import and export of material, but not energy. An example is the recently created 'Biosphere-2', which receives sunlight, but is supposed to function with no other exchanges. It is a hermetically sealed glasshouse constructed in the desert of Arizona in an attempt to establish a functioning ecosystem *separate* from the Earth's environment and capable of supporting several people indefinitely. Experiments begun in 1992 have so far had difficulty maintaining a stable, habitable, enclosed ecosystem. The Earth ('Biosphere-1') is in effect a closed system, although it receives dust, larger meteorites and solar radiation.

Open systems

Boundaries allow free exchange of material and energy. Many of the Earth's environmental systems are of this form and may actually be interdependent.

By degree of human disturbance

Park (1980: 42) suggested that environmental systems could be classified as:

1 *Natural systems* unaffected by human interference.
2 *Modified systems* affected to some extent by human interference.
3 *Control systems* human interference, by accident or design, plays a major role in function (includes most agricultural systems).

Environmental limits

Von Liebig's Law of the Minimum states that whichever resource/factor necessary for survival is in short supply is the critical one (the limiting resource/factor). The limiting resource/factor restricts population growth of a species. Limiting resources/factors include: lack of water, space, nutrients, harsh climate, noise, radiation, recurrent fires, a predator, disease, etc. A population of organisms will tend to grow until it encounters a resource limit/limiting factor; the effect may then be gradual or sudden, limited or catastrophic, or cyclic 'boom and bust', but whatever course is taken there will be a cessation of growth.

The limits to human growth have been the subject of debate since Thomas Malthus in the late eighteenth century (see the discussion of limits to growth in chapter 3). Doubtless many rural communities were aware of limits well before that, as they had evolved and enforced restrictions on marriage. There has been speculation as to what the maximum global human population could be without causing serious disruption of the Earth's life-support systems. Miller (1991: 138) suggested that with likely technology and foreseeable economic development, global population might reach 10 billion or even 30 billion (where 1 billion = 1 thousand million). The global population is already more than halfway toward the lower of those two suggestions and, given that growth is exponential, humanity would be well advised to start treating the problem with urgency. Caution is necessary when dealing with estimates of the population that the Earth might support. It may be possible to produce 40 tonnes of food per person for the 1990 global population (roughly 100 times greater than present needs), but will there be investment, environmental and social stability allowing such levels of production in the future? It is also possible that there will be more environmental damage in the future which would depress production. Meadows *et al.* (1992) suggested that the limits have already been exceeded through development and population growth, but that there is still hope of survival and ultimately stabilization at a level offering reasonable lifestyles.

Human population growth is not the only long-term threat: the environmental manager must consider 'over-consumption' by the rich with its associated pollution and demand for resources; nuclear warfare; technological accidents, such as release of radioactivity, chemicals or a dangerous, perhaps 'genetically engineered', organism; and natural catastrophes such as a major volcanic eruption or planetesimal impact (see page 61).

Popular environmental books commonly publish variants of three 'laws of ecology' which it is suggested should guide those concerned with environmental management:

1 We can never do merely one thing. Any intrusion into nature has numerous effects, many of which are unpredictable.
2 Because everything is connected to and intermingled with everything else, we are all in it together, what one person or country does affects others.
3 Any substance that we produce should not interfere with any of the Earth's natural biogeochemical processes.

Resources

A resource is something with utility: the capacity to meet people's perceived needs or wants. In practice, economic and non-economic criteria determine utility although the economic tend to dominate. Non-economic criteria include those things not easily quantifiable in economic units: aesthetic quality, sense of moral duty to conserve wildlife, cultural importance, religious beliefs, etc. A useful definition of a resource is 'something which meets needs or wants'.

An economist might subdivide resources into those with actual value, those with option value (possible use perceived), and those with intrinsic value (no obvious practical value, but there is a will to maintain it). These somewhat utilitarian definitions require a resource to be used or have the potential for use by humans (this may require extraction, processing, etc), and there must also be demand for it. Something may, however, have no actual or perceived utility at present, but there is no guarantee that in the future it will not have; indeed resource demand is subject to change as human perceptions alter, new technology is developed, fashions vary and perhaps new materials are substituted. There are signs of interest in environmental management with non-utilitarian goals (this could be seen as part of the postmodern trend), the resource being valued for its own sake and if need be utilization being forgone.

Resources may be directly available (e.g. air, seawater or soil), others only indirectly after processing, so as to be economically viable and 'safe' enough to be accepted by people (most metals and petroleum products). A classification of resources follows below, but a simpler classification than that would be:

1 those that can be safely 'stretched' by humankind;
2 those that can be 'stretched' with care;
3 those which cannot or should not be 'stretched'.

The 'stretching' of resources might be achieved by such strategies as the alteration of natural vegetation to agriculture; the conversion of slow-growing woodland to fast-growing plantation; the farming of fish rather than fishing wild stocks.

A classification of resource types

All resources apart from perpetual resources are renewed at a rate that makes them finite as far as humans are concerned. Potentially renewable and extrinsic resources are ends of the same continuum and the classes between overlap somewhat.

Perpetual resources (continuous resources)

These are more or less inexhaustible because they are renewed over a short enough timescale to meet human demands. Mismanagement is unlikely to damage supply (Rees 1990: 15 uses the term *non-critical flow resources*), e.g. solar energy; geothermal heat; wind and wave energy.

Non-renewable resources (stock resources)

These may become economically depleted, i.e. too costly to exploit; or totally depleted so that future technological progress and or demand is unlikely to improve supply (e.g. species extinctions, cultural losses). In short, they are consumed by use or capable of being permanently damaged. A metaphor often applied to such resources is that of a lollipop: each 'lick' (usage) is a step towards none in the future.

Theoretically renewable (theoretically renewable stock resource)

Probably any elemental mineral is theoretically renewable; if, not at present, due to costs and technology limitations.

Recyclable resources (theoretically recyclable stock resource)

Non-renewable resources and some potentially renewable resources can be recycled, i.e. collected after use and processed for reuse (e.g. aluminium, steel, some plastics, glass, etc). Recovery may be theoretically possible, but material could be too dispersed and/or mixed with difficult to separate contaminants to be worth recovery efforts.

Potentially renewable resources (flow resources)

In theory they can last forever as they are renewed (within a timespan relevant to humans) by natural processes (e.g. soil, river flows, grassland, forest, fisheries, etc) or in some cases by human agency. However, if poorly managed these resources can become non-renewable. Rees (1990: 15) uses the term 'critical flow resources' to denote those capable of renewal but which become non-renewable if mismanaged, i.e. utilization exceeds the regenerative capacity of the resource. Under growing resource demand a critical point (threshold) or zone will be reached, if use continues to pressure the resource there may be temporary or permanent collapse to non-renewability.

Extrinsic resources

These include human skills (human resources), institutions, management abilities, etc. These can be fickle, prone to breakdown or degradation, but can also be renewable resources if well managed (Riddell 1981: 23). Schumacher (1974: 64) argued that education is the most vital of all resources.

Common property resources

Often nobody is able or willing to control the use of resources which are owned by no single person, yet used by many. Each user helps place the resource under stress; too many users may exceed the critical threshold or zone and cause the resource to fail. The failure can be difficult to forecast and sudden, e.g. grazing may not increase on common land but might happen at a critical time of year resulting in problems.

Common property resources may be divided into *local and regional common property resources* (Berkes 1989) and *global commons*. Examples of the former include many of the world's grazing lands, fisheries, deep ocean mineral deposits, and the Antarctic. Examples of global commons include the world's atmosphere or ozone layer, and geosynchronous orbits. Biogeochemical and biogeophysical cycles are crucial for the maintenance of a number of global commons and are at risk from or are already disrupted by human activity.

It is also possible to divide common property resources into open access, communally owned and state owned.

1 Open access resources belong to no one and have no restriction on who makes use of them. These can be very difficult to manage.
2 Communally owned resources mean that the co-owners can exclude 'outsiders' e.g., in medieval Britain, common grazing land was open only to those groups who were granted or inherited the right of use.
3 State-owned resources might be better managed, although that is not always true.

In practice a common property resource may be a blend of the above sub-types and ownership (common, private etc.) may be less important than incentive to stewardship.

Resource use

The amount of a particular resource believed to exist is the total resource; the term 'identified resource' is applied to that which has actually been mapped and assessed. Reserves or economic resource is that which is extractable, given current technology, economic conditions and civil order. Undiscovered resources are those deemed by specialists to be likely to exist but not proven (e.g. offshore oil deposits around the Falkland Islands). Resources vital to a country are critical resources and those needed to ensure national security are strategic resources. Strategic resources may prove crucial triggers for human conflict, and there is also a tendency for environmental controls or resource use restrictions to be waived when dealing with these.

A comparison of known resource supplies and rates of use yields a depletion rate, typically the time it takes for 80 per cent of known reserves to be used.

A question increasingly being asked is: who should bear the cost of rehabilitation after resources exhaustion and abandonment of exploitation? For example, the Pacific island of Nauru has provided phosphates for some 90 years; now independent, does it have any claim on past colonial powers to remedy mining damage? Nauru has made a claim through the International Court of Justice for damage done before its independence in 1967 (Anderson 1992).

Are there areas of the world particularly prone to environmental problems?

The answer is 'yes', certain parts of the world are more likely to suffer natural disasters: seismically and volcanically active areas, those subject to hurricane or tornado, typhoon or tsunami, land prone to flooding, avalanche, landslide, sudden frost or intense storms. These natural threats can be recorded on a hazards map or database. It is also possible to map things that are especially vulnerable to disruption, say by pollution damage or erosion. Hazardous industry or power generation and potential military targets can also be mapped. These maps can be overlaid to assess risks. Unfortunately efforts at such mapping are often limited.

Given long enough, even the safest area could be subjected to some sort of catastrophe (Huggett 1988; 1989). Onset and severity of problems may be altered by human activity, for example, land may be 'sensitized' by humans, making it more vulnerable to threats like drought or soil erosion. The effects of storms can be magnified if people have removed vegetation cover from watersheds or coastal land. Development may drive people, or they may be attracted, into areas where they themselves cause or they trigger natural processes to cause environmental problems. People often evolve strategies of environmental use: nomadic grazing, shifting cultivation, etc, which minimize environmental problems and risks to their well-being but these strategies may break down as a result of development. For example, international trade and changing economic conditions can cause land use to degenerate (Shrybman 1990). Addressing the problems caused by breakdown of established land-use strategies is one of the more important challenges facing development.

Managing the environment

For decades there have been calls for the integration of environmental thinking with planning and to link conservation with development of resources. Recently the concept of sustainable development has gained popularity and might offer a means of achieving those goals. Two approaches to the pursuit of such aims have emerged, without a clear-cut split between them: natural resources management and environmental management. The former is primarily concerned with management of land and renewable resources and may be described as a process of decision-making, whereby resources are allocated after consideration of environmental, economic and cultural aspects (Omara-Ojungu 1992: 5). Environmental management has a wider brief stretching to encompass global problems and future issues.

Environmental management

Environmental management has been described as the process of allocating natural and artificial resources so as to make optimum use of the environment in satisfying basic human needs at the minimum, and more if possible, for an indefinite future (Jolly 1978). In the early 1980s the UNEP saw environmental management as the control of all human activities which have a significant impact on the environment (Tolba 1982: 25).

The goals of environmental management are to minimize environmental impacts, to husband, and to add to existing resources for the future, in effect sustaining and where possible improving 'quality of life', however this may be locally perceived, for all the world's people (i.e. sustainable development). Environmental management is based on the assumption there is an optimum balance of natural resource uses. The manager must determine where this balance lies and then use planning and administration to reach it. A criticism has been that many efforts at such management consider the environment only after development objectives have been set (Redclift 1985: 8). Another potential problem is a piecemeal approach to development with managers adopting short-term planning horizons and too limited (spatially) a view. Project-by-project development must be replaced by better overall management at policy level (Shramm and Warford 1989).

The environmental manager needs to know the structure and function of the environment and how the human users act. Once the processes of change are known it should be possible to control them better. Certain inputs are crucial: access to consistent, reliable and comparable sets of data on environment and human activity; enforceable rules and laws of environment and resource use; and means of measuring performance in achieving successful environmental management (Ruddle and Manshard 1981: 5). Recently some have advocated an environmental accounts framework to assist with data gathering, storage and use and to attach a value to environment and resources (Pearce *et al.* 1989: 93–119) (discussed in chapter 11). Given the uncertainties involved, the environmental management process should be flexible and adaptive (Holling 1978).

One of the main goals of environmental management, sustainable development (discussed in chapter 4), may involve trade-offs and the loss of immediate benefits from resource and environment exploitation to ensure longer-term utility. For some countries or groups of people such trade-offs will be a cruel choice unless assistance can be given them, so in the future more foreign aid is hopefully going to be linked to the drive for environmental management and sustainable development.

References

Allen, J.P. (1991) *Biosphere 2: the human experiment*. New York, Penguin

Anderson, I. (1992) Can Nauru clean up after the colonialists? *New Scientist*, 135(1830): 12–13

Atkinson, A. (1991) *Principles of political ecology*. London, Belhaven

Bartelmus, P. (1986) *Environment and development*. London, Allen and Unwin

Berkes, F. (ed.) (1989) *Common property resources: ecology and community-based sustainable development*. London, Belhaven

Blaikie, P.M. (1985) *The political economy of soil erosion in developing countries*. Harlow, Longman

Cann, J. and Walker, C. (1993) Breaking new ground on the ocean floor. *New Scientist* 140(1897): 24–9

Carson, R. (1962) *Silent spring*. Boston, Mass., Houghton Mifflin (London, Hamilton)

Chadwick, M.J. and Goodman, G.T. (eds) (1975) *The ecology of resource degradation and renewal*. Oxford, Blackwell Scientific

Clements, F.E. (1916) *Plant succession. An analysis of the development of vegetation* (Publication no. 242). Washington, DC, Carnegie Institute

Elton, C. (1927) *Animal ecology*. Seattle, University of Washington Press

Fraser-Darling, F. (1963) The unity of ecology. *British Association for the Advancement of Science* 20: 297–306

Holling, C.S. (ed.) (1978) *Adaptive environmental assessment and management*. Chichester, Wiley

Huggett, R.J. (1988) Terrestrial catastrophism, causes and effects. *Progress in Physical Geography* 12(4): 509–32

Huggett, R.J. (1989) Superwaves and superfloods: the bombardment hypothesis and geomorphology. *Earth Surface Processes and Landforms* 14(5): 433–42

Johnson, W.H. and Steere, W.C. (eds) (1974) *The environmental challenge*. New York, Holt, Rinehart and Winston

Jolly, V. (1978) The concept of environmental management. *Development Forum* VIII(2): 13–26

Mannion, A.M. and Bowlby, S.R. (1992) *Environmental issues in the 1990s*. Chichester, Wiley

Meadows, D.H., Meadows, D.L. and Randers, J. (1992) *Beyond the limits: global collapse or a sustainable future*. London, Earthscan

Miller, G.T. jnr (1990) *Living in the environment: an introduction to environmental science* (6th edn). Belmont, Calif., Wadsworth Publishing Company

Miller, G.T. jnr (1991) *Environmental science: sustaining the earth* (3rd edn). Belmont, Calif., Wadsworth

Odum, E.P. (1975) *Ecology: the link between the natural and the social sciences* (2nd edn). London, Holt, Rinehart and Winston

Omara-Ojungu, P.H. (1992) *Resource management in developing countries*. Harlow, Longman

O'Riordan, T. (1976) *Environmentalism*. London, Pion

Park, C. (1980) *Ecology and environmental management*. Folkstone, Dawson

Pearce, D., Markandya, A. and Barbier, E. (1989) *Blueprint for a green economy*. London, Earthscan

Pepper, D. (1984) The roots of modern environmentalism. London, Croom Helm

Redclift, M. (1985) *The struggle for resources: limits of environmental 'managerialism'*. Berlin, International Institut für Umwelt und Gesellschaft (IIUG ks 85-5)

Rees, J. (1985) *Natural resources: allocation, economics and policy*. London, Methuen

Rees, J. (1990) *Natural resources: allocation, economics and policy* (2nd edn). London, Methuen

Richerson, P.J. and McEvoy, J. III (1976) *Human ecology: an environmental approach*. North Scituate, Mass., Duxbury

Riddell, R. (1981) *Ecodevelopment – economics, ecology and development: an alternative to growth imperative models*. Farnborough, Gower

Ruddle, K. and Manshard, W. (1981) *Renewable natural resources and the environment*. Dublin, Tycooly International

Sargent, F. II (ed.) (1974) *Human ecology*. Amsterdam, North Holland

Schumacher, E.F. (1974) *Small is beautiful: a study of economics as if people mattered*. London, Sphere Books (first published 1973 London, Blond and Briggs)

Shramm, G. and Warford, J.J. (eds) (1989) *Environmental management and economic development*. Baltimore, Md., Johns Hopkins University Press

Shrybman, S. (1990) Free trade versus the environment: the implications of GATT. *The Ecologist* 20(1): 30–4

Steward, J. (1955) *Theory of cultural change: the methodology of multilinear evolution*. Urbana, Ill., University of Illinois Press

Tansley, A.F. (1935) The use and abuse of vegetational concepts and terms. *Ecology* 16: 284–307
Tolba, M.K. (1982) *Development without destruction: evolving environmental perceptions.* Dublin, Tycooly International
Treshow, M. (1971) *The human environment.* New York, McGraw-Hill
Tudge, C. (1977) *The famine business.* London, Faber and Faber
Watts, D. (1976) *Principles of biogeography.* New York, McGraw-Hill
Wilkinson, R.G. (1973) *Poverty and progress: an ecological model of economic development.* London, Methuen

Further reading

Miller, G.T. jnr (1991) *Environmental science: sustaining the earth* (3rd edn). Belmont, Calif., Wadsworth (London, Chapman and Hall) [Readable introduction to environmental science, resources and problems of the environment with abundant colour illustrations. Intended as textbook for American colleges]
Rees, J. (1990) *Natural resources: allocation, economics and policy* (2nd edn). London, Methuen [Good introduction to natural resources and their use]

CHAPTER 3

The state of the earth: environmental crisis; ecocatastrophe; environmental catastrophe?

What is meant by 'crisis'?

Warnings that humans and the environment face a crisis or crises have been widely articulated, particularly since the 1960s (Arvill 1967; Park 1980: 21; White 1993). Often the cause is identified as people's cavalier use of nature. Crisis may be seen as a turning-point, a last chance to avoid or mitigate or adapt to undesirable, possibly irreversible change, which if bad enough may merit being called a catastrophe. A crisis may also offer a chance for change, something the Chinese have long acknowledged: Johnston and Taylor (1989: 4) noted that the Mandarin character for 'crisis' is a combination of pictographs for 'danger' and 'opportunity' – *wei-chi.*

Crisis has become an over-worked word, a cliché. People's perceptions differ so not all may agree given circumstances constitute a crisis: what is crisis for one may be normal to another. The term is also prone to emotive, journalistic usage (Blaikie 1988). Some, mainly on the political left, suggest that the idea of a crisis is a 'liberal cover-up' to divert attention from 'real problems' like social injustice and poverty, the 'evils of capitalism' (Clark 1975). Young (1990: 142–3) presents a right-wing and a left-wing perception of the environmental crisis. Weston (1986: 4) felt that environmental problems were in the main really due to social problems; he and others would therefore point to a social or ethical crisis (in that ethics and attitudes determine how humans use the environment). Earlier L.K. Caldwell (1990) had suggested that the roots of environmental crisis were in 'mind and spirit'.

Large disasters get noticed, everyday side-effects of development go unnoticed or are accepted: it is these that are often behind major environmental problems.

Is there a crisis or crises?

Identification of crisis may be a mistaken response to a 'patchy problem', reflecting inadequate observation. Objective and careful monitoring is vital to be sure there really is a crisis (Thompson *et al.* 1986; Blaikie and Unwin 1988: 7). Writing on 'rural poverty unperceived', Chambers (1983: 13–27) noted a range of biases which lead to false impressions, e.g. a researcher's tendency to view roadside areas; the majority of studies are made during dry seasons; interviews with unrepresentative groups of people; research that is too short term. Ives and Messerli (1989) discussed the Himalayas, which many identify as in environmental crisis; their studies indicated little firm evidence that this is so, indeed some records show that conditions in certain areas were markedly worse several decades ago and have now improved. Blaikie and Unwin (1988: 13) cited an example of gully erosion in Zimbabwe identified as constituting a crisis, where careful study revealed that only about 13 per cent of total soil loss was from the spectacular gullys, while 87 per cent was from insidious inter-gully sheet erosion. Funds could easily have been spent averting the 'crisis' of gullying (a symptom of the problem) rather than sheet erosion (the actual problem).

There are dangers in adopting a crisis-orientation: that decision-makers will suddenly respond to a problem (crisis management or 'fire-fighting' approach) rather than devote sustained proactive efforts to avoid or solve it (Henning and Mangun 1989: 3). Too often a response focuses on symptoms and not on real causes of problems. It should also be stressed that infrequent or random events can suddenly cause considerable environmental change with no sign of an approaching crisis.

Some environmentalists in the 1960s and 1970s predicted crisis, even 'doomsday', by the present time (Ehrlich 1970; Eckholm 1976). Either they were right and we still do not recognize the threat or they got the timescales wrong, or were needlessly worried; it would be foolhardy to dismiss the warning 'save nature or humankind will die'.

With any complex system there may be crisis of component parts, the breakdown might be relatively insignificant or it might alone or in combination with other factors contribute to overall system failure. The global environment is such a complex system. Decision-makers need to recognize significant crisis points (thresholds), act and avoid problems. An area of mathematics, catastrophe theory, is concerned with the way in which systems can suddenly change by passing through a catastrophe (i.e. crisis point). Catastrophe theory may aid the identification of critical environmental thresholds before they are reached.

In practical terms it matters little whether death is through one or a few major or a multitude of minor cuts: the implication is that environmental management should try not to overlook anything. Sir Crispin Tickell's comments on global loss of flora and fauna could be applied to environmental crisis: 'We can remove one, two, or ten rivets. But at a certain point – it could be the eleventh or the thousandth rivet... things fall apart' (*The Times* 27/4/91: 4).

It is often possible to forecast natural hazards, but difficulties caused fully or in part by human activity may be less predictable. Such problems may cause a crisis for a locality, a group or groups of people, a region, nation, or the Earth as a whole. The causation may not be direct in time or space: it could be the cumulative effect of more than one factor or the result of some unexpected feedback process. Consequently the recognition of a crisis in advance may be difficult. Even with adequate knowledge of the structure and function of nature and human behaviour things are unpredictable (for example, the Cultural Revolution in the PRC led to tremendous damage to infrastructure and agriculture which would have been difficult to predict in advance). Problems could generate 'ecorefugees' who then generate further difficulties, or a resource shortage might generate a war leading to further threats. Accident and terrorism nowadays have the potential to generate severe environmental problems.

Global crisis, megacrisis, supercrisis, or pseudocrisis?

Biogeochemical cycles, some of which sustain life, have been or could be altered (Rees 1990: 242; Thomas 1992). Many commentators at the Rio 'Earth Summit' observed that environmental conditions had deteriorated in the 20 years since the 1972 Stockholm Conference. Authors like Merchant (1992: 17) recognize a global ecological crisis, rooted in social causes, notably the 'clash' of industrial and agricultural production and population increase with nature. However, there is no general agreement on whether and when the world will face a crisis or crises. One may recognize the following categories of perceived or potential crisis from literature or media sources (these are not arranged in any order of importance, nor are they wholly separate groups of problems):

1 Renewable resource depletion and degradation.
2 Global environmental change.
3 Pollution.
4 Until recently nuclear warfare was seen by many as a major threat.

Ekins (1992: 1) identified four 'interlocking crises':

1 Nuclear weapons.
2 Hunger and poverty.
3 Environmental pollution and loss of species diversity.
4 Increasing human repression that deprives many people of their basic rights and standard of living.

A global crisis?

The Brundtland Report (World Commission on Environment and Development 1987) rekindled the warnings made in the 1960s and 1970s, in particular those of global environmental threats. The report called for action as quickly as possible if a global environmental crisis was to be averted. Table 3.1 lists some of the environmental threats identified since the 1960s; a number of these have the potential to disrupt, or are already disrupting processes or affecting organisms on a global scale.

In the mid-1990s accelerated global warming is generally *seen* to pose the greatest threat. Projected population growth rates set against projected per capita availability of key resources: land, water, food and fuelwood also present a worrying picture. Forecasts are that forested area, irrigated land and grazing land per person will decrease alarmingly and there can be little doubt that soil degradation is a very serious problem. While the threat of nuclear war has receded it has not gone. It seems that environmental managers are going to have to deal

Table 3.1 Critical environmental issues identified since 1960

Trend/significance (see key)				
B		□	○	Agrochemical pollution (pesticides, herbicides, fungicides, etc)
B		□	○	Acid precipitation
C		□	○	AIDS (Acquired Immune Deficiency Syndrome) – upset to land use/increased poverty
B		□	○	Biotechnology (see chapter 8)
C		□	◑	Biological warfare (see also genetic engineering accident)
A		□	●	Climate change (see chapter 5)
C		□	⊘	Chemical warfare
C	■		●	CFCs (chlorofluorocarbons – disruption of ozone layer/greenhouse effect) (see chapter 5)
B		□	●	Consumerism – accelerates production of goods increasing resource demand/pollution
B		□	○	Coral reef destruction – loss of biodiversity, loss of fish/shellfish breeding sites, exposure of coasts to storms/ erosion
A		□	●	Degradation of soil – loss of productivity of land (see chapter 6)
C		□	○	Desertification (overlap with previous threat – see chapter 6)
A		□	●	Deforestation (loss of tropical forest especially damaging – rapid loss of species) (see chapter 7)
B	■	□	◑	Depletion of freshwater resources
B		□	◑	Ecorefugees
B	■	□	●	Enclosure of common resources
B	■	□	○	Effluent – sewage
B	■	□	●	Effluent – industrial
B		□	●	Economic stagnation/depression
A		□	●	Greenhouse effect
A		□	●	Genetic loss/erosion (extinction of flora and fauna, loss of biodiversity)
A		□	◑	Genetic engineering accident ('escape'/uncontrolled release of harmful biological agent)
C		□	○	Industrial accident (toxic release/blast)
B	■	□	◑	Nuclear generation/waste disposal accident
B		□	◑	Nuclear warfare
B	■		●	Ozone layer damage (see chapter 5)
A		□	●	Population increase
B	■	□	●	Pesticide use (agrochemical pollution) (see chapter 8)
B		□	○	Poverty
C		□	○	Resettlement – pressure on land from resettled
C		□	○	Technology – promotion of inappropriate technology
A		□	●	Toxic releases (via effluent, accident, warfare, agrochemicals, atmospheric pollution)

Key

Potential solutions/trends
■ good chance of solution
□ threat increasing

Subjective assessment of risk
greatest A medium B least C

Subjective assessment of significance
● global/national significance – active
○ national/regional significance – active
◑ global/national significance – threat
⊘ national/regional significance – threat

with a range of threats, simultaneously or in succession.

The growing number of threats has been seen as indicating a 'progressive loss of ecological stability' (Simonis 1990: 26). Reviewing the state of the world, L.R. Brown *et al.* (1991) noted that 'every major indicator shows a deterioration of natural systems: forests are shrinking, deserts are expanding, croplands are losing topsoil'.

The Brundtland Report suggested a rough timescale: 'Most of today's decision-makers will be dead before the planet feels the heavier effects of acid precipitation, global warming, ozone depletion.... Most of the young voters of today will still be alive' (World Commission on Environment and Development 1987: 8). Another indication of time to crisis can be obtained by reviewing projected human demand for the Earth's primary production. Given the crucial role of photosynthesis for life, any disruption of primary production, whether through pollution, ozone depletion or accelerated global warming, could lead to crisis. In 1992 humankind used about 40 per cent of total primary production of on-land photosynthesis ('usable land plants') for food. In roughly one generation from now human population will have doubled: it can be assumed that this will result in the use of about 80 per cent of terrestrial primary production. Even if climatic change and pollution do not depress photosynthesis and if agricultural productivity improves, the limits are getting close: within two generations crisis seems likely unless population slows or primary production can be hugely improved (Holmberg 1992: 27).

A crisis in the South?

At the 1972 Stockholm Conference on the Human Environment, problems of the environment were widely regarded in LDCs as matters for the DCs, as the LDCs were too poor and beset with development difficulties to afford the luxury of worrying; anyway, the rich were to blame for most pollution. Until the mid-1980s it was common for LDCs to suspect calls to protect the environment of being a ploy to hold back their development and continue their dependency or to withhold aid. Virtually all LDCs would now accept that there are real problems, which require attention. The nature and blame for problems are less clear. Questions often asked are:

1 Are the DCs' and the LDCs' environmental problems similar?
2 Are some or all of the LDCs' problems caused by the DCs (or vice versa)?
3 Are LDCs suffering more environmental damage than the DCs?
4 Are the LDCs more vulnerable to problems?

The LDCs have tremendous diversity of environment, style of government, administration, historical background, degree of poverty, etc. However, two things are common: poverty and mainly low-latitude location. Although some LDCs lie partly or wholly outside the tropics or have extensive highlands with relatively temperate conditions, between the Tropics of Capricorn and Cancer lies 35–40 per cent of the Earth's landsurface where, with the exception of a few city-states, there are virtually no DCs (Ooi Jin Bee 1983: 2). Whether this reflects accidents of history or special handicaps associated with the tropics (e.g. soils that are often infertile and difficult to manage without degradation, more pest and disease problems, and water supply difficulties) has long been debated (Kates and Haarmann 1992).

Adams (1990: 6–8) suggested that the LDCs faced a double-crisis: a crisis of development and a crisis of environment. The first of these involves debt, falling commodity prices and poverty. The second involves

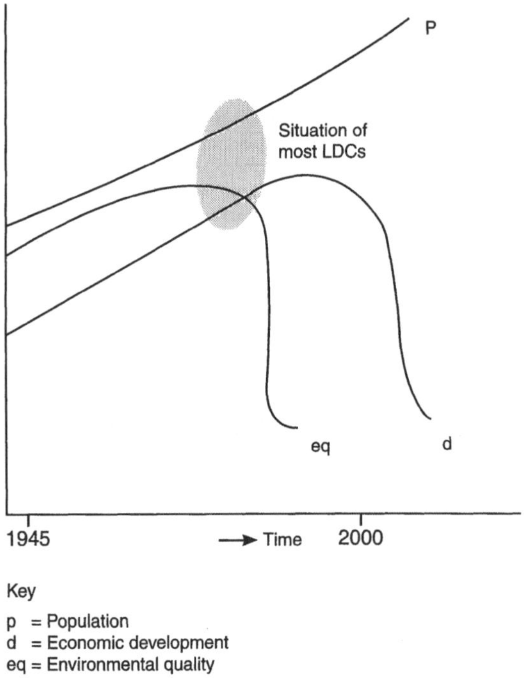

Fig. 3.1 Schematic representation of the predicament of many LDCs.

global change, resource exploitation and energy supplies.

There are authorities who argue that environmental problems are largely due to physical and intellectual duality (humans are bound to the Earth but transcend it in thought and wants) or to a failure to arrive at ways of thinking about environment that deal satisfactorily with physical and social issues (Caldwell 1990: 8). A school of thought with growing support holds that there is a LDC environmental crisis which relates to the disempowerment of local people, i.e. 'locals' can no longer participate in resource management and are losing access to common resources (Ghai and Vivian 1992: 72; The Ecologist 1993).

That the LDCs are facing severe problems is clear. Figure 3.1 is a schematic representation of the predicament of many LDCs, struggling to maintain living standards with little to spend on countering environmental problems. LDC populations are growing; however, their individuals consume far less per capita of the world's resources than DC populations. In an interdependent world both will have to cooperate or conflict and failure to resolve problems will probably follow. Whether global

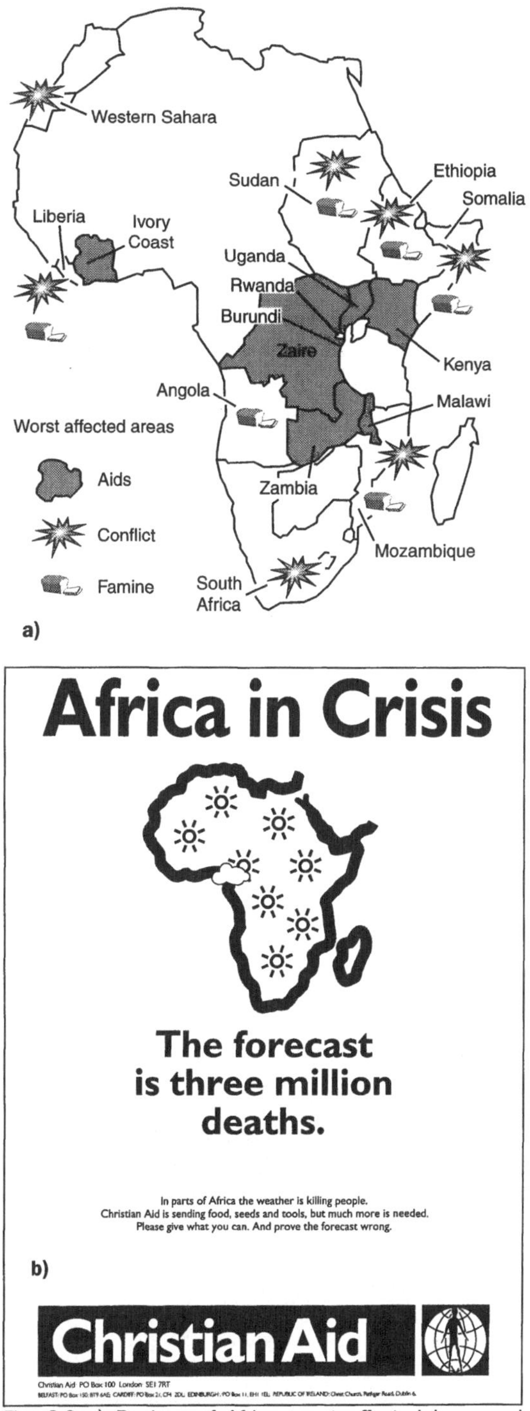

Fig. *3.2* a) Regions of Africa worst affected by armed conflict, AIDS and famine (1991) ⟨*Source: Sunday Times* 9/7/91: 1.23 modified⟩. b) NGO Africa crisis poster 1993 ⟨*Source:* Christian Aid⟩.

environmental change will relatively disadvantage LDCs is difficult to predict; for example, climatic change may have very different effects on even closely neighbouring countries. It would also be easy for desperate LDCs to disrupt the complex workings of DCs if abandoned to any crisis.

An African crisis?

Africa is frequently singled out as having or being close to an environmental crisis, a development crisis or both (Timberlake 1985; Commins *et al.* 1986; Ravenhill 1986; Watts 1989; Davidson *et al.* 1992). Things look particularly grim south of the Sahara (excluding South Africa), consequently many refer to a crisis in subSaharan Africa.

Harrison (1987: 17–26) concluded there was a crisis, particularly an environmental crisis, and that most of the African continent was not developing but was regressing. In his view there were four main factors to blame:

1 There was a food supply crisis, manifest as a decline in per capita food production.
2 Poverty was increasing.
3 There was a debt crisis which got worse as Africa's exports fell in value and imports rose in cost.
4 There was an environmental crisis, getting worse, as vegetation and soils become degraded.

To these difficulties must be added unrest, which may be an important factor: Africa, with less than 10 per cent of the world's population, had almost 50 per cent of the world's refugees in the late 1980s (Harrison 1987: 52) (see Figs 3.2a and 3.2b). The disease AIDS seems to have taken a particular hold in parts of subSaharan Africa and may increasingly contribute to the continent's problems.

Africa's crisis, some claim, results from a particularly unfavourable geology and climate. The continent is ancient and weathered and most of Africa escaped the glacial activity that many other parts of the world experienced in the last few million years, which gave them fresh, often fertile deposits. Consequently, except near the volcanoes of the Rift Valley and Cameroon and in some river floodlands which receive regular deposits of fresh sediment, it is often claimed soils are poor (the volcanic and floodland soils of Rwanda and Burundi support some of the densest populations in Africa).

However, not all agree: Blaikie and Unwin (1988: 20) were sceptical that there was a general African environmental crisis, noting that soil erosion was not spectacular and river silt loads were modest.

Many point to drought as a cause of crisis, yet there is no conclusive evidence that rainfall is less or receipts more variable in recent decades than it has been during the past few thousand years (Holmberg 1992: 225). More likely drought in Africa exposes the weakness of governance: it is a 'litmus of development'.

Food production *is* worrying: in the 1950s most of Africa was self-sufficient; now most states import, and it is the only continent to show a per capita decline in food production. By AD 2000 it has been predicted that 65 of the world's 117 nations will be unable to feed their populations, and at least 30 of those will be in subSaharan Africa (Barrow 1991: 15). Food security for Africa would probably demand a 4 per cent per annum increase of the continent's food production, plus a similar increase in export crops to provide foreign exchange for inputs. Since the early1960: the average growth rate for food production has been about 2 per cent and export crop production has shown a decline (added to which the prices for these crops have fallen making it difficult to buy inputs and to encourage producers) (World Bank 1990: 89–90). There is widespread acceptance that Africa has a severe food problem but little agreement as to why (Rau 1991), although suspicions have been voiced that it reflects widespread traditional communal or state land ownership, and that a move toward individual land ownership (Harrison 1992: 106) or improved communalism might help.

What are the key 'crisis' causes?

Often it is possible to recognize what (natural and human activity-related) might cause a crisis, but tracing *why* these things happen is less easy. It has been suggested that shortcomings in western ethics are frequently the root cause (ethics are disciplined reflection by persons in all walks of life on moral ideas and ideals), and that new ones must be found which consider nature as well as human demands. However, non-western countries also have environmental problems and act in ways likely to lead to more, for example the neglect of education, aid and equality for women can be found in western, westernized and non-western countries. Women are often those who grow food, fetch fuel, draw water, etc., and should be involved in problem avoidance and mitigation (women and environment are discussed in chapter 11). Expenditure on armaments and on conflicts diverts money from environmental management and is likely to lead to disruption of people and physical damage; this too is something found in western and non-western countries.

Table 3.2 Availability (global) of basic natural resources per person in 1990 and projected for AD 2000 (in hectares)

Resource	1990	2000
Grain land	0.13	0.11
Irrigated land	0.04	0.04
Forest land	0.79	0.64
Grazing land	0.61	0.50

Source: Brown *et al.* 1991: 17 (Table 1.3)

Environmental problems might be reduced or even avoided if key difficulties were identified. Identifying threats to 'key' resources (see Table 3.1) is valuable, but is no guarantee that some unforeseen natural catastrophe, human activity or environmental process will not suddenly accelerate degradation or elevate another less obvious threat to criticality. Study, monitoring and development must be sensitive, flexible and wide-ranging to try and cope with the unforeseen. The following sections review possible causes of crisis; in any given situation there may be one or several causes at work (usually one is crucial – the limiting resource or factor).

Population changes

The Malthusian thesis is that human reproductive capacity puts pressure on means of subsistence. A population may be said to have reached a crisis point if either its rate of growth or its size generate demands out of balance with the prevailing environmental, economic and social conditions. In Europe, the arrival of the Black Death (*c.* AD 1348) 'solved' supply problems for a population that had been growing for more than four centuries and which was overstressing natural resources; it also overturned the established socioeconomic order.

It is often argued that population increase places a strain on resources and that there will be a rapid worsening of the position (Myers 1992), in particular stress on food supplies and infrastructure (see Table 3.2). However, that ignores the effect ('Boserüpian thesis') that population increase can have to stimulate agricultural production and advance technology, for example population growth in Europe probably 'drove' farmers to farm clay-lands and shift from long-fallow to annual cropping. For a discussion of

Malthusian and Boserüpian views see Harrison (1992: 11–19). It is too simplistic (Malthusian determinism) to say it is a rule that population growth leads to environmental, economic or social problems: environmental impact is a function of population, standard of living, the technology practised and attitudes. Some people can cause devastation at low population levels and it is probably fair to say that, up to a point, population increase becomes a socioeconomic problem only if food production technology fails to keep up. Population changes may not be synchronous with environmental degradation.

Population change results from imbalance between rates of change in mortality (rate of deaths), fertility (rate of increase) and migration (movement into or out of a given area) (Johnston and Taylor 1989: 78; Wood 1989: 161). Between 1800 and 1990 world population is estimated to have increased six-fold (Kemp 1990: 3); between 1650 and 1992 it increased nearly eleven-fold (see Fig. 3.3a).

In DCs a demographic transition has occurred, whereby death-rates have fallen as a consequence of improved healthcare, nutrition and hygiene and then birth-rates have fallen. Rising 'standards of living' probably help trigger demographic transition, although specific causes are complex and debated and vary from locality to locality (see Fig. 3.3b). In most LDCs death-rates have fallen far more than birth-rates but a demographic transition may be difficult to achieve if economic development is slow or possibly because of sociocultural factors, for example in some regions of Africa reduction of length of breast-feeding may be the cause.

A hoped-for decline in human population growth during the 1980s has not materialized. The world population in 1992 was over 5400 million, more than twice what it was in 1950, and it is projected to grow by over 70 million a year (Young 1990: 106) to around 6260 million in AD 2000 (L.R. Brown *et al.* 1991: 16). About 90 per cent of this increase will be in LDCs (Henning and Mangun 1989: 16). Optimistic predictions suggest that stabilization may occur at around 8000 million (Holmberg 1992: 326); less optimistically 11 500 million (Davidson *et al.* 1992: 61); conservative predictions are that it is unlikely to occur before 12 500 million (World Bank 1992: 26); pessimists suggest 14 000 to 19 000 million by AD 2050. Most agree that the 1990s is the last decade in which control can be humanely achieved in order to avoid catastrophe (Hartshorn 1991: 401).

There have been some promising signs, for example Thailand has reduced its population growth by 50 per cent between 1978 and 1993 (L.R. Brown *et al.* 1993: 19). The impact that HIV/AIDS and other major disease outbreaks will have is unknown.

There is a difference between maximum human population that can be supported with no respect for other organisms and that which can be supported if there is to be adequate conservation. Conservation issues and optimum level of quality of life should be considered in any attempt to compute a target human population for the world; it is probable that such a population would not be much more than today's.

It seems likely that Africa and South Asia will continue to have rapid growth until the early twenty-first century, at which point they will have about half the global population. Latin America and East Asia show some signs of slight decline in growth and by the early twenty-first century they will have about 30 per cent of the global population. Europe, the CIS, North America and Oceania are likely to continue to fall, from about 20 per cent of the global population at present, to around 9 per cent. An increasing proportion of the world's population will be urban: around AD 2000 numbers in cities will start to exceed those in rural areas (World Bank 1992: 28).

Often, population growth has negated or will soon negate significant improvement in food production and, in some areas has begun to degrade agricultural production. These problems may be exacerbated by the influx of migrants. At present most migrants are refugees, displaced by unrest or attracted by hope for better livelihood; in the future environmental degradation may generate large numbers of ecorefugees.

Projected population growth and food production suggest depressing future scenarios, particularly for subSaharan Africa (Harrison 1983; 1984; Barrow 1987: 14–19). Increasingly LDC environmental problems will occur not only in ecologically fragile rural areas and marginal lands, but also around cities (peri-urban areas) where the destitute concentrate seeking housing and livelihoods, and perhaps also where people are vulnerable to rising sea-levels.

There are situations where even at very low population density certain activities (e.g. ranching) lead to severe damage, especially in sensitive environments. Caution is necessary when examining population–environment relationships: for example, there has been little study to check a common assumption that the presence of poor people correlates with environmental degradation (Kates and Haarmann 1992). Population increase might lead to reduced environmental damage if agriculture becomes more intensive, provided the right strategies

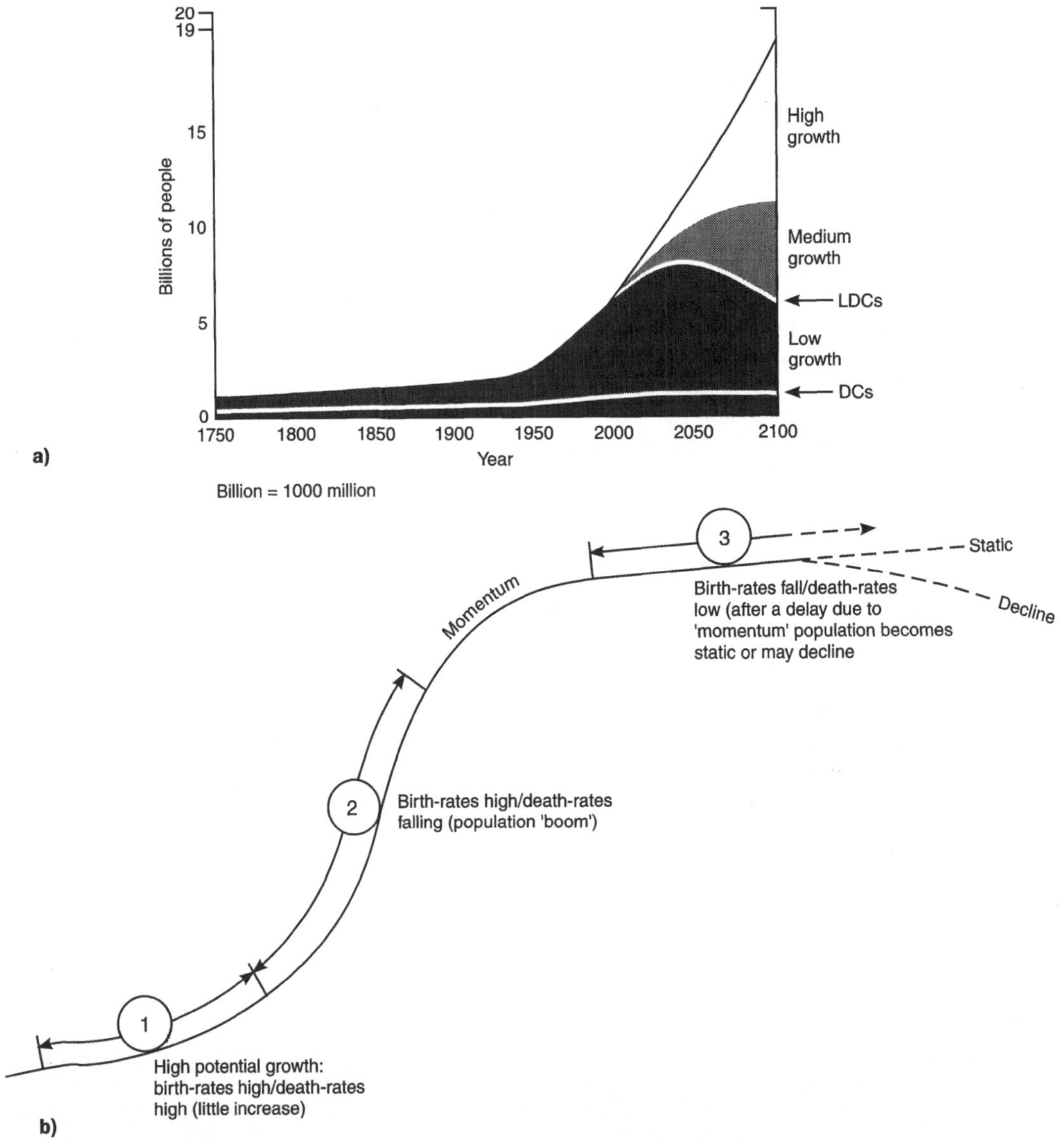

Fig. *3.3* Human population growth. a) Past and projected population growth AD 1700–2100 (*Source:* part based on *Population Reports* (The Environment and Population Growth). Series M, No. 10 (1992) Baltimore, Johns Hopkins University Press: Fig. 1 p. 5). b) Demographic transition: a pattern of growth experienced by western industrialized democracies.

and methods are adopted. Tiffen (1993) discussed the relationship between increasing population density, productivity and environmental degradation in the Machakos District of Kenya and concluded development and environmental conservation occurred even though population density increased and the climate was difficult (drought-prone dryland).

Population stagnation or decline can also lead to environmental problems for example, if neglected pasture develops a scrub cover or erosion control measures are abandoned (Barrow 1991: 17). Population growth projections are therefore not a particularly reliable indicator that environmental problems will increase (Sarre 1991; UNFPA 1991).

Breakdown of traditional agriculture and strategies for coping with adverse conditions

Strategies which have served people for perhaps thousands of years, often in harsh environments, have in some cases broken down in recent years or are under stress. The reasons for this are diverse, but include population increase; spread of commercial agriculture; adoption of new crops; restrictions on movement of people or livestock, etc. Often traditional agriculture has suffered from the 'stigma of backwardness': those in authority scorn it, offer little support, seek to replace it with something modern, and perhaps unsuitable, and blame it for land degradation and poverty. Unfortunately modern agriculture has so far had little to offer agriculturalists in harsh environments that works in practice.

An energy crisis in the South?

There is a growing literature on the 'Third World energy crisis' (Agarwal 1986; Leach and Mearns 1989). Most of the world's poor depend on fuelwood, charcoal, dung or crop residues for cooking and heating; a good deal of small- and medium-scale industry in LDCs also uses such energy sources. When fuel supply becomes difficult, poor people tend to pay a large proportion of their income on obtaining it and may seek to reduce cooking which is not a good thing for health and nutrition under LDC conditions and with most tropical foodstuffs.

Land degradation is often linked to the collection of fuel (Barrow 1991: 168), for example the World Bank (1985) found fuelwood collection was a crucial cause of desertification in parts of the Sahel. The link between land degradation and energy demand is a subject of considerable debate, Foley *et al.* (1984: 11) and Leach and Mearns (1989: 9) suggested fuel often comes from areas undergoing agricultural clearance, i.e. is a by-product, rather than cause of tree loss. Even so, fuel collecting may deliver the *coup-de-grâce* to damaged land. To protect land, people's dependence on traditional fuels may have to be altered or met from sustainable sources.

Assessing how bad the fuel supply situation is can be a problem: attempts to compare estimates of fuel use (seldom accurate or comparable) and projected population growth do not allow for the fact that people adjust their usage and consumption; also supply networks can be complex and secretive (Munslow *et al.* 1988: 14).

Marginalization

People may become marginalized – forced or attracted on to poor quality, perhaps easily degraded land, where they live 'close-to-the-edge' and become progressively more disadvantaged and vulnerable. In 1980 perhaps three-fifths of LDC households had inadequate landholdings – effectively most were marginalized (Harrison 1992: 126). The reasons for marginalization are diverse, and include efforts to escape unrest, the hope for employment or access to farmland, eviction from conservation areas or from the estates of large landowners, altered trade opportunities, changes in labour costs and availability, widowhood, reservoir flooding and so on. Environmental or socioeconomic change or technological innovation can cause people to become marginalized (or de-marginalized) *in situ*, for example drought, disease or pests, pollution, decline in demand for produce due to change in fashion or substitution, labour becoming too costly, communications degenerating, marginal land user attitude changes, the arrival of employment opportunities or crops new to the area.

What tends to happen during marginalization is that the resource exploiters overstress their environment and cause a vicious downward spiral, often closely connected with poverty (see Fig. 3.4), with nowhere to move or no incentive to move, and no means of moving, they become unwilling agents of environmental damage and their own ultimate demise. Marginal land tends to demand inputs and to be less forgiving to users and so is least likely to get investment as aid is directed to where returns are more certain. Poor people may be quite aware of the implications of their action, but have no alternative means of survival.

People forced to move often end up in difficult environments and are also likely to be to some degree traumatized; they are likely, together with many of those who have willingly relocated, to lack the necessary local experience and resources to establish sustainable agriculture. Many practising degenerate shifting cultivation are 'shifted cultivators', people who have been forced to move. Marginal land gets overlooked; better land receives inputs as damaged marginal land can be abandoned with little impact on central government.

Resource access difficulties

Where land tenure or access to some other vital resource is insecure, users will tend to maximize

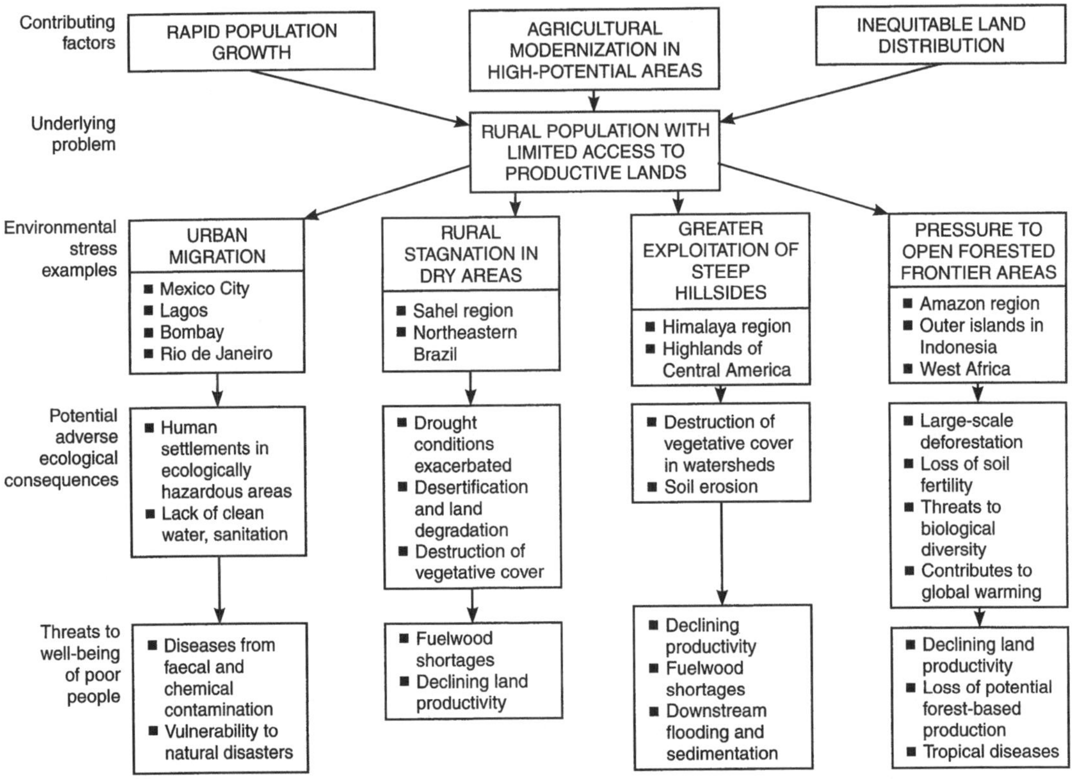

Fig. 3.4 The poverty and environment connection ⟨*Source:* Leonard *et al.* 1989: 7 (Fig. 1)⟩.

short-term gains and will be reluctant or unable to invest in sustaining the quality of the land or resource. Sometimes resource access problems drive people to settle dangerous or easily degraded environments, for example Bangladeshi farmers settle newly exposed mudflats that are at risk of flooding because the land tenure situation denies them other usable land.

As discussed in chapter 2, maintaining the quality of common property resources (commons) tends to be problematic. There are two alternatives for controlling commons, either enclose/'police', and usually develop them, or rely on local control, often through traditional rules. World-wide, the enclosure of common resources is a growing problem. The 'tragedy of the commons' argument has lost supporters, mainly because it missed the fact that individual users of common resources tend to get penalized for misuse and so things are self-policing. There is now more support for a 'tragedy of enclosure' argument: that enclosures can often abuse a resource and escape established community punishment. It has been argued that enclosure has widened in recent times to include labour and intellectual property as well as land and, more recently still, through patent-like laws, genetic material (The Ecologist 1993).

Food production

The struggle to produce more food to supply growing, often more demanding populations, means that there have been several decades of borrowing resources, which it can be argued should be left for the future (L.R. Brown *et al.* 1991: 11). Per capita output of grain (on a world scale) peaked in 1984: one reason for the decline seems to be environmental degradation (L.R. Brown *et al.* 1993: 11). As the 1990s progress there may be reduced per capita food resources, more volatile food prices and greater risk of famine. In spite of good harvests in 1990 major grain producing countries stockpiled little against future shortages and the trend continues.

Those regions of the world that currently produce a food surplus should not be assumed to be invulnerable, greenhouse effect-related changes: the long-term consequences of commercial farming and policy changes could lead to declining production, with no

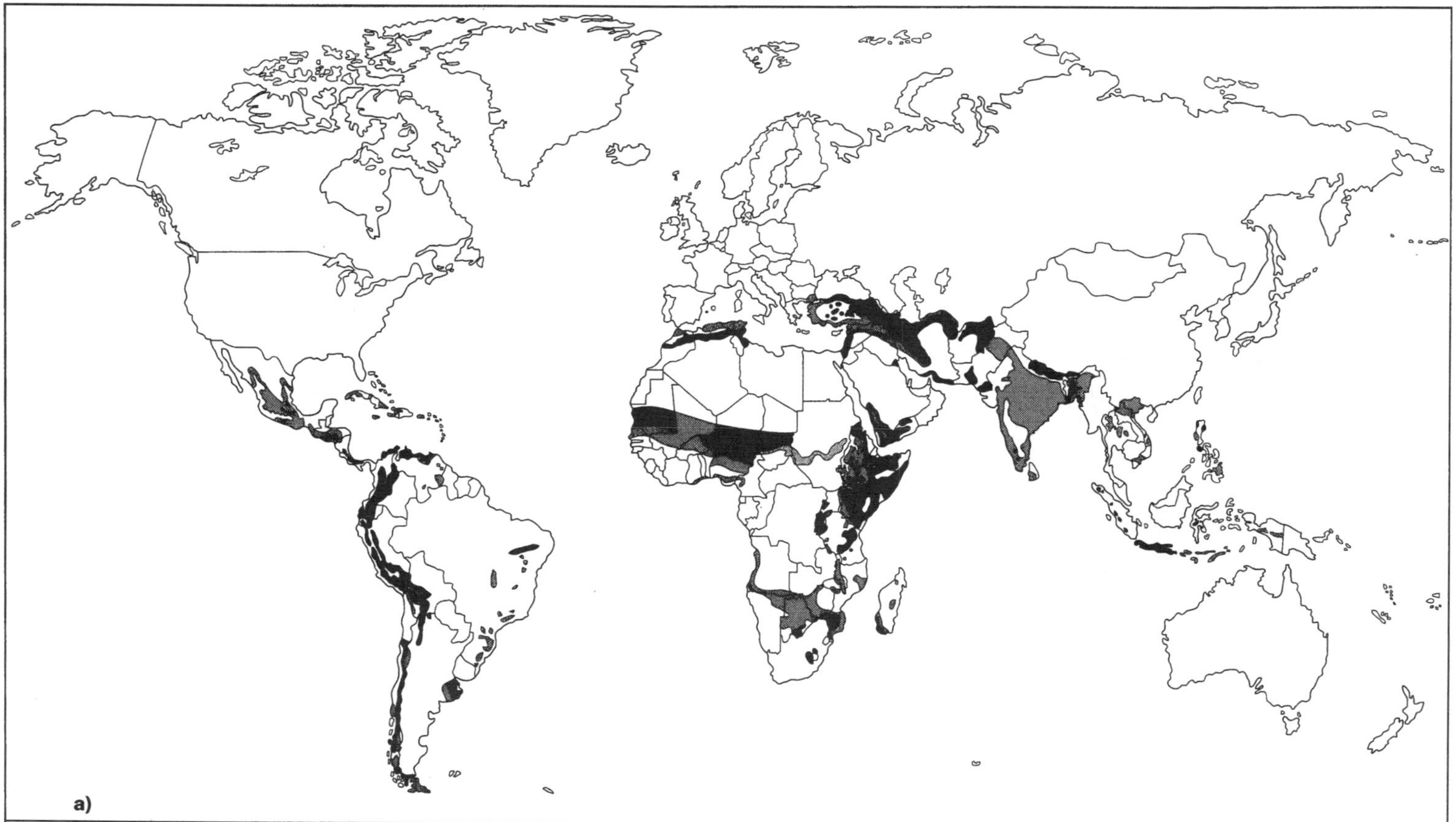

Population exceeds carrying capacity even with high levels of green revolution inputs available

Population exceeds carrying capacity with intermediate levels of inputs available

Population exceeds carrying capacity with low level, traditional inputs available

Note: In 1975, 38 per cent of the total land area was carrying more people than could be fed with low inputs; 22 per cent of the total land area was carrying more people than could be fed even with high inputs.

Fig. 3.5 Exceeding the limits. a) Regions where population exceeds the land's potential carrying capacity ⟨*Source:* based on *Environment* 30(9): 31 (Fig. 1)⟩.

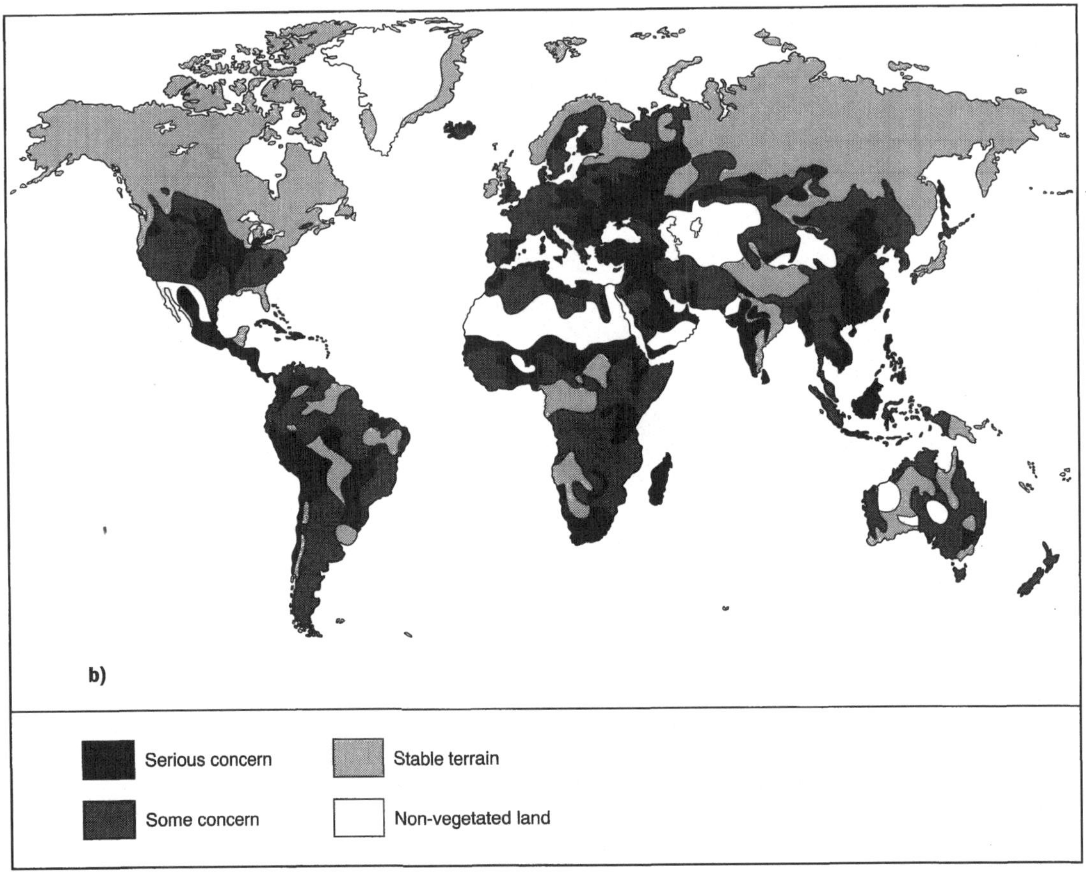

Fig. 3.5 b) Areas of concern for soil degradation (*Source: New Scientist* 16/5/92: 7).

guarantee that alternative regions will make up the shortfall.

Some countries, a huge proportion of which are in subSaharan Africa, are soon likely to have, or already have, populations in excess of their production capacity (using presently affordable agricultural techniques) (Fig. 3.5a). Unless output is raised or trade in food between surplus and non-surplus areas is improved whole countries or regions will suffer.

Poverty

Roughly one-fifth of the world's population, about 1500 million, suffer absolute poverty, i.e. lack enough to eat, and suffer degraded and shortened lives (Davidson *et al.* 1992). In the short term, attempts to remedy and avoid environmental problems may divert funds and efforts from aiding the poor: ideally environmental protection and poverty alleviation should complement each other.

People who for centuries have practised resource exploitation or landuses which have caused no significant long-term damage may be disrupted by development, modify their strategies, cause environmental problems and possibly suffer hardship. Disruption can be caused by even quite minor changes, for example in demands, habits, attitudes, trade, seasonal weather, and so on, particularly if the resource or landuse strategy is poised on a 'knife-edge' (Chambers *et al.* 1981). The terms of trade are often an important root cause of poverty which may drive people to damage the environment and also starves governments of funds to counter such problems (see chapter 11).

It is often observed that the poor inhabit degraded environments and that they are more likely to be found in the tropics (see Fig. 3.5b). The evidence for the former assertions is not clear, although there is some correlation between latitude and numbers of poor people (Barrow 1987: 27;

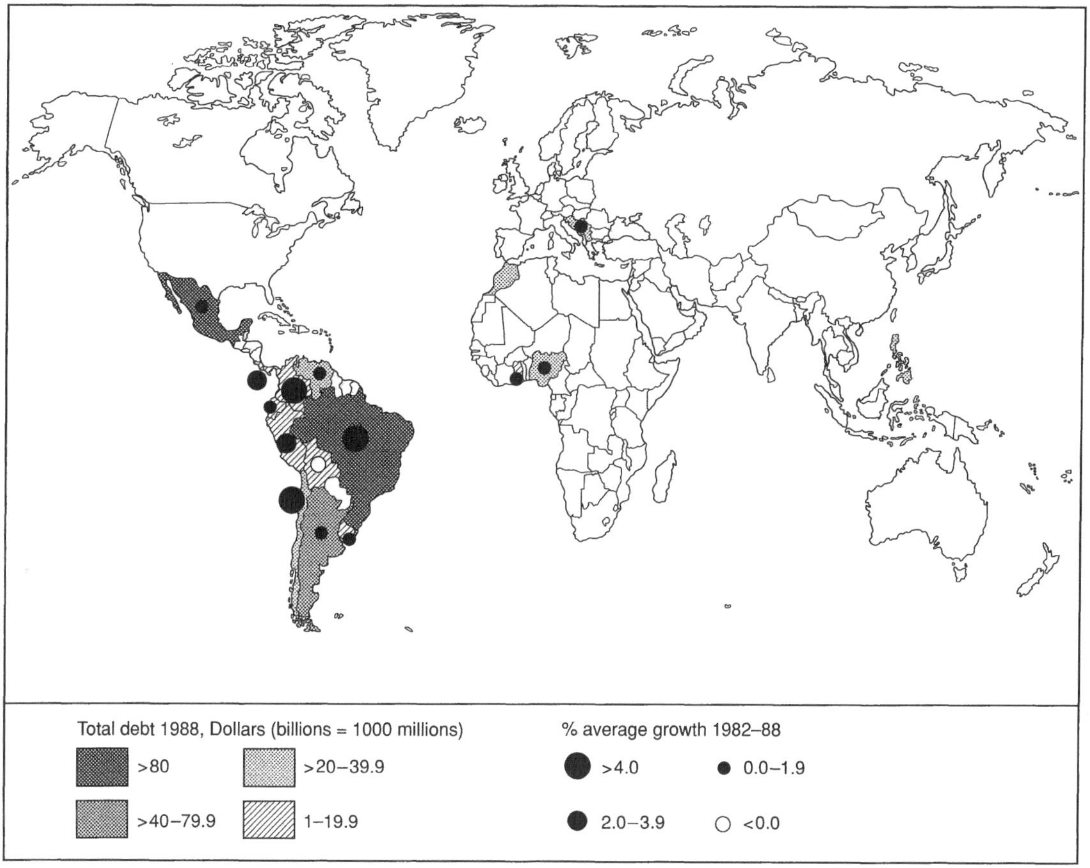

Fig. 3.6 World map of debt (*Source:* Johnson and Taylor 1989: 37 (Fig. 2.7)).

Kates and Haarmann 1992). Study of poverty and environmental degradation suggest that three factors can combine to cause marginalization: (1) rapid population growth; (2) land consolidation and agricultural modernization in fertile agricultural areas; (3) prevailing inequalities in land tenure (Leonard *et al.* 1989: 5). Harrison (1992: 236–44) suggested that the impact of population increase on environment was a function of population increase × consumption × impact per unit of consumption. Technological changes can increase or decrease the impact of a given population, e.g. better engines have helped counter increased numbers of cars, whereas the introduction of non-returnable plastic bottles and aluminium cans has increased the impact of soft drink consumption.

The reason why poor people continue to have many children may be social and economic insecurity: a large family is seen as a defence. Breaking this cycle of poverty linked to rising population and environmental damage will require social reform, particularly adequate welfare support.

Economic causes and inappropriate policies

The South's environmental problems are commonly attributed to exploitation during colonial times and the more recent effects of a world trade situation that has disadvantaged LDCs (Redclift 1984). Yet proposals to 'liberalize' world trade in agricultural and related products might, if not carefully implemented and internationally coordinated, severely compromise the ability of nations to protect their natural resources (Ritchie 1990). For example, increased competition between companies as a consequence of free trade might reduce the will to invest in pollution control. Alteration of subsidies, export quotas and other policies could have considerable environmental impact.

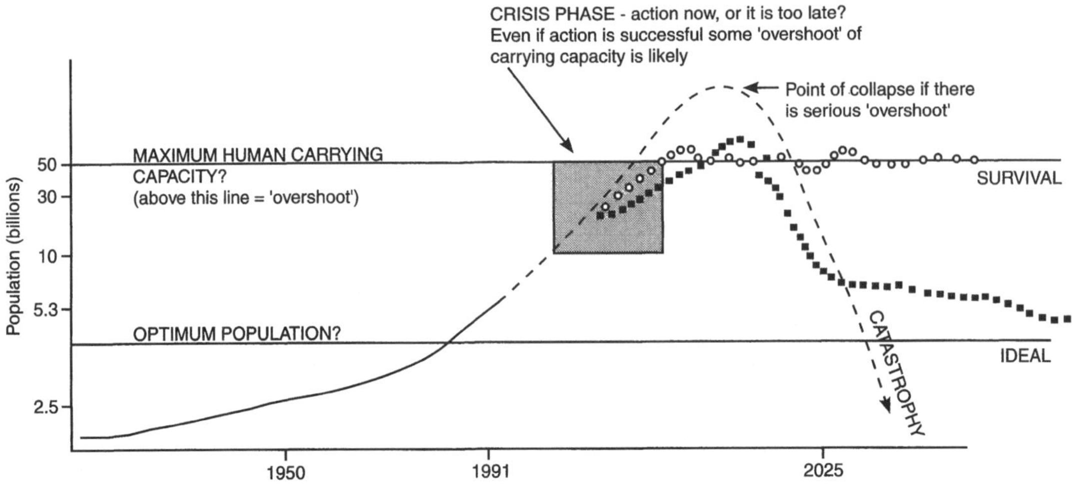

Fig. 3.7 Generalized population growth and control options scenarios.

One problem highlighted by Young (1990: 133) and others is that in general the economic philosophy that controls most modern development regards ethical considerations as beyond its scope. Another problem is that development is often on too grandiose and therefore inflexible a scale, with the developers too careless or arrogant to consider local needs, risks or limitations. Once a decision is made such a development process becomes a juggernaut that cannot easily be modified or halted to meet unforeseen challenges (Adams 1990).

Debt crisis

Foreign debt has become a significant factor in causing environmental degradation (see Fig. 3.6). For a complex of reasons many LDCs and former eastern bloc countries acquired large foreign debts after the mid-1970s, and more recently the USA and the former USSR have developed massive balance of payments problems. By 1982 some of the LDC debtor countries were facing a crisis servicing the interest and debt repayment and a few had to default. The burden of debt has meant many LDC and DC countries have less to spend on environmental management (LDCs especially) and may overexploit resources to obtain foreign exchange to meet their commitments (Carley and Christie 1992: 118).

Inappropriate technology

Concern that development has drawn on inappropriate technology has been voiced since the 1970s (Farvar and Milton 1972). Certainly there are many modern innovations that are not really suitable for LDCs but which are attractive (asbestos roofing, air-conditioning, etc). Often problems arise because western technology has been developed to fit temperate DC environments and requires petroleum inputs, specialist skills and so on. LDCs may have difficulty getting access to some western technology or the system favours the DCs. This is particularly apparent with biotechnology innovations and modern crops. The solution is to develop appropriate technology and many institutes, universities and NGOs are now active in this field, in many cases with spectacular results. With headquarters in the UK, the Intermediate Technology Development Group (an independent charity founded by the late E.F. Schumacher) helps to develop and introduce technologies suitable for LDC rural communities.

Urbanization

Although not an inevitable cause of environmental problems, urbanization has generally led to environmental degradation in the built-up area; air pollution, especially downwind; considerable water pollution if sewage or industrial effluent are discharged, and; pressure on surrounding land (perhaps for hundreds of square kilometres) for fuelwood or recreation use. A more subtle problem is that urban peoples tend to lose contact with the realities of food production, waste disposal and the need to support rural development, and being close to centres of government and often better educated,

Fig. 3.8 One of Brazil's fastest growing cities: Marabá (Pará State, eastern Amazonia). Population in 1992 c.250 000, Marabá has grown about 20-fold since the early 1970s following creation of roads.

the demands they voice are heeded even if it leads to problems outside the cities (the urban bias discussed earlier). Urban populations are increasing rapidly and this must surely stress city facilities and surrounding regions, certainly in LDCs. For example, Mexico City has grown nine-fold between the mid-1940s and the mid-1990s and now has severe pollution problems; some cities have even more rapid growth (see Fig. 3.8 – the problems related to urbanization, particularly pollution, are discussed in chapter 10).

Overconsumption

Populations of DCs and the middle classes of LDCs are making increasingly unsustainable demands on the world's resources. International trade in agricultural commodities, industrial products and minerals expanded at about 4 per cent per year during the 1980s (L.R. Brown *et al.* 1991: 7). Increases in resource consumption will probably occur as the eastern bloc, South Asia, the PRC and other countries develop and demand refrigerators, televisions, and other goods and services; DCs are also likely to have considerable expansion of vehicle ownership and, even if exhaust pollution is reduced, there will be problems of congestion, resource demand for construction of cars and roads and scrap disposal and recycling needs. India, the PRC and a number of other LDCs seem likely to expand their use of coal which, unless adequate emission controls and sequestering measures are achieved, will mean increasing carbon levels in the atmosphere and the risk of enhanced greenhouse warming.

Recycling of materials and efforts to discourage 'planned obsolescence' and the attitude that associates status with consumption will help reduce, but not prevent the huge increases of demand on resources and pressure on the environment resulting from pollution during manufacture and energy demand to run new consumer items.

Norman Myers (1992) has observed that the UK with an annual population growth rate of 0.2 per cent was not much less of a threat than Bangaldesh with 2.4 per cent, given that each Bangladeshi consumes roughly 80 times less fossil fuel.

Pollution

Pollution associated with power generation, industrial development, disposal of sewage, agricultural development, warfare and land clearance now poses a global threat. Warnings of pollution-related global warming, like that of Hartshorn (1991: 398), which suggests a one in two chance of a 6°C rise in mean global surface temperature by AD 2050, and the threat to public health from pollutants, are serious enough to demand very careful attention from decision-makers. Crucial needs are to recognize the critical thresholds for pollutants in natural systems and organisms, to monitor levels and to act if such limits are approached. Given the sheer diversity and often insidious build-up of pollutants, control is unlikely to be easy, cheap or fully effective.

Loss of biodiversity

Loss of biodiversity poses a serious threat to the long term well-being of the environment and humans. Species loss has accelerated; in the last few decades it has been greater than at any time since the end of the Cretaceous 65 million years BP (perhaps 100 to 10 000 times normal levels of extinction, although there have been natural mass extinctions at several points in the Earth's history). There is little time to correct this loss, and no second chance to avoid the extinction of flora and fauna (Wilson and Peter 1988). Biodiversity is being or will be lost through:

1 direct destruction (hunting, logging, etc);
2 habitat destruction, e.g. destruction of coral reefs;
3 pollution;
4 introduction of exotic, competitor species;
5 subtle selective pressures related to human activity;
6 global environmental change.

Because tropical forests are particularly rich in species, rapid, extensive deforestation and associated genetic resources loss is particularly worrying. At the Rio Earth Summit (UNCED) in 1992 a number of nations signed the UN Convention on Biodiversity, intended to provide LDCs with funds to take protective measures. There is unfortunately usually quite a difference between numbers signing and numbers ratifying such treaties, and actual progress.

Energy crisis?

Between AD 1800 and 1990 global energy consumption was estimated to have increased six-fold (Kemp 1990: 3). There is now less concern about energy supplies than there was in the 1970s. Alternatives to petroleum seem feasible, and are gradually being developed or revived; nevertheless, the world still depends on oil and much comes from regions vulnerable to conflict. An 'energy crisis' could thus arise long before resources were exhausted and could trigger warfare with associated environmental damage (not only pollution from burning oilfields as in Kuwait following the Gulf War, but also more widespread war damage to nations competing to retain access to resources). The need to control the pollution resulting from combustion of hydrocarbons (coal, oil, gas, wood) will hopefully encourage a change to alternative energy sources and energy conservation.

Warfare

Humankind has long had the capacity to cause catastrophic damage to the environment as a consequence of conventional, nuclear, chemical or biological warfare (Harwell and Hutchinson 1989; Pittock *et al.* 1989). The end of the 'Cold War' (*c.*1989) and a start made towards reduction of more horrific armaments by the superpowers may signal a decreased threat, but only provided superpower relations remain amicable and the proliferation of the more devastating weaponry to LDCs can be controlled. Some weapons pose a threat even if they are not used in earnest, through leakage in storage, the need to dispose of aged warheads or an accidental or terrorist release.

Wars generate refugees who are generally slow to re-establish sustainable land use. In 1991 (and not counting those from the Balkan conflict) the UN High Commission on Refugees estimated that the world had 17 million refugees (a large proportion in Africa), those counted being only the ones who crossed national borders; if in-country refugees were added the figure would be much higher.

Topsoil degradation

DCs and LDCs alike suffer from soil degradation. Given the problems that the USA and USSR encountered in the 1930s and 1950s respectively, one might have expected more concern. Restoration of lost or degraded topsoil is in many situations impossible, at least at reasonable cost and at a fast enough speed to avoid problems.

The scale and rate of degradation coupled with the lack of opportunity to resolve the problem once it has happened makes topsoil degradation one of the most serious environmental threats. Estimates of topsoil loss suggest that each year world-wide an area equivalent to Australia's wheat lands are going out of use. Added to this, there is considerable loss of non-arable land as rangelands suffer overgrazing (L.R. Brown *et al.* 1991: 7). Unfortunately, attention has been diverted from the problem, partly by concern for energy supplies and global warming, and because soil damage is often insidious and missed. Modern agriculture has also helped conceal the problem, but for considerable areas, there must come a time when application of fertilizer or new crops fail to boost or even maintain yields because insufficient depth of topsoil remains for plant roots. Soil conservation should be given a much higher priority. Doing so should present no great *technical* challenges as soil protection measures are well developed and can often be spread using appropriate technology and local peoples.

AIDS

A 1991 estimate put the number with fully developed Acquired Immune Deficiency Syndrome (AIDS) in Africa alone at 6 million (*Sunday Times* 9/7/91: 22). There are suggestions that, given the number of people carrying the Human Immunodeficiency Virus (HIV) but without yet developing AIDS, large parts of Africa could suffer severe workforce shortages which in turn might lead to abandonment of farmland (PANOS 1989; Barnett and Blaikie 1992).

It is also clear that in a number of LDCs the middle-classes and professionals are being decimated, leading to skills shortages. Statistics are still incomplete or inaccurate so AIDS predictions should be treated with caution (*The Times* 21/3/93: 1.2).

Biotechnology

Biotechnology offers both tremendous potential, some already realized, for boosting food and commodity crop production, combating pollution, diseases and reducing fossil fuel demand, and serious threats if misused or managed without care. There is a risk, unknown before the 1960s, of a mistaken or accidental uncontrolled, perhaps catastrophic, release of artificially created or modified (genetically engineered) organisms. This might damage any one of innumerable organisms or natural processes, say, something crucial like photosynthesis in the oceans or on land or some other vital biogeochemical cycle. Control of releases may prove difficult in future as the technology involved is likely to be much more widespread and more difficult to monitor than that concerned with nuclear weaponry.

Natural catastrophe

There are sufficient indications of past catastrophic events that we should be concerned about: large earthquakes, tsunamis, planetesimal (asteroid, meteorite, comet or bolide) strikes, and so on (Huggett 1990). It is wise to site nuclear waste repositories, atomic power stations, gene banks, etc, where they are as safe as possible from natural catastrophes: often this has not been the case.

Planetesimal strikes may be infrequent by human time-scales, but nevertheless seem to have nearly wiped out almost all lifeforms more than once in the past (notably at the end of the Permian and at the end of the Cretaceous: see pages 61–2). A 'near-miss' by asteroid 1191BA on 1 January 1991 prompted popular interest in protection against celestial objects (*The Times* 2/7/92: 3); currently a large camera-telescope on Kitts Peak (USA) is engaged on assessing the threat and the US Congress have recommended a warning system estimated to cost US$40 million (*The Observer* 20/10/91: 3; *The Times* 20/11/91: 14). The collision of comet Shoemaker Levy-9 with Jupiter in 1994 prompted further interest.

Reacting to environmental problems or crisis

Before the mid-1980s environmental problems were seen as side-effects of development which could be dealt with later once standards of living had improved. Now it is generally accepted that development problems and environmental problems are closely interrelated and to wait to cure them is a mistake: development efforts should be environmentally appropriate.

Environmental problems and the threat of crisis seem to have increased faster than the ability or willingness to counteract them. Inertia in responding to environmental problems has not been confined to twentieth-century humans. Easter Island, isolated in the Pacific, apparently had a forest cover when settled about 1200 years ago. The settlers, perhaps assisted by seed-eating rodents that they introduced, totally destroyed the palm forests by about AD 1600, and in so doing triggered soil erosion and lost their means of escape – wood, without which ocean-going rafts or boats were impossible (Flenley *et al.* 1991; Bahn and Flenley 1992). It is probable that much of the forest was cut for cultural purposes, to transport stone

Table 3.3 Ranking of dilemmas for the 1980s and for possible development paths for the year AD 2030

Pathway	Very serious	Moderately serious	Not serious
1980s		Soil acidification Transport growth Summer traffic-related air pollution	Water management Forestry wood supply Marginalized land Sea-level rise Coastal pollution Chemical time bombs Non-point toxics Urbanization
Present trends continuing Europe and elsewhere	Soil acidification Transport growth	Water management Forestry wood supply Marginalized land Sea-level rise Coastal pollution Chemical time bombs Non-point toxics Summer traffic-related air pollution	Urbanization
High-growth economy/low environmental concern Europe and elsewhere	Water management Soil acidification Forestry wood supply Sea-level rise Chemical time bombs Non-point toxics Transport growth Summer traffic-related air pollution	Marginalized land Coastal pollution Urbanization	
Environmentally friendly economy Europe and elsewhere		Transport growth	Water management Soil acidification Forestry wood supply Marginalized land Sea-level rise Coastal pollution Chemical time bombs Non-point toxics Urbanization Summer traffic-related air pollution
Environmentally friendly economy Europe but not elsewhere	Water management Forestry wood supply Sea-level rise	Marginalized land Chemical time bombs Transport growth Urbanization Summer traffic-related air pollution	Soil acidification Coastal pollution Non-point toxics

Source: Stigliani *et al.* 1989: 7 (Table 2)

statues, rather than for survival. Unable to escape and with a growing population the Islanders 'must surely have realized that their very existence depended on the limited resources of a small island. Yet they were unable to devise a system that allowed them to find the right balance with their environment' (Ponting 1991: 7). Let us hope that modern humans learn from Easter Island.

Once a crisis is recognized its resolution is likely to require the parties involved to accept trade-offs; there is also the danger that 'crisis management' will adopt a short-term, poorly coordinated, 'fire-fighting' approach. Far better is an objective study, with identification of causes and understanding of processes rather than rash treatment of what may be only symptoms. The avoidance of 'crisis management' requires proactive monitoring, research, money and expertise, stable socioeconomic conditions and institutions.

Science fiction crises are generally resolved by nature, good luck or hero(ine), often in a white lab-coat; in reality environmental managers are going to have to work hard at monitoring and researching potential problems, decision-makers will have to implement suitable policies, appropriate legislation and planning approaches will be needed and, *above-all*, the 'common people' will have to change their individual ways. Everyone will probably have to accept unwelcome trade-offs and the avoidance of one problem is no guarantee that others will not arise. The alternatives are worse. Technology may prolong the period of population and industrial growth, and will be vital to get us out of trouble, but it has not removed the ultimate limits to growth.

Modelling crisis situations

In his 1798 and 1803 essays, Malthus (1970) indicated that the relationship between population growth and efforts to raise food production was likely at some point to result in a crisis. However, he presented a more complex message than many acknowledge. The main axiom of his 1798 essay was that whereas population grows geometrically, food production grows only arithmetically (at best) so that, unless numbers are reduced by pestilence or warfare (effective family planning not being available then) resources will be outstripped (O'Riordan 1976: 42; Dupâquier and Grebenik 1983). Malthus was saying that the Earth limited social and economic development, but he was writing after the French Revolution (1789) with intellectuals like Condorcet and Godwin advocating social utopianism and anarchy, so he was probably also warning that the Poor Laws (primitive British 'social security' enacted in 1796) would not lead to social reform, indeed might prove harmful.

The Limits to Growth revived Malthusian concern, suggested a time-scale, and, through modelling, explored ways that the crisis caused by continued worldwide economic and population growth might be avoided. The message presented (Meadows *et al.* 1974: 23) was that 'If present growth trends in world population, industrialization, pollution, food production, and resource depletion continue unchanged, the limits to growth on this planet will be reached sometime within the next hundred years.'

Models dealing with growth and ecocatastrophe, which appeared from the early 1970s, including those of Meadows *et al.* (1974), were hindered by poor data, untested modelling methods, inadequate computing, incomplete knowledge of the variables, and sometimes lack of objectivity (GIGO – 'garbage in/garbage out' – became a widespread criticism). Furthermore, the public had little awareness of environmental problems to really judge the warnings (Gribbin 1979; Park 1980: 23). More recent modelling draws upon better data, much better computing facilities and a more objective approach (e.g. that of Stigliani *et al.* 1989: see Table 3.3). Meadows *et al.* have published a sequel to their 1970s work, in which they argue that they see a challenge, not necessarily, 'doom'; the limits they warned about in the 1970s were exceeded by the 1990s, but a sustainable society was still 'technically and economically possible' (Meadows *et al.* 1992: xvi).

References

Adams, W.M. (1990) *Green development: environment and sustainability in the third world.* London, Routledge

Agarwal, B. (1986) *Cold hearths and barren slopes: the woodfuel crisis in the third world.* London, Zed

Arvill, R. (1967) *Man and the environment: crisis and the strategy of choice.* Harmondsworth, Penguin

Bahn, P. and Flenley, J.R. (1992) *Easter Island, earth island.* London, Thames and Hudson

Barnett, T. and Blaikie, P. (1992) *AIDS in Africa: its present and future impact.* London, Belhaven

Barrow, C.J. (1987) *Water resources and agricultural development in the tropics.* Harlow, Longman

Barrow, C.J. (1991) *Land degradation: development and the breakdown of terrestrial environments.* Cambridge, Cambridge University Press

Blaikie, P.M. (1988) The explanation of land degradation in Nepal. In J. Ives and D.C. Pitt (eds) *Deforestation: social dynamics in watersheds and mountain ecosystems.* London, Routledge

Blaikie, P.M. (1989) The use of natural resources in developing and developed countries. In R.J. Johnston and P.J. Taylor (eds) *A world in crisis? Geographical perspectives* (revised edn). Oxford, Basil Blackwell

Blaikie, P.M. and Unwin, T. (eds) (1988) *Environmental crisis in developing countries* (IBG Developing Areas Research Group, Monograph no. 5). London, Institute of British Geographers

Brown, W. (1990) Trade deals blow to the environment. *New Scientist* 128(1742): 20–1

Brown, L.R. *et al.* (1991) *State of the world 1991* (Worldwatch Institute report on progress toward a sustainable society). New York, W.W. Norton

Brown, L.R. *et al.* (1993) *State of the world 1991* (Worldwatch Institute report on progress toward a sustainable society). New York, W.W. Norton

Caldwell, L.K. (1990) *International environmental policy: emergence and dimensions* (2nd edn). Durham, N.C., Duke University (1st edn 1984)

Carley, M. and Christie, I. (1992) *Managing sustainable development.* London, Earthscan

Chambers, R. (1983) *Rural development: putting the last first.* Harlow, Longman

Chambers, R., Longhurst, R. and Pacey, A. (eds) (1981) *Seasonal dimensions to rural poverty.* London, Frances Pinter

Clark, R. (1975) *Notes for the future: an alternative history of the past decade.* London, Thames and Hudson

Commins, S. *et al.* (eds) (1986) *Africa's agrarian crisis: the roots of famine.* Boulder, Colo., Lynne Rieinner

Davidson, J., Myers, D. and Chakraborty, M. (1992) *No time to wait: poverty and the environment.* Oxford, OXFAM

Eckholm, E.P. (1976) *Losing ground: environmental stress and world food prospects.* Oxford, Pergamon

Ehrlich, P.R. (1970) *The population bomb.* New York, Ballentine

Ehrlich, P.R. (1992) The value of ecodiversity. *Ambio* XXI(3): 219–26

Ekins, P. (1992) *A new world order: grassroots movements for global change.* London, Routledge

Farvar, M.T. and Milton, J.P. (eds) (1972) *The careless technology: ecology and international development.* Garden City, NY, Natural History Press

Flenley, J.R., King, A.S.M., Jackson, J. and Chew, C. (1991) The late Quaternary vegetation and climatic history of Easter Island. *Journal of Quaternary Science* 6(2): 65–115

Foley, G., Moss, P., and Timberlake, L. (1984) *Stoves and trees – how much wood would a wood stove save if a wood stove could save wood?* London, Earthscan

Gerasimov, I.P. and Armand, D.L. (1971) *Natural resources of the Soviet Union: their uses and renewal* (first published in Moscow 1963). San Francisco, Calif., Freeman

Ghai, D. and Vivian, J.M. (eds) (1992) *Grassroots environmental action: people's participation in sustainable development.* London, Routledge

Gribbin, J. (1979) What future for futures? *New Scientist* 84(1186): 21–3

Gunder Frank, A. (1980) *Crisis in the world economy.* Farnborough, Gower

Harrison, P. (1983) Earthwatch: land and people the growing pressure. *People* 13: 1–8

Harrison, P. (1984) Population, climate and future food supply. *Ambio* XIII(3): 161–7

Harrison, P. (1987) *The greening of Africa: breaking through in the battle for land and food.* London, Paladin

Harrison, P. (1992) *The third revolution: population and a sustainable world.* Harmondsworth, Penguin

Hartshorn, G.S. (1991) Key environmental issues for developing countries. *Journal of International Affairs* 44(2): 393–402

Harwell, M.A. and Hutchinson, T.C. (1989) *Environmental consequences of nuclear war – vol. 2: Ecological and agricultural effects* (SCOPE 28). Chichester, Wiley

Henning, D.H. and Manguin, W.R. (1989) *Managing the environmental crisis: incorporating competing values in natural resource administration.* Durhan, N.C., Duke University Press

Holmberg, J. (ed.) (1992) *Policies for a small planet.* London, Earthscan

Huggett, R. (1970) The bombarded earth. *Geography* 75(2): 114–27

Ives, J.D. and Messerli, B. (1989) *The Himalayan dilemma: reconciling development and conservation.* London, Routledge

Johnston, R.J. and Taylor, P.J. (eds) (1989) *A world crisis? Geographical perspectives* (revised edn). Oxford, Basil Blackwell

Kates, R.W. and Haarmann, V. (1992) Where the poor live: are the assumptions correct? *Environment* 34(4): 4–11, 25–8

Kemp, D.D. (1990) *Environmental issues: a climatological approach.* London, Routledge

Komarov, B. (1981) *The destruction of nature in the Soviet Union.* London, Pluto

Leach, G. and Mearns, R. (1989) *Beyond the woodfuel crisis: people, land and trees in Africa.* London, Earthscan

Leonard, H.J. (with Yudelman, M., Stayker, J.D., Browder, J.O., De Boer, A.J., Campbell, T. and Jolley, A.) (eds) (1989) *Environment and the poor: development strategies for a common agenda.* Washington, DC, Overseas Development Council

Malthus, T.R. (1970) *An essay on the principle of population as it affects the future improvements of society* (original edn 1798) (1970 edn, ed. A. Flew). Harmondsworth, Penguin

Malthus, T.R. (1803) *The principles of population, a view of its past and present effects on human happiness* (original edn 1798). London

Meadows, D.H., Meadows, D.L., Randers, J. and Behrens, W.W. III (1974) *The limits to growth* (USA edn 1972). London, Pan

Meadows, D.H., Meadows, D.L. and Randers, J. (1992) *Beyond the limits: global collapse or a sustainable future?* London, Earthscan

Merchant, C. (1992) *Radical ecology: the search for a livable world.* London, Routledge

Munslow, B. Katerere, Y., Fert, A. and O'Keefe, P. (1989) *The fuelwood trap: a study of the SADCC region.* London, Earthscan

Myers, N. (1992) Population/environment linkages: discontinuities ahead. *Ambio* XXI(1): 116–18

Ooi Jin Bee (1983) *Natural resources in tropical countries.* Singapore, Singapore University Press

PANOS (1989) *AIDS and the third world* (PANOS Dossier 1). London, PANOS Institute

Park, C. (1980) *Ecology and environmental management.* Folkestone, Westview/Dawson

Pittock, A.B., Ackerman, T.P., Crutzen, P.J., MacCracken, M.C., Shapiro, C.S. and Turco, R.P. (1989) *Environmental consequences of nuclear war – vol. 1: physical and atmospheric effects* (SCOPE 28). Chichester, Wiley

Ponting, C. (1991) *A green history of the world.* London, Sinclair-Stevenson

Rau, B. (1991) *From feast to famine: official cures and grass roots remedies to Africa's food crisis.* London, Zed

Ravenhill, J. (1986) *Africa in economic crisis.* London, Macmillan

Redclift, M. (1984) *Development and the environmental crisis: red or green alternatives?* London, Methuen
Rees, J. (1990) *Natural resources: allocation, economics and policy.* (2nd edn.) London, Routledge
Ritchie, M. (1990) GATT, agriculture and the environment. *The Ecologist* 20(6): 214–20
Sarre, P. (ed.) (1991) *Environment, population and development* (for Open University). London, Hodder and Stoughton
Schofield, A. (1984) *The crisis in economic relations between north and south.* Farnborough, Gower
Simonis, U.E. (1990) *Beyond growth: elements of sustainable development.* Berlin (WZB), Edition Sigma
Smil, V. (1983) *The bad earth: environmental degradation in China.* London, Zed
Stigliani, W.M., Brouwer, F.W., Munn, R.E., Shaw, R.W. and Antonovsky, M. (1989) *Future environments for Europe: some implications of alternative development paths.* Laxenburg, Austria, International Institute for Applied Systems Analysis
The Ecologist (1993) *Whose common future? Reclaiming the commons.* London, Earthscan
Thomas, C. (1992) *The environment in international relations.* London, Royal Institute of International Affairs
Thompson, M., Hatley, T. and Warburton, M. (1986) *Uncertainty on a Himalayan scale.* London, Ethnographica
Tiffin, M. (1993) Productivity and environmental conservation under a rapid population growth: a case study of Machakos District. *Journal of International Development* 5(2): 207–24
Timberlake, L. (1985) *Africa in crisis: the causes, the cures of environmental bankruptcy.* London, Earthscan
UNFPA (1991) *Population and the environment: the challenges ahead.* New York, United Nations Population Fund
Watts, M.J. (1989) The agrarian crisis in Africa: debating the crisis. *Progress in Human Geography* 13(1): 1–41
Weston, J. (ed.) (1986) *Red and green: the new politics of the environment.* London, Pluto Press
White, R.R. (1993) *North, south, and the environmental crisis.* Toronto, University of Toronto Press
Wilson, E. and Peter, F. (eds) (1988) *Biodiversity.* Washington, DC, National Academy of Sciences
Wood, R. (1989) Malthus, Marx and population crisis. In R.J. Johnston and P.J. Taylor (eds) *A world crisis? geographical perspectives* (revised edn). Oxford, Basil Blackwell
World Bank (1985) *Desertification in the Sahelian and Sudanian zones of West Africa.* Washington, DC, World Bank
World Bank (1990) *Sub-Saharan Africa: from crisis to sustainable development.* Washington, DC, World Bank
World Bank (1992) *World development report 1992: development and environment.* Oxford, Oxford University Press
World Commission on Environment and Development (1987) *Our common future* (Brundtland Report). Oxford, Oxford University Press
World Resources Institute (1992) *World resources 1992–93.* Oxford, Oxford University Press
Young, J. (1990) *Post environmentalism.* London, Belhaven

Further reading

Harrison, P. (1987) *The greening of Africa: breaking through in the battle for land and food.* London, Paladin (Grafton Books) [Review of Africa's problems, development successes and potential]
Lean, G., Hinrichsen, D. and Markham, A. (1990) *Atlas of the environment.* London, Arrow [Colourful maps plus articles on the state of the world]
Meadows, D.H., Meadows, D.L. and Randers, J. (1992) *Beyond the limits: global collapse or a sustainable future?* London, Earthscan [Sequel to *Limits to Growth*, warns of global collapse and suggests ways of avoiding it]
Myers, N. (ed.) (1985) *The gaia atlas of planet management: for today's caretakers of tomorrow's world.* London, Pan [Lively presentation of facts on the state of the world 1985 and options for the future, with colour illustrations]
Ponting, C. (1991) *A green history of the world.* London, Sinclair-Stevenson [Readable review of how humankind got into their present predicament – available as 1992 Penguin edn]

CHAPTER 4 Stable environment? Sustainable development?

How stable are environments?

The concept of ecosystem stability has provoked much debate, is still not fully resolved, and has an inadequate terminology (Hill 1987). Nevertheless, theorists have tried to apply the concept of ecosystem stability to development, at local, regional and global scales.

On the whole, biogeochemical and biogeophysical processes tend to resist change and are self-regulating within limits. However, natural ecosystems are rarely static, and the best one can expect is a sort of dynamic equilibrium, not a fixed stability. Some global cycles, environments and organisms (and groups of people) are more sensitive to change than others and consequently enjoy less stability. Stability is in large part a function of sensitivity and resilience to change. Sensitivity may be defined as the degree to which a given ecosystem undergoes change as a consequence of natural or human actions or a combination of both. Resilience may be used to refer to the way in which an ecosystem can withstand change; originally it was proposed as a measure of the ability of an ecosystem to adapt to a continuously changing environment without breakdown. It would be misleading to give the impression that these concepts – stability and resilience – are straightforward.

Stability (some prefer to use constancy) is often invoked by those interested in establishing whether conditions will remain steady or will return via a predictable path to something similar to the initial 'steady state' after disturbance. It is widely held that ecosystem stability is to a significant degree related to biological diversity: the more variety of organisms there are in an ecosystem, the less likely is there to be instability. However, it is quite possible that a change in some parameter could have an effect on all organisms regardless of diversity: thus diversity may help ensure stability, it does not guarantee it.

An ecosystem may not be stabilized when disturbed, it may be close to a 'starting-point', or it could be undergoing cyclic, more or less constant or erratic change. Return to a pre-disturbance state is therefore uncertain. Resilience is often measured by the speed of recovery of a disturbed ecosystem, but can refer to how many times a recovery can occur if disturbance is repeated (Holling 1973). An ecosystem may return to stability after several disturbances, but fail to after a subsequent upset for a variety of reasons.

The concept of resilience has been applied to human ecology: some societies absorb or resist social change and continue with traditional skills and landuses or develop satisfactory new ones, other societies fail and their resource use and livelihood strategies degenerate (Burton *et al.* 1977). Some resources such as topsoil have little resilience; once fertile soil is lost it may well take centuries to develop a new cover.

Referring to sensitivity and resilience, Blaikie and Brookfield (1987: 11) suggested a simple classification of land, which may be modified to be applicable to ecosystems in general:

1. *Ecosystems of low sensitivity and high resilience* These suffer degradation only under conditions of poor management or natural catastrophe. Generally these are the best ecosystems to 'stretch' to improve production of food or other commodities.
2. *Ecosystems of high sensitivity and high resilience* These suffer degradation easily but respond well to management and rehabilitation efforts.
3. *Ecosystems of low sensitivity and low resilience* These initially resist degradation but, once a

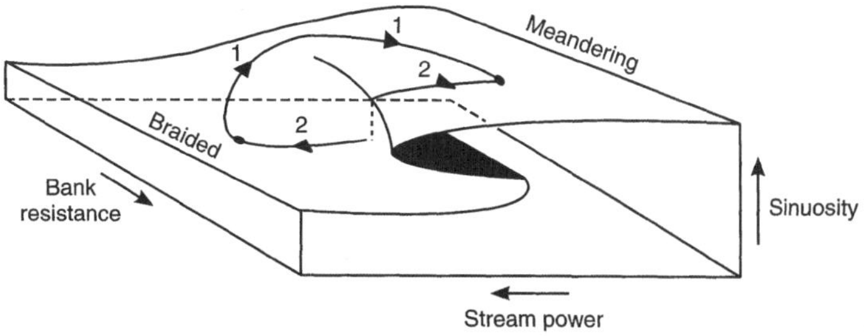

Fig. *4.1* Application of catastrophe theory to predict changes in a system (stream channel behaviour) ⟨*Source:* based on Huggett 1989: 7⟩.
Note: A cusp catastrophe surface relating the response variable of channel sinuosity (which includes braided and meandering channel patterns) to the control variables of stream power and bank resistance. Path 1 shows smooth and gradual change from a braided channel to a meandering channel owing to a reduction in both stream power and bank resistance. Path 2 shows an abrupt, discontinuous switch from a meandering channel to a braided channel owing to an increase in stream power at constant bank resistance.

threshold is passed, it is very difficult for any management and restoration efforts to save things.

4 *Ecosystems of high sensitivity and low resilience* These degrade easily and do not readily respond to management and rehabilitation efforts. It is probably best either to leave such ecosystems alone or to radically alter them, for example, forest might be converted to rice paddy-field and suffer less degradation than were it converted to rain-fed rice or treecrops.

Development managers or researchers often wish to establish in advance of, sometimes after, a disturbance what the consequences will be (not in order of importance):

1 Will the ecosystem re-establish its initial state?
2 Will there be a shift to a new state?
3 If re-establishment takes place how rapid will the recovery be and how complete?
4 What path does the recovery take?
5 How often can recovery occur?
6 Will the same recovery path be always followed?
7 Will successive, similar disturbance have the same effect?
8 Would change still occur if there were no disturbance?
9 Can the recovery path be predicted?

Ecosystem stability: concepts and caveats

Concepts

Dynamic equilibrium supporters argue that individual parts of an ecosystem change, but the overall function remains 'stable', thanks to homoeostatic processes. Some ecosystems, however, are in constant non-equilibrium or frequent flux rather than in a stable state at or near carrying capacity.

The behaviour of an ecosystem (physical or human) can be modelled; however, there is often such complexity that outcome is difficult to reliably predict with simple systems analysis. The cliché about the flapping of a butterfly's wing over say Africa triggering a chain of events affecting weather over Europe is often cited – the 'butterfly effect'. To cope with these challenges; analysts are turning to studies of chaos and to catastrophe theory (see Fig. 4.1). Catastrophe theory is better developed and allows examination of systems where there is a process that changes gradually, possibly with little serious effect, but then reaches a threshold, after which if change progresses further, there is a sudden 'flip' to a marked effect. An example is a river, which over a given length of course can render pollution harmless, provided the amount of pollutants do not exceed the threshold beyond which the water and biota ceases to cope. Figure 4.1 shows a similar 'flip' from a braided (multi-channel) stream pattern to a single meandering channel pattern. The channel meandering pattern (sinuosity) is a 'steady-state' function of stream power and bank resistance. A smooth transition can take place with braided gradually giving way to meandering (path 1 in Fig. 4.1 represents this), or the system may suddenly 'flip' (path 2 in Fig. 4.1) to meandering.

There is disagreement as to whether an ecosystem evolves in the long term toward a steady-state with equilibrium of its biota through slow, gradual evolution of species (phyletic gradualism) or generally steady slight and slow evolution punctuated by occasional sudden catastrophes and extinctions, after which there may comparatively rapid and considerable biotic change (punctuated equilibrium – supported for example by Steven J. Gould and Niles Eldridge) (Gould 1984; Goldsmith 1990). Whatever the process, the end result is held to be a 'climax stage', reached via a succession which may involve more or less transient successional stages (at which succession is held by some parameter) (Clements 1916).

Succession may not be as reliable and useful a concept as some hope; in particular, ecological change may not be as predictable as might be wished. For example, in some environments heavy grazing leads to increased scrub cover, a reduction of grazing might be expected to lead to a reduction of the scrub; in practice it sometimes causes a thickening of the woody vegetation (Barrow 1991: 24). Some plant communities do not exhibit succession as a directional change but follow a cyclic fluctuation about a mean (the classic case being bog or tundra hummock formations) (Kershaw 1973: 65–84).

Ecologists have developed a range of concepts and parameters, many of which have been adopted (sometimes modified) by those seeking to manage the environment or understand human society. The more widely borrowed concepts and parameters are now discussed.

Primary productivity The rate at which organic matter is created, usually by photosynthesis (in some situations by other metabolic processes). The total energy fixed by a plant community is termed gross primary production.

Maximum sustainable yield The fraction of primary production (as organic matter) in excess of what is used for metabolism (net primary production), that it is feasible to remove on an ongoing basis without destroying the primary productivity, i.e. 'safe harvest'. Under US law, maximum sustainable yield would be defined as maintenance in perpetuity of a high-level of annual or regular periodic output of renewable resources (Henning and Mangun 1989: 110).

Carrying capacity Definitions vary and can be imprecise, examples include the maximum number of individuals that can be supported in a given environment (often expressed in kg live weight per km^2); the amount of biological matter a system can yield, for consumption by organisms, over a given period of time without impairing its ability to continue producing; the maximum population of a given species that can be supported indefinitely in a particular region by a system, allowing for seasonal and random changes, without any degradation of the natural resource base. For human society, there can be a maximum or an optimum carrying capacity.

Assimilative capacity An environment has a capacity to purify pollutants up to a point where the pollutant(s) hinder or wholly destroy that capacity. Assimilative capacity can vary over time and from locality to locality.

Evolution The theory of evolution deals with the development of complex organisms from pre-existing simpler organisms over the course of time. Charles Darwin proposed a mechanism for explaining how change in genetic composition of a population exposed to new environmental conditions took place, i.e. the process of evolving new species; the idea had been part suggested much earlier by Erasmus Darwin (grandfather of Charles) in the 1790s and in 1809 by Jean-Baptist de Lamarck (who published his own views on inheritance and evolution roughly 20 years after Charles Darwin). Darwin's mechanism was natural selection ('survival of the fittest'), proposed in 1858 (published with his other theories in *On the Origin of Species by Means of Natural Selection* in 1859); Darwin acknowledged that A.R. Wallace was an independent discoverer of natural selection. Natural selection seeks to explain how organisms with genetic traits best adapted to their environment on average produce more offspring over time and so tend to replace less successful organisms (in the long run, there is no such thing as a final optimum adaption because the environment changes: evolution is a process). Darwin was not the only formulator of evolutionary theory, but as he was the most influential, the label 'Darwinism' is often applied to evolutionary theories developed in the second half of the nineteenth century and first half of the twentieth century.

Lamarckism Proposed by Jean-Baptiste de Lamarck from 1809. Holds that organisms respond to environment and make use of organs which then adapt, e.g. a giraffe that can reach further survives and somehow passes on a longer neck to offspring, or a fish forced to lie on its side in shallow water becomes a flat-fish. The problem is explaining how such behavioural and physical adaptions of individuals are passed on to be inherited by offspring. The concept was adopted in the USSR during the Stalinist era as social Lamarckism (notably by Lysenko, who carried out 'experiments' which claimed to show that plants could be modified and would pass on new characteristics without need for conventional breeding). Soviet geneticists, like Vavilov, who rejected Lamarckism, were purged, hindering USSR genetics and crop breeding until well into the 1950s.

Creationism Various groups (e.g. supporters of creative evolution, special creation, etc). Some recognize a long-term purpose or guiding principle behind evolution, i.e. are teleologists (e.g. supporters of orthogenesis), others see no separation of divine plan from nature.

Neo-Darwinism Different schools of thought soon arose over the way organisms changed from generation to generation. Attacking the ideas of Lamarck on inheritability of acquired traits, neo-Darwinists like August Wisemann tried to show experimentally that such heredity was impossible. Above all neo-Darwinists deny physical surroundings can give rise to new species (i.e. they hold that acquired traits are not inheritable). Seeks to take account of modern genetics, biochemistry, etc. Emphasizes individual natural selection is the 'motor' of evolution. A recent proposal (by Gerald Edelman) – neural Darwinism – suggests that there is an intimate link between functioning of the human brain and selection of the fittest, i.e. genetics may not be a major determinant of mental ability.

Neo-Lamarckism Claims evolution is much influenced by environment. Like neo-Darwinism seeks to explain how traits are passed on to successive generations.

Caveats

Environmental managers and those considering the concept of sustainable development should be aware that some of the aforementioned concepts and parameters have faults. If an error is made in calculating maximum sustainable yield (or some environmental change occurs) a resource could be over-exploited. An example is the 1972–3 El Niño event, whereby sudden physical changes in the Pacific destroyed the Peruvian anchovita fishery although the catch had remained at the level it had been at for some years. Maximum sustainable yield calculations may give resource users a false sense of security.

A given plot of land can have more than one carrying capacity, depending on the intensity of landuse, the technology used, etc. The biogeophysical carrying capacity may differ from the behavioural carrying capacity, such that organisms could be fed, but feel crowded and suffer stress sufficient to limit their numbers or survival before food is a limiting factor. The more people that the Earth supports, the lower is the standard of living that they can enjoy; also too high a human population will mean environmental damage and biodiversity loss. Adequate conservation and environmental quality might require a human population on Earth of 5000 million or fewer, and that has already been exceeded.

Carrying capacity (at the regional or national level) can be stretched by means of trade or technology and military power which ensures tribute from elsewhere. Some organisms adjust to their environment through 'boom and bust' strategies, feeding and multiplying during good times and in bad times suffering population decline, migrating or hibernating. The timing of resource use may be important, for example, rangeland might support a certain population of grazers provided grazing is restricted for a few critical weeks (at times when plants are setting seed, becoming established or are otherwise temporarily vulnerable), if timing is wrong much less grazing may be possible and there could be serious land damage.

The Gaia hypothesis

Our perspective on the interrelation between humans and environment determines how we approach development. Since the 1860s Darwinian concepts of 'survival of the fittest' and slow adaption of organisms to the environment have held sway; though still generally accepted today, Darwinian evolutionary theory is undergoing considerable reconsideration (Goldsmith 1990).

The Gaia hypothesis, proposed in 1969 by James Lovelock, offers a way of thinking about the environment and development relationship. Similar views had been expressed by 'the father of modern geology' James Hutton: as early as 1785 he, and later Pierre Teilhard de Chardin, a Jesuit geologist, archeologist and palaeontologist, suggested well before Lovelock that the biosphere acted as a self-directing entity. In spite of Lovelock's impressive research record (he did pioneering work on CFCs in the upper atmosphere and was one of the NASA team who designed the biological sensors on the Viking Martian landing craft), the hypothesis received little support before the late 1980s; indeed, it was and still is often ridiculed. Critics point out that it is difficult to see how evolution leads to gaian 'global altruism', whereby organisms act to benefit other organisms.

There are several variations of the Gaia hypothesis proposed (Schneider 1990: 8), but whichever variant is accepted, it runs counter to the long-held attitude in the west that humans can exercise what controls they want over the Earth (Lovelock 1972; 1979; 1988; 1992; Joseph 1990; Schneider 1990; Watson 1991). Whether or not they accept the hypothesis, many have been stimulated by it to think carefully about environment and development issues, for example it has helped provoke valuable research into the global carbon cycle. The Gaia hypothesis also provides a framework for people–environment study that is holistic.

The hypothesis suggests that life on Earth might not have simply adapted to the conditions it encountered, but may have altered, and still controls, the global environment to keep it habitable in spite of disruption from things like long-term changes in solar radiation or occasional planetesimal strikes. The hypothesis seeks to explain the survival of life on Earth by treating organic and physical environment as two parts of a single system ('Gaia') in which biotic components act as regulators so it can control and repair itself (this is not seen by Lovelock as a conscious process, nor is there implied a 'design' or 'purpose') (Sagan and Margulis, 1983). Temperature and composition of the atmosphere according to the hypothesis are regulated by biota, the evolution of which is influenced by the factors regulated: life at least partly initiates environmental conditions. Without Gaian regulation the suggestion is that average global temperatures would be more extreme, perhaps as hot as Venus (+474°C compared with Earth's +16°C mean surface temperature), or as cold as Mars (−55°C), and reactive gases like oxygen would probably be locked up in rocks and not be present as it is in the atmosphere.

In effect the Earth is seen as a super-organism, a single homoeostatic system with feedback controls that maintain global temperature, atmospheric gases and availability of nutrients. The controls involve a number of biogeochemical cycles, notably those of carbon dioxide, nitrogen, oxygen, sulphur, carbon, and phosphorus. An example of a global feedback is given in Fig. 4.2, and is one with which humans have interfered. The system functions in the 'interests' of the whole physical environment and biota: the whole is greater than the sum of the parts. The message is: we are part of a complex system, fit in, obey the limits or be cut out. If humans upset Gaian mechanisms, there could be sudden, possibly catastrophic, runaway changes which might remove us and other higher lifeforms, nature will not save humans from their mistakes.

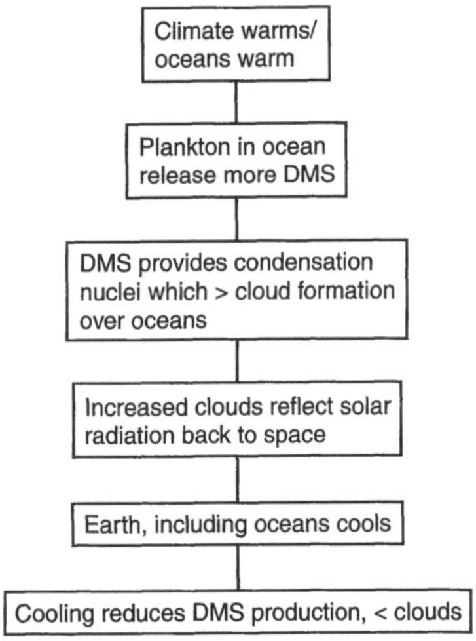

Fig. *4.2* Suggested Gaian feedback process (DMS = dimethyl sulphide derived from dimethyl sulphoniopropionate)*.
Note: * methyl iodide can be released by plankton which may form methyl chloride in the stratosphere and reduce ozone, possibly with warming effects.

Hazards, catastrophe, contingency and periodic environmental change

A hazard is a threat to life or property (Whittow 1980: 19; Bryant 1991). A disaster or catastrophe (the latter is of greater scale) is the realization of a hazard, perceived or otherwise. Catastrophe denotes a sudden 'overturning' or conclusion (cataclysm implies sudden violent deluge, although other events are often described as cataclysmic). Insurance companies usually divide catastrophes into natural ('acts of God') and human-made. Some places are especially prone to hazards and the art of hazard mapping is available to warn of risk and assist avoidance and mitigation (e.g. Degg 1992: 205). Human activity may render a locality or a group of people more likely to suffer disaster or catastrophe or may increase their vulnerability, so assessment and mapping should be regularly updated.

The scale of catastrophes varies from global to local and even down to the molecular scale. Timing of such events varies from relatively frequent to an interval of hundreds of millions of years and may or may not have a predictable pattern of recurrence. Catastrophe may be sudden, or of the 'creeping' form discussed earlier where a system is stressed and changes until a threshold is reached, whereupon there may be sudden drastic alteration (Huggett 1990a: 8).

Given time, organisms may adapt to their environment. Classical Darwinian evolutionary theory sees organisms as undergoing steady, gradual evolution. Those who support punctuated equilibrium hold that long periods of stasis are broken by sudden bursts of change perhaps in response to environmental stimuli (this may be true for organisms and landscapes). Given long enough, chance events probably affect survival at least as much as evolution – the 'process' has been described as contingency (Gould 1984). Louis Agassiz had similar ideas in the nineteenth century that 'the fittest are those who have survived by whatever accident'. Changes and events that challenge life but give insufficient time for adaption would allow some organisms to prevail for quite fortuitous reasons (rather than a 'survival of the fittest'). The environmental manager might note the comment made by Gould (1989: 320–1): 'The divine tape player holds a million scenarios, each perfectly sensible. Little quirks at the outset, occurring for no particular reason, unleash cascades of consequences that make a particular future seem inevitable in retrospect.'

Early earth scientists invoked catastrophic events such as Noah's Flood to explain erosive landforms, prehistoric extinctions and geological unconformities. The idea of a world-wide flood (diluvial theory) was supported by late-eighteenth-century catastrophists like the geologist William Buckland and the palaeontologist Georges Cuvier (Thomas Huxley probably coined the term 'catastrophism' in 1869). With the publication of *The Principles of Geology* in 1830, Charles Lyell helped uniformitarianism (the idea of continuing gradual change, involving processes operating in the past that operate today) to prevail over catastrophism (Simmonds 1989: 25).

Since the mid-nineteenth century there have been various attempts to revive catastrophism to explain landforms, for example the 'channelled scablands' in northwestern USA, or features in eastern Scotland where tsunami waves within the last 7000 years may have affected land 19 metres above sea-level (Smith and Dawson 1990; Ager 1993); this is something that those siting nuclear power stations or waste repositories should note. There are indications that eastern Australia may have been struck by a tsunami at least 50 metres high about 105 000 BP (this and other large tsunamis may be generated by submarine landslips of silt down continental shelf slopes, or by earthquakes or small planetesimal strikes).

A number of palaeontologists and other scientists recognize 'sudden' mass extinctions, perhaps four major and eleven significant, in the last 600 million years: at the end of the Ordovician 440 million years BP, during the late Devonian 390 million years BP, during the Permian 220 million years BP and at the close of the Cretaceous (the 'K/T boundary event') 65 million years BP (Raup 1988; 1992). The cause of the assumed K/T and other mass extinctions is debated: there is some question whether there really is adequate evidence for sudden catastrophes, suggesting instead more gradual non-catastrophic loss of species (Bowler 1990; for a review of extinction episodes see Wilson 1992: 22–30).

In the early 1980s Walter Alvarez noted the widespread occurrence of iridium (a rare metal), glass spherules and 'shocked quartz' grains in a thin clay layer of K/T boundary age (Alvarez and Asaro 1990). This and 'tsunami beds' around the Gulf of Mexico have been widely interpreted as evidence of a planetesimal (of roughly 10 km diameter) impacting with the Earth (Kerr 1972). Others suggest the cause was one or more very large volcanic eruption(s), Courtillot (1990) has linked a number of mass extinctions with the Deccan Traps (Deccan Plateau Basalts) of India, which are about the right age for the K/T extinction, and several other large volcanic outpourings such as the Karoo and Paraná Basalts. These are by no means the only causes cited by supporters of the K/T mass extinction, others include climatic cooling, sea-level falls, reduction of atmospheric oxygen levels, disease, increased UV-radiation, and so on. Perhaps more than one cause were responsible, if a mass extinction actually occurred. In order to gain support for planetesimal strike-related extinction catastrophists need evidence of impact craters. Possible planetesimal strike craters have been identified, one about 300 km in diameter at Chicxulub in Yucatán, Mexico, that was dated to 65 million BP would perhaps link with K/T 'tsunami beds'.

Whether or not a planetesimal strike 'terminated' the dinosaurs and many of their associates, and caused earlier and subsequent disruptions of life, there is enough evidence of impacts to indicate a threat that planners should very seriously consider (Fig. 4.3). Over 100 ancient craters, a few of more

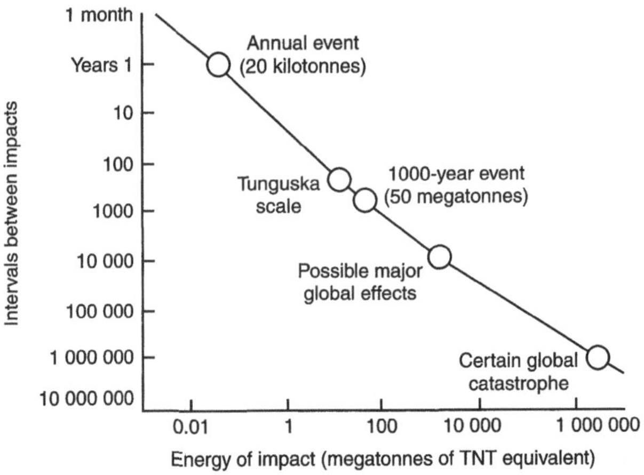

Fig. *4.3* Risk of planetesimal strikes (*Source:* based on *New Scientist* 14/11/92: 4).

than 100 km diameter, are known on Earth and some are as recent as 1500 BP (Huggett 1990b). A small meteorite (estimated at 100 metres diameter or about 50 000 tonnes) probably exploded about 8 km up in the atmosphere at Tunguska, Siberia, in AD 1908 and was the cause of the blasting down of 1200–2200 square kilometres of *taiga* forest. A similar strike may have occurred in South Island, New Zealand *c.*800 BP (Hecht 1991), and a blast (of about 100 kilotons yield) in the South Atlantic in 1978 may have been caused by a planetesimal (Lewin 1992). Anything over 100 metres in diameter is a serious threat to urban areas; a 1 km diameter body could threaten civilization (even if it did not trigger early warning systems to initiate a mistaken nuclear war); bodies over 5 km diameter are enough for mass extinction of organisms. Chapman and Morrison (1994) consider a major strike to be no less a risk than natural disasters we do plan for. Schneider and Boston (1991) speculated that a 'Shiva' hypothesis of repeated catastrophes, extinctions and diversification of survivors could compliment the Gaia hypothesis.

Planetesimals could be watched for with early-warning optical or radar systems to be intercepted and deflected with simple rocket impacts if up to 100 metres diameter, or if larger, with some of the world's many nuclear weapons (200 to 2000 megatons should be enough for asteroids of up to 3 km diameter) or a large plastic-film mirror could be deployed in space to vaporize surface material generating enough thrust to cause them to miss Earth (a mirror of 0.5 km diameter flown parallel for several months should be capable of deflecting a 2.2 km diameter planetesimal). The technology is more or less available (NASA was investigating and costing such possibilities in 1992 and the Moscow Institute of Dynamics of Geospheres have made proposals). A detection and deflection system could also assist a little in the problem of post-Cold War re-employment of military and space agencies staff (Ahrens and Thomas 1992; Lewin 1992; *The Times* 4/11/93: 10). Caution will be needed to ensure that such measures did not violate the 1966 Treaty on Peaceful Use of Outer Space or the 1972 Antiballistic Missile Treaty which prohibit nuclear warheads in Earth-orbit. To take such measures makes sense and the level of threat justifies the investment: the difficult question is who should pay for such a planetary defence system?

Volcanic eruptions can be locally devastating (e.g. Pompeii and Herculaneum in AD 79; Hekla, Iceland, in AD 1636; Tambora, Indonesia, in AD 1815; Krakatoa, Indonesia, in AD 1883) and some have global impacts: large outpourings of lava or eruptions of acidic ash, gases and aerosols into the stratosphere could alter climate. Smaller eruptions like Mt Agung, Indonesia (1963), El Chichon, Mexico (1982) and Mt Pinatubo, Philippines (1991) caused temporary lowering of global temperatures. Palaeoecologists and archaeologists have assembled considerable evidence for a link between past eruptions, acid deposition in

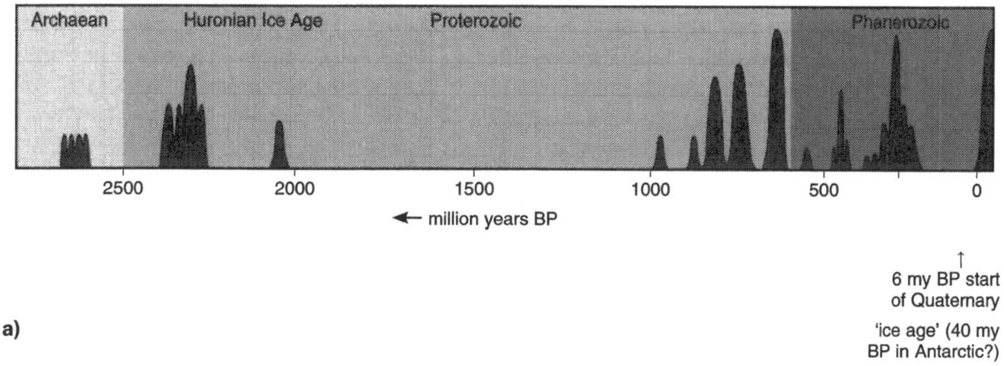

a)

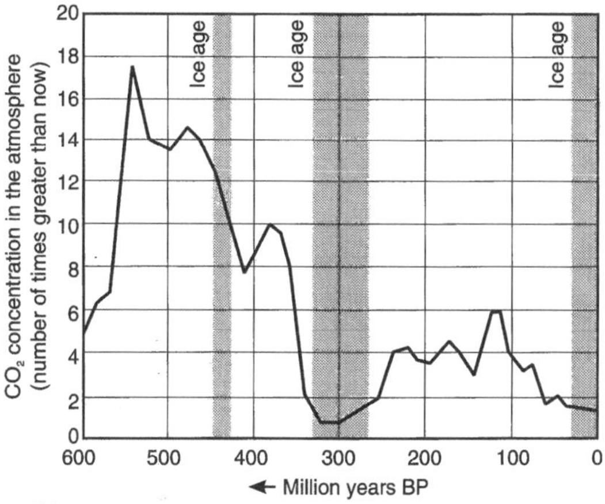

b)

Fig. *4.4* Natural climatic change. a) Ice ages during the last 3000 million years. b) Ice ages and atmospheric carbon dioxide levels during the last 600 million years (*Source:* author from various sources).

Greenland ice and sufficient alteration of climate to have affected human fortunes in Europe.

The recurrence of catastrophic events may not be random, planetesimal strikes, variation in Earth's solar radiation receipts (which it is claimed display a periodicity of 21 000, 41 000, 100 000 and 400 000 years) and perhaps volcanicity and seismic activity may be more likely at certain points in the orbit of the planets or galaxy, perhaps as the solar system passes the galactic plane every 26 to 33 million years. Velikovsky (1950; 1952; 1955) suggested that planetary alignments within the solar system and planetesimal strikes could be blamed for catastrophic events; however this, and a modern variant, the 'nemesis' hypothesis, have won limited support. The nemesis hypothesis suggests that a hidden companion star to the sun in an eccentric orbit periodically affects the solar system enough to alter climate and possibly trigger volcanic activity.

Windelius and Tucker (1991) warn of the threat of global cooling (rather than greenhouse warming) and possibly increased seismic activity and volcanicity during periodic falls in solar radiation. There have been attempts to establish the existence of periodic climatic changes related to the movements of the sun and moon – so-called luni-solar or atmospheric tides. Such atmospheric tides exist, but it is uncertain whether they are strong enough to cause the approximately 19-year periodicity of climate attributed to them (Currie 1988).

'Ice ages' are periods of Earth history consisting of cold (glacial) phases alternating with warmer interglacial or less cold interstadial phases; they have happened at several points during the Earth's history (see Fig. 4.4). During the glacials ice extends further from the poles and to lower altitudes on high ground. The most recent cooling began 40 million BP, became more pronounced from about 15 million and reached maximum chill in the last 2 million years.

In the last 1.8 million to 2.4 million years (the Quaternary Era) an 'ice age' with roughly 20 major warm/cold oscillations occurred (Greenland ice-core data suggest that climate during the last 250 000 years has fluctuated more often and more markedly than was previously assumed). The major interglacials lasted between 10 000 and 20 000 years and the glacials roughly 120 000 years. The peak of the last interglacial was about 132 000 to 120 000 BP and the last glacial maximum was about 18 000 BP. The post-glacial began quite fast around 13 000 BP in Europe and ice had retreated to broadly present limits worldwide by around 10 000 BP (between 7000 and 3000 BP average conditions may have been as much as 2°C warmer than today). Bearing in mind the sub-optimal conditions of the last few thousand years, one might ask whether there is a threat of global cooling and if so whether accelerated global warming as a result of human activity is such a bad thing?

There is considerable support for the Milankovitch theory (first proposed in the 1920s) that shifts in tilt and precession ('wobble') of the Earth's spin-axis and in the shape of the Earth's orbit, possibly due to cumulative irregularities in rotation or some external influence, cause 'ice ages' to come and go (Broeker and Denton 1990). Some are still unwilling to accept the Milankovitch theory, and cite recent palaeo-environmental studies at Devil's Hole (Nevada, USA) which might show that ice advances do not match astronomical causes. Recently the 'plateau theory' has attracted supporters (Ruddiman and Kutzbach 1991; Paterson 1993). These are just some of the many proposed glacial/interglacial causes (the list is by no means exhaustive):

1 Milankovitch effects: orbital causes.
2 Volcanic ash/dust/aerosol/gas emissions. Palaeoecologic/archaeological/modern observations indicate that marked cooling soon follows some volcanic eruptions. Some suggest that El Niño – Southern Oscillation (ENSO) oceanic changes are triggered by eruptions.
3 Solar radiation fluctuation (cause within Sun): sunspot cycles. The Little Ice Age (Europe *c*.AD 1500–1850) might correlate with solar activity; however, some palaeoclimatologists do not accept that there was a Little Ice Age. Periods of low sunspot activity could be indicative of weaker electromagnetic radiation and heating. There may be a double *c*.2 × 11 year periodicity.
4 Variation in biotic activity, particularly oceanic plankton photosynthesis.
5 Gradual lock-up of carbon dioxide, methane or other atmospheric gases by biotic or inorganic processes.
6 Plateau theory: uplift of Tibetan Plateau and Himalayan Mountains 40 million years BP led to increased rainfall which took more CO_2 out of the atmosphere and weathered carbon and carbonates from litter, soil and rocks; all this carbon was sequestered in the sea initiating cooling.
7 Evolution: new or more abundant organisms alter atmospheric composition.
8 Astronomical effects: dust in space reduces Earth's solar radiation receipt.
9 Ocean current changes/thermohaline circulation altered.
10 Cloud cover changes (effect likely to depend on altitude of clouds): CFC-like compounds in atmosphere might alter trophosphere/stratosphere relationship; warming might mean more water vapour in the atmosphere leading to more cloud, reflection of radiation and cooling.
11 Continental drift brings land to polar area and might cause ocean current changes.
12 Release of methane from Earth triggers warming sufficient to increase moisture in the atmosphere, causing snow accumulation at poles and on mountains leading to greater albedo and reflection back to space of more solar radiation.
13 Planetesimal strike scatters dust and acids into atmosphere causing cooling.
14 Tectonic (earthmovement) activity alters sea-level or affects atmospheric circulation.
15 Forest fires reduce lock-up of carbon.

While cause may be disputed, glacials and interglacials clearly occurred. There are also well-established links between glacial conditions, low levels of carbon dioxide in the atmosphere (approximately 25 per cent reduction compared with the present), low levels of methane in the atmosphere, and low sea-levels (which may drop perhaps 140 metres less than today's during glacials). During warm interglacials, it is estimated that carbon dioxide and methane in the atmosphere were high and sea-levels rose (perhaps 40 metres above present levels).

According to the generally accepted concept of uniformitarianism, processes that operate today operated in the past, so knowledge of the past can help forecast the future. In 1988 the US Geological Survey began to coordinate a programme of research into environmental conditions and distribution of organisms between 5.3 million and 1.6 million years BP. During that period, the Pliocene, the continental configuration was similar to today's but conditions were warmer. This Pliocene Research Interpretation and Synoptic Mapping (PRISM) project could help modelling of future climatic change and prediction of the likely effects of global warming on plants and animals (Hecht 1990). Another attempt at 'backcasting' has been the Climate, Long-range Investigation, Mapping and Prediction (CLIMAP) programme. Begun in 1970, this has used deep-sea bottom sediment core dates to test the 'Milankovitch theory'.

Drought in Africa, South East Asia and other parts of the world and the patterns of monsoonal rainfall have been linked to atmospheric and oceanic changes which show some sign of periodicity or quasi-periodicity. Particular attention has focused on the El Niño–Southern Oscillation (ENSO). ENSO is believed to function in the following manner: a low-pressure, high-temperature weather system lies over Indonesia, while thousands of miles away over the southwestern Pacific is a related high-pressure, low-temperature system. It has been established that if pressure in one increases, it falls in the other. These pressure differences cause the southeast trade winds to blow steadily and move water away from the western coast of South America. This causes upwelling of nutrient-rich cold seawater, on which wildlife depends. Every year in spring and autumn there is a weakening, even cessation of the trade winds, peaking in the middle of the austral summer (around Christmas, hence El Niño – 'the Boy-Child') and if fully manifest, the eastern tropical Pacific can warm markedly (Diaz and Markgraf 1992). ENSO events tend to cause increased rain along the Pacific coast of South America, and later drought in Brazil, Australia and Australasia and reduced austral summer rainfall and cloud cover in South Africa (even Europe may be affected) (Diaz and Markgraf 1992; Hamlyn 1992). Recent El Niños occurred in 1982–3, 1986–7 and 1992–present. Study of the 1986–7 ENSO event allowed rough prediction of the conditions mentioned above nine months in advance.

Infrequent events pose a threat to humankind and we would be well advised to devote more resources to provide early warning, defence or mitigation measures. For example, submarine mapping might identify and allow monitoring of underwater mudslide risk; these slides appear to be responsible for some of the catastrophic tsunami in the past and might also release large amounts of methane from oceanic mud to the atmosphere, endangering shipping and threatening global warming. Perhaps slides could be triggered in a controlled fashion like avalanches in alpine environments, and so rendered less harmful? Humankind is Earth's first organism with the capacity to examine 'the playground of contingency', identify and perhaps counter threats. Environmental management should try to plan for catastrophes, even planetesimal strikes, as well as more 'everyday' threats and ongoing processes.

'Greener' development

By the 1870s some foresters had begun to try to practise sustained forest yield strategies and ecologists had begun to study the dynamics of natural

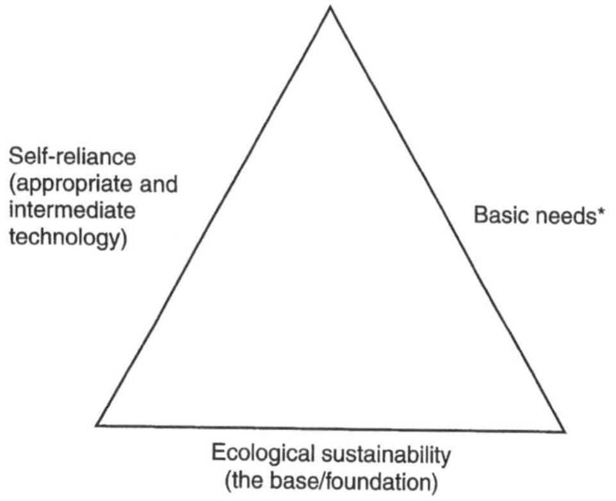

Fig. *4.5* The concept of ecodevelopment put forward by UNEP and published at the 1975 Cocoyoc Declaration on Environment and Development ⟨*Source:* Dasmann 1985: 215; Redclift 1987: 35; Turner 1988: 38⟩.
Note: * Unless poor people can be assisted to a better standard of living, sustainable development is unlikely to be possible.

ecosystems before 1914 (Cowles 1901; Clements 1916). Range managers, foresters and applied ecologists gained experience in predicting and managing natural resource responses to the demands of development, but it was not really until the late 1960s that environmentally informed development was widely advocated.

Today fewer people see quantitative economic growth as the only, or even main, objective of development (Max-Neef 1986; 1992). 'Sustainable development' has become not only a sincere goal but also a placebo and a cliché used by environmentalists, development theorists, aid agencies and politicians (Redclift 1987: 2; Adams 1990: 3). The challenges of the 1990s are to integrate environmental concern with economic policies and to make socioeconomic objectives compatible with the environment. Sustainable development seems to be an attractive vehicle with which to achieve these aims but to get it will require a shift in accepted ethics (Engel and Engel 1990: 35).

Ecodevelopment

The concept of ecodevelopment was articulated at the 1968 Founex (Paris) Conference, and at the 1972 Stockholm Conference (Dasmann *et al.* 1973: 17; Sachs 1979; Riddell 1981; Glaeser 1984; Adams 1990: 52). After 1972 Maurice Strong, the UNEP and Raymond Dasmann of the IUCN (Dasmann *et al.* 1973) promoted ecodevelopment (see Fig. 4.5) which has been defined as development at the regional and local levels consistent with the potentials of the area involved, with attention given to the adequate and rational use of the natural resources, and to applications of technological styles and organizational forms that respect the natural ecosystems and local sociocultural patterns (Bartelmus 1986: 45). Put crudely, ecodevelopment seeks to 'stretch' ecosystems to provide more food and commodities for human development without causing a breakdown of natural systems – a symbiosis of humankind and environment. Living 'within one's environmental means', as implied by ecodevelopment, might result in inequalities between countries being continued because development is not as much a goal as it is in the sustainable development approach.

Sustainable development

Sustainable development in theory

Much of the world is in a 'treadmill situation' with efforts at best simply keeping living standards from regressing and doing little to protect the environment. Often things end up in a worse state than predevelopment. Many concerned with agricultural development had accepted by the 1980s that raising crop yields was not enough; this was flagged by Conway (1985), who saw sustainability as one of four key qualities of agricultural systems: productivity (of yield or income); stability (of yield or income); sustainability (of yield or net income); equity or equitability (in terms of distribution of equitability) (Tisdell 1988: 375).

The concept of sustainable development (Barbara Ward is credited with the first use of the term in the mid-1970s) was widely disseminated by the *World Conservation Strategy* (IUCN, UNEP, and WWF 1980), which sought to stimulate a more focused approach to the management of living resources and to provide policy guidance on how to do so. Sustainable development was seen as a way of shifting support for conservation and environment to make it also play a part in improving human welfare. The three main aims stated in the *World Conservation Strategy* Executive Summary are as follows:

1 To maintain essential ecological processes and life-support systems (the latter being ecosystems natural and human-modified necessary for food production and other aspects of human well-being).
2 To preserve genetic diversity (i.e. maintain wild and domesticated plants and animals to ensure as few as possible become extinct).
3 To ensure the sustainable utilization of species and ecosystems.

The *World Conservation Strategy* called for international and national responses; an example of such a response came from the UK in 1983 (Programme Organizing Committee of the Conservation and Development Programme for the UK). The *World Conservation Strategy* has been criticized as superficial, for failing to integrate environment and economics and for giving limited consideration to social aspects (Pearce *et al.* 1989: xii); nevertheless, it drew attention to a concept which many feel should be elevated to a global ethic if humans and nature are to survive. The three bodies which produced the *World Conservation Strategy* recently released principles for achieving sustainable living (IUCN, UNEP and WWF 1991).

The Brundtland Report, *Our Common Future* (World Commission on Environment and Development 1987); placed environmental issues on the political agenda, rekindling interest that had died down after the mid-1970s; it also stressed the interdependence between environment and development and the need to deal with world poverty and inequality. In many respects it was a development of concepts put forward in the Brandt Report and related publications (The Independent Bureau on International Development Issues 1980; The Independent Commission on International Development Issues 1983), but it went further in calling for 'a new holistic ethic in which economic growth and environmental protection go hand-in-hand'. That may be one of the most crucial calls made in the twentieth century.

Little of the literature on ecodevelopment and sustainable development before 1987 moved from theory to consider the real world, although there were limited attempts to come to terms with the politics of development (Adams 1990: 57). Today, advocates of sustainable development tend to place emphasis on improving the livelihood of the poor. This demands development strategies which must be environmentally and socially sustainable, i.e. maintain, and if possible enhance the natural and human resources development is based on (Tidsell 1988).

Since 1987 sustainable development has become a catch phrase, some might say a cliché or shibboleth, but many see it as the social and development

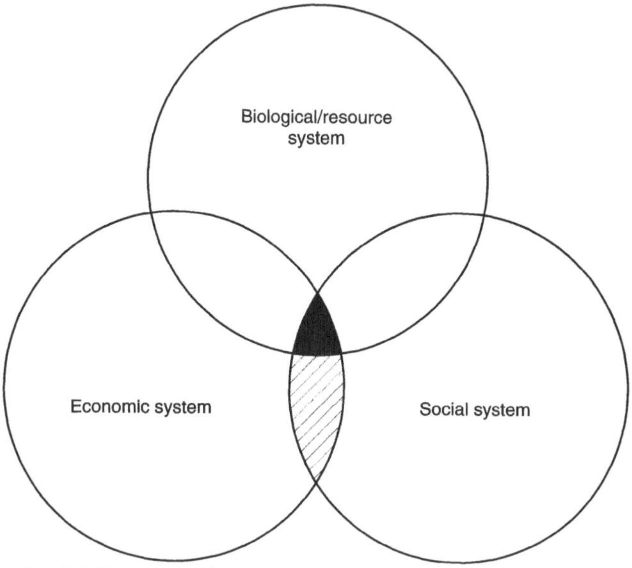

Fig. 4.6 Representation of the concept of sustainable development (*Source:* based on several sources).
■ shaded area = maximization of goals across the three systems and, in effect, represents sustainable development.
▨ = Marxist economics.
Notes
Goals of biological system: keep genetic diversity, maximize productivity.
Goals of economic system: supply basic needs, improve equity, improve goods and services.
Goals of social system: sustain institutions, improve social justice, improve participation.
Conventional economics maximizes economic system and social system.

paradigm of the 1990s (Engel and Engel 1990: 14; Lélé 1991: 607; Elliott 1994). Probably the best thing is to view it as a goal. There is much debate over the exact definition of sustainable development, and it is unlikely that there will ever be a precise meaning; indeed, there is value in a broad, even vague concept that allows people with perhaps otherwise irreconcilable positions to reach common ground.

Redclift (1987: 32) suggested sustainable development meant that lessons from ecology should be applied to economic processes; certainly to achieve it there must be ecological, social and economic sustainability. There are those who use sustainable development to indicate a sustained rate of growth of per capita real income, without depleting the national capital or environment (in effect sustainable economic development); an increasing number view it as a more radical departure from conventional growth-orientated society and development. For the latter, conservation becomes *the* basis for defining criteria with which to judge desirability of development options. Figure 4.6 is a representation of the sustainable development concept; some definitions of sustainable development are as follows:

- Development that meets the needs of the present without compromising the ability of future generations to meet their own needs (World Commission on Environment and Development 1987: 43).
- The kind of human activity that nourishes and perpetuates the historical fulfilment of the whole community of life on Earth (Engel and Engel 1990: 10–11).
- A concept which draws on two frequently opposed intellectual traditions: one concerned with the limits which nature presents to human beings, the other with the potential for human material development (Redclift 1987: 199).

- A vague concept, but it is as powerful as many other vague concepts such as liberty, equality and justice (Timberlake 1988: 61).
- A change in consumption patterns towards environmentally more benign products, and a change in investment patterns towards augmenting environmental capital (Pearce *et al.* 1989: xiv).
- A development strategy that manages all assets, natural resources, and human resources, as well as financial and physical assets, for increasing long-term wealth and well-being. It rejects policies and practices that support current living standards by depleting the productive base . . . to leave future generations with poorer prospects and greater risks than our own (Reppetto 1986: 15).
- A new era of economic growth, one that is forceful, global, and environmentally sustainable, with a content that enhances the natural base rather than degrading it (PANOS 1988).
- A . . . handrail that guides us along as we proceed toward development (Tickell 1991).

Pearce *et al.* (1989: 173–85) provide a list of definitions of sustainable development. Since its appearance three dimensions have generally been added to sustainable development: first, awareness of the possibility of serious global problems; second, new ethical stances, such as the concept of rights of nature; and third, the recognition that sustainable development requires suitable, institutions, understanding of social relations accounting procedures and methods of analysis which treat the environment as natural capital and measure its depletion or enhancement (Turner 1988: 35). Environmental accounting procedures were advocated by Pearce *et al.* (1990) in *Blueprint for a Green Economy*, which also called for:

- Valuation of the environment: attachment of monetary values to things that often don't have a market price.
- Extension of the planning horizon: ideally the present generation should pass to the next at least as much as they inherited.
- Provision for less advantaged people.

Sustainable development in practice

If the concept of sustainable development is accepted, then development and environmental issues cannot be separated, attempts to protect the environment are tied to poverty alleviation, world-wide trade, and so on. However, to sustain economic growth, and integrate economic, social and environmental goals is not at all easy, if indeed ever fully possible. Some of the critical objectives that should be sought and approaches that must be adopted if the concept of sustainable development is accepted (according to the Brundtland Report: World Commission on Environment and Development 1987: 49) are:

- reviving growth;
- changing the quality of growth;
- meeting essential needs for jobs, food, energy, water and sanitation;
- ensuring a sustainable level of population;
- conserving and enhancing the resource base;
- reorientating technology and managing risk;
- managing environment and economics in decision-making.

Furthermore, the pursuit of sustainable development requires (World Commission on Environment and Development 1987: 65; Pearce *et al.* 1990: 11):

- A political system that secures effective citizen participation in decision-making.
- An economic system that is able to generate surpluses and technical knowledge on a self-reliant and self-sustained basis.
- A social system that provides for solutions for the tensions arising from disharmonious development.
- A production system that respects the obligation to preserve the ecological basis for developments.
- A technological system that fosters sustainable patterns of trade and finance.
- An administrative system that is flexible and has the capacity for self-correction.
- Justice in respect of socially disadvantaged people.
- Justice between generations.
- Justice toward nature.
- Aversion to risk arising from our ignorance about the nature of interactions between environment, economy, and society, and from the social and economic damage arising from low margins of resilience to external 'shocks' such as drought, or to 'stress' such as soil erosion.
- Economic efficiency.

In general, the requirement is for an 'anticipate-and-avoid' approach to development, rather than a 'react-and-mend' approach (Barbier 1987; *Futures* 1988; Munn 1988).

Carley and Christie (1992: 78) divided environmental ethics into four general viewpoints:

1 *Technocratic* resource-exploitative, growth-orientated. (Continued on p. 70.)

Box. 4.1 Case Study: sustained agriculture in the colony of Tomé-Açu, Amazonia

Attempts by small farmer settlers to sustain agriculture in non-floodland environments of Amazonia have seldom been successful; one exception has been the Japanese colony at Tomé-Açu and its satellite settlements (about 150 km from the city of Belém at the mouth of the Amazon). Some of the land had been cultivated and then abandoned by Brazilian settlers as 'unfarmable'. The colonists set up a mixed agricultural cooperative in 1929, producing rain-fed rice and cocoa (Tanaka 1957). In 1931 they expanded into vegetable production for the city of Belém and after 1945 began black pepper (*Pimenta nigrum*) production. The pepper profits allowed the improvement of the cooperative (CAMTA); by 1955 they were exporting the crop to the USA and Europe and had gained municipality status. By 1961 there were 116 families farming, each working between 20 and 80 hectares.

The appearance of pepper fungus blight prompted CAMTA to disperse production to minimize disease transmission and to switch to other crops like citrus, papaya, passion-fruit, guaraná (*Paulinia cupana*), cocoa and a maize-bean/cowpea rotation. The colony had been linked with the rest of Brazil by road since the mid-1970s; the production of eggs, poultry and pigs therefore expanded and was integrated with cropping, the pigs being fed on chicken slaughtering waste and the pig manure being spread on the fields (Barrow 1990; author's field notes).

The key to success at Tomé-Açu appears to have been the support given by a close-knit cooperative group that paid attention to, and invested in, agricultural experimentation, market research and marketing. While the colony is relatively close to a large city (Belém) and has quite good road links with the rest of Brazil, there seems to be no reason why similar efforts should not work elsewhere.

Fig. *4.7* Sustainable agriculture. a) Bust of the founder of the Tomé-Açu colony, Pará Brazil. Members of the Japanese colony formed an agricultural cooperative in the late 1920s and their offspring are still working the land today.

Fig. 4.7 b) Tomé-Açu: black pepper vines (to right of Japanese colonist), young cocoa trees among pepper.

2 *Managerial* resource-conservative, orientated toward limited or sustainable growth.
3 *Communalist* resource-preservationist, orientated toward limited or zero growth.
4 *Bioethicist or deep ecological* extreme preservationist, anti-growth.

The more moderate viewpoints (2 and 3) seem to offer the most promising foundation from which to work for sustainable development.

Sustainable development embraces contradictory goals (O'Riordan 1985; Redclift 1987), because development has always been associated with economic growth and yet there must be 'Malthusian limits' in a finite world. A distinction needs to be made between sustainable development and sustainable exploitation, it may be possible to maintain production at a given level (exploitation), but if development is interpreted to mean providing for rising numbers of people with increasing expectations sustaining development may be difficult.

Sustainable development can be viewed on two levels: first, global sustainable development, aimed at ensuring that world-wide catastrophe does not happen, where priority should be given to ensuring development does not cause runaway global problems; and second, local and regional sustainable development. There is unlikely to be a single route to sustainable development, given differing environments, aspirations and problems (see Box 4.1 and Figs 4.7a, 4.7b and 4.7c). Trade-offs required to achieve sustainable development will vary from location to location and through time. The most promise of achieving sustainable development probably lies in localized 'grass-roots' efforts (Davidson *et al.* 1992: 27; Ghai and Vivian 1992). A 'grass-roots', 'bottom-up' approach to sustainable development (originally advocated by the IIED) which has much to commend it is Primary Environmental Care (PEC), which involves:

1 Satisfying basic needs.
2 Optimum use and protection of environment.
3 Empowerment of groups and communities (empowerment refers to the process of granting power to individuals or groups to enable them to

Fig. 4.7 c) Smallholder who had worked the same family plot on the levee of the Tocantins River for 14 years. Every year the floods cover the land and deposit fresh sediment so that there is no need for input of fertilizer. Given extension support and careful management (especially strict control of pesticide use) many thousands of hectares of Amazonian floodland could be worked sustainably with acceptable environmental impacts.

secure and defend their basic rights and manage their own affairs: Illich 1970).

PEC ensures that local people are major agents of change and not just recipients of aid. Some of the foundations of PEC were developed in primary health care (for a discussion of PEC and some examples see Davidson *et al.* 1992: 27–38).

It will be difficult to get sustainable development in practice (Fisk 1991); clearly it is not a costless strategy, keeping future options open, as has already been stressed, may involve forgoing present-day benefits. 'Investment' in the future has to be weighed against other possible investments. Sustainable development will require longer planning horizons. In Victorian times it was not unusual for investors, engineers and planners to adopt such a long-term view, but since the 1940s shorter planning horizons have usually been adopted, possibly driven by what has been termed a NIMTO ('not in my term of office') attitude, i.e. not done unless the politician or decision-maker benefits from it. Individual and corporate behaviour will have to change if there is to be a transition to sustainability. Already policy proposals are appearing, for example the proposals of Von Weizsäcker and Jessinghaus (1992) for providing incentives and penalties through ecological tax reform or the latest additions to the *Blueprint for a Green Economy* series (nos 2, 3 and 4) (Pearce 1991; Pearce *et al.* 1993; Pearce and Barbier 1994). The UK Government has published (1994) a Command Paper on the British strategy for sustainable development, which goes little way beyond discussion of theory and principles; however, a more useful contribution has been the Town and Country Planning Association's report *Planning for a Sustainable Environment.* The latter is the product of a charitable body and charts a way forward for the UK, setting policy goals and putting forward policies with which to reach them (Blowers 1993a; for a review see Blowers 1993b).

It is possible for a nation, region or group of people to achieve 'sustainable' development at a cost to others, by importing resources and exporting pollutants. But at some point other territories and peoples may become unwilling or unable to support this, or there may be disruptive global environmental change affecting everyone (a number of countries are effectively in this situation today).

Successful environmental management requires accurate comparison, ranking and quantification of attributes; a means of achieving this is environmental accounting (or environmental valuation). France has been developing National Heritage Accounts (Comptes du Patrimoine Natural) since 1978, which are sets of accounts measuring and describing economic, ecological and social functions and environmental assets. Norway has a similar system (Pearce and Warford 1993: 88), The Netherlands published a long-range National Environmental Policy Plan in 1989 (Carley and Christie 1992) and the World Bank has promoted the approach (Ahmed *et al.* 1990). However, there are many who feel that this approach is flawed: Ghai and Vivian (1992: 29) were sceptical of the value of such resource accounting approaches for LDCs, and perhaps DCs, arguing that a change in development attitude and ethics is required, not just a revised 'stocktaking' approach (see also Gillis *et al.* 1992). Stirling (1993) reviewed environmental evaluation and noted many shortcomings; nevertheless planning and management tools must be found and it is unrealistic to expect them to be perfect.

To summarize, there are clear indications that the Earth is, at least in the long term, subject to marked environmental changes. In our approach to sustainable development we need to acknowledge that uncertainty and seek to incorporate flexibility and options for adaption. Even then, there will still be risks of catastrophes, humanity could make contingency plans or even develop safety measures against these – if those in power and the people accept the need. Sustainable development is an attractive goal, but reaching it is not going to be easy, nor will it be easy maintaining it.

Failure to accept that the Earth is a closed system that can be easily disrupted by human activity has already led to potentially threatening changes which must now be considered along with natural hazards. Humankind has made its game of survival more difficult. Chapter 5 examines atmospheric changes that occur naturally plus those caused by human activity which appear to pose considerable danger.

References

Adams, W.M. (1990) *Green development: environment and sustainability in the third world*. London, Routledge

Ager, D. (1993) *The new catastrophism: the importance of the rare event in geological history*. Cambridge, Cambridge University Press

Ahmed, Y., Elseraty, A. and Lutz, E. (eds) (1990) *Environmental accounting for sustainable development*. Washington, DC, World Bank

Ahrens, T.J. and Thomas, A.W. (1992) Deflection and fragmentation of near-Earth asteroids. *Nature* 360(6403): 429–33

Allaby, M. and Lovelock, J. (1983) *The great extinction*. London, Paladin

Alvarez, W. and Asaro, F. (1990) What caused the mass extinction? An extraterrestrial impact. *Scientific American* 263(4): 76–84

Barbier, E.B. (1987) The concept of sustainable development. *Environmental Conservation* 14(2): 102–2

Barrow, C.J. (1990) Environmentally appropriate, sustainable small-farm strategies for Amazonia. In D. Goodman and A. Hall (eds) *The future of Amazonia: destruction or sustainable development?* London, Macmillan

Barrow, C.J. (1991) *Land degradation: development and the breakdown of terrestrial environments*. Cambridge, Cambridge University Press

Bartelmus, P. (1986) *Environment and development*. London, Allen and Unwin

Blaikie, P. and Brookfield, H. (1987) *Land degradation and society*. London, Methuen

Blowers, A. (ed.) (1993a) *Planning for a sustainable environment: a report by the Town and Country Planning Association*. London, Earthscan

Blowers, A. (1993b) A world in balance. *The Times Higher Education Supplement* 17/9/93: 48

Bowler, S. (1990) Three steps on the road to extinction. *New Scientist* 127(1773): 37

Broecker, W.S. and Denton, G.H. (1990) What drives glacial cycles? *Scientific American* 262(1): 42–50

Bryant, E. (1991) *Natural hazards*. Cambridge, Cambridge University Press

Burton, I., Kates, R. and White, G. (1977) *The environment as hazard*. Oxford, Oxford University Press

Carley, M. and Christie, I. (1992) *Managing sustainable development*. London, Earthscan

Chapman, C.R. and Morrison, D. (1994) Impacts on Earth by asteroids and comets: assessing the hazards. *Nature* 367 (6458): 33–9

Clark, W.C. and Munn, R.E. (1988) *Sustainable development of the biosphere*. London, Cambridge University Press

Clements, F.E. (1916) *Plant succession. Analysis of the development of vegetation* (Publication no. 242). Washington, DC, Carnegie Institute

Conway, G.R. (ed.) (1985) *Agricultural systems research for developing countries*. Canberra, International Agricultural Research

Courtillot, V.E. (1990) What caused the mass extinction? A volcanic eruption. *Scientific American* 263(4): 85–92

Cowles, H. (1901) The physiographic ecology of Chicago and vicinity: a study of the origin, development and classification of plant societies. *Botanical Gazette* 31: 73

Currie, R. (1988) Lunar tides and the wealth of nations. *New Scientist* 120(1637): 52–5

Dasmann, R.F. (1985) Achieving the sustainable use of species and ecosystems. *Landscape Planning* 12: 211–19

Dasmann, R.F., Milton, J.P. and Freeman, P.H. (1973) *Ecological principles for economic development*. London, Wiley

Davidson, J., Myers, D. and Chakraborty, M. (1992) *No time to waste: poverty and the global environment*. Oxford, OXFAM

Degg, M. (1992) Natural disasters: recent trends and future prospects. *Geography* 77(3): 198–209

Diaz, H.F. and Markgraf, V. (1992) *El Niño: historical and palaeoclimatic aspects of the Southern Oscillation*. Cambridge, Cambridge University Press

Eden, M.J. (1990) *Ecology and land management in Amazonia*. London, Belhaven

Elliott, J.A. (1994) *An introduction to sustainable development: the developing world*. London, Routledge

Engel, J.R. and Engel, J.G. (eds) (1990) *Ethics of environment and development: global challenge, international response*. London, Belhaven

Fisk, D. (1991) Growth without tears. *The Times Higher Education Supplement* 13/9/91: 16

Futures (1988) Special issue on sustainable development. 20(6): 595–678

Ghia, D. and Vivian, J.M. (eds) (1922) *Grassroots environmental action: people's participation in sustainable development*. London, Routledge

Gillis, M., Perkins, D.H., Roemer, M. and Snodgrass, D.R. (1992) *Economics of development*. New York, W.W. Norton

Glaeser, B. (ed.) (1984) *Ecodevelopment: concepts, projects, strategies*. Oxford, Pergamon

Goldsmith, E. (1990) Evolution, neo-Darwinism and the paradigm of science. *The Ecologist* 20(2): 67–73

Gould, S.J. (1984) Toward the vindication of punctuational change. In W.A. Berggren and J.A. van Couvering (eds) *Catastrophes and earth history: the new uniformitarianism*. Princeton, NJ., Princeton University Press

Gould, S.J. (1987) *Time's arrow, time's cycle*. Harmondsworth, Penguin

Gould, S.J. (1989) *Wonderful life: the Burgess Shale and the nature of history*. Harmondsworth, Penguin

Hall, A. (1989) *Developing Amazonia: deforestation and conflict in Brazil's Carajás Programme*. Manchester, Manchester University Press

Halperin, D.T. (1980) The Jarí Project: large-scale land and labor utilization in the Amazon. *Geographical Survey* 9: 13–21

Hamlyn, M. (1992) El Niño fathers distant drought. *The Times* 29/5/92: 12

Hecht, J. (1990) Global warming: back to the future. *New Scientist* 128(1745): 38–41

Hecht, J. (1991) Asteroid 'airburst' may have devastated New Zealand. *New Scientist* 132(1789): 19

Henning, D.H. and Mangun, W.R. (1989) *Managing the environmental crisis: incorporating competing values in natural resource administration*. Durham, N.C., Duke University Press

Hill, A.R. (1987) Ecosystem stability: some recent perspectives. *Progress in Physical Geography* 11(3): 313–33

Holling, C.C. (1973) Resilience and stability of ecological systems. *Annual Review of Ecology and Systematics* 4: 1–23

Hoppe, A. (1992) The Amazon between economy and ecology. *Natural Resources Forum* 16(3): 232–4

Huggett, R. (1990a) *Catastrophism: systems of earth history*. London, Edward Arnold

Huggett, R. (1990b) The bombarded Earth. *Geography* 75(2): 114–27

Illich, I. (1970) *Deschooling society*. London, Harper and Row

IUCN, UNEP and WWF (1980) *World conservation strategy: living resource conservation for sustainable development*. Gland, Switzerland, International Union for Conservation of Nature and Natural Resources

IUCN, UNEP and WWF (1991) *Caring for the earth: a strategy for sustainable living*. London, Earthscan

Johnston, R.J. and Taylor, P.J. (eds) (1989) *A world in crisis? Geographical perspectives* (2nd edn). Oxford, Basil Blackwell

Joseph, L.E. (1990) *Gaia: the growth of an idea*. London, St Martins Press

Kerr, R.A. (1972) Huge impact led to mass extinction. *Science* 257(5072): 878–80

Kershaw, K.A. (1973) *Quantitative and dynamic plant ecology* (2nd edn). London, Edward Arnold

Lélé, S.M. (1991) Sustainable development: a critical review. *World Development* 19(6): 607–21

Lewin, R. (1992) How to destroy the doomsday asteroid. *New Scientist* 134(1824): 12–13

Lovelock, J.E. (1972) Gaia as seen through the atmosphere. *Atmospheric Environment* 6: 579–80

Lovelock, J.E. (1979) *Gaia: a new look at life on earth*. Oxford, Oxford University Press

Lovelock, J.E. (1988) *The ages of gaia: a bibliography of our living earth*. Oxford, Oxford University Press

Lovelock, J.E. (1992) *Gaia: the practical science of planetary medicine*. London, Gaia Books

Lovelock, J.E. and Margulis, L. (1973) Atmospheric homoeostasis by and for the biosphere: the gaia hypothesis. *Tellus* 26: 2

Lyell, C. (1830) *Principles of geology, being an attempt to explain the former changes of the earth's surface, by reference to changes now in operation* (3 vols). (1830–1834)

Macedo, D.S. and Anderson, A.B. (1933) Early ecological changes associated with logging in the lower Amazon floodplain. *Biotropica* 25(2): 151–63

Munn, R.E. (1988) *Towards sustainable development: an environmental perspective*. Paper presented to Conference on Environment and Development, Milan, 24–26 March 1988 (*mimeo*)

Max-Neef, M.A. (1986) *Economía descalza: señales desde el mundo invisible*. Stockholm, Nordan

Max-Neef, M.A. (1992) *From the outside looking in: experiences in 'barefoot economics'* (originally published in 1982 by Dag Hammarskjold Foundation). London, Zed

O'Riordan, T. (1985) Future directions in environmental policy. *Journal of Environment and Planning* 17: 1431–46

PANOS (1988) *Towards sustainable development*. London, PANOS

Paterson, D. (1993) Did Tibet cool the world? *New Scientist* 139(1880): 29–33

Pearce, D. (ed.) (1991) *Blueprint 2: greening the world economy*. London, Earthscan

Pearce, D. and Barbier, E.B. (1994) *Blueprint 4: sustaining the earth*. London, Earthscan

Pearce, D. and Warford, J.J. (1993) *World without end: economics, environment and sustainable development*. Oxford, Oxford University Press

Pearce, D., Barbier, E.B. and Markandya, A. (1989) *Blueprint for a green economy* (Pearce Report). London, Earthscan

Pearce, D., Barbier, E.B. and Markandya, A. (1990) *Sustainable development: economics and environment in the third world.* London, Earthscan

Pearce, D., Barbier, E.B. and Markandya, A. (1993) *Blueprint 3: measuring sustainable development.* London, Earthscan

Raup, D.M. (1988) Diversity crisis in the geological past. In E.O. Wilson (ed.) *Biodiversity.* Washington, DC, National Academy Press

Raup, D.M. (1993) *Extinction: bad genes or bad luck?* (1st edn 1992, Norton, London). Oxford, Oxford University Paperbacks

Redclift, M. (1987) *Sustainable development: exploring the contradictions.* London, Methuen

Rees, W. (1990) The ecology of sustainable development. *The Ecologist* 20(1): 18–23

Repetto, R. (1986) *World enough and time.* New Haven, Conn., Yale University Press

Riddell, R. (1981) *Ecodevelopment: economics, ecology and development: an alternative to growth-imperative models.* Farnborough, Gower

Rudderian, W.F. and Kutzback J.E. (1991) Plateau uplift and climatic change. *Scientific American* 264(3): 42–50

Sachs, I. (1979) Ecodevelopment: a definition. *Ambio* XVIII(2/3): 113

Sagan, D. and Margulis, L. (1983) The Gaian perspective of ecology. *The Ecologist* 13(2): 160–7

Schneider, S.H. (1990) Debating gaia. *Environment* 32(4): 5–9

Schneider, S.H. and Boston, P. (eds) (1991) *Scientists on gaia.* Washington, DC, MIT Press

Smith, D. and Dawson, A. (1990) Tsunami waves in the North Sea. *New Scientist* 127(1728): 46–9

Stirling, A. (1993) Environmental valuation: how much is the emperor wearing? *The Ecologist* 20(3): 97–103

Tanaka, K. (1957) Japanese immigrants in Amazonia and their future. *Kobe University Economic Review* 3: 1–23

The Independent Bureau on International Development Issues (1983) *Common crisis North–South: cooperation for world recovery.* London, Pan

The Independent Commission on International Development Issues (1980) *North–South: a programme for survival* (the 'Brandt Report'). London, Pan

The Programme Organizing Committee of the Conservation and Development Programme for the UK (1983) *The conservation and development programme for the UK: a response to the world conservation strategy.* London, Kogan Page

Tickell, C. (1991) *Speech to UNEP–UK seminar, London* (9th October 1991)

Timberlake, L. (1988) Sustained hope for development. *New Scientist* 119(1620): 60–3

Tisdell, C. (1988) Sustainable development: differing perspectives of ecologists and economists, and relevance. *World Development* 16(3): 373–84

Turner, R.K. (ed.) (1988) *Sustainable environmental management: principles and practice.* London, Belhaven

UK House of Commons (1994) Command Paper 2426 *Sustainable development: the UK strategy.* London, HMSO

Velikovsky, I. (1950) *Worlds in collision.* London, Victor Gollancz

Velikovsky, I. (1952) *Ages in chaos.* London, Victor Gollancz

Velikovsky, I. (1955) *Earth in upheaval.* London, Victor Gollancz

Von Weizsäcker, E.U. and Jesinghaus, J. (1992) *Ecological tax reform: a policy proposal for sustainable development.* London, Zed

Watson, A. (1991) Inside science no. 48: 'Gaia'. *New Scientist* 131(1776): 4.

Whittow, J. (1980) *Disasters: anatomy of environmental hazards.* Harmondsworth, Pelican

Wilson, E.D. (1992) *The diversity of nature.* Harmondsworth, Penguin (pp. 22–30).

Windelius, G. and Tucker, P. (1991) The sun, sovereign ruler with chilling power: an assessment of the potential impact of solar activity on future climate. *Unasylva* 41(163): 15–21

World Commission on Environment and Development (1987) *Our common future* ('the Brundtland Report'). Oxford, Oxford University Press

Further reading

Hill, A.R. (1987) Ecosystem stability: some recent perspectives. *Progress in Physical Geography* 11(3): 313–33 [Good overview]

Independent Commission on International Development Issues (1980) *North–South: a programme for survival.* London, Pan [Brandt Report, one of the first popular books to stress LDC/DC mutual interdependence]

IUCN, YNEP and WWF (1991) *Caring for the Earth: a strategy for sustainable living.* London, Earthscan [Sequel to *World conservation strategy*]

Joseph, L.E. (1990) *Gaia: the growth of an idea.* London, St Martins Press [Review of the Gaia hypothesis]

Pearce, D. (ed.) (1991) *Blueprint 2: greening the world economy.* London, Earthscan [Application of economic analysis to world-wide dangers]

Pearce, D., Markandya, A. and Barbier, E.B. (1989) *Blueprint for a green economy.* London, Earthscan [Pearce Report presents practical proposals for financing a sustainable environment – application of economic analysis to national environmental problems]

Redclift, M. (1987) *Sustainable development: exploring the contradictions.* London, Methuen [Critical discussion of sustainable development]

Roberts, N. (ed.) (1994) *The changing global environment.* Oxford, Basil Blackwell [Especially chapters 1–9 – provides a clear account of the nature of change in the Earth's environment]

Tisdell, C. (1988) Sustainable development: differing perspectives of ecologists and economists, and relevance. *World Development* 16(3): 373–84 [Introduction to the concept]

Watson, A. (1991) Inside science no. 48: 'Gaia' (4 pp.). *New Scientist* 131(1776) [Concise introduction to Gaia hypothesis]

CHAPTER 5

Atmospheric changes due to natural and human causes

Climate change and development

Climatic trends are inadequately established and it is not certain whether human activity will counter or reinforce natural changes. Planners and administrators have tended to ignore the possibility there will be climatic change; the consequences of such neglect could be disastrous. Environmental managers need to weigh up the threats, establish contingency plans and (if appropriate) avoidance strategies. The speed with which the issue of enhanced global warming has been accepted on the political agenda is striking: before 1985 there was widespread scepticism, but by 1992 it was a major issue at the Rio 'Earth Summit'. Perhaps discussion will now lead to necessary action.

Global cooling or warming may progress via transient steps, rather than steady change, and there may be different effects from region to region, at least for a time, as say, air mass movements or ocean currents change. Thus, global warming could lead to cooler conditions in some places or in general (warming just might trigger an 'ice age'). (The term 'equilibrium warming' is used to indicate warming reached some time after it has been triggered.) There is abundant evidence of natural climatic changes in the past; similar changes still pose serious threats to human well-being.

Since the 1750s a range of pollution-related global climatic threats have been increasing. Warnings about these threats have been issued for decades, and from about 1988 there has been sufficient scientific consensus to argue that governments and international agencies would be foolish not to prepare, support research and initiate control or mitigation strategies (Gribbin 1990). To wait for proof, which is what governments tend to do, may well mean delaying avoidance or remedial measures until they cost much more (in real and human terms) or become ineffective (Gleick 1989; Grubb 1990; Mohnen *et al.* 1991; Thompson 1991; World Bank 1992: 160). Faced with uncertainty and a threat, governments should act: no insurance company allows a client to wait until a problem happens to take out a policy and everyone accepts that. Foley (1991: 68) suggests that 'insurance' should be a 'no regrets' approach. Therefore if no threat materializes there is still benefit, for example energy conservation; if climatic change proved no threat it would still have improved environmental quality and saved on energy costs; forestry offers timber and a healthier environment, even if it proves ineffective at locking up carbon or if the need does not arise.

Many consider that enhanced global warming (accelerated greenhouse effect or anthropogenic warming) is the most serious challenge to be faced in the twenty-first century (in my view it is unwise to focus only on one threat; there are others such as natural climatic change and soil degradation). Today's decision-makers may well determine the course of world climate change for generations to come (Leggett 1990; De Freitas 1991). Earth for much of its history (before the last 2 million years) has been warmer and more biologically productive than it is now. Idso (1991) therefore saw future warming as an opportunity for a 'rebirth of the global biosphere', provided change is not too sudden, and humans do not hinder natural adaptions (change may well be rapid and adaptation may be upset). At the other extreme some fear that human activities or natural processes, or a combination of both, could cause a sudden, catastrophic 'runaway' change of atmospheric gas mix or climate.

'Natural' climate change

Climate might be affected by many factors (Broecker and Denton 1990; Williamson and Gribbin 1991). Palaeoecology indicates considerable, apparently related, variation in global mean temperature and atmospheric gas composition, and in oceanic and atmospheric circulation patterns (regional weather conditions) before human beings began to pollute the world. Norse settlements established in Greenland in the tenth century AD failed during cooler wetter conditions between roughly AD 1342 and 1500 as they were slow to adapt and were also overgrazing their cattle pastures. Between roughly AD 700 and 1300 (the 'Little Ice Age') and AD 1550 to 1700 conditions in western Europe are believed by many to have been unusually cool. The Earth was probably warmer between roughly 5000 and 9000 BP than now (the climatic optimum or hypsithermal).

It might be possible to predict future trends from what happened in the past, but in doing so allowance must be made for three possibilities:

1 The possibility that human pollution may counter or reinforce natural changes.
2 The possibility that unpredictable events such as volcanic eruption may also affect climate.
3 The possibility that climate will change faster than in the past.

Speed of change may be more crucial than magnitude, because wildlife, agriculture and other human activities may be able to adapt if they have long enough. People with entrenched attitudes may not adapt. The medieval Greenlanders; the Tiwanaka culture of the Peruvian Andes, who failed to respond to cooling and drought around AD 1000 and many others provide twentieth-century man with a warning.

The global system is complex, for example warming may initiate greater snowfall in higher latitudes as more moisture gets into the atmosphere, this could then cause cooling (as sunlight is reflected from snowcover) or delay the onset of warming (Davidson 1992). There are many potential 'feedbacks', positive (which amplify a change) and negative (which counteract a change), these make forecasting difficult (see Table 5.1). Some feedbacks are allowed for in climate models, but others may still be unknown.

Some natural climatic changes might be periodic, others random. There are many explanations offered for the ice advances (glacials) of the Quaternary 'Ice Age' (approx. 2.4 million–1.8 million BP to approx. 11 000 BP) and it may well be that more than one operate (see p. 64). There is a possibility that the world is presently in an interglacial and that a glacial might return within 4000 to 5000 years, or would do so were it not for anthropogenic warming (the enhanced 'greenhouse effect').

The end of the last glacial (*c.*10 000 BP) may have been quite swift, perhaps a roughly forty-year 'warm-up' (Gribbin 1990), possibly even less. Human activity might delay or prevent such return, might trigger a return, or might warm the Earth enough to cause problems or even a catastrophic 'runaway'

Table 5.1 Some feedbacks (+ and −) which might be associated with atmospheric and climatic change

+	Cold oceans absorb more dissolved inorganic carbon to form carbonates, locking up carbon, if it becomes warmer then < lock up, and more warming. G
+	Industrial pollution controlled – reduction of sulphates in atmosphere, loss of their cooling effect, > warming. G
−	Warming alters ocean currents > carbon lock-up, > cooling. G
+	Warming changes ocean mixing < carbon lock-up, > warming. G
+	Warming > water vapour in atmosphere, > warming. G
+	Warming thaws tundra and leads to release of methane, > warming. B
−	Warming dries soils cutting methane release, < warming. B
−	Warming > water vapour in atmosphere, increases clouds that reflect sunlight > cooling. G
−	Warming > water vapour, more snowfall, reflection of radiation, > cooling. G
−	Sunlight increases ocean plankton production of DMS → condensation nuclei → greater cloud cover → reduction of solar radiation reaching surface > cooling. B
−	Warming/rise of CO_2 → Increased plankton and terrestrial plant activity ('fertilizer effect') → greater fixation of carbon from atmosphere → reduction of 'greenhouse' warming. B
−	Warming/rise of CO_2 → increased activity of sulphur-emitting bacteria in soil (sulphur emissions might help reduce stratospheric ozone) this generates DMS → both processes tend to reduce global temperature. B
+	Warming releases methane hydrates in permafrost/cold seas, accelerates warming. B

Note: B = biogeochemical; G = geophysical/geochemical. DMS = dimethylsulphide (formed by decomposition of a 'precursor': dimethylsulphonisopropionate). Leggett (1990: 30) lists 23 possible feedbacks, and concludes that on balance positive feedbacks are more likely to be felt.
Source: Krause *et al.* 1990; Leggett 1990: 30; Idso 1991: 180; Williamson and Gribbin 1991

heating. Greenland ice cores and evidence of rapid sea-level change in the Bahamas indicate considerable short-term fluctuation during the Quaternary long before human influence: the climate is probably more changeable than we have tended to accept. Sudden global warming or cooling as a consequence of a natural catastrophe is another possibility.

Orbital and astronomical factors as possible causes of climatic change

The Milankovitch theory

This theory postulates a link between climatic changes and alterations of the orientation of the Earth in its orbit around the Sun. Supporters of the hypothesis claim three main cycles which they say cause climatic fluctuation; these cycles are of 22 000, 41 000 and 100 000 years' duration.

Extraterrestrial causes

These can be dust in space, planetesimal strikes, solar, lunar and planetary alignments, and solar energy output variation. Solar activity does vary, and at some points in the '11-year' 'sunspot cycle' (actually 8 to 14 years) the Earth does appear to cool. Cycles of roughly 460, 180, 80 and 11 years have been recognized, and there is some correlation with evidence from tree-rings and historical records. The possibility is that anthropogenic warming has now reduced any such cooling (Roberts 1989).

Volcanic activity

Volcanoes can eject dust and ash, gas and aerosols. Ejected sulphur compounds, especially sulphur dioxide, appear to cause global cooling, other gases (including water vapour) and dust might enhance greenhouse warming. The effect of volcanic eruption depends on what and how much is ejected and to what altitude it rises. It has been possible to link some volcanic eruptions with marked climatic changes. Some claim volcanic activity is influenced by factors like Earth–Sun–Moon alignments. At present there is little possibility of forecasting much in advance when volcanic activity will occur nor the exact character of an eruption. Material ejected into the stratosphere will spread further and stay aloft longer, having greater effect on climate. Dust and aerosols liberated into the troposphere (lower atmosphere, below about 15 km altitude) tend to be 'rained-out' within days. Because the troposphere–stratosphere boundary is lower at higher latitudes, volcanic activity in parts of the world such as Alaska, northern Asia or southern South America will tend to have greater impact on global conditions. Volcanic dust appears to be less likely to affect global climate ('dust veil effect') than sulphate compounds, which cause at least short-term cooling (effective over a few years following an eruption). The indications are that eruptions rich in sulphur compounds, particularly those ejecting material high into the atmosphere, and especially those at high latitude can cause quite rapid global cooling. Ice cores from Greenland, and Antarctica provide a record of repeated acid deposition caused by eruptions over the last 250 000 years or more. Volcanoes liberate many 'greenhouse gases' (see pp. 79–83), whereas sulphur compounds may lead to cooling; other gases are capable of global warming, but links between eruptions and global warming are not clear-cut (Kemp 1990: 108). It is also possible that volcanic activity could affect the Earth's ozone layer leading to climatic changes.

There are several examples of volcanic activity which appear to have reduced mean global temperature (Kemp 1990: 108; other sources):

- Santorini (Greece, 1628 BC) – probably cooling effect.
- Hekla III (Iceland, 1159 BC) – long eruption (about 20 years) – acidic aerosols/ash – Greenland ice cores record eruption – tree-rings in UK, Scandinavia, Eire correlate and indicate cold and wet conditions – catastrophic crop failures in Bronze Age Scotland linked to eruption.
- Hekla I (Iceland, 1104 BC) – cooling?
- Tambora (Indonesia, AD 1815) – huge eruption, ejected acidic material/dust to great height – probably caused cool summer conditions, *c.*1°C below average, and harsh winter following in Europe and USA in 1816 (Europe had catastrophic harvests in that year).
- Krakatoa (Krakatau) (Indonesia, AD 1883) – some global cooling in few years following eruption (even in high latitude north) – effects due to high altitude clouds of dust and possibly acidic sulphur compounds?
- Mt Agung (Indonesia, AD 1970) – spread acidic material/dust into the stratosphere and so world-wide – may have caused some cooling.
- Mt St Helens (USA, AD 1980) – dust, but not rich enough in sulphur compounds or ejected high enough to really cause much cooling.
- El Chichón (Mexico, AD 1982) – acidic materials – possibly *c.*0.3°C reduction of mean global

temperatures in 1983 and 1984. El Chichón's outgassing of sulphur dioxide also reduced stratospheric ozone.

- Mt Pinatubo (Philippines, AD 1991) – appears to have had enough cooling effect to distort the trend toward global warming. More effective at cooling than El Chichón, triggered up to *c*.0.4°C reduction of mean global temperature. Ejected sulphur dioxide/dust/acidic aerosols high into stratosphere. Cooling effect most felt in continental interiors.

Other possible natural causes of climatic change

The oceans may well play a role in climatic change (e.g. the ENSO-type events which seem to relate to droughts – discussed in chapter 6), whether they initiate it or are a link between, say an astronomical cause and terrestrial cooling or warming is so far unclear. Change might be initiated by some process of 'turnover' relate to salinity changes oceanic current variations or alteration of photosynthetic activity by oceanic plankton. The latter could change the composition of the atmosphere leading to greenhouse cooling or warming, and might result from natural pollution (of say, volcanic origin), from increased predation of plankton, variation of ocean currents or mixing affecting supply of nutrients, increased receipts of iron compounds in ocean surface waters (perhaps as wind-blown dust from land). To help understand the role of plankton the Global Ocean Flux Programme (1989–99) will, among other things, try to assess their fixation of carbon (Williamson and Gribbin 1991).

'Anthropogenic' climatic change

Dust, soot and aerosols as a possible cause of climatic change

These materials are mainly generated by burning forests and grasslands; farming activity; wind action in drylands, especially if there has been grazing or tillage; industry; combustion of fuel for power generation, by vehicles or home heating; and warfare. Dust from farming can travel great distances, for example winter ploughing in northern PRC generates dust that is registered in Hawaii. There have been increasing reports of dirty snow and 'winter haze' in the Arctic shown to result from industrial pollution in North America, northern Europe and high latitudes of Asia. However, so far it is more a problem of pollution of wide areas with undesirable compounds than one of altered albedo and climatic change so far. Probably because these materials are relatively heavy and generally do not get into the stratosphere, they have caused limited increase of atmospheric turbidity (increase in suspended particulate matter) and probably little cooling or warming (Kemp 1990: 112–13).

Large-scale, particularly nuclear warfare, might raise dust, soot and aerosols into the stratosphere, block sunlight, cause serious cooling ('nuclear winter' effect), and stratospheric ozone damage. Fears voiced in the 1970s that high-flying supersonic transport aircraft would eject sufficient aerosols and gases into the stratosphere to affect climate or the ozone layer are unfounded at the moment, given the very limited number of such craft.

Reduction of natural 'lock-up' of greenhouse gases as a possible cause of climatic change

The cutting of forests and pollution of oceans has probably cut the rate at which atmospheric gases that are responsible for global warming are rendered inactive by photosynthetic lock-up or deposition as carbonates.

'Greenhouse gases' as a possible cause of climatic change

The atmosphere consists mainly of nitrogen (78 per cent), oxygen (21 per cent) and argon (0.9 per cent); the remaining 0.1 per cent includes trace gases that are active in causing global warming or 'forcing'. The amounts of these 'greenhouse gases' in the atmosphere varied before humankind appeared, due to volcanic emissions, or because of fluctuations in biogeochemical cycles or because climate changed. However, during roughly the last two centuries, human activity has been increasing the atmospheric concentration of some greenhouse gases at rates which probably pose a significant threat to global climate.

Global warming

Even if there is no variation in Earth's solar radiation receipts, changes in the concentration of the component gases and amounts of particulate matter drifting through the air alter the atmosphere's 'greenhouse effect' and so global temperatures. The idea that the concentration of greenhouse gases like water vapour or carbon dioxide controls global temperature is not new, suspicions were voiced in 1827 (by Jean-Baptiste Fourier), in 1861 (by John Tyndall) and in 1896 and 1903 by Svante Arrehenius, who published calculations of the effects of doubling atmospheric carbon dioxide. Warnings that humans were upsetting the climate have been voiced since the 1940s (staff of the Scripps Institute of Oceanography did so in 1957); probably the first to attract serious attention were those of Wallace Broecker in the early 1970s and the World Meteorological Organization, which issued an alert in 1974 (Gribbin 1988a; McKay 1991).

The 'greenhouse effect' operates broadly as follows: the Sun's energy passes through the Earth's atmosphere with adsorption of some wavelengths by water vapour, gases and dust, and some reflection from clouds. Most of this incoming radiation reaches the surface where roughly 70 per cent is re-radiated at longer wavelength back to space. Passing back out of the Earth's atmosphere, some of this longer wavelength radiation is 'trapped', mainly in the lower atmosphere below 14 km, mostly by water vapour and also 'greenhouse gases', so maintaining global temperatures.

Roughly 90 per cent of natural greenhouse effect is due to water vapour, but there are 20 or so other 'greenhouse gases', the main ones being carbon dioxide (CO_2); nitrous oxide (N_2O); methane (CH_4); carbon monoxide (CO); hydrogen sulphide (H_2S); a range of chlorofluorocarbon gases (CFCs) and related compounds like halons; and low altitude (tropospheric) ozone (O_3). Greenhouse gas releases are not constant: the sources are diverse and they vary a lot in their ability to force global warming and in the time they remain active once released. Molecules of methane are about 20–30 times more effective at trapping radiation than carbon dioxide, and CFCs as much as 20 000 times as effective as carbon dioxide (World Resources Institute, UNEP and UNDP 1990: 12). Methane is active in the atmosphere for far less time than CFCs. Climatic change is due to a 'cocktail' of gases, the mix of which will change over time. Attention today is focused on carbon dioxide (Fig. 5.1a); in the future, methane, CFCs, carbon monoxide and nitrogen oxides are likely to become more important. Figure 5.3 (see p. 90) shows the relative warming effect (forcing) of the main 'greenhouse gases' in the mid-1980s.

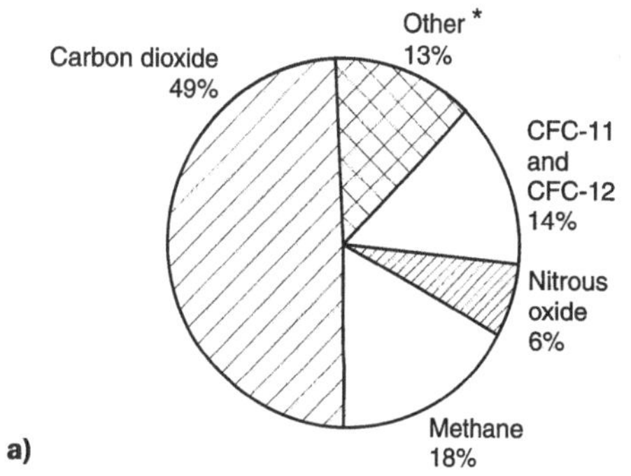

Note: * Other = tropospheric ozone; halons; stratospheric ozone; water vapour

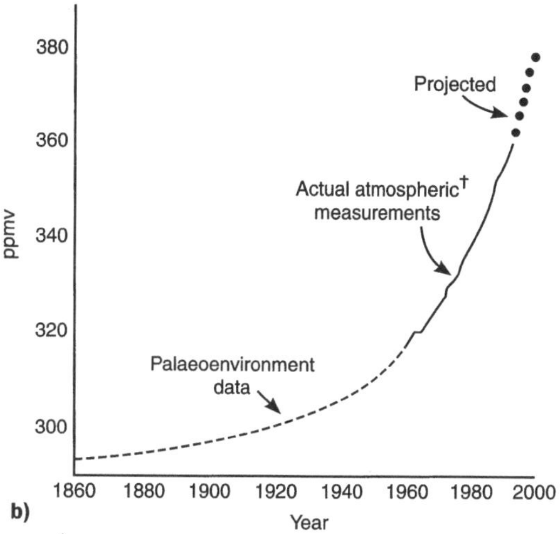

Note: † Seasonal fluctuation reflecting biotic activity – (curve is averaged)

Fig. 5.1 Greenhouse gases. a) Approximate contributions to the 'greenhouse effect' by the main 'greenhouse gases' in the 1980s (*Source:* Boyle and Ardill 1989: 27 (Fig. 5); Pearce 1989: 39; other sources). b) Carbon dioxide concentration in the atmosphere 1860–2000 (*Source:* various sources).

Greenhouse gases

Carbon dioxide

Currently most greenhouse gas-related climatic change (forcing) is due to carbon dioxide – about 55 per cent in 1993; in the future this percentage contribution is likely to fall (Mohnan *et al.* 1991: 111). Levels have been continuously monitored since 1957 (initiated during the International Geophysical Year) at Mauna Loa (Hawaii) and at the South Pole and show a rising trend from the 1750s until recently. Gas bubbles in ice taken from cores through icecaps in Greenland and Antarctica (Vostock Base) show atmospheric composition back to 160 000 BP or earlier, covering the last glacial and interglacial. High levels of carbon dioxide correlate with warm conditions and low with cold conditions and other evidence confirms these findings. When the last glacial reached its greatest chill, roughly 18 000 BP, the concentration of carbon dioxide in the atmosphere was 190 to 210 ppmv (parts per million by volume) and the global mean temperature was about 4°C less than today's (Boyle and Ardill 1989: 25; Leggett 1990). During the last interglacial, when temperatures were a little above those of today carbon dioxide levels were around 300 to 350 ppmv (Foley 1991: 6). Since the industrial revolution atmospheric concentrations of carbon dioxide have risen: from round 250–90 ppmv in 1750 to roughly 315 ppmv in 1957, to roughly 346 ppmv in 1985, and to roughly 350 ppmv in 1989 (Boyle and Ardill 1989: 25).

In the last 100 years carbon dioxide has increased to a level 20–25 per cent higher than at any point in the last 16 000 years, and there are predictions that if no remedial action is taken the level will be 560 ppmv or more by AD 2000 (Barbier 1989: 20). By 2030 there may well be twice the level of carbon dioxide present at the start of the industrial revolution, even if countries like the PRC and India do not increase coal consumption, which is likely. Raised carbon dioxide, plus increasing amounts of other 'greenhouse gases', and altered cloud cover are thought likely to cause significant warming. Already greenhouse gases other than carbon dioxide are believed to contribute roughly equally to climatic forcing and their relative importance will probably increase (Bolin *et al.* 1986: xxi).

The Earth had carbon dioxide levels higher than those of the last few million years for most of its history between roughly 600 million BP and 4 million BP. In the late Cambrian (roughly 570 million BP) there was probably about eighteen times today's carbon dioxide level; in the Carboniferous (around 300 million BP), twice today's level and in the mid-Cretaceous (around 100 million BP) five to six times today's level (Hecht 1991) (Fig. 4.4b). In the past, life has survived more carbon dioxide than pollution is likely to give in the next hundred years. However, we cannot afford to be complacent because humans, and many other organisms, are adapted to present conditions and even an increase or decrease in global mean temperature of only 2°C would significantly affect food supply and raise sea-level.

Most of the carbon dioxide increase is due to combustion of fossil hydrocarbon fuels (coal, oil, gas) and the burning of vegetation (Levine 1991; Crutzen and Goldammer 1993). Biomass burning probably contributes about 40 per cent of annual CO_2 increase (Levine 1991). Heating limestone, used as a flux in metal smelting and as a raw material for cement and agricultural lime, also liberates carbon dioxide. Loss of vegetation, which has the capacity to act as a carbon 'sink', means less carbon dioxide is taken out of the atmosphere.

Roughly half of the carbon dioxide produced by human activity seems to be absorbed by natural 'sinks' (Gribbin 1988a: 2, Bolin *et al.* 1986). These are not fully understood, but carbon dioxide easily diffuses into the oceans to be precipitated as carbonate deposits or is fixed by plankton through photosynthesis and by organisms which form shells or skeletons from carbonates (Broecker and Denton 1990: 47) (by convention carbon dioxide emissions are commonly cited in terms of weight of carbon contained in the gas). Coal, oil, peat, soil and plant tissue lock-up carbon. Chemical weathering also removes carbon dioxide from the atmosphere, vegetation cover plays a part in this by acidifying rainwater enough to speed up the process and warming is likely to assist – a probable negative feedback acting against global warming (Schwartzman and Volk 1991).

Lovelock (1979; 1990) and others have discussed the possibility that plankton play a major role in controlling climate through carbon fixation and release of compounds like dimethyl sulphide and methyl iodide (Fig. 4.2). Considerable effort is being made to discover how well ocean plankton act as sinks for carbon dioxide, for example by the International Geosphere Biosphere Programme (1992–2002) which is investigating the ocean carbon cycle. The huge northern (boreal) forests also play a key role in locking-up carbon (see chapter 7).

Methane

Estimates vary a good deal, however, methane (CH_4) levels probably rose from roughly 800 ppbv (parts per billion by volume) in the 1750s to about 1700 ppbv by the late 1980s (Pearman *et al.* 1986; Foley 1991: 10). Atmospheric monitoring began in the 1960s and the rate of increase in the early 1990s was about 16 ppbv per year, according to Warrick and Farmer (1990: 7).

This represents an increase of between 1 and 2 per cent per year, which projected gives an AD 2050 level of roughly 2500 ppbv (Kemp 1990: 154). However, although there has been a three-fold increase in atmospheric concentration since 1945, there were signs in late 1993 of a marked reduction in methane increase.

Methane is more effective at raising global temperatures than carbon dioxide (some suggest between 3 and 25 times as effective, others 63 times), also it can affect ozone formation in the troposphere (Leggett 1990: 39). Past concentrations of methane can be assessed from bubbles trapped in ancient ice, and like carbon dioxide, these appear to increase when the climate is warm and decrease during glacials.

Sources and sinks are not fully understood but efforts are being made to model the increase and its effects (Rotmans *et al.* 1992). Most emissions probably result from decomposition of organic matter by micro-organisms under anaerobic conditions, for example in paddy-fields, swamps, coal-mines, landfill refuse tips, sewage and livestock slurry, in the digestive tracts of animals, especially cattle and probably termites in the tropics (Bolin *et al.* 1986: 161; Lee *et al.* 1987; Bouwman 1990a). Inefficient combustion of hydrocarbon fuels, charcoal making and leakage of natural gas from oil and gas wells and pipelines may comprise roughly one-fifth of anthropogenic methane emissions (some suggest leakage related to hydrocarbon exploitation is a major source – and could be easily controlled). Estimates have suggested roughly 6 per cent of UK gas supplies leak between the North Sea and use – enough to virtually cancel out the advantage gas should have over coal use in terms of Britain avoiding causing 'greenhouse warming' (McKenzie 1990). There may also be natural emissions from the Earth (Bouwman 1990b), e.g. some volcanic eruptions, notably cold, mud volcanoes, release methane.

A significant proportion of anthropogenic methane emissions are related to food production and this could increase in future as efforts are made to boost agriculture to feed the world's growing population, especially through increased irrigation, fertilizer use and growing numbers of livestock. Bouwman (1990b) suggested paddy-field emissions could already account for 20 per cent or more of total annual anthropogenic methane increase. Nilsson (1992) suggested that swamps and other wetlands were the main sources; rice-paddies the second; livestock enteric emissions the third; escapes from oil/natural gas wells the fourth; termites the fifth; landfill refuse decay the sixth; coal-mines the seventh.

Methane sinks include fixation by soil micro-organisms and deposition in frozen soil, cold and deep seas as methane hydrates (clathrates). Fears have been voiced that as global warming occurs, methane hydrate deposits might suddenly release large quantities of methane leading to a positive feedback strong enough to cause 'runaway' global warming. Ultimately (after 8 to 11 years) a good deal of the CH_4 in the atmosphere breaks down to CO_2 and water.

Carbon monoxide

Carbon monoxide (CO) is the second most abundant 'greenhouse gas' after CO_2. Levels are probably increasing at 1–5 per cent per annum, mainly as a result of incomplete combustion of hydrocarbon fuel (Rambler *et al.* 1989) and from biomass burning (roughly 32 per cent of total annual emissions, according to Levine 1991; a similar estimate is given in *Nature* 359{6398}: 776).

Nitrous oxide

Nitrous oxide (N_2O) is an effective 'greenhouse gas' and plays a part in the formation of tropospheric ozone (the latter also acts as a 'greenhouse gas'). The gas is released naturally through nitrification by micro-organisms in soil and by forests and grasslands. Human beings have raised emissions mainly through biomass burning (about 21 per cent of the total annual emissions, according to Levine 1991) and the use of artificial nitrogenous fertilizers, but even unfertilized soil when cultivated can release nitrous oxide. Combustion of hydrocarbon fuels and biomass, emission from sewage, livestock slurry and from marine plankton (and possibly algae) also contribute to rising levels of atmospheric N_2O.

Knowledge of releases and the global budget for this gas is incomplete so it is difficult to forecast. Before 1950 the levels were about 285 ppbv, by 1989 it was at 305 ppbv and increasing at roughly 0.2 to 0.3 per cent per annum (Boyle and Ardill 1989: 30; Rambler *et al.* 1989: 70). Nitrous oxide breaks down slowly so even limited emissions can accumulate and remain effective for a long time.

Nitrogen dioxide

Nitrogen dioxide (NO_2) acts as a greenhouse gas, is highly toxic, and plays a part in photochemical smog formation (urban air pollution). Catalytic converters on motor vehicles help to reduce emissions.

Nitric oxide

Nitric oxide (NO) is often grouped with the two aforementioned gases (N_2O and NO_2); the three are often referred to as NO_x (this is more usual when dealing with air pollution and is not an ideal practice).

Ammonia

Ammonia (NH_3) is a very effective greenhouse gas. Anthropogenic emissions derive from sewage, livestock dung, breakdown of nitrate fertilizers, biomass burning and some industrial sources. Livestock and fertilizer usage probably accounts for 80–90 per cent of the emissions.

Tropospheric ozone

Attention has focused on the role of ozone (O_3) in the stratosphere as a filter against ultra-violet radiation; however, in the atmosphere below roughly 15 km (the troposphere) ozone also acts as a potent greenhouse gas (and as a pollutant). Tropospheric ozone is formed by the action of ultra-violet radiation on molecules of nitrogen dioxide, mainly formed when biomass or hydrocarbon fuel is partially burnt (this accounts for about 38 per cent of annual emissions, according to Levine 1991). The gas is not stable and soon breaks down. Concentrations vary in time and spatially according to topography and factors like wind patterns so it is difficult to forecast future climatic effects. As well as affecting climate the gas is toxic, damages crops and other vegetation, higher organisms and some construction materials (Le Fohn 1992; see p. 267).

Water vapour

Water vapour is the most important atmospheric constituent in terms of forcing (warming). Most is invisible and as global temperatures rise there will tend to be more of it in the atmosphere. Raised levels of water vapour enhance cloud formation; whether this increases global warming or reflects solar radiation to cool the Earth seems to depend on the type of cloud and altitude. It is unlikely that human activity will alter water vapour levels directly, although warming due to other gases will probably cause levels to rise.

Chlorofluorocarbons and halons

Chlorofluorocarbons (CFCs), halons and halocarbons are synthetic chemicals containing chlorine or bromine, fluorine and carbon. Known since the 1930s and developed mainly after 1951 as 'safe', inert aerosol-can propellants, refrigeration/air-conditioning fluids, de-greasing agents and fire extinguishers (peak production was in 1974). CFCs are extremely strong 'greenhouse gases', and also act to destroy the stratospheric ozone layer (see Box 5.1). Once CFC molecules are in the atmosphere they have a long, active life (emission controls will therefore not stop or prevent some future increase of CFC-related problems and it will be decades before levels fall). CFCs have been largely replaced as aerosol-propellants but there are worries that demand for refrigeration, air-conditioning and industrial usage will increase emissions, particularly in developing countries (PANOS 1991).

CFCs like the 'freons' CFC-11 and CFC-12 are still widely used in refrigeration, electronics manufacture, etc. Attention is focused on getting a shift to 'safe' substitutes that do less damage to the stratospheric ozone layer but care should be taken to see these are not active greenhouse gases or pose a pollution threat (one possible substitute, CFC-113, apparently does). Substitutes must do less damage to stratospheric ozone, have less warming effect and still be inert, effective and cheap. Hopes that HCFCs would suffice have been dashed, and the Montreal Protocol has

Box 5.1 Some widely used CFCs and halons

CFC-11 ($CFCl_3$) plastic foamer[a]
CFC-12 (renamed HCFC-22) (CF_2Cl_2) refrigerant/aerosol propellant[b]
CFC-113 ($C_2F_3Cl_3$) semiconductor de-greasing
halon 1211 (CF_2ClBr) fire extinguisher halon 1301[c] and 1302 (CF_3Br) fire extinguisher
methyl chloride (CH_3Cl)
methyl bromide crop and soil fumigant[d] (pesticide)

Note:
a lasts about 65 years in atmosphere
b lasts about 130 years in atmosphere
c roughly ten times as active in destroying ozone as CFC-11
d about 33 per cent of world production is reputedly produced by one chemicals complex in Israel (perhaps causing 3 per cent of world ozone thinning – *The Times* 22/8/93: 10)

For a list of CFCs, their application and ozone depletion potential see Porteous (1992: 61–4)

since 1991 phased these out (notably HCFC-22); the plan now is that HFCs (which have no ozone-scavenging chlorine) will ultimately replace CFCs. Already domestic and vehicle air-conditioning manufacturers are starting to change to R134a, a non-CFC which is effective but much more expensive to use.

Some CFCs are roughly 100 000 times more effective than carbon dioxide in forcing and, although the amount in the atmosphere is relatively small, were already responsible for as much warming as methane by 1990 (Gribbin 1990: 145; Warrick and Farmer 1990: 7).

Sulphur dioxide

There are indications that anthropogenic emissions of sulphur dioxide (SO_2) have increased ten-fold during the last century. The main effect seems to be to increase cloud cover, and so cool the Earth, especially in the northern hemisphere, where concentrations are greater. The gas also plays an important role in acid deposition and efforts to control its emission are largely prompted by concern for that. Emission control may prove relatively easy but, if achieved before other 'greenhouse gases' are managed, the loss of cooling effect might help trigger more warming. The main anthropogenic source is the burning of hydrocarbon fuels; however, some volcanoes release large quantities.

Hydrogen sulphide

There are indications that hydrogen sulphide (H_2S) can cause global cooling. Naturally the gas is released by decay processes, by living vegetation, and volcanic eruptions can release large amounts. Human activities are raising natural levels; indeed they probably account for over 90 per cent of what there is in the atmosphere. The main anthropogenic sources are oil refining, woodpulp production, the combustion of hydrocarbons, and breakdown of sewage.

Is global warming really occurring?

If warnings of accelerated global warming are to be taken seriously, two questions need to be considered. Is there indication of warming? Is change recorded during roughly the last 150 years a deviation from what would have been natural patterns and therefore due to anthropogenic causes?

Over the last 140 years the trend does seem to be one of warming (Pearce 1991). P.D. Jones and Wigley (1990) suggested that there had been a 0.5°C warming over the last 100 years. Global temperatures recorded in the 1980s and 1990s were appreciably warmer than the average. However, weather records may not be proof enough, as some meteorological recording sites might have got warmer, due to the heat island effect of growing cities (see p. 246) or clearance of vegetation and measuring techniques have changed over the years. There have been reports that US and UK submarines have logged thinner ice at the poles in the late 1980s (Leggett 1990). Attempts to compare Australian wheat yields with known carbon dioxide increases are said to indicate warming (Skiba and Cresser 1988). Tree-ring evidence from Tasmania tends to confirm that there has been rising temperatures, as do signs of tropospheric warming and stratospheric cooling (Gribbin 1991). Warming has probably taken place and will continue in 'jumps'; from region to region conditions may vary a lot, with some areas perhaps even cooling after volcanic eruptions or shifts in ocean currents or airmass movements.

A rising trend in meteorological records may not yet be obvious. There are indications of warming but irrefutable proof is difficult to find (De Freitas 1991: 12). The suggestion is that since the 1860s there has been an 'average global warming' of 0.5°C to 0.7°C (R.A. Houghton and Woodwell 1989: 19). Intergovernmental Panel on Climatic Change (IPCC) studies concluded that world average ocean surface temperatures have increased between 0.3°C and 0.6°C between 1861 and 1989 (Foley 1991: 29). Given that the thermal inertia of the oceans will slow any realized warming, the previous figures may offer a conservative indication of change taking place.

Establishing if warming is a deviation from what would have happened naturally presents challenges. Extrapolating from past patterns of climatic change and study of how conditions have altered after certain past volcanic eruptions or known changes in solar activity gives some indication. Modelling the effects of greenhouse gas increase provides further information. While there is not absolute proof, the general consensus is that a significant amount of warming is anthropogenic (although sudden cooling after volcanic eruptions or caused by warming increasing the snowfall is still possible).

What are the likely impacts of global warming?

Precise prediction of impacts is impossible because there are so many uncertainties and gaps in the data and understanding of how environmental processes operate. On the whole it is scenarios that are being suggested and discussed, the 'what if...?' speculation. The models used to identify these scenarios and the data fed into them are not perfect. Nature is also capable of springing surprises.

Impacts could include climatic change, sea-level rise, altered conditions for agriculture and wildlife. And there could be indirect impacts, for example hindrance of agriculture, property development problems, health and pest control, or industrial development problems as a consequence of altered environmental conditions, restrictions and costs associated with 'greenhouse gas' emission controls. Some important impacts may be cumulative, the result of combined effects of more than one, possibly insignificant, change, which makes accurate forecasting difficult (Barbier 1989; Houghton and Woodwell 1989; Leggett 1990).

It might be that without 'greenhouse warming' the Earth would have slipped back to conditions like those of the last glacial; if so, any coming problems of warming are almost certainly the lesser of two evils. The literature tends to focus on negative impacts, but there are a few, like Ince (1991), who see warming as offering opportunities. However, they should not forget that life has adapted to the cold of the last 2 million years (Krause *et al.* 1990), and development may limit future response and adaption to changes.

The harm caused by warming depends on how fast as well as how much it happens and on the quality of long-term planning and policy making: if policy making is poor, opportunities may be lost and damage magnified. Development planners must not assume that environmental conditions are going to remain constant and should exploit opportunities and meet challenges. Maunder (1989: 27) suggested a 'disaster' for flora, fauna and human development could be a 10 per cent decrease in cloud cover over North America, a 1°C increase in summer temperature over the South Pacific and a 3°C decrease in summer temperature over the CIS: apparently minor changes, but enough to seriously disrupt the world's main grain-producing regions.

Climate models

General circulation models (GCMs) are being developed to try to simulate (often in 3-D) what happens if atmospheric gas-mix alters (for a review see Leggett 1990: 44–67). GCMs are seriously hampered by an incomplete database; gaps in understanding of environmental processes, especially greenhouse gas sinks and the way clouds propagate and reflect sunlight, and inadequate computers. No GCM is yet satisfactory but a few can be run to show recent known past climatic changes ('backcasts') or present seasonal climatic change quite well and so might give reasonable forecasts. In a complex environment with many potential feedbacks, crucial process can be missed by a model: a minor thing may lead to a significant development, what has been termed the 'butterfly effect' (i.e. insignificant more or less random input ultimately affecting global systems). Present GCMs make coarse predictions (yet decision-makers seek regional or local details) and fail to adequately show the nature and timing of changes, also there may well be sudden shifts rather than gradual modification of conditions, giving little time to plan a response (De Freitas 1991: 4; Mohnen *et al.* 1991).

There are at least fourteen centres that have developed GCMs; attention has focused on five of them:

- National Center for Atmospheric Research (USA);
- NASA Goddard Institute for Space Studies (USA);
- Princeton University, Geophysical Fluid Dynamics Laboratory (USA);
- Moscow Hydro-Meteorological Centre (CIS);
- UK Meteorological Office (UK).

Some of these assume little change in human habits – 'a business as usual scenario' (no real breakthrough in control of emissions, no great change of habits) – some seek to allow for a realistic scenario of change.

Although conditions today are very different from those at the peak of the last glacial, global-mean surface temperatures are probably only about 3°C higher: a 'minor' change in global temperature can thus have marked effect. It should also be remembered, repeating the point made earlier, that unpredictable natural events like volcanic eruptions have the potential to suddenly upset established trends, accelerating warming or counteracting it.

Predicted changes in climate

In 1988 the Intergovernmental Panel on Climatic Change (IPCC) was established by the UN General Assembly to try and assess likely greenhouse effect-related global climatic change, its environmental and socioeconomic impacts and possible policy responses and actions required to cope (J.T. Houghton *et al.* 1990; 1992; Jäger and Fergson 1991). The establishment of the IPCC was the first real step by the international community to address the issue of development-related climate change. In 1990 the IPCC reported its findings to the World Climate Conference (Intergovernmental Panel on Climatic Change 1991).

A doubling of CO_2 by AD 2030 is widely accepted, and predicted to lead to a 1.5°C to 4.5°C rise of global-mean surface temperature. Most would suggest a figure below +2°C (Gribbin 1988b; Warrick and Farmer 1990; Krause *et al.* 1990); there are some, however, who suggest rises as high as 6.5°C or even 11°C within a century or so (Idso 1991). It probably makes most sense to plan for around 2°C increase by 2030 (Barbier 1989: 21).

What effect such changes would have on regional and local conditions is difficult to assess. Changes will be uneven, with probably more effect felt at higher latitude. There will probably be some lag in warming due to the thermal inertia of the oceans; there may even be some initial cooling in higher latitudes as warmer air delivers more moisture to land to form snow. Scenarios may therefore refer to 'actual' and 'committed' change, such that in AD 2030 conditions may have become x°C warmer but even if there is no more pollution it will still rise to x° + y°. For example, the IPCC suggest an actual (realized) change of +1°C (above 1990 temperatures) in AD 2030 with a committed rise of +3°C fully felt by the end of the twenty-first century.

Warming may well increase what are today seen as extreme weather events: droughts, floods and storms, partly because of warmer seas. Climatic belts are likely to shift toward the high latitudes, with the effect felt more in the northern hemisphere land areas away from moderating oceanic influences (there is less land in the southern hemisphere). Continental interiors may become hotter and drier. Warmer conditions at high latitudes may reduce winter heating needs and expand growing seasons and may perhaps reduce contrast between day and night temperatures. One consequence may be improved conditions for winter shipping along the northern coasts of Eurasia.

Impacts are unlikely to be simple: warming may assist a temperate country like the UK by improving the growing season and reducing winter heating needs, but it might also increase storms and summer droughts.

Predicted oceanic changes

Sea-level rises are likely, due to thermal expansion of the oceans as temperatures rise and, to a lesser extent, through melting of land ice and sea ice (Hekstra 1989). Rises in sea-level will probably be quite slow and delayed as oceans respond slowly to warming. There is a possibility that the Greenland and West Antarctic ice sheets may break free and cause a sudden large (over 5 metres) rise in sea-level. Warrick and Farmer (1990: 12) think that this is unlikely to happen until there has been at least a 4°C rise, which may not come for at least 100 years, and that even then, global sea-level rise would not be very sudden. But there have been recent reports (*Nature* 359{6398}: 816) which suggest that the East Antarctic ice sheet melted about 3 million years BP in the mid-Pliocene (when temperatures may not have been much higher than now). More likely are rises by AD 2030 of between 17 and 30 cm (the lowest estimate is 5 cm and the highest, realistic prediction is 44 cm) (J.T. Houghton *et al.* 1990). A 2°C rise would probably give about 30 cm sea-level rise (Ince 1991), and would have serious consequences. A conservative forecast for 2030 would be a realized rise of 20 cm and a committed rise of a further 30–100 cm by the end of the twenty-first century.

Much is uncertain, warming could increase snowfall in high latitudes and offset thermal expansion of the oceans, or it might help provoke a surge and subsequent melting of the West Antarctic ice sheet. A possibility is that global warming might suddenly alter ocean currents or the 'turnover' of deeper cool waters. Such occurrences could be difficult to reverse and could suddenly cool or warm whole continents like Europe. Northern cold seas are believed to act as significant sinks for carbon; warming might have a positive feedback by reducing the carbon fixing. Both aforementioned alterations could have catastrophic impacts on marine life which might also trigger a positive feedback – a reduction of carbon fixation and runaway warming.

Predicted sea-level rise impacts

Sea-level is not a constant: there were rises, before pollution became significant, e.g. in late Roman times, which are known to have flooded extensive areas of Britain, Europe and other parts of the world. The rate of sea-level rise is important; slow change allows flora, fauna and people some chance to adapt. Bowler (1991) suggested that natural environments might be able to cope with 1°C change provided it took place at no more than 0.1°C per decade; he felt that wetlands and coral reefs could cope with a sea-level rise of around 2 cm per decade. Provided sea-level rise is not too rapid, coral reefs may grow fast enough to keep pace (corals can grow at 1 to 10 mm per year and predicted sea-level rises are as much as 6 mm per year); however, there is some concern that increasing sea-temperatures may already be causing coral bleaching and dieback in some areas (Goreou and Hayes 1994). The predicted 2°C warming at a rate of 0.2°C per decade, together with a 20 cm sea-level rise (which authorities like the IPCC predict for 2030) would mean serious disruption.

In the Pliocene (5.3 million to 1.6 million years BP), with global mean annual temperatures perhaps 2°C to 4°C above those of today, sea-levels rose 30 to 35 m above the present mean; during the last glacial a fall of as much as 120 m may have taken place (Tooley and Shennan 1987). A complete melt of the Earth's ice caps might raise sea-levels by 70 metres.

Sea-level change is difficult to predict, as a consequence of local tectonic subsidence or emergence, changes in currents, storm patterns and prevailing winds; for example, in 1981–2 reduction of prevailing easterly winds during an El Niño event led to a rise of 35 cm along parts of the west coast of South America and similar rises along the Pacific coast of the USA (De Freitas 1991: 13). Low-lying deltas, islands with little high ground and coastal lands are especially at risk. There are also likely to be difficulties with seawater penetration into freshwater aquifers and estuaries, for small islands or regions where there is no alternative water supply replacement of fresh groundwater with salt groundwater would be a serious blow. Vulnerability to sea-level rise and storm or tidal extremes is increased by tectonic subsidence (Bell 1993: 4), reef damage and destruction of mangroves (reducing their 'breakwater' effect), coastal erosion, excessive removal of groundwater and building (which may lead to lands sinking). The Mississippi Delta region, for example, is naturally sinking by roughly 1 metre per century, according to Warrick and Forester (1990), as is southeastern UK, although more slowly.

Vulnerable areas include eastern Bangladesh, the Nile Delta, Guyana, islands like Kiribati and Tuvalu (Pacific), the Maldives (Indian Ocean), many major cities and The Netherlands (for a review of threatened areas see Barrow 1991: 40–3). The economic impact of the threat of rising sea-levels has already been faced by The Netherlands. Having recently completed 20 years of coastal defence construction, the Dutch have embarked on a new programme (estimated to cost US$6000 million at 1986 prices) to protect themselves against a possible 50 cm sea-level rise (Hekstra 1986; 1989). Few other countries will be willing or able to pay such sums.

If there is significant sea-level rise vulnerable areas may become a source of 'ecorefugees'. The Maldives have 200 000 people living within a couple of metres of present sea-level. A 50 cm rise would probably force Egypt to relocate about one-sixth of its population. The worst potential ecorefugee situation is in Bangladesh, where a rise of 50 cm could swamp 14 per cent of territory and affect 10 per cent of population; a rise of 1 metre might hit 17 per cent of territory and, were it to occur by 2025 (predicted population 346 million) it might necessitate relocation of 34.6 million; a 1-metre rise might displace 112 million people (Titus 1990: 140; Foley 1991: 41; UNEP 1991: 161; Harrison 1992: 226). In many cases the land lost to sea-level rise would be some of the best farmland available, together with cities and much infrastructure.

Predicted vegetation changes as a consequence of global change

Plants and animals flourished under much higher temperatures and levels of carbon dioxide than today's before roughly 6 million and especially before 40 million years BP; indeed, today's conditions in comparison might well be described as sub-optimal (Kemp 1990: 156; Ince 1991). Whether or not present conditions are sub-optimal, flora, fauna and humans are adapted to it and will almost certainly suffer considerable upset if there is anything other than slow, gradual warming. There has been speculation that increased carbon dioxide levels may raise plant productivity (a 'fertilizer effect'); together with warming, this might mean one-third more biomass by the mid-twenty-first century assuming CO_2 concentrations resulting from an IPCC 'business as usual' energy-use scenario (Fajer and Bazzaz 1992). Whether this will be of good quality, and thus a boost to agriculture, or simply more material of less value remains to be seen. There is some chance that if biomass production does increase it will act as a

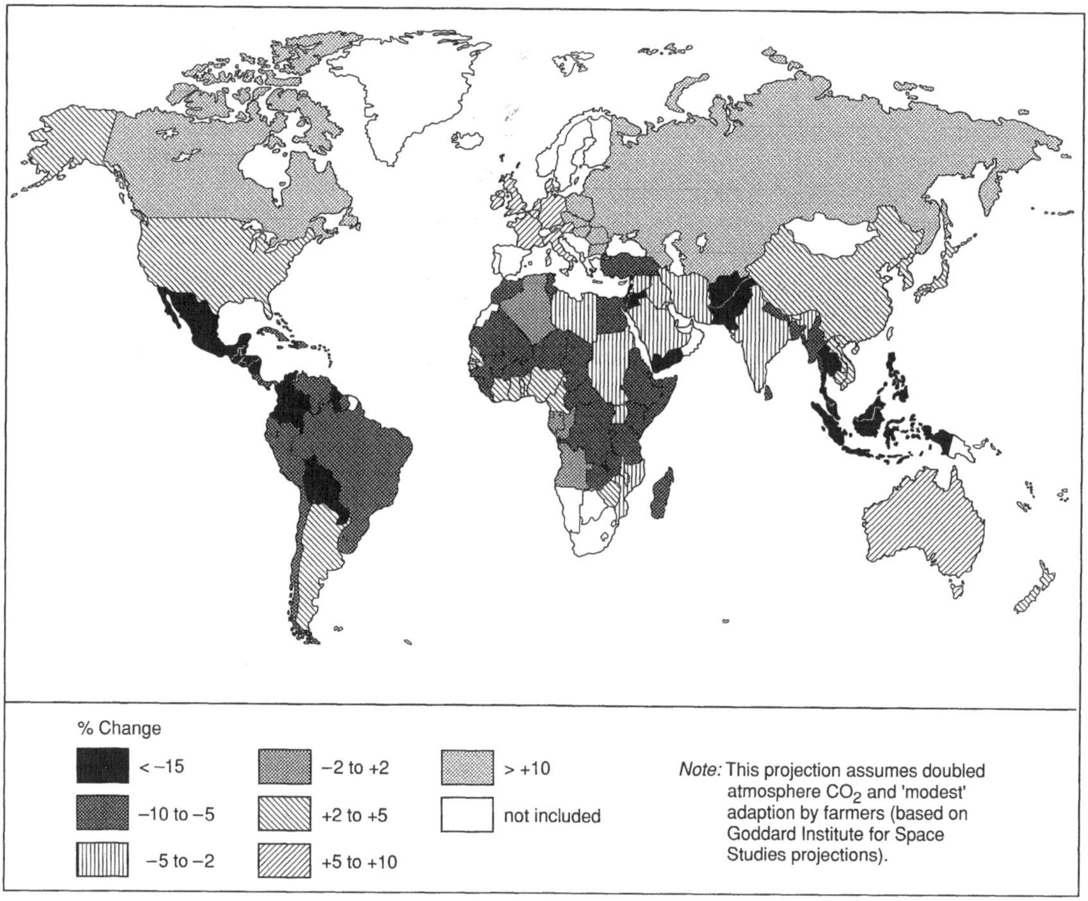

Fig. 5.2 Global warming and predicted agricultural production changes (*Source: IIASA Annual Report 1992:* 12; other sources).

negative feedback, locking up more carbon, to help counter warming.

Efforts have been made to predict the agricultural impacts of increased greenhouse gases and global warming (e.g. Warrick 1988; Fajer 1989; Parry *et al.* 1988; Parry 1990) and the effects on wetlands (Titus 1991) (Fig. 5.2). There are predictions that rice crops might improve, particularly in regions like north Japan (Joyce 1991). Unfortunately, the lands which produce the world's traded cereals could be upset by global warming. A northward shift of cereal growing may not be the simple outcome, because suitable soils will probably not be found where climatic conditions become favourable. Developing new crop varieties and agricultural strategies takes time; crop-breeders should be trying to assess likely needs now so as to start preparations for conditions twenty years or more ahead. The need for such proactive measures has been questioned, one argument being that many farmers face considerable year-to-year environmental changes and so may adapt to gradual global warming without serious difficulty (Reifsnyder 1989; De Freitas 1991: 13): the assumption, perhaps not a sound one, is that global change will be gradual.

Higher plants have two main metabolic pathways controlling their photosynthesis: the C_3 and C_4 (a third, the CAM group, are less widespread). The C_3 group includes crops like wheat, soya, rice, barley, oats, cotton and many trees. C_4 plants include maize, sugar cane, sorghum and several other important tropical crops. Broadly, C_3 crops may benefit more from raised carbon dioxide levels; C_4 crops, which tend to be those of the LDCs, will tend to be disadvantaged. A further complication is that fields tend to contain a mix of crops and weeds: where crops are C_3 and weeds C_4; prospects may be good. Where crops are C_4 and weeds C_3, there may be more weed control problems with increased carbon dioxide. Many weeds, especially in warmer climates, are C_3 (Wellburn 1988: 170; Skiba and Cresser 1988: 144).

Raised carbon dioxide levels may lead to reduced transpiration and so less demand for moisture; this would be a boon for much of the world's agriculture. There is some indication that transpiration rates have been changing from plant specimens at the British Museum of Natural History that were collected in the nineteenth century, before modern increases in carbon dioxide. These appear to have more stomata (leaf pores) than modern equivalents, suggesting that increased carbon dioxide results in reduced transpiration.

If global changes increase plant cover this would help counter soil erosion in some regions, make soil formation more rapid and perhaps reduce the levels of pollution in surface and groundwaters because of increased microbial activity. In some situations warming might increase leaching of pollutants from soil leading to more groundwater and surfacewater contamination.

Weeds, pest insects, rodents and crop diseases may be affected by warming and raised carbon dioxide (and by altered ultra-violet receipts). If change favours the pests then problems could arise. Even if global change favours crop metabolism, warmer and wetter winters may favour increased fungal diseases that depress harvests. Aphids can seriously damage crops and might flourish thanks to global changes. Increased movements of goods and people may make pest and disease control more difficult than in the past. Monitoring is therefore a wise investment.

Forests take time to mature: many trees now growing or being planted will still be alive in 2030 or later but may not then flourish or regenerate. It is necessary to plant species or varieties of trees today which will suit likely future conditions and to ensure conservation areas and other forests are able to adapt. In order to survive environmental change, plants and animals may have to move; where development has made this impossible it may be feasible to provide 'corridors' to alternative areas or to collect and relocate flora and fauna. It may be important to plan for increased fire risk or storm damage. Global warming may shift forest belts northward, perhaps into presently treeless tundra. Vegetation zones in highland areas will also tend to move upslope if there is warming (due to the Massenherebung effect there may be altitudinal variation from side to side of mountains and between mountain ranges and isolated peaks).

Predicted changes affecting soils

Soil micro-organisms and soil forming processes will alter if climate and atmospheric composition change. Some of these changes could act as important feedback effects, for example increased or decreased soil carbon fixation will affect climate (Bouwman 1990b).

Health impacts as a consequence of global warming

Environmental change may alter the distribution of disease vectors and possibly their numbers. There may also be changes in the disease resistance of organisms and in transmission of illness because of different climatic stresses and, for higher organisms, changes of diet and habit. Human diseases most likely to cause problems include malaria (and other insect- or snail-borne infections like yellow-fever and leishmaniasis), schistosomiasis, hookworm infestation and polio. Malaria causes particular concern because it could spread into many areas that are presently free of it. Rodents may increase and with them the transmission of a number of serious diseases. It should be possible to predict malaria risk areas and in Africa to map those likely to become newly infested or which will lose tsetse flies (carriers of human and livestock 'sleeping sickness' – trypanosomiasis – which have quite a specific temperature range). Areas now free of tsetse which have high human populations could encounter difficulties, such as the Kenyan Highlands.

Socioeconomic impacts of global change

Predictions at this point are difficult (see Parry 1993). However, there may well be large numbers of ecorefugees and possibly less security of food production which might lead to conflict (Maunder 1989). Tourism is vulnerable to environmental change, in particular, alpine resorts which have invested in infrastructure to support skiing may suffer if there is a reduction of winter snowfall. The need to control emissions and to adapt to change might involve huge costs for national economies. Growth and development could be hugely affected, however: the only reasonably firm prediction that can be made is that costs and benefits will not be spread evenly.

Bodies like the OECD have begun to assess the economic impacts of predicted climatic change and the benefits of abatement efforts (T. Jones 1992; OECD 1992). Insurance companies will be among the first to register the effects of environmental change as storm and severe weather damage or summer forest fires increase. Between 1988 and 1994 there were unprecedented claims for severe weather damage in Europe, North America and the Caribbean to the extent that insurance companies found it difficult to get reinsurance to provide cover (Bagnall 1993). Already underwriters are discussing how they might respond to global change. Land and road drainage may have to be modified if there are more intense storms and irrigation infrastructure may become inadequate if climate changes. Given time, building practices and lifestyles will adapt to change but the process of adaption could be painful.

Altered precipitation and evapotranspiration

Climatic change will affect streamflow, groundwater, plant water demand, the level of many lakes and the character of wetlands. Where human water use is high there may be difficulties, for example in southwest USA and where limited river flows are shared by states (e.g. the Blue Nile).

Responding to global climatic change

A wide body of opinion accepts that there is a real risk of global climatic change; the responses might include: ignore the problem, seek to adapt to it, seek to counter it, mitigate it, remedy it, or try to avoid it. If warming is happening, countermeasures should reduce difficulties, but not prevent (perhaps considerable) environmental change because effective and long-active greenhouse gases have already been released into the atmosphere. Adaption should therefore accompany counter-measures. Counter-measures and adaption will require:

1 Understanding of the process of change and rate of change.
2 Establishment of who is emitting what.
3 Development of a control strategy.
4 Deciding how to fund and enforce control.
5 Monitoring.
6 DCs will probably have to assist LDCs with emissions controls and perhaps adaption to warming.

The costs of avoiding, countering and adapting to global climatic change could be huge, but this has to be weighed against the costs that will accrue from not acting. Policy makers are being forced to consider this problem before they are fully convinced that there are going to be dangerous changes, i.e. are obliged to follow a 'precautionary principle'. The most sensible response would be to adopt a 'no regrets' approach: counter-measures offer benefits even if warming fails to materialize, for example, carbon-sink afforestation still yields timber and amenity value. While there may be improved emission controls, many forecasts assume the worst and base their predictions on a 'business as usual scenario'.

Efforts are being made to assess the policy implications of global climate change (e.g. National Academy of Sciences 1991; Pearman 1991; Hymes and Smith 1993). Geopolitics, international law and cooperation will be severely tested in seeking controls.

Who is responsible for 'greenhouse gas' emissions?

In 1990 estimates suggested that, of the total carbon (carbon equivalents) added to the Earth's atmosphere: 63 per cent was through CO_2 emissions, 14 per cent as CH_4 and 23 per cent as CFCs (Boyle and Ardill 1989: 34; Sandbrook 1991: 413–14).

While it is possible to assess the general sources and total amount of 'greenhouse gases' in the atmosphere (Fig. 5.3), it is less easy to assess exactly who are emitters and how much they emit, but without accurately doing so control will be difficult or impossible (Fig. 5.4). Emissions data are fragmentary (for example, methane emission estimates are poor), the processes involved are not fully understood, and the effectiveness and time of residence in the atmosphere of the gases are only partly known.

According to Grubb (1990: 71) projected emission estimates varied by a factor of ten or more! The 'cocktail' of greenhouse gases is not static: in some countries there are likely to be future increases in coal use and therefore carbon dioxide release, elsewhere emissions of other more powerful 'greenhouse gases' like CFCs or methane might rise.

Estimates generally give little attention to the past, yet there are nations which became major emitters long before others and some of these may now have cut back (see Table 5.2). For example, UK emissions of CO_2 peaked in 1973 and have since fallen markedly; many other DCs show a similar trend. Some argue that for any comparisons to be made

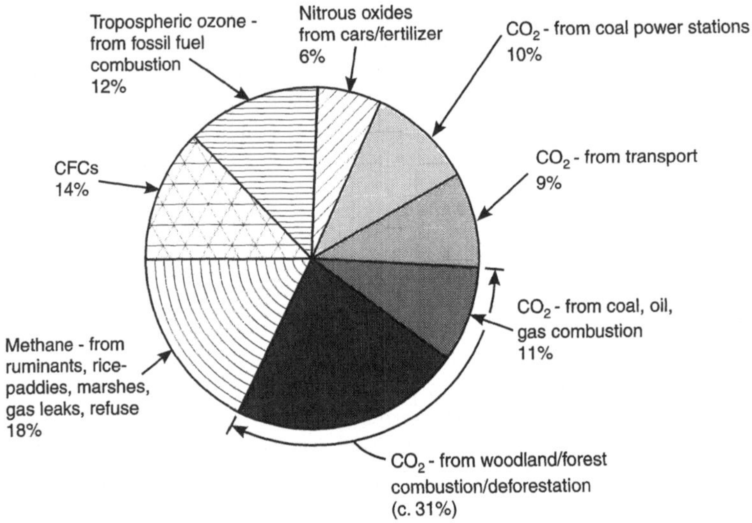

Fig. 5.3 Approximate percentage contribution of 'greenhouse gases' to global warming in the late 1980s (and their sources) ⟨*Source:* Carley and Christie 1992: 34; Waterstone 1993: 148⟩.

between countries' emissions they should be assessed on a per capita basis (others disagree, for reasons discussed on p. 92). Attempts to assess non-industrial emissions have tended to focus on tropical deforestation, particularly in Amazonia – but that has not been constant so assessments are debated. If Amazonian deforestation continued to accelerate as it had up to 1989, it would equate to as much as 70 per cent of the fossil fuel combustion emissions forcing in the early 1990s; however, such high rates seem unlikely. Tropical grasslands probably fix as much carbon as forests, have been widely burnt, and have been given inadequate attention (Boyle and Ardill 1989: 34; De Groot 1990). Acid deposition may increase emission of carbon from soils and dying vegetation, and if a projected 10 per cent of the world's deciduous forests are affected by acidification this could seriously affect lock-up of carbon.

Efforts have tended to concentrate on carbon dioxide or carbon equivalents (CO_2 output is roughly a function of population size, wealth per capita, type of energy used and efficiency of use). A better measure of each nation's responsibility might be a greenhouse index (or global warming potential index), accurately representing the contribution to warming for each nation's mix of emissions (past and present), plus other relevant factors. Getting such an index and getting it right is not easy, given that there is a changing cocktail of greenhouse gases, some of which may remain in the atmosphere for a long time, others only briefly and that they all vary in ability to force warming. Ideally present emissions should be weighed against past emission 'national debts' – in the case of DCs acquired through past industrial activity and for some LDCs through extensive burning of forests or grassland (for an example of calculation of such indices see *Environment* 23{2}: 44). Equitable calculation will be difficult: in some countries people have had to burn fuel to keep warm and it has not really contributed to their becoming DCs; also it could be claimed that DCs' emissions in part led to technical advances now available to LDCs who have not had to perfect them.

Efforts to assess blame and develop indices have been made by a number of bodies, for example the Intergovernment Panel on Climatic Change (IPCC), the World Resources Institute (WRI), the East–West Institute (Hawaii), the Centre for Science and Environment (CSE) (India) and the Organization for Economic Cooperation and Development (OECD). Other research groups have also tried to make estimates (Mohnen *et al.* 1991: 113). Not only is calculating an index complex, but also some of the bodies involved are accused of having vested interests (Brookfield 1992: 93; Pachauri *et al.* 1992). For an assessment of the situation see Hymes and Smith (1993: 51–69; 70–96).

WRI estimates have attracted criticism, notably from the CSE, for seeking to apportion too much

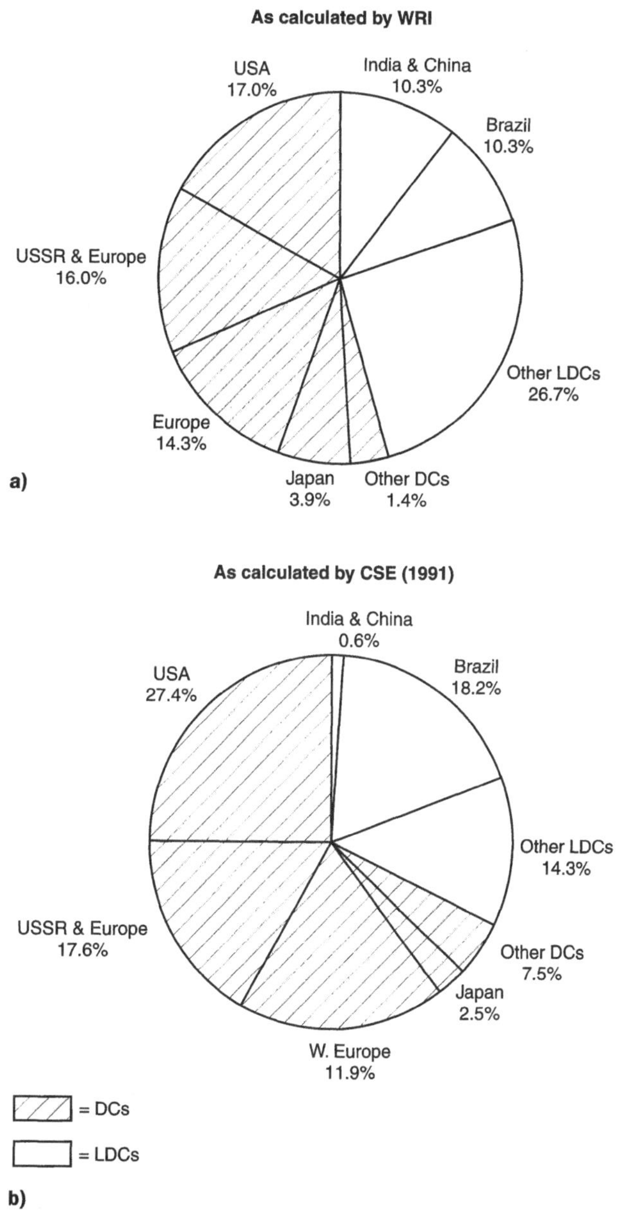

Fig. *5.4* Percentage responsibility for net emissions of 'greenhouse gases'. a) As assessed by the World Resources Institute, Washington, DC. b) As assessed by the Centre for Science and Environment, New Delhi ⟨*Source:* Agarwal and Narain 1991: 15 (Fig. 5)⟩.

blame on LDCs, for poor data, inappropriate extrapolations and flaws of methodology (Agarwal and Narain 1991; Hammond *et al.* 1991; Hurtado 1991; Ahuja 1992; Thery 1992). The WRI (1990: 3) estimated that the 30 main industrial countries emit 55 per cent of total 'greenhouse gases', but of the ten largest emitters half are non-industrial. In effect the responsibility was seen as DCs roughly 55 per cent

Table 5.2 Fifty countries with highest per capita greenhouse gas net emissions (1987)

Rank	Country	Metric tons per capita
1	Laos[a]	10.0
2	Qatar[b]	8.8
3	United Arab Emirates	5.8
4	Bahrain[b]	4.9
5	Canada	4.5
6	Luxembourg	4.3
7	Brazil[a]	4.3
8	Ivory Coast[a]	4.2
9	United States	4.2
10	Kuwait[b]	4.1
11	Australia	3.9
12	German Dem. Rep.	3.7
13	Oman[b]	3.5
14	Saudi Arabia[b]	3.3
15	New Zealand	3.2
16	Netherlands	2.9
17	Denmark	2.8
18	Costa Rica	2.8
19	Singapore	2.7
20	United Kingdom	2.7
21	Fed. Rep. of Germany	2.7
22	Finland	2.6
23	Ireland	2.5
24	Belgium	2.5
25	USSR	2.5
26	Switzerland	2.4
27	Nicaragua	2.4
28	Colombia	2.3
29	Trinidad & Tobago	2.3
30	France	2.2
31	Austria	2.2
32	Czechoslovakia	2.1
33	Israel	2.1
34	Ecuador	2.1
35	Italy	2.1
36	Norway	2.1
37	Greece	2.1
38	Poland	2.0
39	Mynmar	2.0
40	Bulgaria	1.9
41	Spain	1.9
42	Japan	1.8
43	Iceland	1.8
44	Liberia	1.7
45	Portugal	1.7
46	Sweden	1.7
47	Guinea-Bissau	1.6
48	Malaysia	1.6
49	Cameroon	1.6
50	Venezuela	1.5

Note: [a] due to high rate of deforestation
[b] Gulf states high because of gas flaring (before Gulf conflict)
Source: The World Resources Report 1990–91 (Table 24.2)

and LDCs roughly 45 per cent (DCs represent only about 20 per cent of global population). Figure 5.4(a) is the WRI ranking of nations with the highest 'greenhouse gas' emissions: some countries are included because of carbon releases related to forest clearance and agricultural development rather than combustion of hydrocarbons (see Table 5.2).

Other criticisms of WRI estimates include, first, that it estimated Brazilian deforestation would be greater than it probably is, and second, it used a limited Indian case study to estimate methane emissions from paddy-fields. The following wide range of estimates (from McCully 1991a: 162) of annual methane emission from paddy-fields shows the need for further research:

WRI	*c*.18 million tonnes year
CSE	*c*.3–9 million tonnes year
USA	*c*.30 million tonnes year
PRC	*c*.30 million tonnes year

Some critics see the WRI estimates as 'neo-colonial' (McCully 1991a) (Figs 5.4a and 5.4b contrast CES and WRI estimates). The CES has argued for estimates to be on a per capita basis, arguing that every person has a moral right to air; others suggest a per capita income (World Bank 1992: 165) or unit GNP basis. The UNEP (1991: 162) published a 'greenhouse index' (carbon heating equivalents per capita 1988) for the world's nations. Such indexes are liable to distortion because of exchange rates, and compliance may require a degree of economic development that most LDCs have yet to reach (Grubb 1990: 75; 84). Estimates based on per capita emissions have been seen as favouring countries with large, mainly young populations and as a disincentive for population control, a compromise giving some degree of 'equitability' might be possible: to use per capita emissions, but impose a cut-off by means of which persons below a certain age, say 18 years old, would have no allowance (Grubb 1990: 84). This would be more acceptable to countries with slow-growing populations (on the whole the DCs) as well as meeting some of the wishes of LDCs.

Individual countries should be able to estimate, and have ratified, their own national emissions under the 1992 (UNCED) Convention on Climate Change (Brown and Adger 1993 look at how this could be done for the UK).

The next step after calculating emissions indices is to agree a taxation or emissions quota or permit system.

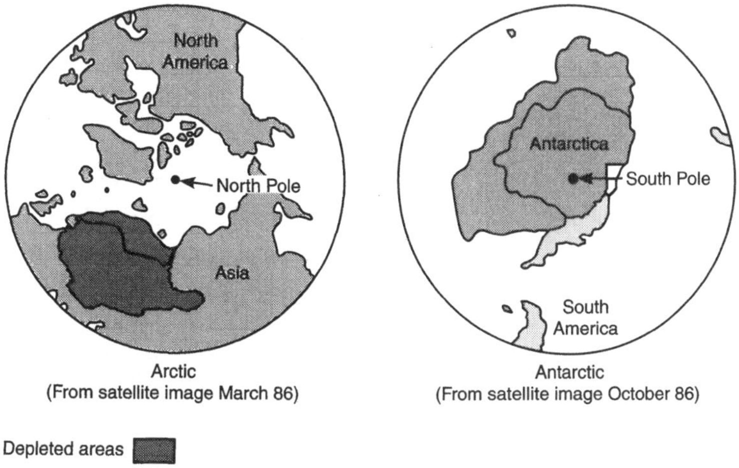

Fig. 5.5 Ozone depletion over the Arctic and Antarctic in the late 1980s (*Source:* based on various sources including *The Daily Telegraph* 4/10.88: 2; Barrow 1991: 51).

Table 5.3 Estimated relative contribution to forcing by greenhouse gas and sector (1980–2030)

Sector	(% forcing by each gas)					
	carbon dioxide	methane	ozone	nitrous oxide	CFC	sector
Energy	35	4	6	4	0	49
Deforestation	10	4	0	0	0	14
Agriculture	3	8	0	2	0	13
Industry	2	0	2	0	25	24

Source: based on UNEP and Beijer Institute data of 1989 (Table 2.1); Bouwman 1990a: 10

Measures to control emissions by agreement

The 1988 Toronto Conference on the Changing Atmosphere called for a 20 per cent reduction (on 1988 levels) of carbon dioxide emissions by AD 2005. Since 1988 some countries have tried to meet, or in a few cases, better that goal; many have not. The Second World Climate Conference was held in Geneva in 1990 with one main aim: to agree a policy leading to the world Climate Change Convention signed at the Rio 'Earth Summit' in 1992 (Hymes and Smith 1993: 3–18). The problem is to arrive at controls agreed to by all nations (Krause *et al.* 1990; Morrisette *et al.* 1991; Pearman 1991). The situation was described by Grubb (1990: 75) as a global problem of countries with different levels of development and different resource endowments seeking economic development, with populations expanding all combined with hydrocarbon fuel use. The situation is akin to Hardin's (1968) 'tragedy of the commons' scenario with nations exploiting the atmosphere with little regard for the problem of ultimate degradation or rights of other parties involved. Unusually LDCs have some power in international negotiations on global environmental change, for if they fail to cooperate with DCs adequate emission controls will be impossible.

For emission control to work there must be accurate, impartial monitoring by a body or bodies with the authority to enforce emission controls. Such a body might be the UN Security Council, or the Sustainable Development Commission set up by the UN General Assembly at the 1992 Rio 'Earth Summit', but there is still a long way to go. Whatever is selected, control should start soon if there is to be significant mitigation or avoidance of warming.

A Climate Change Convention, intended to curb greenhouse gas emissions, was agreed at the 1992 Rio 'Earth Summit'; unfortunately signatories are not bound to much firm commitment. The hopes had been for agreement to a global reduction that would hold each country at its 1990 levels of emissions by AD 2000. For many DCs this should be no problem as their emissions are already falling (although it might be argued that some have a large 'debt' of past emissions), but for countries like India and the PRC with growing populations and expanding use of coal it looks less promising.

Fuel use tax

It is possible to discourage hydrocarbon fuel use, and consequent carbon emissions through a tax on fuel (some call this a carbon tax, but it is not the same as that discussed further on this page). The proceeds could be channelled to funds to counter disasters, aid LDCs, counter warming and perhaps compensate oil-producers. It may prove easier to impose such a tax than adopt other approaches, although administering the funds generated could be a challenge. Oil-producing countries might not welcome such taxes, nor might poor nations with little choice of energy sources. McCully (1991b) suggested that the OECD countries could take the initiative, and that to do so would be in line with the 'polluter pays principle' (see p. 251) which these nations profess to support in other pollution control agreements. The USA (at the time of writing this) is reported to be reluctant to adopt such policy but OECD action might 'pressure' it into support.

A fuel tax might encourage energy-efficient and sustainable technology as well as a reduction of greenhouse emissions, whereas an emissions tax would probably lead only to reduction of emissions (Loske 1991). Efficient energy use should mean that a country actually buys less fuel and this helps to pay for emission control.

All this should be balanced against the problem that emissions other than carbon dioxide, for example methane and CFCs, are becoming more important in forcing warming: a carbon tax would not address these. The effects of such taxes on economies is uncertain, although Denmark has retained heavy fuel taxes introduced during the 1970s fuel crisis and it seems to have encouraged good energy management without unacceptable checks to economic growth.

In 1993 the UK Government, which alone in the EU opposed carbon taxes, announced an intention to phase in a value added tax on domestic fuel and electricity. Recent estimates suggest that this would generate income for the state but cut carbon emissions by only about 3 per cent. A carbon tax directed at industry and domestic users would probably make more sense, as would improving the UK's generally inadequate domestic heat insulation to cut energy waste and encouraging use of energy-efficient light bulbs, and so on.

Carbon emission tax

Emissions could be controlled by calculating carbon content of various fuels, assessing pollution released by use in power stations, cars, burning of vegetation, etc, and then levying a tax on import, production or processing of such fuels (Helm 1990). An alternative is to try and monitor what is actually emitted – which presents difficulties.

If levied at a low level across the world the negative economic impacts of such a tax should be relatively minor (Grubb 1990: 80). Carbon emission tax fits in with the 'polluter pays principle' and should be relatively easy to police and collect. Finland, Sweden and The Netherlands have carbon emission tax and the European Union seems likely to gradually apply it to oil after 1993.

Tradable emissions quotas

The advantage of tradable emissions quotas or permits (TEQ) is that they are flexible, may prove relatively easy to negotiate and are two-edged in that they give some encouragement to low emitters to stay low and prompt high emitters to try and reduce. TEQ would give each country quotas or permits which if unused could be sold, and which if exceeded, could be extended at a cost. Extensions would normally be by purchase of unused TEQ available on the market or the country exceeding its permitted emissions could be fined, and those fines paid to a fund for controlling environmental problems or aiding poor nations (Grubb 1990; Agarwal and Narain 1991: 19–21; Holmberg 1992: 314). A disadvantage of TEQs is that they do little to encourage reafforestation that would help lock up carbon.

Other problems with TEQs are the agreement of indices on which they are based and their management. Among the management problems are the risk that countries might hoard surplus TEQs or that weak nations might be pressurized to sell to stronger. An impartial international overseeing body and periodic TEQs re-issue would limit those difficulties. A country consuming inefficiently or doing little to control pollution can make reductions relatively easily compared with one that has already made efforts to improve. When assessing TEQs it should be borne in mind that some countries have already made efforts to improve energy use efficiency (e.g. Japan did so following the 1973/74 oil crisis). Countries with hydroelectric generation potential or other non-polluting energy sources should find it less disadvantageous to comply than those committed to coal or oil.

Population control

Unless population increase is controlled, some or all emissions' reduction and mitigation measures will be reduced. However, improved living standards, even without population increase, tends to mean increased energy consumption and almost certainly increased pollution although technology is tending towards being 'cleaner'.

Mitigating, counteracting and avoidance measures

Control of biomass burning

A good deal of the world's emissions of greenhouse gases results from the burning of vegetation. Levine (1991) estimated that this was responsible for roughly 40 per cent of annual carbon dioxide releases, 32 per cent of carbon monoxide releases, 21 per cent of nitrogen oxide and 38 per cent of tropospheric ozone. Control of burning forests, woods and grasslands would thus have a significant impact.

Physical interference with climate

Global warming might be countered by altering the Earth's albedo to reflect more solar radiation back to space, possibly by dispersing some suitable dust or gas in the stratosphere. There could well be problems of excessive chilling with this strategy if the reflection proved too effective or if a volcanic eruption suddenly added to it.

It may be that industrial pollution, notably sulphur dioxide, counters global warming, this counteraction could be lost when there is better pollution control. The suggestion has been made that sulphur dioxide could be released from aircraft to counter global warming, even if it worked it would also cause acid deposition.

Sink enhancement

Removing about 3000 million tonnes of carbon from the atmosphere each year and locking it up would mitigate the worst of expected global warming. It should be noted that CO_2 emissions from some countries including the UK, Germany, Japan, USA and Canada have already peaked, and in some cases have been declining for some years.

There are a number of already available potential sink enhancement routes, e.g. temperate afforestation; tropical agroforestry; reafforestation worldwide; irrigated carbon sequestering crops. Most attention has focused on forestry as a means of locking-up carbon (e.g. Andrasko 1990; R.A. Houghton 1990; Dixon *et al.* 1993). There are two approaches: conservation of forests and afforestation (planting). One estimate is that to counter present increases in atmospheric carbon would require the planting of roughly 1200 million to 1600 million ha of growing temperate forest each year (in 1992 the USA had a total of 300 million ha of forest – World Bank 1992: 163). Foley (1991: 63) cautioned that it would require planting of about 70 million ha (roughly the area of Turkey) with trees to absorb 10 per cent of the 1991 carbon dioxide emissions each year for 3 to 40 years.

Claims for the effectiveness of forestry in carbon lock-up should be treated with caution, for rates of fixation vary from tree species to species; they also differ with local environmental conditions and fall as a tree matures and its growth slows or if it is unhealthy. Tropical forests generally require less area to absorb a given amount of carbon than temperate environments; Mohnen *et al.* (1991: 117) estimated that 120 million to 250 million ha of tropical forest could deal with about 20 per cent of present global emissions of carbon dioxide from fossil fuel combustion. Forestry is clearly a useful strategy (especially for countries like the USA, Canada and the CIS, which have space available), but alone is probably not enough, even for controlling carbon dioxide-related warming, let alone the whole 'cocktail' of greenhouse gases. Bekkering (1992), after reviewing the value of tropical forestry for sequestering carbon, suggested that 15 countries had potential, but that afforestation should be concentrated in 11 and that results would be 'at best moderate'. It would make sense to pursue carbon sequestering forestry in a way that gives other benefits such as recreation and timber. Another practical route would be to establish seawater-irrigated fields of halophytes (salt-tolerant plants) like *Salicornia* spp. in little used arid areas, which would lock-up carbon and the biomass could be used for fuel, paper, and so on.

There have been proposals to modify oceanic biogeochemical cycling. One approach would be to enhance biotic fixation of carbon, perhaps by the addition of iron compounds to the ocean to increase plankton or macro-algae activity. This is speculative: it is unlikely that it would work and there is a chance it might get dangerously out of control (Victor 1991).

Carbon extraction and sequestration

It might be possible to extract carbon dioxide from the atmosphere, liquify or solidify it and pump it into underground reservoirs or sink it to great depth in the sea (at least 700 metres depth and possibly deeper than 2000 metres) where it should remain for centuries. Extraction from the atmosphere requires power, which would have to come from a non-polluting source (such as solar or nuclear power) if the process were not to become self-defeating. Photovoltaic electricity (solar cells) might be harnessed in the tropics to precipitate calcium carbonate from seawater using a cathode grid; this biogenic carbonate precipitation by electrolysis might form useful artificial coral reefs as well as lock-up carbon (Hilbertz 1992). The challenge is to find a safe, cost-effective means of carbon sequestering; in the future this might even prove to be a profitable industry.

Proposals have been made for hydrocarbon fuel use with removal of greenhouse gases from the exhausts for safe disposal as liquid CO_2 or as dry ice possibly deep in the sea, in suitable natural or constructed underground reservoirs or insulated stores above ground. One suggestion is to moor shipboard power stations over marine oil/gas-fields (like the North Sea fields), the exhaust would be pumped down into the reservoir rocks to keep it from the atmosphere and where it would help maximize full recovery of the gas or oil reserves. Sending power to land by cable may be cheaper than constructing pipelines for hydrocarbons. Whether it proves cost-effective for onshore power stations to capture emissions and send them via a pipeline network to suitable oilfield reservoirs or to ocean deeps remains to be seen. If such CO_2 capture, compression and sequestration proves cost-effective, storage must be carefully sited: the gas may be inert but the 1986 Lake Nyos (Cameroon) natural CO_2 emission disaster shows what a cloud of it can do – over 1700 people, large numbers of cattle and great swathes of vegetation were killed (Freeth 1992).

There have been proposals for power stations to capture CO_2 and pipe it to pools of microalgae which would capture the carbon, and which could then be exploited for fuel – in effect a partially closed cycle. The main requirements are space for the algae ponds and sunlight; also there would be some loss of power in capturing the gas.

Reduction of greenhouse gas emissions through energy conservation, recycling resources, improved technology and alternative energy sources

Energy use could be very much reduced through improved insulation of buildings and fridges, heating and air-conditioning and more efficient transport. Many other energy uses could also be made much less wasteful or non-polluting energy sources could be used. Cultures must be educated to reduce energy lost in inefficient and unnecessary travel or wasteful lifestyle. Not only would that reduce greenhouse gas emissions, but also it could reduce pollution in general and may save money for other uses. Countries like the USA and the UK have huge opportunities, given the presently wasteful practices. Real savings will come from grassroots changes of habits – for example if all citizens fitted low-energy lightbulbs and cut down a little on heating – but the move will need to be promoted by more effective propaganda. Governments can offer tax incentives to energy-conservers and public transport users or subsidies to cut the prices of public transport, assuming there has been adequate investment to provide cheap, reliable, efficient and clean public transport (in the UK there has not been for many years). Promoting public transport, reducing hydrocarbon fuel consumption or the popularity of private vehicles cuts smog, volatile organic compounds (VOCs – e.g. ethelene, propylene, benzene, acetone), and ozone pollution as well as greenhouse gas emissions.

Development of clean and renewable energy sources – hydroelectricity, wind-power, wave-power, solar-power, pollution-free fuels like hydrogen and minimal net emission biomass crops (e.g. rape-seed oil, reeds, willow or poplar plantations) – can help reduce greenhouse emissions (Victor 1991). A willow- or poplar-fuelled electricity generation plant supplying 5000 UK homes could be sustained by 15000 ha of arable coppice, harvested at 12 months after planting and then about every 3 years. Plans are advanced for such a plant in southwest UK as a pilot project to prepare for wider adoption. The net releases of carbon dioxide and other pollutants will be minimal.

There has been progress with alternative energy use but the potential remains huge. California had over 15 000 large wind-power generators supplying its electricity grid by 1992; in the UK plans had been lodged by mid-1993 for 4000 generators producing a total of about 1100 MW (*c.*1. per cent of total energy). Denmark makes widespread use of biomass, urban waste and agricultural waste to generate heat and power (biomass releases carbon when burnt but fixes it again as it grows so there is no net release). A growing number of communities harness geothermal heat for domestic heating, electricity generation, desalination or greenhouse heating. Photovoltaic cells are rapidly falling in cost and a range of thermal solar generation equipment has potential well beyond warm areas like California (where costs are already less than nuclear generation). In the future desert areas of the USA, Asia, Spain, etc., might be producing power for other regions, either as electricity or as solar-generated hydrogen.

According to Foley (1991) Denmark had actually cut energy consumption to 70 per cent of its 1970 level by 1991 without loss of economic growth. Between 1973 and 1985 Japan had a roughly 46 per cent increase in GDP while energy use remained more or less static. Energy conservation and alternative power sources offer benefits like saving foreign exchange. Some countries are increasing recycling which reduces consumption of 'virgin' materials and energy to produce or refine them; there will, however, be some LDCs keen to export minerals for many years as they have invested a lot in developing mines and need foreign exchange. Progress is patchy but the potential for improvement is good.

Agricultural strategies to reduce greenhouse emissions

Increasing use of fertilizers, irrigation and rising numbers of livestock generate gases like methane. In the future these sources will become more and more important. It may prove necessary to restrict fertilizer type and use (nitrate and phosphate pollution will also prompt this – see chapter 8), to develop new crop varieties, to modify cultivation and to breed 'low-emission' livestock varieties and modify feeding practices (and perhaps use growth hormones) to reduce enteric emissions.

The prospects for global climatic change problem avoidance

Some countries prefer energy conservation and carbon sink enhancement; a number have begun to adopt emission controls ahead of agreement of international accords. There are some promising individual country initiatives, for example Sweden aims to reduce its carbon emissions within 20 years to those it had in 1986 and phase out its use of nuclear power. This will be done partly through the use of biomass as fuel (there are some risks of soil nutrient depletion associated with this unless composted refuse or sewage sludge can be spread on the plantation areas), and by improving equipment and insulation of buildings, whilst still seeking 1.9 per cent economic growth. In India, Karnataka State has a long-range plan for carbon emissions reduction without hindering economic development; this depends on an energy strategy that includes nuclear power.

Controlling greenhouse gas emissions involves long-range planning and expenditure on threats that are not proven when there are pressing present-day demands for funds and difficult issues of sovereignty and enforcement of agreements (Waterstone 1993). Progress with international agreements on nuclear weapons testing and proliferation has been slow, as have full controls on whaling and on trade in ivory. Ozone controls are a little more promising. In 1992 at UNCED a Framework Convention on Climate Change was agreed (see chapter 12), so there has been some progress towards international agreement. However, even if greenhouse gas controls were reached by 2000, it is likely that there would be warming due to the gases already in the atmosphere.

Depletion of stratospheric ozone

Until recently natural processes maintained a tenuous, sometimes discontinuous, layer of ozone in the stratosphere between roughly 15 and 50 km altitude, the ozone formation normally roughly balancing destruction (it is probable that life on Earth remained in the oceans until such a layer formed). There is reduced ozone formation when sunspot activity periodically decreases, roughly every 11 years (the next sunspot 'low' is around 1997), and possibly as a consequence of biogeochemical processes, and there could be serious destruction following those volcanic eruptions which eject ozone-scavenging chlorinated molecules or fluorine compounds into the stratosphere: according to Brasseur and Granier (1992), Mt Pinatubo did.

Since the 1930s there have been releases of powerful, long-lived, human-made ozone-scavenging compounds (or precursors to such compounds), notably the CFCs, halons and halocarbons described earlier, plus nitric oxide, carbon tetrachloride, hydrogen oxides, methyl chloroform, chlorine, oxides of nitrogen, and bromine (Molina and Rowland 1974). Anthropogenic ozone-scavenging compounds may themselves do enough damage to stratospheric ozone to cause problems and also increase its vulnerability to unpredictable volcanic emissions. Not all ozone damage is caused by CFCs and related compounds. Increasing numbers of ruminant livestock emit enough methane to be a significant cause of ozone loss. The denitrification of fertilizer by bacteria is not fully understood but may lead to stratospheric ozone destruction, as may methane emissions from irrigated land, especially paddy-fields.

The control of ozone-damaging emissions like methane may mean restrictions on agricultural development as well as control of industrial pollution. Agriculture increasingly depends on the use of nitrogenous fertilizers and irrigation: if these lead to serious ozone depletion alternatives will have to be found.

The process of ozone destruction is broadly as follows: little of the gas is formed or destroyed over the polar regions during winters due to the lack of sunlight and cold. Such conditions also favour the accumulation of ozone-scavenging compounds in polar stratospheric clouds. When polar spring arrives scavenging is rapidly activated, leading to ozone layer thinning or even holes (see Fig. 5.5).

By 1991 the world as a whole had lost between 4 and 8% of its stratospheric ozone layer, the effect being less noticeable at the Equator and low latitudes (Foley 1991: 17). By 1992, the Antarctic and high southern latitudes thinning and 'holes' from time to time reached populated areas like Punta Arenas (Argentina), Ushuaia (Chile) and the Falkland Islands. Estimates suggest that in the Southern Hemisphere stratospheric ozone levels were falling at about 5 per cent per decade by 1990. The most marked loss up to 1993 was between 10°S and 20°S. Antarctic thinning has so far been more advanced than that seen in the Northern Hemisphere because southern global circulation is less complex and perhaps because volcanoes like Mt Erebus (*c.*4000 metres altitude) – lying near the middle of the Antarctic hole – emit chlorine compounds high enough to reach the stratosphere (this is debated).

In the Arctic and over northern high latitudes the greatest thinning, up to 1993, has been recorded between 10°N and 60°N latitude. Northern Hemisphere 'holes' were reported in early 1993 (which is worrying given the greater population in the northern high latitudes). Parts of Europe have had periods of significant ozone layer reduction: a loss of about 8 per cent was reported for the belt between 30°N and 48°N in 1990 (Rowlands 1990: 365).

Over the Arctic the winter atmosphere is more dynamic than it is above Antarctica; while there has been thinning by as much as 30 per cent in some regions (Austin *et al.* 1992: 223), there has so far been a lot less thinning than over Antarctica and the southern high latitudes. There are fears that when atmospheric carbon dioxide doubles (predicted to occur around 2030), tropospheric-stratospheric climatic change will lead to more polar stratospheric clouds over the Arctic and then marked thinning and holes like those of the Antarctic might develop with serious impacts on human and other life in the Northern Hemisphere (Anon 1992). Given the influence of factors like solar wind, volcanic emissions and the orientation of the Earth to the Sun, ozone thinning may never be equal in both hemispheres (at the time of writing, the WHO were hopeful that ozone loss would be controlled).

What are the likely impacts of stratospheric ozone depletion?

Biological impacts of ozone depletion

Stratospheric ozone reduction exposes terrestrial and marine organisms and construction materials to more ultra-violet radiation, especially UV-B wavelengths (see Box 5.2). Ozone levels monitored in the Swiss Alps fluctuate seasonally and periodically but appear to show a roughly 3 per cent decrease between 1931 and 1970. Ozone reduction may have biological and climatological impacts. Biological impacts are at first likely to be greatest at high altitude and high latitude. Regions with frequent cloudy and polluted atmosphere will probably suffer less.

The most worrying impacts are possible reduction of food crops; human health damage; loss of wildlife; upsets to global biogeochemical cycles. Some UV is beneficial in assisting with the production of vitamin D in the skin of higher organisms so slight increases may not be especially harmful. More intensive and longer exposure particularly to UV-B (see Box 5.2) damages DNA and proteins, leads to increased mutation, decreased growth rate and possibly a depressed immune system (depressed immunity makes organisms more vulnerable to infections). It is probable that life on Earth was restricted to the seas until an ozone layer formed so the loss of UV-shielding is potentially serious.

Box 5.2 Ultra-violet radiation striking the Earth

This is subdivided by wavelength into:

UV-A (320–400 nm) Most damaging to organisms in terms of disrupting DNA, causing cell death and mutation.

UV-B (290–320 nm) The wavelength most likely to increase with ozone thinning. Causes mutagenesis, slow organic growth, eye damage. Damage to construction materials like rubber or PVC.

UV-C (<290 nm) Almost totally filtered out by normal ozone levels. Less damaging to organisms.

Ozone concentration units A frequently used unit of measurement is the *Dobson unit* (DU). The DU has been developed from spectroscopic measurement techniques, and can best be comprehended by making the following hypothetical assumption: all ozone dispersed overhead is somehow brought down to sea-level pressure and a temperature of 0°C so the thickness can be measured – one DU would be equivalent to one-thousandth of a cm of ozone measured in that manner. Between the 1950s and 1970s typical austral winter-time Antarctic readings would have been *c.*300 DU).

Note: Ozone absorbs shorter wavelengths best. The angle of the sun, altitude and clarity of the atmosphere affect exposure: high sun, high altitude and clean, cloud-free air is the worst scenario; low sun, low altitude and polluted, misty air means reduced UV receipts.

With present levels of thinning, humans and animals are likely to suffer increased cataracts and retinal damage (eye injuries), cancer, especially skin cancer (melanomas), ageing of the skin and allergic reactions to various pollutants (Russel-Jones and Wigley 1989). Occurrence of skin cancer in New Zealand has increased in recent years to such an extent that the country has one of the world's highest incidents of the disease; this might relate to ozone reductions (but recreational and clothing habits may also be to blame – people expose themselves to sunlight more now than in the past). In 1992–3, when ozone levels fell over Europe and North America, radiation levels did not rise as much as had been expected; one suggestion is that pollution and cloud cover provided UV-shielding. There may be indirect impacts as some species better withstand or adapt to increased UV and out-compete others (this might have considerable effect on marine plankton).

Plants exposed to too much UV-B suffer inhibition of photosynthesis (Le Fohn 1992) and seedling survival may be reduced. Certain plants seem more vulnerable than others: tomatoes, lettuce and members of the pea and bean family are particularly sensitive. It has been suggested that as many insects see toward the short wavelength end of the visible spectrum, increased UV might be enough to hinder pollination. Temperate zone plants are generally less well adapted to withstand UV-B than many tropical species: that might counter relatively greater benefits from global warming in higher latitudes.

Although marine plankton vary in sensitivity to ultra-violet, UV-B may penetrate down to 20 metres depth, so that a large part of the ocean plankton could be affected. As the ocean's primary producers, plankton are the base of the marine food-web and play an important role in a number of global biogeochemical cycles. Damage to plankton could have serious impacts on oceanic carbon fixation, on fisheries, on wildlife and perhaps the Earth's atmospheric composition.

Climatic impacts of ozone depletion

There is a lot of uncertainty over the climatological effects of increased UV penetration of the atmosphere; it will depend on how much the ozone decreases and at what altitude this takes place. Increased penetration of UV through photochemical reactions in the troposphere could produce enough (low altitude) ozone to damage organisms and might trigger a process leading to global cooling (Kemp 1990: 137). Such cooling might occur if tropospheric ozone increases enough to cause the lower atmosphere to heat up, this would probably allow atmospheric circulation to extend upward leading to high altitude clouds that would reflect away more solar radiation (Gribbin 1990: 147). Stratospheric ozone helps warm the stratosphere, reduction means high altitude cooling and probably significant impacts on global wind systems.

Stratospheric ozone protection

Progress so far with international emission controls

There has always been periodic natural thinning of the ozone layer; radio transmissions have long been affected by ionispheric conditions associated with periodic solar activity that is believed to cause ozone depletion. Fears that human activity could deplete stratospheric ozone were voiced in the USA by 1971. Although there were suspicions about CFCs in the 1970s, the danger was seen to be emissions from supersonic aircraft, and it was not until 1976 that CFCs were regarded as a threat (Molina and Rowland 1974; Kemp 1990: 132; Morrisette *et al.* 1991).

The UNEP had begun to seek CFC controls well before Antarctic ozone depletion had been discovered in 1985 by Farman and his colleagues (Farman *et al.* 1985). The peak of CFCs production on a world scale was probably 1974. In 1978 the USA introduced a partial ban on CFCs as propellants for hair-sprays and other aerosols; Canada followed, banning CFC-pressurized aerosol sprays in 1980, and the EC acted similarly in 1989. By the late 1970s Norway and Sweden had banned the use of CFC for aerosol spray-can propellants.

The first international move (apart from bans on CFCs as propellant for aerosol spray-cans) to protect stratospheric ozone was made by the UNEP in 1985 – the Vienna Convention for the Protection of the Ozone Layer. The Convention was mainly concerned with encouraging research and exchange of information, rather than CFC control, and did not commit signatories to reduce use of ozone-damaging compounds.

Antarctic ozone depletion was confirmed by atmospheric research and by satellites like NASA's Total Ozone Mapping Spectrometer (TOMS) in 1986, following this a greater sense of urgency developed, although the link between stratospheric chlorine, CFCs and ozone damage was not really proven until 1988 (Dütsch 1987). The Montreal Protocol on Substances that Deplete the Ozone Layer (also a UNEP initiative) was signed by many, but not all, industrial nations in late 1987. The Protocol called for an agreement by each signatory to cut total CFCs consumption by 1990 to its 1986 level and to freeze consumption of listed CFCs and related compounds (a further cut of 20 per cent of 1986 levels is to come into force in 1994, and in 1999 a reduction to 50 per cent of 1986 levels).

The Protocol has been amended a number of times since 1987 (e.g. in 1988, 1990 and 1992), for example, carbon tetrachloride controls have recently been added. As it stood in 1992 the Protocol called for signatories to phase out CFCs manufacture and use by AD 2000, with LDCs given 10 more years to make that reduction and ultimately access to a US$240 million Interim Multilateral Fund to help them adopt alternatives (Haas 1991). In 1989 a meeting at Helsinki led to the agreement by 81 nations to total elimination of CFC production by AD 2000. In late 1992 there were still gaps in the agreements, for example methyl bromide (a fumigant used to rid warehouses, ships and soil of pests) was added to the Protocol with the aim of freezing emissions by 1995. Given that the compound probably causes about 10 per cent of global ozone loss, the pace of control agreement seems rather sedate.

If the Protocol is adhered to by enough nations it will slow but not halt ozone destruction due to the persistence of CFCs already in the stratosphere. A fall to near natural CFC levels is unlikely before the end of the twenty-first century. However, the Protocol is widely seen as a 'landmark' in terms of international agreements to protect the environment. Before the Montreal Protocol there had been few international agreements on pollution control, apart from the 1979 Convention on Long-Range Transboundary Air Pollution and the 1985 Vienna Convention on the Protection of the Ozone Layer – the latter designed to improve cooperation in monitoring, research and information exchange rather than control.

Alternatives to ozone damaging CFCs and reduction of emissions

Because there are relatively few producers of CFCs and related compounds, and because some of the compounds are about to have their patent protection expire, there will probably not be too much resistance from industry to alternatives. Manufacturers of CFCs like DuPont and ICI have been seeking safe replacements. It may not be necessary to develop expensive compounds for things like refrigeration and air-conditioning: propane gas is an effective and very cheap alternative. However, it may be unusable under some countries' fire-precaution laws, although it seems irrational to ban sales of a refrigerator with the same quantity or less of propane in it than can be found in any table-top cigarette lighter, which is freely sold. Water can also be used as a working fluid for refrigeration. There is scope for more use of absorption-type refrigeration to avoid CFCs, and for better building and ventilation to reduce the need for air-conditioning.

Some chemical companies, specialist pollution control companies and a number of governments have instituted measures for collection, storage and safe disposal of CFCs from old refrigeration and air-conditioning equipment, etc. Consumer goods manufacturers have also begun to adopt codes of practice aimed at avoiding CFCs. For example, in 1993 the Saab car company's UK marketing publicity gave a high profile to CFC-free air-conditioning units and companies like Ford are taking a similar route. After 1995 manufacture and importation of refrigerators using CFCs will be banned within the EU.

What are realistic goals?

To keep greenhouse gas emissions at 1990 levels, Foley (1991: 51), drawing on IPCC findings, suggested that releases of carbon dioxide should be cut by 60 per cent, methane by 5–20 per cent, nitrous oxide by 70–80 per cent and CFC-11 by 70–75 per cent. More realistic goals, which would reduce but not prevent global change, would be to eliminate CFC use and production by 1995 (ensuring that substitutes were not effective as greenhouse gases); halt deforestation by AD 2000 and start 'extensive' reforestation; reduce carbon dioxide emissions from fossil fuel combustion to 30 per cent of the 1991 level by AD 2020; and cut annual rise in methane and nitrous oxide to 25 per cent of the 1991 level (Foley 1991: 52).

In 1992 at the Rio 'Earth Summit' 154 countries signed the Framework Convention on Climate Change. That this will prove to be like Article IV of the Non-Proliferation Treaty – which many countries signed but which has done little to prevent the spread of nuclear weapons – is a distinct possibility. Members of the EU have shown some interest in moving toward an energy tax to try and hold emissions of carbon at about 1990 levels after AD 2000 and the USA collects taxes on CFCs and halons. Even if enough countries can cooperate to prevent disaster, there is always a risk of 'free-riders', those which benefit from the abatement efforts made by others without bearing the costs.

References

Agarwal, A. and Narain, S. (1991) *Global warming in an unequal world: a case of environmental colonialism*. New Delhi. Centre for Science and Environment, 807 Vishal Bhwan, 95 Nehru Place, New Delhi

Ahuja, D.R. (1992) Estimating national contributions of greenhouse gas emissions: the CSE–WRI controversy. *Global Environmental Change* 2(2): 83–7

Andrasko, K. (1990) Global warming and forests: an overview of current knowledge. *Unasylva* 41(163): 4–6; 12–14

Anon (1992) Global change: a looming Arctic ozone hole? *Nature* 360(6401): 209–10

Austin, J., Butchart, N. and Shire, K.P. (1992) Possibility of an Arctic ozone hole in a double-CO_2 climate. *Nature* 360(6401): 221–5

Bagnall, S. (1993) Global warming sends cold shivers through insurers. *The Times* 5/4/93: 34

Barbier, E.B. (1989) The global greenhouse effect: economic impacts and policy considerations. *Natural Resources Forum* 13(1): 20–32

Barrow, C.J. (1991) *Land degradation: development and breakdown of terrestrial environments*. Cambridge, Cambridge University Press

Bazzaz, F.A. and Fajer, E.D. (1992) Plant life in a CO_2-rich world. *Scientific American* 266(6): 18–25

Bekkering, T.D. (1992) Using tropical forests to fix atmospheric carbon: the potential in theory and practice. *Ambio* XXI(6): 414–19

Bell, E.C.F. (1993) *Submerging coastlines: the effects of sea level on coastal environments*. Chichester, Wiley

Benedick, R.E. (1991) *Ozone diplomacy: new directions for safeguarding the planet*. Cambridge, Mass., Harvard University Press

Bolin, B., Döös, Bo. R., Jäger, J. and Warrick, R.A. (eds) (1986) *The greenhouse effect, climate change and the ecosystem* (SCOPE 29). Chichester, Wiley

Bouwman, A.F. (ed.) (1990a) *Soils and the greenhouse effect*. Chichester, Wiley

Bouwman, A.F. (1990b) Land use related sources of greenhouse gases. *Land Use Policy* 7(2): 154–65

Bowler, S. (1991) The politics of climate: a long haul ahead. *New Scientist* 128(1740): 20–1

Boyle, S. and Ardill, J. (1989) *The greenhouse effect: a practical guide to the world's changing climate*. London, Hodder and Stoughton

Brasseur, G. and Granier, C. (1992) Mount Pinatubo aerosols, chlorofluorocarbons, and ozone depletion. *Science* 257(5074): 1239–42

Broecker, W.S. and Denton, G.H. (1990) What drives glacial cycles? *Scientific American* 262(1): 43–50

Brookfield, H.C. (1992) 'Environmental colonialism', tropical deforestation, and concerns other than global warming. *Global Environmental Change* 2(2): 93–5

Brown, K. and Adger, N. (1993) Estimating national greenhouse gas emissions under the Climate Change Convention. *Global Environmental Change* 3(2): 149–58

Bryson, R.A. (1989) Will there be a 'global greenhouse' warming? *Environmental conservation* 16(2): 97–9

Cannell, M.G. and Hooper, M.D. (eds) (1990) *The greenhouse effect and terrestrial ecosystems in the UK*. London, HMSO

Carley, M. and Christie, I. (1992) *Managing sustainable development*. London, Earthscan

Crutzen, P.J. and Goldammer, J.G. (eds) (1993) *Fire in the environment: the ecological, atmospheric and climatic importance of vegetation fires*. Chichester, Wiley

Davidson, G. (1992) Icy prospects for a warmer world. *New Scientist* 135(1833): 23–6
De Freitas, C.R. (1991) The greenhouse crisis: myths and misconceptions. *Area* 23(1): 11–18
De Groot, P. (1990) Are we missing the grass for the trees? *New Scientist* 125(1698): 29–30
Dixon, R.K., Ninjum, J.K. and Schroeder, P.E. (1993) Conservation and sequestering of carbon: the potential of forest and agroforest management practices. *Global Environmental Change* 3(2): 159–73
Dütsch, H.U. (1987) The Antarctic ozone hole and its possible global consequences. *Environmental Conservation* 14(2): 95–6
Fajer, E.D. (1989) How enriched carbon dioxide environments may alter biotic systems even in the absence of climatic change. *Conservation Biology* 3(3): 318–20
Fajer, E.D. and Bazzaz, F.A. (1992) Is carbon dioxide a 'good' greenhouse gas? Effects of increasing CO_2 on ecological systems. *Global Environmental Change* 2(4): 301–10
Farman, J.C., Gardiner, B.G. and Shanklin, J.D. (1985) Large losses of total ozone in Antarctica reveal seasonal ClO_x/NO_x interactions. *Nature*, 315(6016): 205–10
Foley, G. (1991) *Global warming: who is taking the heat?* London, PANOS
Freeth, S. (1992) The deadly cloud hanging over Cameroon. *New Scientist* 135(1834): 23–7
Gleick, P.H. (1989) Climatic change and international politics: problems facing developing countries. *Ambio* XVIII(6): 333–9
Goreou, T.J. and Hayes, R.L. (1994) Coral bleaching and ocean 'hotspots'. *Ambio* XXIII(3): 176–80
Gribbin, J. (1988a) The greenhouse effect. *New Scientist* 120(1635): special supplement (Inside Science no. 13), 4 pp.
Gribbin, J. (1988b) *The hole in the sky: man's threat to the ozone layer*. London, Corgi
Gribbin, J. (1990) *Hothouse earth: the greenhouse effect and gaia*. London, Bantam
Gribbin, J. (1991) Climate and ozone: the stratospheric link. *The Ecologist* 21(3): 133
Grubb, M. (1990) What to do about global warming? *International Affairs* 66(1): 67–87
Haas, P.M. (1991) Policy responses to stratospheric ozone depletion. *Global Environmental Change* 1(3): 224–34
Hammond, A.L., Rodenburg, E. and Mooman, W.R. (1991) Calculating national accountability for climate change. *Environment* 33(1): 11–15; 33–5
Hardin, G. (1968) The tragedy of the commons. *Science* 162(3859): 1243–8
Harrison, P. (1992) *The third revolution: population, environment and a sustainable world.* Harmondsworth, Penguin
Hecht, J. (1991) Monitoring ancient greenhouse effect. *New Scientist* 132(1797): 21
Hekstra, G.P. (1989) Global warming and rising sea levels the policy implications. *The Ecologist* 19(1): 4–15
Helm, D. (1990) Who should pay for global warming? *New Scientist* 128(1741): 36–9
Hilbertz, W.H. (1992) Solar-generated building material from seawater as a sink for carbon. *Ambio* XXI(2): 126–32
Holmberg, J. (ed.) (1992) *Policies for a small planet: from the International Institute for Environment and Development.* London, Earthscan
Houghton, J.T., Jenkins, G.J. and Ephramus, J.J. (eds) (1990) *Climate change: the IPCC scientific assessment.* Cambridge, Cambridge University Press
Houghton, J.T., Callander, B.A. and Varney, S.K. (eds) (1992) *Climate change 1992: the 1992 supplement report to the IPCC scientific assessment.* Cambridge, Cambridge University Press
Houghton, R.A. (1990) The future role of tropical forests in affecting the carbon dioxide concentration of the atmosphere. *Ambio* XIX(40): 204–9
Houghton, R.A. and Woodwell, G.M. (1989) Global climatic change. *Scientific American* 260(4): 18–26
Hurtado, M.E. (1991) raising the heat in the greenhouse. *Panoscope* 25 (July 1991): 2–3
Hymes, P. and Smith, K. (eds) (1993) *The global greenhouse regime. Who pays? (Science, economics and north–south politics in the Climate Change Convention).* London, Earthscan (for UN University)
Idso, S.B. (1991) Carbon dioxide and the fate of the Earth. *Global Environmental Change* 1(3): 178–82
Ince, M. (1991) *The rising seas.* London, Earthscan
Intergovernment Panel on Climatic Change (1991) *Climatic change: the IPCC scientific assessment.* Cambridge, Cambridge University Press
Jäger, J. and Ferguson, H. (1991) *Climatic change: science, impacts and policy.* Cambridge, Cambridge University Press
Jones, P.D. Wigley, T.M.L. (1990) Global warming seems to be occurring. *Scientific American* 263(2): 66–73
Jones, T. (1992) The economics of climatic change. *OECD Observer* 179: 22–5
Joyce, C. (1991) The world's rice crop vulnerable to climatic change. *New Scientist* 129(1751): 34
Kemp, D.D. (1990) *Global environmental issues: a climatological approach.* London, Routledge
Krause, F., Bach, W. and Kooney, J. (1990) *Energy policy in the greenhouse, from warming fate to warming limit.* London, Earthscan
Lee, T.H. *et al.* (eds) (1987) *Methane age.* Dordrecht, Kluwer
Le Fohn, A.S. (1992) *Surface-level ozone exposures and the effects on vegetation.* Boca Raton, Fla., Lewis
Leggett, J. (ed.) (1990) *Global warming: the Greenpeace report.* Oxford, Oxford University Press
Levine, J.S. (1991) *Global biomass burning: atmospheric, climatic and biospheric implications.* Cambridge, Mass., MIT Press
Loske, R. (1991) Ecological taxes, energy policy and greenhouse gas reductions: a German perspective. *The Ecologist* 21(4): 173–6
Lovelock, J.E. (1979) *Gaia: a new look at life on Earth.* Oxford, Oxford University Press
Lovelock, J.E. (1990) Hands up for the gaia hypothesis. *Nature* 344(6262): 100–2
McCully, P. (1991a) Discord in the greenhouse: how WRI is attempting to shift the blame for global warming. *The Ecologist* 21(4): 157–65
McCully, P. (1991b) The case against climate aid. *The Ecologist* 21(6): 244–51
McKay, G.A. (1991) The changing atmosphere: a review. *Natural Hazards* 4: 353–72
McKenzie, D. (1990) Leaking gas mains help warm the world. *New Scientist* 127(1735): 24

Mainguet, M. (1991) *Desertification: natural background and human mismanagement*. Berlin, Springer-Verlag
Maunder, W.J. (1989) *The human impact of climatic uncertainty: weather information, economic planning and business management*. London, Routledge
Mohnen, V.A., Goldstein, W. and Wei-Chyung Wang. (1991) The conflict over global warming: the application of scientific research to policy choices. *Global Environmental Change* 1(2): 109–23
Molina, M.J. and Rowland, F.S. (1974) Stratospheric sink for chlorofluoromethanes: chlorine atom-catalysed destruction of ozone. *Nature* 249(5460): 810–12
Morrisette, P.M., Darmstadter, J., Plantinga, A.T. and Toman, M.A. (1991) Prospects for a global greenhouse gas accord. *Global Environmental Change* 1(3): 209–23
National Academy of Sciences (1991) *Policy implications of greenhouse warming* (Synthesis Panel of Committee on Science, Engineering, and Public Policy) Washington, DC, National Academy Press
Nilsson, A. (1992) *Greenhouse earth* (SCOPE). Chichester, Wiley
OECD (1992) *Global warming: estimating the economic benefits of abatement*. Paris, OECD
Pachauri, R.K., Gupta, S. and Mehra, M. (1992) A reappraisal of WRI's estimates of greenhouse emissions. *Natural Resources Forum* 16(1): 33–8
PANOS (1991) *Who is taking the heat? Global warming and the third world*. London, PANOS
Parry, M. (1990) *Climate change and world agriculture*. London, Earthscan
Parry, M. (1993) *Climatic change and the world economy*. London, Earthscan
Parry, M., Carter, T. and Konijn, N. (eds) (1988) *The impact of climatic variations on world agriculture* (vol. 1, *Predictions for cooler environments*; vol. 2, *Predictions for warmer environments*). Dordrecht, Kluwer
Pearce, F. (1989) Methane locked in permafrost may hold the key to global warming. *New Scientist* 121(1654): 28
Pearce, F. (1991) Warmer winters fit the greenhouse model. *New Scientist* 129(1752): 20
Pearman, G.I. (ed.) (1991) *Limiting greenhouse effects: controlling carbon dioxide emissions*. Chichester, Wiley
Pearman, G.I., Ethridge, D., De Silva, F. and Fraser, P.D. (1986) Evidence of changing concentrations of atmospheric CO_2, N_2O and CH_4 from bubbles in Antarctic ice. *Nature* 320(6059): 248–50
Porteous, A. (1992) *Dictionary of environmental science and technology* (revised edn). Chichester, Wiley
Rambler, M.B., Margulis, L. and Fester, R. (eds) (1989) *Global ecology: towards a science of the biosphere*. London, Academic Press
Reifsnÿder, W.E. (1989) A tale of ten fallacies: the sceptical inquirer's view of the carbon dioxide/climate controversy. *Agricultural and Forestry Meteorology* 47: 349–71
Roberts, L. (1989) Global warming: blaming the sun. *Science* 240(4933): 992–3
Rotmans, J., Den Elzen, M.G.J., Krol, K., Stewart, R.J. and Van der Woerd, H. (1992) Stabilizing atmospheric concentrations: towards international methane control. *Ambio* XXI(6): 404–13
Rowlands, F.S. (1990) Stratospheric ozone depletion by chlorofluorocarbons. *Ambio* XIX(6–7): 281–92
Ruddiman, W.F. and Kutzbach, J.E. (1991) Plateau uplift and climatic change. *Scientific American* 264(3): 42–50
Russel-Jones, R. and Wigley, T. (eds) (1989) *Ozone depletion: health and environmental consequences*. Chichester, Wiley
Sandbrook, R. (1991) Development for people and the environment. *Journal of International Affairs* 44(2): 403–20
Schwartzman, D. and Volk, T. (1991) When soil cooled the world. *New Scientist* 131(1777): 33–6
Skiba, U. and Cresser, M. (1988) The ecological significance of increasing atmospheric carbon dioxide. *Endeavour* (new series) 2(3): 143–7
Thery, D. (1992) Should we drop or replace the WRI Global Index? *Global Environmental Change* 2(2): 88–9
Thompson, P. (ed.) (1991) *Global warming: the debate*. Chichester, Wiley
Titus, J.G. (1990) Greenhouse effect, sea-level rise and land use. *Land Use Policy* 7(2): 138–54
Titus, J.G. (1991) Greenhouse effect and control wetland policy: how Americans could abandon an area the size of Massachusetts at minimum cost. *Environmental Management* 15(1): 39–58
Tooley, M.J. and Shennan, I. (eds) (1987) *Sea-level changes* (IBG Special Publication no. 20). Oxford, Basil Blackwell
UNEP (1991) *Human development report 1991*. Oxford, Oxford University Press
Victor, D.G. (1991) How to slow global warming. *Nature* 349(6309): 451–6
Von Weizsäcker, E.U. and Jesinghaus, J. (1992) *Ecological tax reform: a policy proposal for sustainable development*. London, Zed
Warrick, R.A. (1988) Carbon dioxide, climatic change and agriculture. *Geographical Journal* 154(2): 221–33
Warrick, R. and Farmer, G. (1990) The greenhouse effect, climatic change and rising sea-level: implications for development. *Transactions of the Institute of British Geographers* (new series) 15: 21–34
Warrick, R.A., Barrow, E.M. and Wigley, T.M.L. (1990) *The greenhouse effect and its implications for the European Community*. Brussels, Commission for the EC (ECU 5)
Waterstone, M. (1993) Adrift in a sea of platitudes: why we will not resolve the greenhouse issue. *Environmental Management* 17(2): 141–52
Wellburn, A. (1988) *Air pollution and acid rain: the global threat of acid pollution*. London, Earthscan
Williamson, P. and Gribbin, J. (1991) How plankton change the climate. *New Scientist* 129(1760): 48–52
World Bank (1992) *World development report 1992: development and the environment*. Oxford, Oxford University Press
World Resources Institute, UNEP and UNDP (1990) *World resources 1990–91* (Report by the UNEP). Oxford, Oxford University Press

Further reading

Agarwal, A. and Narain, S. (1991) *Global warming in an unequal world: a case of environmental colonialism*. New Delhi. Centre for Science and Environment, 807 Vishal Bhwan, 95 Nehru Place, New Delhi [Challenge to the WRI suggested allocation of responsibility for emissions]

Foley, G. (1991) *Global warming: who is taking the heat?* London, PANOS [Simple, concise, introduction]
Gribbin, J. (1990) *Hothouse earth: the greenhouse effect and gaia.* London, Bantam [Readable overview]
PANOS. (1991) *Who is taking the heat? Global warming and the third world.* London, PANOS [Lively introduction with focus on LDC impacts]
Parry, M. (1990) *Climate change and world agriculture.* London, Earthscan [Examines impact of likely change on world agriculture]
UNEP (1983) *Selected multilateral treaties in the field of the environment: vol. 1.* Cambridge, Grotius [Details of environmental agreements]
UNEP (1991) *Selected multilateral treaties in the field of the environment: vol. 2.* Cambridge, Grotius [Details of environmental agreements, including those relating to atmospheric pollution/climate]

CHAPTER 6

Desertification and soil degradation

Desertification

Desertification: helpful concept or misleading catch-phrase?

In drylands, tundras, cold uplands, wherever land is periodically deprived of adequate moisture, where soils are infertile, or poor drainage leads to saline conditions, crusts or pans, if vegetation is present it will probably be easy to disturb and slow to re-establish. These difficulties are most common in semiarid and dry subhumid environments, but can also be found in some tropical and subtropical humid and temperate environments of both LDCs and DCs (Fantechi and Margaris 1986). Because the end-product often resembles desert the process has been termed desertification (vegetation and soil degradation are usually components) (see Fig. 6.1).

The USA 'Dust Bowl': monocropping marginal land

The 'Dust Bowl' prompted considerable concern for land degradation avoidance in the USA. The short-grass prairies of the southern midwest USA were settled after the late nineteenth century when the development of heavy multi-horse-drawn ploughing enabled the tough sod to be broken, bison had been hunted to virtual extinction, and rail communications had opened up access. From the late 1880s to the 1930s drought-resistant wheat growing expanded. During wetter years, especially when grain prices were high, there were profits to be made. But this was essentially semiarid land, with a history of drought (e.g. in the 1880s–90s). Continuous dry farming involving regular tillage and fallows with little ground cover, combined with drought, fine loess soils and strong wind led to soil erosion and severe dust storms (the dust was so bad that people suffered respiratory distress over a wide swathe of the USA). By the mid-1930s the midwest Great Plains (northwest Kansas, northwest Oklahoma, southeast Colorado, northern Texas, north and south Dakota, parts of Nebraska and northeast New Mexico) had been degraded enough to earn the nickname 'Dust Bowl' (Bonnifield 1979; Worster 1979; Russel 1988).

Farmers were either too slow to perceive the land degradation risk or were trapped by debts acquired in a time of severe economic depression by buying land, agricultural machinery or inputs, and their continued attempts to try and produce led to misery and land damage. Falling grain prices drove farmers to try to squeeze more out of the land. The dust storms became so bad that by 1935 they were darkening the sky over Chicago and Washington, DC. An estimated 80 million ha of grain-producing land were destroyed and about 20 million ha were abandoned (Kassas 1987: 393). The human misery was vividly captured by John Steinbeck in his novel *The Grapes of Wrath*.

The response of the US Government was practical and utilitarian: loans were made available for soil conservation and farming inputs, public works employment schemes were established to give off-farm jobs, relief was made available, wind-breaks were planted, etc. National resources were poured into the depressed region: according to Heathcote (1983: 184), in the year 1934–35 alone US$375 million were requested. A Soil Conservation Service and a Civilian Conservation Corps were established, to develop soil protection techniques and advise farmers, and the Taylor Grazing Act was passed to try and check the considerable damage caused by private grazing on federal lands. By 1942 the worst of the land degradation was over.

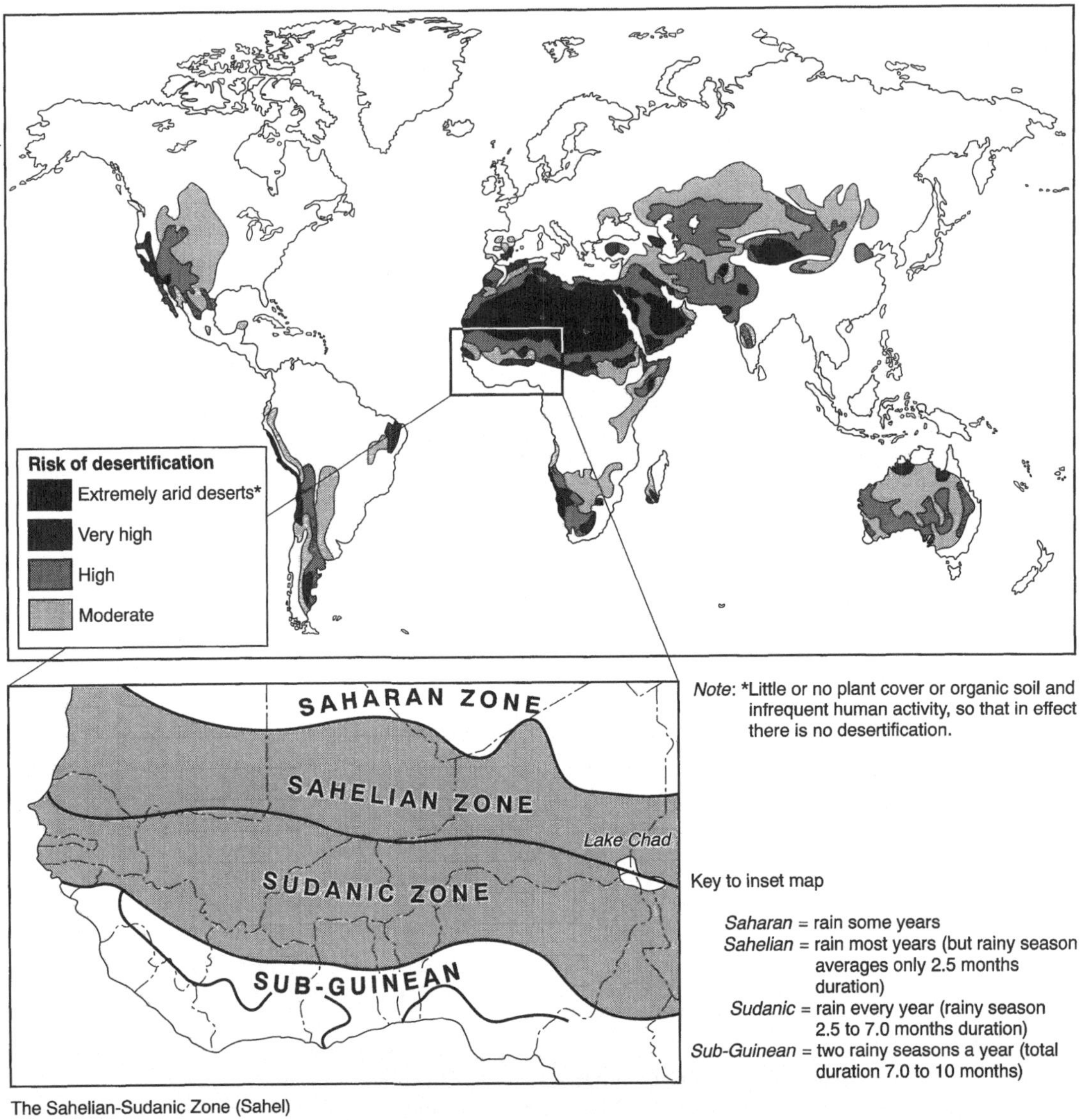

The Sahelian-Sudanic Zone (Sahel)

Fig. *6.1* Deserts and areas subject to desertification ⟨*Source:* Barrow 1987: 40–1⟩.

The Dust Bowl is not unique: there have been similar experiences in the Mallee wheatlands region of western Australia, in the 'Virgin Lands' of Kazakhstan and the southern Ukraine, in parts of the Sahel, South Africa, Argentina and Nigeria. Commercial exploitation of marginal lands is risky but exploiters are attracted by 'average' or better than average conditions and new technology (e.g. heavy 'sod-busting' ploughs in the Dust Bowl, 'stump-jump' ploughs in the Australian Mallee region), only to suffer when there is below average rainfall or some other set-back.

The Sahelian drought

Concern was stimulated again from the late 1960s by the Sahelian drought and famine and associated land degradation (see Fig. 6.1 inset). The UN Commission on Desertification (UNCOD) met in Nairobi in 1977 and began to try and better understand the problem and counter it by means of a Plan of Action to

Combat Desertification (PACD). Perhaps because the PACD efforts had so little effect, perhaps because concern for global warming diverted interest after the early 1980s, desertification has had less attention since then. Interest has recently revived: action was taken at the 1992 'Earth Summit' in Rio, to initiate a second phase of international response to desertification when the International Negotiating Committee on Desertification started preparing for a Convention to Combat Desertification – due to be signed in June 1994 (although the decision to work for a desertification treaty was part of *Agenda 21* agreed at Rio, disagreements on giving priority to combatting African desertification were delaying progress by late 1993).

Terminology

The perception has often been that dryland degradation is getting worse and is a process that can be attributed to specific causes and perhaps countered. Because the problem is more common in drylands and the end product is desert-*like* it has been common to speak of desert spread, desert advance or desert encroachment. Images of fields, roads or houses being engulfed by dunes have helped to reinforce the false idea that desertification mainly spreads out of deserts, particularly in time of drought. Various terms have been applied to the process of increasing desert-like conditions: desertification, desertization, aridization, aridification, xerotization. Desertification has come to be the most widely used of these terms, and was probably originated by Aubreville (1949). Unfortunately, desertification has been given many different definitions by researchers or commentators (Barrow 1991: 142). Some recent comments on and definitions of desertification include the following:

> Desertification revealed by drought is *caused* by human activities in which the carrying capacity of the land is exceeded, it proceeds by exacerbated natural or man-induced *mechanisms*, and is made manifest by intricate steps of vegetation and soil deterioration which *result*, in an irreversible decrease or destruction of the biological potential of the land and its ability to support population (Mainguet 1991: 4 – her italics).

> a process whereby an ecosystem loses its capacity to maintain and repair itself... desertification is a litmus of development.

> spread of desert-like conditions in arid, semiarid or dry subhumid areas due to man's action or, climate change.

> the impoverishment of ecosystems indicated by reduced biological productivity and accelerated degradation of the soil.

> the process of environmental degradation by which productive land is made non-productive and desert-like.

> land degradation in arid, semiarid and dry subhumid areas, resulting from various factors, including climate variations and human activities.

> catch-phrase to encapsulate a series of poorly understood problems, usually in semiarid lands.

> a complex process of land degradation in arid, semiarid and dry subhumid areas resulting mainly from adverse human actions.

> the main cause and mechanism of global loss of productive land.

> the diminution or destruction of the biological potential of the land, and can lead ultimately to desert-like conditions. It is an aspect of the widespread deterioration of ecosystems, and has diminished or destroyed the biological potential; i.e. plant and animal production. (Resolution no. 7 of the 1977 UN Conference on Desertification, Nairobi).

Of various desertification definitions those that involve simply spreading or encroaching desert are least satisfactory, for the process often takes place *in situ* and can be well away from desert margins. After a thorough review of definitions Mainguet (1991: 17) rejected the term in favour of 'land degradation'. Pearce (1992: 42) felt that desertification was a catch-phrase that hinders as much as helps understanding; Thomas (1993) warned that it is an emotive term subject to misuse and often misunderstood and Thomas and Middleton (1994) argue that desertification is not as great a problem as has been claimed, nor is it a global problem.

The perception of desertification

Awareness that desertification poses a problem depends on the outlook of various groups of people involved; there may be abundant indications, but these may go unnoticed or unheeded (Heathcote 1980). People need to be cautious about recognizing desertification without adequate proof (Helldén 1991).

Extent, trends and rate of desertification

There are clearly areas of land that have become degraded; establishing the extent and trends are less easy. The failure to standardize what is being measured and lack of widespread unambiguous measures has made monitoring difficult and often unreliable. One researcher may map vegetation in a wet year, another in a dry year; the location of measurements may not be fixed, the component plants measured may differ. Comparison of one year's data or one area's data with another may thus be little indication of real change. There is often no accurate information about the original state of degraded land, making it difficult to ascertain if there has been long-term improvement or degeneration.

In drylands boundaries are seldom static: they shift considerably from year to year and from site to site, especially if there are localized rainstorms. Conditions may be subject to cyclic variation but in the long term show little overall change. It is worth note that degradation can be caused by an activity that may cause little harm normally but which takes place at some crucial moment, e.g. grazing or burning during a few crucial weeks when plants set seeds.

Even apparently objective measurements may prove misleading, satellite imagery of quantity of biomass showing change may be interpreted as sign of no desertification. In reality the effectiveness of plant cover in protecting soil from erosion, for providing fodder or for genetic conservation depends on species composition, and changes may not be reflected in biomass measurements. Land invaded by thorn-scrub could be losing grass and herbs and so is less use as pasture and may suffer soil erosion, but remote sensing may show no loss, perhaps even a gain of biomass. Clearly the choice of indicator of desertification is vital and their monitoring and interpretation of the results requires skill and caution if a false picture is not to be gained.

There are three main indicators of desertification: physical, biological (vegetation and animal) and social/economic (Reining 1978: 5–8, 10; Kassas 1987: 391).

Physical indicators of desertification

- decrease in soil depth
- decrease in soil organic matter
- decrease in soil fertility
- soil crust formation/compaction
- appearance/increase in frequency/severity of dust/sandstorms/dune formation and movement
- salinization/alkalinization
- decline in quality/quantity of groundwater
- decline in quality/quantity of surface water
- increased seasonality of springs and small streams
- alteration in relative reflectance of land (albedo change).

Biological indicators of desertification: vegetation

- decrease in cover
- decrease in above-ground biomass
- decrease in yield
- alteration of key species distribution and frequency
- failure of species to reproduce successfully.

Biological indicators of desertification: animal

- alteration in key species distribution and frequency
- change in population of domestic animals
- change in herd composition
- decline in livestock production
- decline in livestock yield.

Social/economic indicators of desertification

- change in land use/water use
- change in settlement pattern (e.g. abandonment of villages)
- change in population (biological) parameters
- demographic evidence, migration statistics, public health information
- change in social process indicators
- increased conflict between groups/tribes, marginalization, migration, decrease in incomes, decrease in assets, change in relative dependence on cash crops/subsistence crops.

Measuring and mapping desertification

As with deforestation, statistics on desertification are subject to misuse:

1. Inaccurate information may be cited again by successive workers who have lost sight of its origins and limitations.
2. Agencies, countries or NGOs may use dubious statistics to make a case that a region is threatened and in need of aid.
3. Countries may suppress evidence of desertification to appear more successful or stronger, or may overlook evidence.

Roughly 35 per cent of the world's landsurface can be classed as drylands (over 6.1 billion hectares) and about 15 per cent of the world's landsurface is semiarid: the latter has sufficient population and the environment is vulnerable enough to face the greatest

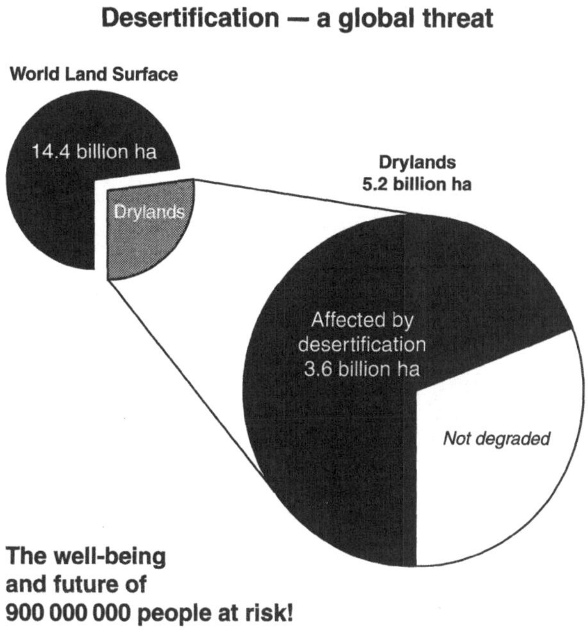

Fig. 6.2 UNEP view of desertification 1993 (*Source: Ecoforum* (UNEP) 17(1/2): 12 (UNEP 17/5/93)).

desertification risks. Drylands have between 17 and 20 per cent of the total world human population, much of it in the semiarid regions, Middleton and Thomas (1992: iv) suggest that about one-sixth of the world population was threatened by the effects of desertification. Thomas (1993; Thomas and Middleton 1994) warned that the degree of threat may be exaggerated because desertification assessments have overestimated the world-wide extent by as much as a factor of three – one reason being the failure to distinguish natural fluctuations from which there should be recovery without permanent degradation.

Semiarid areas have tended to have marked increase of human and livestock populations in recent decades, possibly due to improved veterinary care and healthcare, especially human and livestock immunization (UNEP and Commonwealth of Australia 1987: 4; Dixon *et al.* 1989).

Figures 6.2 and 6.3 present data released in 1993 by a UN agency keen to stress the importance of fighting land degradation. It has been claimed that roughly 7 per cent of the world's soils are more 'arid' than would be expected, given their climate, and that about 9 per cent of world vegetation cover is 'out of balance' with climate: the interpretation might be that approximately 7–9 per cent of land has become desertified, possibly through human activity (Eckholm 1976: 60; Heathcote 1983: 16). Since the 1970s there have been several attempts to map the extent of desertification and areas at risk; for example the UN published a *World Map of Desertification* in 1977 and Middleton and Thomas (1992) a *World Atlas of Desertification* (for the UNEP). In the preface of the latter it is suggested that Asia has suffered most, and that North America and Africa currently have the worst desertification. Scoging (1993) felt that rangelands were probably the worst affected land-type.

Concern has been expressed that desertification might affect global climate and therefore deserves international attention and funding, i.e. it is or will be a global problem, not just a matter of foreign aid allocation.

Causes of desertification

Many statements about desertification and drought in the literature are either weakly connected shopping lists of 'causes' unrelated to one another or are unduly hypothetical (plausible observations not supported by case-specific data) (Mortimore 1989: 119). There are two main categories of arguments put forward to explain why desertification takes place: natural causes and human causes.

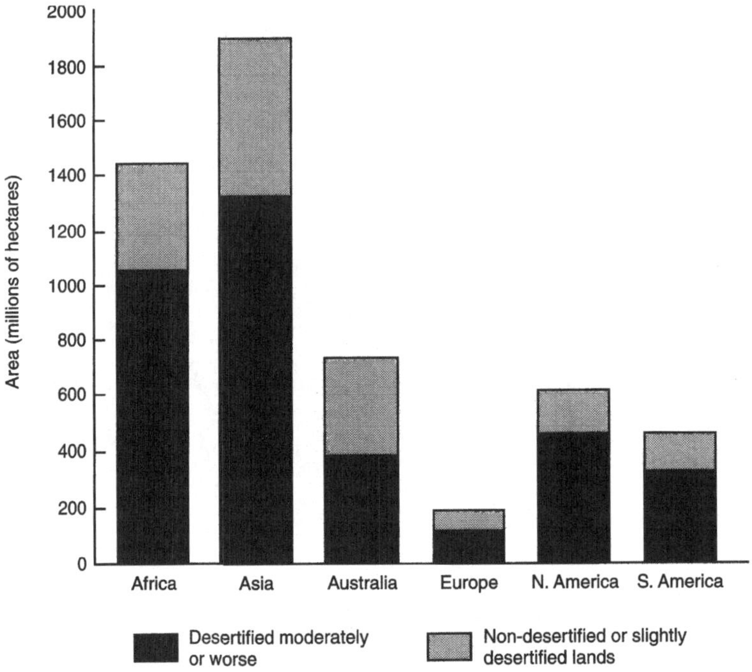

Fig. 6.3 Global status of desertification/land degradation within the world's drylands (1993 UNEP view) ⟨*Source: Ecoforum* (UNEP) 17(1/2): 15 (UNEP 17/5/93)⟩.

Natural causes of desertification arguments

Desertification is held to be primarily due to physical events: drought, climatic change, pest or disease outbreak. If so, inevitable causes prevail and there are fewer opportunities to avoid or mitigate desertification, although it might be possible to forecast trends and modify development to minimize impact.

Human causes of desertification arguments

First, population increase: desertification is blamed on the growth of human population or livestock populations (this may not always be the case: see Tiffen *et al.* 1994).

Second, structural arguments: social and economic structures and relations (patterns of ownership, rights of use and control of resources) are blamed.

Third, political and economic arguments: political or economic factors which may be local or non-local (terms of trade, LDC debt, etc) are blamed. The former might be countered by local people, the latter will require action outside the area suffering desertification.

Fourth, human fallibility arguments: desertification results from stupidity, greed, ignorance, short-sightedness of local and/or non-local people (peasants, herdsmen, governments, developers), or the adoption of inappropriate technology or approaches (ill-informed 'do-gooders' may be a cause).

Fifth, resource exploitation: people are attracted into drylands to exploit the vegetation or soil for crops or livestock. However, there are situations where drylands are degraded for other reasons (e.g. nuclear weapons testing, mineral exploitation, etc).

Whatever the cause(s), the process of degradation or desertification generally begins with damage to the vegetation cover. In very arid areas, there is little or no vegetation to destroy so that in a sense there can be no desertification, although erosion of soil or rock may be started or accelerated by the passage of humans, animals and vehicles.

Desertification can be the consequence of a complex mix of hidden and apparent causes leading to cumulative or indirect impacts. Even though there may appear to be similar processes and causes at work in different regions, the reality may be different and interpretation needs caution (Eckholm and Brown 1977: 17; Spooner and Mann 1982: 5; Dregne 1983: 96). Another trap for those seeking the causes of desertification is that there is often a web of causation–feedback–causation. This means that a physical cause might lead to a human feedback and that to further physical causes, or vice versa.

Natural causes of desertification

In drylands plants often grow near their limits of tolerance, so even slight changes in environmental factors or a little disturbance can lead to more bare ground, species loss and degradation (De Vos 1975: 134).

Drought Droughts are common in tropical drylands. They may be short-lived and localized or widespread, or persistent and localized or widespread; they may be recurrent with a clear pattern or apparently random.

A drought occurs when a region has insufficient moisture to meet the demands of plants, people, livestock or wildlife. The moisture shortage may be due to less than normal precipitation for the period in question (meteorological drought), or it may result if natural vegetation or crops are altered, becoming more moisture-demanding (an agricultural drought), or it may be due to changes that reduce the absorption and storage of moisture in the soil (hydrological drought). Drought could occur where precipitation receipts and groundwater remain unaltered and where crops are unchanged, because farmers alter their practices, i.e. due to human behavioural changes.

Drought can lead to desertification, but desertification may occur without there being drought. Drought need not be a cause of desertification, but it often highlights or triggers clear signs of desertification (it can be like litmus showing change in an otherwise transparent liquid or a catalyst assisting other causes) (Dregne 1983: ix, 7). Vegetation may suffer due to drought or drought may arise because vegetation has been damaged.

Whatever the cause of drought, there are a number of theories as to how it might lead to further, possibly more intense and more persistent drought and desertification. These (usually biogeophysical) 'feedback' processes may involve:

- reduced infiltration or water retention, with reduced soil moisture less cloud forms over the land, surface temperatures rise and airmass movements are affected (Charney *et al.* 1975; Mortimore 1989: 148);
- degeneration of soil fertility and/or decline in humus content, soils then erode easily and plant cover decreases;
- generation of enough airborne dust to affect air temperature and movements;
- clearing of woodland or overgrazing resulting in altered albedo and reduced evapotranspiration, airmass movements may alter and rainfall downwind may decrease (Otterman 1974; Arnon 1981: 113).

Such a desertification process has been recognized for the Rajaputna Desert (India), where seventh-century AD overgrazing supposedly altered vegetation and led to regional climatic change that has meant more or less permanent desert conditions ever since.

To assess the role of drought in desertification the following questions must be addressed:

1. Is there a pattern of droughts? Can it be predicted?
2. What are the causes? Are all droughts in the region due to the same causes? Is human action making droughts worse?
3. Are human activities making the land more vulnerable to drought? Are cropping or livestock-rearing strategies more or less vulnerable to drought?
4. Can drought be avoided or mitigated?
5. Are there feedback mechanisms operative?

A number of researchers have examined the pattern of drought occurrence for Africa (e.g. Schove 1977; Glantz and Katz 1985: 335). (A table of drought occurrence for Africa, the Americas, Asia, Europe and Oceania, based on 1986 WMO data, can be found in UNEP 1987: 321–3.) Until better data are available, separating meteorological droughts from those due to human activity will be uncertain. Some droughts may be initiated by Milankovitch-type orbital variations; altered dust and aerosols levels in the atmosphere (especially from volcanic eruptions); changes in general atmospheric airmass circulation due to oceanic temperature variation; or by amount of snow-cover over Eurasia and North America. Although some mechanisms operate at a global scale, they might be affected by human activity; if this is the case, evidence of past occurrence of droughts will have to be carefully interpreted for forecasting future patterns.

There have been claims that the Sahelian-Sudanic Zone (SSZ) of Africa has suffered more drought in recent decades; whether there is a long-term deterioration or just the normal swings to be expected in erratic rainfall areas remains to be proven (Thomas 1993). These changes might be due to southward shifts of the Inter-Tropical Convergence Zone, perhaps in response to oceanic temperature changes (WMO 1986; UNEP 1987: 303), or variations in snow-cover over Eurasia (heavy spring snows might delay the rise of air pressure over Eurasia which 'drives' rain-bearing monsoons to India and other parts of the world, so drought is ultimately felt in the SSZ) (Gee 1988). Other explanations include global warming or volcanic eruptions (Charney *et al.* 1975; Bryson and Murray 1977: 101; 111–14). If the mechanism can be established drought forecasting might be improved. There are indications that rainfall fluctuations in some regions are linked to large-scale ocean–atmosphere interactions like the El Niño–Southern Oscillation (ENSO) (Lockwood 1984; 1986; Allan 1988) (see chapter 5). In 1982–3 drought in Australasia was believed to relate to an earlier ENSO event, and in 1986 and 1987 ENSO events were used to make rough forecasts of SSZ drought up to 18-months in advance (Anon 1987; Land 1987).

Other natural causes Wild animals and feral livestock may degrade vegetation enough to initiate desertification. Some animal species respond to environmental opportunities with a population 'boom', and drastically decline in numbers as a result of overgrazing or some other environmental constraint. Termites (common in many semiarid environments) may aggravate heavy grazing by wild herbivores or domesticated livestock and 'tiger stripes' (vegetation/bare ground) may sometimes result from such damage (De Vos 1975: 148).

Locust or grasshoppers might denude vegetation cover enough to trigger desertification. Introduced animals unchecked by natural predators or diseases, like the rabbit in Australia or red deer in New Zealand, have caused severe land degradation.

Plants may play a role in soil degradation: in parts of the USA and Mexico 'salt cedar' (*Tamarix* spp.) are blamed for reduced soil moisture and for raising salt from deep below ground. A plant species may be introduced to a region or spread in naturally, and by transpiring more than the pre-existing vegetation, lower groundwater thus causing desertification. Spread of plants with different transpiration than established species, could raise or lower groundwater: this might improve vegetation cover (if salts have accumulated because of a high water-table and this falls, or if more moisture accumulates without soil salinization) or damage it (if groundwater rises too much and topsoil becomes salinized or groundwater falls too far).

Human causes of desertification

Human activity may play a role in land degradation or there may be situations where if things remained undisturbed there would have been natural degradation but much less pronounced and slower to develop.

Population increase Desertification is often blamed on too many people for the environment to support. But there are also regions with severe desertification, for example Australia and large parts of subSaharan Africa, where there are fewer than 16 people per square kilometre; for comparison, at the time of writing the PRC's average population density was about 100 people per square kilometre and India's was about 225 people per square kilometre (such figures are misleading, for most of the population in, say, the PRC is concentrated in much less than the total land area). Why is degradation occurring where human populations are low? There are a number of possibilities:

1. A few landowners graze large numbers of livestock.
2. Degradation may take place because there is labour shortage, so roads cannot be constructed or maintained, agriculture cannot be sufficiently intensive and the economy cannot be developed. Intensive land use is unlikely and extensive land use causes degradation (Franke and Chasin 1980: 117).
3. Bush-rotation (shifting cultivation) farming can be disrupted by population growth or arrival of cash-cropping: fallows are reduced and the land suffers.

It is seldom easy to establish a definite population increase–land degradation relationship. There are degraded dryland regions where demographic growth has been rapid and marked, for example in Rajasthan (India) there was a rise from 36 000 to 96 000 in 70 years (Ruddle and Manshard 1981: 116) and in the SSZ there is reputed to have been at least a 2.5 per cent per year increase, which means that by AD 2110 it will be double what it was when the 1968–73 drought began (Grainger 1982: 53). As already noted, in some circumstances too low a population will prevent intensive landuse and may mean land degradation. The problem is to establish what the 'critical population' (human and livestock) is for each ecosystem; this will reflect technology, social, economic and other factors as well as number of people.

Overgrazing Overgrazing is reported to have a range of effects and to result from a complex of causes:

1 *Reduction of vegetation cover and trampling* Depending on the type of livestock, may cause compaction and reduced infiltration or loosening of the soil surface which increases erosion. The effects may resemble a drought, even if precipitation has remained unchanged.
2 *By reducing savanna grass cover, grassland fires become less frequent and woody vegetation increases* A false impression may be given, that things are flourishing. However, the reduction of grasses and herbs under the scrub may allow greater soil erosion when it rains and there is probably also reduction of infiltration, dry matter accumulates under the bush and trees where livestock no longer graze and when fire finally starts it is more intense than were previous grassland bushfires and does far more damage to vegetation and soil (Bell 1987: 84).
3 *Unpalatable, spiny or difficult to graze plants tend to survive, other plants tend to decrease* The effect may be patchy: where there are few unpalatable plants vegetation may be stripped; elsewhere where grazing is unattractive there is little degradation.
4 *Alteration of wood vegetation regeneration* Scattered savanna trees/shrubs or forest may be prevented from regeneration (in Africa *Acacia albida* is vulnerable to seedling damage through excessive grazing). However, some savanna/woodland species will not successfully regenerate unless their seeds pass through the digestive tract of a grazing animal, either to kill insect larvae or to break dormancy (this includes some *Acacia* spp.). Some plant species' seeds germinate and survive better where existing vegetation is disturbed, for example by overgrazing or livestock wallowing. Overgrazing may thus favour regeneration of some species and hinder others. The result of overgrazing can be replacement of grassland by bush or scrub or the opposite.
5 *Overgrazing may raise dust* This may be sufficient to cause more stable atmospheric conditions with subsiding air masses, the result of which could be a reduction of precipitation. It has been suggested that this process operates in Rajasthan (India) and the SSZ (Bryson 1967).

There does seem to have been an increase in stock in many dryland regions. Between 1955 and 1976 cattle numbers in LDCs are estimated to have risen by 34 per cent, and sheep and goats by 32 per cent (Arnon 1981: 111). Increased numbers of livestock means overstocking only if these animals exceed the carrying capacity of the land (the latter is a concept which should be applied with caution when dealing with drylands).

'Mismanagement' has been blamed for overgrazing, particularly where communal land ownership and individual herd ownership coincide. Another possible cause is breakdown of traditional grazing controls as a result of 'modernization'. Pastoralists may overgraze pastures because their traditional grazing-damage reduction or drought-avoidance strategies have broken down. Breakdown may be the result of:

1 Hindered migration (border restrictions, stock fences, conservation areas, warfare).
2 Use of trucks to take livestock to formerly less accessible areas.
3 Government efforts to 'sedentarize' the people and their stock.
4 The sinking of wells or the provision of watering places, food-aid distribution centres, trading centres, which tend to attract and hold pastoralists too long in a given locality. A teardrop-shaped area of overgrazing and trampling, the head of which faces into the prevailing wind, is likely to form around the attraction.
5 Commercialization of production, which causes pastoralists to increase stock or attracts in large-scale ranching.
6 Health and/or veterinary care, which has led to increases in human and/or livestock populations.
7 Loss of traditional administration of grazing rights.
8 Loss of youth and able-bodied males (absentee migrant labour, drift to cities, AIDS), elderly people, women and children left to farm/herd and are forced to neglect traditional pastures and keep herds near village where land is degraded.

In practice accurate identification of causes of overgrazing can be difficult. Western ideas of 'correct' grazing-level, 'resilience' and 'overgrazing' may not

Fig. *6.4* Heavily (goat) grazed pasture and dry farming (mid-right) with scattered olive trees. Southern Cyprus.

be appropriate in drylands. In drylands annual variation in vegetation productivity is likely to be greater than progressive decline of vegetation. To pastoralists, the basic unit is not productivity per animal, it makes sense to them to maximize production in good rainfall times and to lose production in bad. Migration, if possible with livestock, traditionally helped reduce bad-times losses (Mortimore 1988: 64; 1989). There may actually be incentives for pastoralists to 'overgraze'; by doing so they can sometimes reduce the risk of damaging bushfires and control taller grasses harbouring ticks, which carry diseases. Studies of pastoralism have tended to see stable production as desirable, implying stocking restrictions matched to the average or the worst possible rainfall conditions; studies have also tended to assume that one or a few years' data describe every year or even the average, whereas they are unlikely to do so.

Sedentarization has been cited as one cause of desertification; the motives vary from a desire by authorities to make it easier to provide modern services to scattered peoples, to a wish to establish closer rule. Rezah Shah of Iran attempted the latter in the 1920s, when he tried to sedentarize the pastoralists of the Zagros Mountains to curb their political power (Beaumont 1989: 181).

Goats are often blamed for dryland degradation (Fig. 6.4). It is more likely they (or their herders, who actually strip the trees) deliver the *coup-de-grâce* after cattle grazing, bushfire or fuelwood collection has already caused ruin. That free grazing goats are more damaging than cattle seems unproven, Dunbar (1984: 33) suggested that because goats prefer bush and tree vegetation they open up some savanna woodlands, increase the groundcover and so might actually *reduce* soil erosion. Goats are the livestock of the poor: poor people and their livestock tend to get blamed for land degradation.

Undergrazing Reduced grazing, by wild animals or domestic stock, can result in lower herb and grass species diversity and woody species may become more common. As woody vegetation increases the groundcover below thins or becomes more rank in growth, the result can be increased soil damage, less pasture value and more intense bushfires.

In Africa there is evidence that in some regions human and livestock populations have fallen in the past, probably due to diseases, and scrub became more widespread. It is too early yet to assess the impact of the spread of AIDS in rural Africa (or other rural lands); however, there are indications that in East and Central Africa the labour force may be sufficiently weakened to upset agriculture and allow bush encroachment (*The Times Higher Education Supplement* 4/3/88: 12). In North America, settlers killed off buffalo (*Bison bison*) in the late nineteenth century, causing vegetation changes; shrubs covered English downlands after myxomatosis reduced rabbit populations in the 1950s. Poaching or disease can reduce wildlife grazing and livestock grazing may decline following drought, herd restriction legislation and falling market prices.

Fuelwood, forage and dung collection When fuelwood is scarce peasants or merchants collecting for urban consumers may so damage treecover that people have to gather animal dung and crop residues, accelerating soil degradation initiated by thinning the woody vegetation. Many LDC cities have surrounding 'fuelshed' zones depleted of wood and dung for as much as 300 km from urban limits.

In some situations where fuelwood collection is seen as the *cause* of de-vegetation, it may actually be the *result* of de-vegetation which has left dead trees and little other source of livelihood from the land (Harrison 1987: 206). Where fuelwood, dung, forage and crop residue collection are causing de-vegetation alternatives that can be afforded by the people must be found (Prior and Tuohy 1987; Whitney 1987: 115–43).

Social, political and economic structures and relationships that may trigger desertification Land degradation is sometimes the result of social and economic relationships. But often attempted solutions are designed after focusing on the physical symptoms. Effective control will require the accurate identification of social and economic causes.

At the root of the problem may be conflict between public interest in the long term and private, short-term resource exploitation (World Bank 1985: iii). Where tenure is insecure there is likely to be little incentive for the individual to minimize degradation.

Breakdown of established authority can cause desertification, for example loss of respect for heads of extended families due to cultural change; decline of traditional systems of tribute or tribal domination; inappropriate intervention by central government, such as the imposition of an outsider as a decision-maker. The new administrator is unlikely to achieve what the traditional leader(s) could because the local people do not respect and obey and generally lose incentive to tackle problems. Governments increasingly tend to enforce national or state boundaries, attempt to sedentarize nomadic and semi-nomadic peoples and develop dryland regions, all of which tends to hinder migration and drought or overgrazing avoidance. In various countries scattered populations have been relocated by force or encouragement, into new villages (the motives vary: a state may wish to offer services like schools and healthcare more cost-effectively; it may wish to subjugate and tax people more easily; it may simply see nomadic peoples or scattered smallholders as a 'stigma of backwardness'). 'Belts' of degradation can form near such settlements, for example in Tanzania around *ujamaa* villages.

In Niger and Burkina Faso, the Tuareg once raided and 'taxed' savanna farmers: this was controlled and arable farming (e.g. that of the Iklan people in Burkina Faso) increased, reducing fallows and putting land under greater stress. At the same time the Tuareg were deprived of a vital link in their drought-avoidance strategy (Franke and Chasin 1980: 68). Similar events have taken place in Kenya (Mabbutt and Wilson 1980). The decline of trans-Saharan caravan routes led to impoverishment in some regions and land damage as people depended more on what they could win from the soil.

The spread of cash-cropping is often cited as a cause of desertification, but it is one thing to chronicle the process of ecological change and another to demonstrate its relationship with the impact of capitalism (Franke and Chasin 1980: 15). Urban-biased policies seldom give much support to peasant cash-croppers: market prices are allowed to fluctuate or tend to be kept so low (to satisfy the city-dweller voters) that the producers get inadequate, insecure returns and may over-exploit the land to survive. Things are little better for many commercial livestock producers as taxes and middlemen take a large portion of their profits in good years, so little is available for improving land management or for hard times. In remote areas herders may have little incentive to sell livestock; cash is of little use if there are virtually no things on which to spend it, so herders' beasts are a more realistic and 'concrete' form of wealth and are unlikely to be reduced in numbers.

There is a risk that cash-croppers may be forced by falling market prices to step up production effort; greed or a failed crop may have a similar effect: a 'spiral' where the producer has to produce more to pay for inputs or compensate for falling produce prices until the land suffers. In the SSZ it is likely that smallholder groundnut production has followed this pattern. Where farmers and herders are not integrated, the expansion of arable cash-cropping may force the latter to abandon traditional grazing lands and overgraze what is left to them by farmers (or larger livestock producers).

If high market prices for agricultural produce coincide with above average rainfall, or if there is an innovation in seed or other aspects of production (e.g. a fall in rural wages) which reduces costs, boost yields or makes cultivation or herding seem more attractive, people are likely to make greater use of marginal land. Any set-back, such as a fall from above average rainfall, produce price slumps or wage rises, can damage agricultural production and set off land degradation in marginal areas.

There are two phenomena that may influence landuse leading to degradation and desertification:

1 *Social differentiation* the process of growth of inequality between sections of local, regional or national society; the result is marginalization.
2 *Marginalization* the retreat of farmers or herders to areas of poor productivity or some other disadvantage(s), areas which are often vulnerable to land degradation.

The two phenomena (or processes) can be due to appropriation of land by richer landowners; by the creation of state reserves, conservation areas, game parks; by the redistribution of land to groups favoured by the state; warfare or unrest; threat of disease, pests or natural disaster; simple lack of available land. In some situations people seek out marginal land to escape debt, to pursue new opportunities or to exercise some vague 'frontier spirit'.

In some LDCs the response to rising population has been to attempt to resettle people in 'vacant lands', often at great cost, and seldom with sustained success. Such programmes have been tried in Indonesia; Amazonian Brazil, Peru and Ecuador, Bolivia; Malaysia and various African countries; they have generally caused deforestation and land degradation. Sometimes settlers fail to adapt to their new environments and resort to land-damaging form of shifting cultivation.

There are regions where the active male population is taken by military service or lured away from the land by opportunities for wage labour in mines, city industry and services or tourism, or simply because the towns are attractive to youth or offer better healthcare, etc. At best they become migrant workers, with elderly people, women, and children left to try and cope with the farming for at least part of the year. Agriculture is likely to become less intensive and land degradation may develop.

Human error as a cause of land degradation Unrest and warfare are all too common causes of land degradation, either directly or because human populations are uprooted and disorientated. Unrest also involves expenditure that might have been used for anti-degradation work and hinders control of pest insects, monitoring of erosion, etc. Unrest may take the form of tribal conflict, terrorism, full-blown warfare or inadequate rule of law over those who would exploit others or the land. Disruption may lead to trauma and loss of livestock, seeds and tools; people may become refugees, in all probability malnourished, poor and in an unfamiliar and possibly hostile environment. At the very least, good, sustainable agricultural practice is unlikely. Even if hostilities cease, there are usually unexploded munitions, especially anti-personnel mines which can kill and maim peasant labour for decades.

Greed, ignorance or fashion may result in excessive demands for resources or potentially damaging innovations. A change in taste can quickly lead to abandonment of traditional crops, possibly for new, less robust, less nutritious and more land-damaging ones. Veterinary care may well improve livestock survival but with no complimentary provisions for culling and marketing excess livestock benefits are short-lived.

Though restricted mainly to drylands and coastal dunefields of richer countries, increased use of off-road vehicles (cross-country motor cycles, 'dune-buggies', jeeps, etc) is a locally significant problem. The problem is most marked in California and other dryland areas of the USA, Australia and game parks in Africa.

Desertification monitoring and control

There are a number of vital steps in monitoring desertification:

- recognize desertification indicators;
- establish standards;
- set-up reliable measurement of indicators (ensuring that the time is noted and measurement is comparable between stations).

It is pointless to measure plant-cover in a dry season in one locality and in a wet season at another and then attempt comparison. Even with caution, obtaining comparable data is difficult where rainfall varies a lot from year to year and site to site. Governments are often reluctant to gather data on 'negative' development like desertification; this, coupled with the shortage of personnel and funding, further hinders monitoring.

UNESCO, FAO, UNEP and IUCN are some of the bodies active in trying to map and monitor desertification. Between 1950 and 1962 UNESCO ran an Arid Zone Research Programme which provided valuable information on drylands, this was continued under the Man and Biosphere Program from 1974. In 1973 the UN established the UN Sudano-Sahelian Office (UNSO) to coordinate efforts to aid the 19 SSZ countries. UNSO's brief was later expanded to include desertification avoidance and control. In 1983 the UNEP selected the Environmental Systems Research Institute to create a geographical information system for monitoring and mapping world desertification hazard and, in 1986, the Desertification Control Programme Activity Centre of UNEP was set up to establish a database on desertification.

In 1973 eight SSZ governments (Mali, Mauritania, Niger, Senegal, Burkina Faso, Chad, The Gambia, Cape Verde) established the Permanent Inter-State Committee for Drought Control (Comité Interétat pour la Lutte Contre la Sécheresse dans le Sahel – CILSS). This aimed to attack SSZ drought and desertification. Under the aegis of the OECD various western aid agencies formed the Club du Sahel (in 1976) to organize and administer donor-to-country aid programmes, and to improve cooperation between SSZ countries and between donor countries and the SSZ (the main aim was to get regional food self-sufficiency rather than simply fight desertification). In 1977 the OECD published a *Strategy and Programme for Drought Control and Development in the Sahel*, and in 1986 IUCN and the Commonwealth of Australia released a SSZ desertification rehabilitation strategy. In Southern Africa, the Southern African Coordination Conference (SADCC), and in Latin America, the Andean Pact have made provisions for desertification control.

Responding to the crisis situation in the SSZ in 1977 the UN General Assembly called for a Conference on Desertification; this Conference (UNCOD) was held in Nairobi (Kenya) and issued a Plan of Action to Combat Desertification (PACD). The UN Environment Programme (UNEP) was charged with administration of the PACD, and UNCOD widened its concern to cover other regions of the world and began this by circulating a questionnaire to check the desertification status of various countries; the results were patchy and unsatisfactory.

The PACD was a broad-based document, in large part more concerned with rural development than control of desertification, in effect virtually any development assistance (aid) in a dryland region has tended to be called 'anti-desertification' (hence many of the figures on costs or extent of control measures are misleading). The UNEP set up three groups to assist in administering the PACD:

1 The Desertification Branch of UNEP. Its role was carrying out programmes (i.e. implementing the PACD) and acting as 'catalyst' and coordinating body. By 1984 it had caused about 30 countries to assess their desertification status and prepare control proposals for submission to DESCON (see following).
2 The Consultative Group for Desertification Control (DESCON). Its role was to secure funding for projects and to screen anti-desertification projects.
3 Inter-agency Working Group on Desertification (IAWGDC). This was charged with facilitating cooperation among UN agencies and to develop a plant of short- and long-term objectives for the PACD. IAWGDC was also to prepare an annual report on progress with PADC.

Apart from generating acronyms, what did these efforts achieve? The PADC stated what *should* be done, but not *how* it could be achieved. The general conclusion is that it failed to generate adequate political support, although its proposals were probably quite sound. UNCOD was underfunded: by the end of 1983 it had received country donations of less than US$50 000 (virtually all from LDCs). In 1983/84 (after seven years' activity) UNCOD called for a General Assessment of Progress with the PADC. Questionnaires were sent out to over 100 countries: the returns were poor. At a UNEP special meeting in 1984 it was admitted that PACD had been an almost total failure. In 1984 only two of the hundred nations recognized to be affected by desertification had proposed a National Plan of Action to Combat Desertification; in 1987 this had grown to just six (these Plans were deemed essential by UNCOD).

Had desertification control been made the responsibility of the FAO or UNDP, rather than the UNEP, which was comparatively underfunded and understaffed, efforts might have had more results. In 1986 the UNEP and the East–West Center (Hawaii) and the Australian Government, in an attempt to initiate renewed anti-desertification effort, held a conference in Canberra which focused on the economic significance of the problem. A Strategy for Rehabilitating the Sahel (and an Action Plan to facilitate this) was published by the IUCN (1986). The proposed strategy was to work over 15 years and generate money at the local level.

There have been some successes in controlling desertification in parts of the USA, PRC, Israel and CIS; Iran made some progress in the 1970s and Libya and some other North African countries have applied dune stabilization measures or have afforested considerable areas of drylands. There is no shortage of techniques for combating dryland degradation (Goudie 1990; Skujins 1991); however, since the mid-1980s debt problems have made desertification control less easy for many nations and there has always been a temptation for governments to spend on regions with promise not 'throw away' money on less productive and degraded land.

Funding desertification control

Most of the world's desertification could probably be reversed (with presently available technology). Provided enough soil remains, plants can recover or there can be re-seeding. In practice the amount of control and rehabilitation will be determined by economics and institutional factors. The UNEP and Commonwealth of Australia (1987: 4) suggested that US$4.5 billion a year would *if effectively spent* avoid roughly US$26 billion lost productivity, that is the costs are huge but should be weighed against the costs and consequences of no expenditure. Forse (1989) noted that less than 10 per cent of the aforementioned US$4.5 billion had been raised in 1988/89 and there is heated debate as to whether what had been collected was spent wisely. Ahmad and Kassas (1987) suggested how funding *might* be generated, and clearly international backing will often be needed.

Identifying priority problems

The major priorities should be:

1 to improve landuse planning in drylands;
2 to train personnel;
3 to overcome political and institutional constraints;
4 improve inter-regional and inter-state cooperation;
5 raise adequate funding.

There has been quite a lot of work on desertification control techniques, but establishing the necessary institutions, management, and funding has had less attention. Since the late 1980s there have been developments within fields of community development, farmer and peasant education and 'grassroots' planning that could help desertification control efforts.

Overgrazing and trampling might be countered by developing feedlot-rearing or stall-feeding rather than free grazing. Beaumont (1989: 213) stressed the need to prevent overgrazing at *critical times*, in particular when seed is being set, once seeds have been dispersed (or bulbs and tubers developed), then overgrazing poses much less threat. Possibilities for diverting investment in livestock to other alternatives should be investigated where there is a risk of overgrazing.

Appropriate forestry, fuelwood-supply plantations, dryland cropping strategies and range management now received more attention than in the past.

The challenges are now to:

1 assemble appropriate 'packages' of crops, livestock, soil management, possibly alternative livelihoods, drought insurance and avoidance, etc., to suit each region;
2 disseminate and 'tune' such 'packages' to local needs and constraints and sustain them by providing necessary inputs and stable, appropriate institutions able to ensure efficient management;
3 fund the rehabilitation of degraded land where it is worth doing so and
4 prevent further desertification.

Soil degradation

Soil degradation is generally a component of desertification; however, it may take place without leading to particularly desert-like conditions. Soil degradation might be defined as *a reduction of the current and/or future capability of soil to produce (in terms of quantity or quality)* (Higgins 1988: 2). Soil degradation can be both quantitative (loss of soil due to erosion, mass-movement or solution) or qualitative (decline in fertility; reduction of plant nutrients; structural changes; changes in aeration or moisture content; change in trace elements, salts, alkaline compounds; pollution with some chemical compound; change in soil flora or fauna). Soil degradation can be a natural phenomenon, for example leaching of glacial soils, pan formation, activity of wildlife, laterite or plinthite formation, or it may be due to human activity.

Fig. 6.5 More obvious soil erosion forms. a) Severe gullying and sheetwash of an area of vegetation cleared for housing development.

Soil is usually formed at a very slow rate (typically a few mm per century), but removal can easily take place at a rate of several centimetres per year or even per hour. Soil renewal depends on a complex of factors that can easily be disrupted. Therefore, although potentially a renewable resource, soil can become a non-renewable resource if mismanaged (some soils require much more careful management than others). The process of soil degradation may be difficult (and very costly) to halt or reverse once it gets under way, particularly if vital seeds, fungi and soil organisms are lost (and with such loss there is also likely to be altered microclimate and structural alterations of the topsoil).

Soil degradation is one of the crucial problems today: it is occurring world-wide afflicting DCs and LDCs from the Equator to the poles and is getting worse (Pimental 1993: 277). The world was quick to recognize an 'oil crisis' in 1973/74, yet, in spite of the fact that it affects all people, millions of them directly, and was a serious problem in many regions long before the 1970s, soil degradation has had much less attention. Ironically, if oil reserves were depleted, soil resources would become more and more valuable because, first, with less energy and artificial fertilizers, agricultural strategies would in many regions have to be revised (global and regional pollution may also force revision), and second, it may be necessary to grow crops to substitute for the energy and industrial raw materials presently obtained from petroleum as well as produce food crops.

Study and control of soil degradation is beset by difficulties, many similar to those of desertification: data have generally been, and often still are, inadequate; the processes of erosion may vary over time and monitoring has seldom been satisfactory. Periodic, frequently in accurate measurements give a distorted view and make comparisons difficult and more research is required if better land management is to be achieved (Thornes 1989). Steady insidious soil damage may fail to attract attention, spectacular gully erosion and landslips (Figs 6.5a and 6.5b) may be dealt with yet the former unspectacular problem may be crucial.

Fig. 6.5 b) Landslip in an area logged and burnt. Malaysia.

For years, over wide areas of the globe, soil degradation has been masked by improving agricultural yields; these improvements have frequently been accomplished at the expense of accelerated soil damage. There may come the time where soil damage reaches a critical point in areas vital for food or commodity production, and technology may not be able to further boost yields to compensate. Some claim there are already signs that technology and crop improvement have reached a 'plateau' (Brown *et al.* 1984: 54; Canter 1986: 1). Soil degradation can involve feedback mechanisms, such that damage to soil may set in motion a chain of causation leading to further, greater decline.

At present soil degradation is mainly associated with land use – farming, grazing, logging, mining, etc. Increasingly soil degradation will become a much more general risk, affecting untouched, natural landscapes as well as those used by humans through acid deposition, pesticide, radioactivity and other pollutants.

Soil degradation can occur even before there has been any erosion (indeed can predispose a soil to erosion), as a consequence of acid deposition, compaction, salts or alkali accumulation, burning, oxidation of organic matter, etc. It is difficult to give a precise definition of soil erosion; a reasonable working definition might be the removal ('eating-away') of soil material by water or wind at rates in excess of soil formation. Soil erosion is generally the culmination of a degradation process, and could be regarded as a 'disease' of landscape not just of soil, as it relates to vegetation, climate, etc (Zacher 1982).

Added to erosion are the effects of mass-movement (the downhill movement of surface materials, including solid rock, under the influence of gravity, frequently assisted by buoyancy of particles due to rainwater) and solution (material dissolved-away by moving moisture). Mass-movement can occur rapidly (landslides, rockfalls, etc) or much more slowly (as 'creep'). As a result of mass-movement surface materials come to rest in a steady-state that depends on the material's character, the slope of underlying terrain, rainfall, and gravity (for a review of soil erosion terminology see Zacher 1982: 27–45).

Non-erosive soil degradation

Compaction and loss of nutrients through burning vegetation cover

If forest or savanna are cleared there may be considerable variation in the subsequent degree of degradation depending on how the clearance was conducted. Much depends on how much compaction occurs, and how much the plant nutrients are lost after vegetation is burnt. When vegetation is burnt a lot of nutrients are carried away in ash and smoke; much of the remaining ash may be washed away if there are heavy rather than gentle rains soon after the fire. If the burn has been particularly fierce and hot, soil micro-organisms and soil organic matter may be destroyed and there may be chemical changes in the surface layers. The timing of the burning relative to seasonal rains is important, both in affecting the fierceness of the burn and in determining how long may elapse before heavy rain strikes the burnt ground.

Even on the same soil the following clearance strategies could have different consequences:

1 *hand removal of vegetation* little soil compaction, some selected trees might be left to control erosion;
2 *use of bulldozer* much compaction of soil, little control over what vegetation is left; generally more destructive than hand-cutting and burning; a single farm tractor pass may well cause a 70 per cent reduction of soil porosity;
3 *bulldozer-clearing* plus burning of cut vegetation is particularly likely to result in soil degradation.

In the UK farm machinery typically weighs between 10 and 15 tonnes and can cause serious long-term compaction, especially on heavy clay soils (Wolman and Fournier 1987: 145). Trampling by livestock or animals used to till the land can cause compaction, but this is likely to be less than that resulting from mechanized cultivation.

Decline of soil fertility and loss of soil organic matter

Soil fertility is a function of a wide range of factors, but most crucial are the availability of plant nutrients and adequate, but not excessive, moisture. There are many ways of assessing the fertility status of a soil, ranging from electronic probes and chemical soil tests to checking crop yields or looking for sensitive 'indicator species' of weeds (e.g. in Africa the appearance of witchweed, *Striga hermothica*, generally shows a fertility decline is taking place).

The loss of soil organic matter and the nutrients used by plants can occur in virtually any environment, but is most dramatic in dryland regions. The consequences are reduced moisture and nutrient retention, degeneration of soil structure, changes in population of soil micro-organisms – essentially loss of fertility. As organic matter is concentrated near the surface this valuable material is generally the first to be lost. Once the organic content falls below about 2 per cent organic carbon content a soil is likely to be easily eroded (Kirkby and Morgan 1980: 286). Loss of soil organic matter therefore should make a good indicator for those wishing to monitor land degradation and desertification; surprisingly, it has seldom been measured on an ongoing basis.

Loss of organic matter may result from removal of crops, fodder, fuelwood, dung and bushfires or stubble-burning or overgrazing without the return of any material to compensate. It may also result from alteration of soil drainage or tillage that accelerates oxidation of organic matter. Peaty soils tend to suffer oxidation of organic matter if drained and can shrink alarmingly as a consequence and sometimes spontaneously combust. In drier climates the loss of soil organic matter generally leads to a reduction in retention of soil moisture and with this a decline in vegetation cover (crops or natural plants), which in turn leads to increased erosion. A feedback can thus arise, as organic matter and moisture in soil falls, so plant cover decreases and thus renewal of organic matter in soil further declines.

Tropical moist forests produce a lot of organic matter (litter and root debris), soil micro-organisms rapidly break this down and it is re-absorbed by the vegetation. If the forest is cleared the breakdown of organic matter generally remains high, but the supply of material ceases so that a rapid decline of soil organic matter takes place. The soil exposed to sunlight is warmed more than it might have been under tree cover and may dry out more.

It is common for a crust or plinthite deposits (duricrusts or cuirassic deposits) to develop after forest clearance for various reasons (Nye and Greenland 1960; Lewis and Berry 1988: 133).

People may use dried animal dung for fuel, which might cause soil organic matter to decline. Agarwal (1986: 26) pointed out that use of cattle dung has a high opportunity cost in terms of lost agricultural output had the dung been used to manure the soil (as well as soil damage, it is likely to lead to increased area of cultivation to get the same crop). Agarwal estimated that for every tonne of cattle dung burnt in India roughly 50 kg of potential food grains were lost. India is a particularly heavy user of cattle dung, consuming at least 73 million tonnes per year. Dung is also much used in the Peruvian Altiplano and highland Ethiopia. In Australia there were virtually no indigenous insects capable of incorporating livestock dung into the soil before it was oxidized, washed or blown away. The solution has been to import suitable dung-burying scarab beetles (Jackson *et al.* 1984).

Structural changes

Compaction and shrinkage through oxidation are not the only causes of structural change. Soil may suffer suffosion, i.e. the washing-out (or the washing down-profile) of soluble salts like gypsum or common salt. Loess-type soils are vulnerable, and the problem may arise if land is irrigated (in the San Joaquin Valley, California, a considerable area subsided over 9 metres through suffosion between 1925 and 1975). Some soils suffer a 'sealing' of soil fissures if they are cultivated (or mis-cultivated) – without pores and fissures to hold air and moisture the soil becomes less fertile.

Acid-sulphate soil problems may arise where land that has been waterlogged (this leads to the reduction of sulphate ions to sulphides under anaerobic conditions) is drained or vegetation is altered causing a fall in water-tables. If there is insufficient calcium in the soil to neutralize the reformed sulphate ions it becomes very acid. Acid-sulphate chemical and structural problems afflict an estimated 12.6 million ha world-wide, much of it lowland peat swamp or mangrove swamp. In Vietnam defoliation during the war formed large areas; in Indonesia the efforts of some transmigration settlers to establish new ricelands have had similar effects. Whether formed in the tropics or temperate environments acid-sulphate soils are very infertile and can be difficult to rehabilitate (Barrow 1987: 193; 215; Lal *et al.* 1989: 66).

A pan is a hard concretionary layer formed at or beneath the soil surface. Often pans form just below the cultivation depth: these are sometimes called plough soles. They can restrict roots of vegetation, making crops or natural cover vulnerable to drought and trees vulnerable to wind-throw. They can also affect drainage leading to waterlogging and salinity or alkalinity problems. Once crusts or pans develop soil moisture recharge declines, vegetation finds it difficult to root, and sheetwash and gully erosion may increase as the land fails to absorb precipitation.

Laterite and laterization are not very precise terms but, generally speaking, refer to the product and process of leaching leading to the concentration, consolidation and hardening of silicates (Al_2O_3 or Fe_2O_3) to form layers, sometimes of great thickness. It is common (though not as widespread as some would claim) in the humid and subhumid tropics. Depending on the nature of the materials involved, some authorities recognize laterite, fericrete, silcrete, calcrete (or *caliche*) (involving aluminium, iron, silica and calcium compounds respectively). The formation of laterites, plinthites, pans and crusts can be very rapid: some tropical soils (often those beneath moist tropical forest or savanna) may harden as soon as they are drained or exposed to the air by tillage, road-cutting, etc. The effect of duricrust formation can be striking: the resulting scattered bush with bare patches of ground has been termed 'pedological leprosy'.

Acidification

Acid deposition through air pollution is already a problem in or downwind of areas of industrial development or widespread use of coal for fuel. Such acidification is spreading to more and more areas as trans-boundary pollution increases. In addition to air pollution, global warming, change of crops and landuse may also cause soil acidity problems, as may careless application of fertilizers. In Australia, for example, the creation of leguminous pastures (with exotic white clover or alfalfa) has been known to cause soil acidification.

The accumulation of toxic compounds, salinization, alkalinization and pest organisms

Salinization and alkalinization are major causes of soil degradation. As well as salts many toxic compounds may build up in soils naturally or as a consequence of development. Boron and selenium are naturally abundant in some soils; human activity may bring them to the surface or increase the level of contamination (if say poor quality irrigation water is used on land). Heavy metals, pesticides, herbicides, disease and pest organisms, radioactive compounds and many other contaminants may degrade soil through human activities ranging from warfare to sewage effluent disposal. Some of these remain or are slowly accumulating as potential chemical time bombs threatening sudden release in the future if some environmental threshold is reached.

Erosion

Erosion, together with mass-movement and solution, acts (in varying combinations) to remove material from a site; these three processes are responsible for quantitative soil degradation (Kirkby and Morgan 1980: 2) and are natural, world-wide components of soil erosion. It is possible to divide soil erosion causes into abiotic causes (due to inanimate processes) and biotic causes (relating to activity of living things). In any given situation one or both groups of causes may operate (but not necessarily at the same time). Of the abiotic causes, water and wind are the main agents; human activity has come to dominate the biotic causes, and indeed may, at least regionally, dominate all causes of soil erosion.

Some refer to human activity-related erosion as anthropogenic erosion, others use secondary erosion (as opposed to natural or primary erosion). Potentially confusing is the use of the term 'accelerated erosion' to imply 'resulting from human activity'. While humankind has often accelerated soil loss well beyond natural rates, there are natural processes or events that can suddenly alter long-established 'natural' rates of erosion, for example earthquakes, tsunamis, major storms and severe drought.

Once topsoil is removed, the subsoil may be more or less vulnerable to erosion. Often the poor organic matter content and unfavourable drainage of subsoil makes it difficult to re-establish a protective vegetation cover. Erosion-resistant soil layers (e.g. pans, stony-layers, clay-bands) will slow and possibly stabilize soil erosion. Thus erosion seldom proceeds in a steady manner unless there is a very thick and homogeneous soil or a slow rate of erosion.

Not all places have erosion in excess of soil formation; there are localities that accumulate material, sometimes as a result of human-induced degradation elsewhere, sometimes due to natural causes. Floodwater or wind may deposit sediment, volcanic ash or loess (deposits of fine wind-blown dust); an area with such accumulation may sustain agricultural production even though the standard of management may be low. There are also situations where rock or subsoil readily form soil or allow plant growth, so that there may be fast recovery from erosion (some clays and shales may support tree cover with virtually no topsoil present). Some of the world's food and commodity crop production regions are areas where accumulation of material matches or exceeds erosion (for example the Nile Valley, lowlands adjoining the Ethiopian and Sudanese uplands, the Amazonian floodlands or *várzeas*). In the case of the Nile Valley and many of the major Indian valleys, upland farming and grazing may have dislodged sediment that helps sustain lowland cultivation.

The amount of erosion that will occur in a given circumstance is determined by erosivity, erodibility, cover and management.

Erosivity is the capacity (potential) of precipitation to cause erosion in given circumstances, i.e. 'aggressiveness' of climate. Erosivity is a function of intensity, duration, timing, and amount of precipitation.

Erodibility is the vulnerability of soil to erosion, i.e. a soil property – its liability to have particles detached and then transported away. Erodibility is dynamic, and changes during a storm, during the year or from year to year. Soils can vary in moisture content and with this resistance to erosion. For example, infiltration can alter if soil structure changes (this may be due to seasonal activity of burrowing animals, swelling, shrinking or saturation of a soil during a rainfall episode, trampling or compaction). Infiltration can alter if vegetation cover alters. Surface water flow causes much soil erosion, subsurface flow can affect surface flows.

Cover is natural vegetation, crop or cover-crop that protects the soil.

Management is landuse, i.e. crop, cropping method, cropping pattern, tillage method, use of mulch.

At various times of year, or over a period of time, erosivity, cover and management are likely to alter. Unless there is disruption, erodibility is more likely to remain stable. If cover is removed, even briefly, erosion can increase considerably.

Solution

The amount of solution depends upon three factors:

1 The character of the soil/rock.
2 The movement of moisture: amount and timing of flows.
3 The quality of the moisture, acid precipitation may act directly on rocks or may lead to acid soil moisture. Soil moisture may become contaminated with salts or other contaminants that accelerate or slow solution, either by altering chemical reactions or because they affect the permeability of the soil or rock and thus movement of moisture.

Vegetation cover

The relationship between erosion and vegetation cover is complex. If cultivation of eroding land ceases, or if there is some other landuse change, the system might move toward more or less erosion. It is important to be able to determine in advance which route will be taken (there may be situations where rehabilitation efforts are a waste of time but prevention is worthwhile). Thornes (1989: 48) presented a model based on differential equations describing change in rate of erosion and growth of vegetation (in practice other factors would need to be added, such as grazing pressure, plant water demand, etc), which could give some idea of the likely vegetation cover–erosion relationship. It is apparent that beyond certain limits the vegetation cover–erosion relationship can suddenly 'flip' from a non-problem to a problem state.

Wind erosion

Wind erosion can be considerable where airflows are not slowed by vegetation, topography, etc., especially if the soils dry out and are fine-grained, incohesive and form no protective crust. Arable farming is likely to lead to seasonal peaks of wind erosion that coincide with periods when crop cover is reduced, soil is driest and most disturbed and wind most erosive. Typically such periods are after tillage or follow late summer harvest before the arrival of winter snows or rain. Nearer the Equator wind erosion increases during dry seasons or if there is drought. Erosive episodes may be associated with winds that occasionally blow from some direction other than the usual prevailing wind.

Monitoring soil degradation

There are a number of ways in which soil erosion may be measured.

Direct measurement at site

Measurements are made of the time to remove a given depth of soil, or the depression of surface in a given time.

Modelling (analogue models)

Rainfall can be simulated by spraying water on to a test bed and measuring the effect. A plot can be shielded with material like mosquito-gauze and the effect compared after a given time with an unshielded plot to judge effect of precipitation impact. A plot may be protected from trampling or grazing by an enclosure (fence): levels within this plot can be compared with those outside; alternatively metal pins or precise levelling can be used to check losses from a site.

Monitor stream sediment loads

This has limited accuracy, as not all sediment will reach a stream; there may be sudden above-average releases (monitoring may miss these, especially if it is not continuous or if severe flood damages equipment); not all sediment reaching a stream will reach the monitoring point (stream sediment transport is discussed on pp. 124 and 126).

Radioactive tracers

There may be opportunities to mark with radio-nucleides (such as Cs-137 or Pb-210) or make use of fall-out from atomic accident or bomb-testing, to establish common reference points, trace where eroded sediment goes and whether any location escapes erosion.

Empirical estimates

Tolerance level (T-factor) may be defined as the loss equal to the rate of new soil formation. Obviously if this is exceeded there will be a net loss of soil. Kirkby and Morgan (1980: 45) defined it as the maximum rate of soil erosion that permits a high level of productivity to be sustained. It is not unusual in the USA to have soil losses of twice the tolerance level. The Universal Soil Loss Equation is discussed on p. 125.

Stream sediment transport

Streams transport sediment as suspended load and as bedload and these vary over time and from place to place along the stream, depending upon discharge, flow velocity, stream gradient, channel morphology, bed roughness, physical characteristics of the fluid and of the particles carried (these variables interact with each other making things even more complex). In the long run climate and geology are the dominant factors. Ephemeral streams in dryland regions probably carry much more sediment than geomorphologists have expected and sediment yield equations presently predict.

Sediment load

A combination of transport rate and mean sediment concentration, it increases with discharge, so that it is likely to be greatest after a storm or at peak snowmelt (Fig. 6.6a). A stream carries suspended particles with settling velocities that are less than the buoyancy generated by turbulent flow, i.e. the faster and more turbulent the flow, the bigger the particles in the suspended load. Bedload consists of particles rolling, sliding or jumping along the bed, i.e. the load generally not suspended in the fluid (and is difficult to accurately measure).

Sediment yield

The mean sediment load carried by a stream gives some measure of rate of erosion in a catchment (and is expressed as weight per unit area).

Sediment discharge ratio

The ratio of total stream discharge to discharge of transported load is expressed as a sediment rating curve (see Fig. 6.6b).

Soil degradation concepts

Soil erosion

This is the gross amount of soil moved by raindrop detachment or runoff.

Soil loss

This is soil moved off an area of land.

Sediment yield

This is soil delivered to a point under evaluation.

Loss on crops

This is material adhering to harvested roots (especially significant with potatoes, beet, yams and other root-crops).

Universal Soil Loss Equation (USLE)

This is a predictive 'tool' which uses a range of parameters to estimate average annual soil loss. It was developed in the 1930s by the US Soil Conservation Service, was improved in 1954 by the US Agriculture Research Service and was further improved in 1978 by the US Department of Agriculture. Much used by consultants and planners world-wide to predict soil loss and select appropriate agricultural practices and crops, it was developed in mid-western USA (Kirkby and Morgan 1980: 16–62; Hudson 1981: 179–96), it has now been adapted for use in other environments (particularly as a result of efforts by the International Institute of Tropical Agriculture, Ibadan, Nigeria), but without modification and caution it can be misleading (Pimenel 1993: 266).

Problems arise when the USLE is applied in areas where data are imprecise or unavailable; unfortunately expediency often obliges such risks. Estimating the losses due to wind is more complicated than applying the USLE to water-related erosion (there is also likely to be rainfall loss as well). For any given soil conditions, the amount of soil which will be blown away depends upon (1) wind velocity and (2) roughness of the soil surface.

Typical form of the USLE:

A = (0.224) RKLSCP

Where :

A = soil loss kg m^2 s^{-1}

R = the rainfall erosivity factor

K = the soil erodibility factor

L = the slope length factor

S = the slope gradient factor

C = the cropping management factor

P = the erosion control practice factor

Typical empirical attempt to assess wind erosion of soil – a Wind Erosion Equation:

E = f(I, C, K, L, V)

(Hudson 1981: 258–9)

Where :

E = soil loss by wind

I = erodibility factor – vulnerability to wind erosion

C = factor representing local wind conditions

K = soil surface roughness

L = width of the field in direction of prevailing wind

V = measure of vegetation cover

Studying degradation

There is a need to be able to obtain a long-term picture of degradation in a given region, to be able to assess the current position, predict possible future trends and try and compare one locality with another. Soil erosion is not often constant: it is more likely to take place as 'episodes' with periods of relative

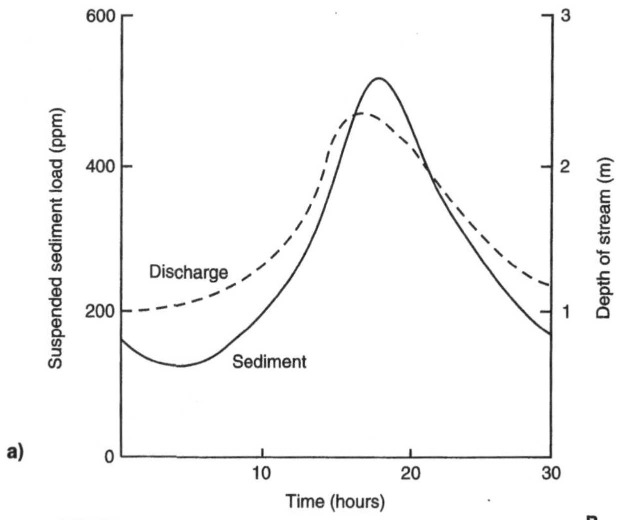

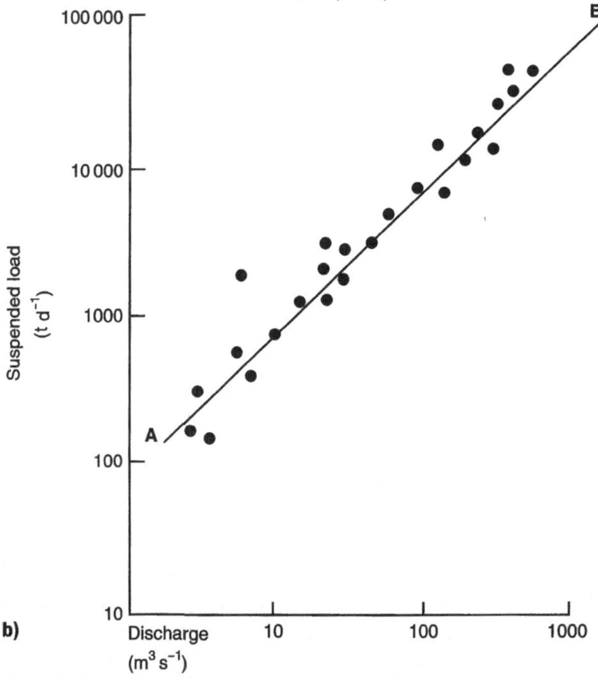

Fig. 6.6 Stream sediment load concepts. a) suspended sediment load and flood discharge relationship (*Source:* author from various sources). b) sediment rating curve for a stream (*Source:* author from various sources).

Note: Suspended sediment discharge correlates with flow, it is plotted on a logarithmic scale to give straight line A–B using the formula $M = kQ^n$ where:

M = rate of sediment movement discharge

Q = discharge

k = } empirical constants which differ from
n = } stream to stream

quiescence. For example, in the UK for some time before and during the winter of 1985–6, conditions were very wet consequently there was more severe erosion than usual. The present position may not be a true reflection of what an environment could once support, and the present rates of erosion may not be maintained. Soil is seldom homogeneous, so soil degradation varies from place to place even over short distances and in the same place as time goes by (especially as different soil horizons are exposed that vary in resistance). Vegetation strongly influences erosion. Usually it protects soil, but there are situations where trees shed large, erosive droplets and have little ground cover beneath: in such circumstances erosion might actually be enhanced, which has been a problem in teak and rubber plantations (in the past when regular weeding was the norm). Landuse can change, for example growing autumn-sown cereals has become more common in the UK since the 1960s and this favours erosion.

Evidence of past erosion is provided by palaeo-erosion studies, for example evidence from lake or marine cores. Common reference points to link different sites are often possible using volcanic ash layers, tree-ring (dendrochronology) or radio-isotope dating. Some DCs and some LDCs have set up soil erosion monitoring organizations but practices vary and often monitoring has been only recently established: the UK (Ministry of Agriculture Fisheries and Food) started to monitor erosion only in 1982.

Given that soil erosion may be continuous, seasonal or episodic, it may be slow, moderate or rapid. The crucial question is: what is the critical rate of soil loss? Crude rates of soil loss have been calculated (in $10^9\,t - 1\,y^{-1}$) for the world by Clark and Munn (1987: 56) and show a clear upward trend:

pre-agricultural = 16.4

AD 1860 = 46.3

AD 1978 = 91.1

The UNEP has a World Soils Policy, which aims to develop methodologies to monitor global soil and land resources. One problem is to collate and correlate national and regional data. To assist, the UNEP prompted the International Society of Soil Sciences (ISSS) to produce a global soil degradation map (1 : 10 000 000-scale published in 1990) and to establish a database. In late 1987 the UNEP, the ISSS and the International Soil Reference and Information Centre (ISRIC) agreed to a project – Global Assessment of Soil Degradation (GLASOD). This was seen as a first step towards global soil degradation assessment.

Extent and rate of soil degradation

As remote sensing, international cooperation and databasing improve, there are better chances for accurately assessing the extent and rate of soil degradation. Problems will still remain; some damage may be readily visible, but much is insidious, for example an erosion loss of $50\,t\,ha^{-1}\,y^{-1}$ amounts to only about 3 mm a year off a soil profile: difficult to recognize, but enough to affect agriculture in quite a short time if the soil is shallow. If one measures the natural rate of soil loss, and then checks it against actual, a mismatch indicates that there might be a chance of erosion control by altering landuse. It should be emphasized that even without change in landuse, soil erosion may vary if, for example, a region starts to get higher rainfall or there is a progressive acidification. The worst phase of erosion may have passed and now there may not be much change, or obvious indication that bursts of erosion took place in the past. If a 'steady-state' or established rate of deterioration has been reached; change of environment or land use will mean change in land degradation rates.

Some generalizations follow which serve to emphasize the scale and widespread nature of soil erosion. One recent global assessment concluded that between 1945 and 1990 about 12.2 million square kilometres of agricultural land suffered serious loss of productivity – equivalent to roughly half the area of Latin America (Harrison 1993: 117). Europe has one of the highest percentages of total area damaged; Africa is also high, followed by Asia and Latin America. Africa suffers a lot of erosion in seasonally dry regions as a consequence of arable farming or grazing livestock. Soils with less than 2 per cent organic matter content and soils with less than 5 per cent clay content are most vulnerable to erosion. Latin America seems to have more of a problem of decline of fertility and increasing soil acidity. A large part of South Asia has been badly affected by soil erosion: El-Swaify *et al.* (1985: 3) estimated that around 175 million ha of India's total area of 328 million ha were 'degraded'. The PRC is increasingly afflicted by soil erosion: Forestier (1989) estimated that roughly one-sixth of the total land was affected and in valleys and lowlands silt-laden rivers appear to be flooding more severely and frequently. Australia, in spite of a relatively low rural population and only a few hundred years of European settlement, has probably got about 38 per cent of its non-arid (i.e. usable) land significantly eroded (Chisholm and Dumsday 1987), and possibly as much as half of Australia's farmland needs remedial treatment (*The Times* 30/8/88: 5). It was estimated in 1987 that 44 per cent of the UK's arable land soil was at risk from erosion. It has been suggested that in the last 200 years the USA has probably lost at least one-third of its topsoil and in the 1990s about 44 per cent of the USA was losing soil faster than it was formed (Harrison 1992: 119). By the late 1970s about one-fifth of the world's cropland had been lost to soil erosion and about one-half of the world's cropland was losing soil at a debilitating rate (Brown 1977: 18; Brown *et al.* 1984: 5; World Commission on Environment and Development 1987: 25; Pimentel 1993). Soil degradation could well result in movements of 'ecorefugees' out of subSaharan and North Africa.

According to Moran (in Little and Horowitz 1987: 69–73) in many parts of the world the problem of soil degradation is not one of erosion or loss of organic matter, rather one of increase in acidity and decline in what is often already poor fertility. Moran suggested:

- 81 per cent of Latin American soils were acidic and nutrient poor;
- 56 per cent of Africa's soils were acidic and nutrient poor;
- 38 per cent of Asia's soils were acidic and nutrient poor;
- 33 per cent of humid tropical Asia has soils with moderately high base status;
- 12 per cent of tropical Africa has soils with moderately high base status;
- 7 per cent of humid tropical America has soils with moderately high base status;
- 61 per cent of tropical America's soils have aluminium toxicity problems;
- 53 per cent of tropical Africa's soils have aluminium toxicity problems;
- 41 per cent of tropical Asia's soils have aluminium toxicity problems;
- 90 per cent of Amazonia's soils are badly deficient in phosphorus.

What are permissible rates of soil erosion?

Natural rates of soil erosion vary greatly and are, in large part, a function of vegetation cover. Undisturbed, 'natural, semiarid regions possibly lose around 1.0 μm y^{-1}' (Kirkby and Morgan 1980: 3). In temperate UK in sites where vegetation is undisturbed (i.e. is 'natural'), erosion seems seldom to exceed 1.0 μm y^{-1} (Kirkby and Morgan 1980: 3).

To judge whether soil degradation is a threat, depth of soil may be compared with soil loss rate and rate of soil formation at a given site (e.g. a 1 mm y^{-1} loss from a soil 1 metre deep with no measurable soil renewal and, assuming crops need a minimum of 40 cm to root, would give roughly 600 years of production before the critical point is reached; if soil depth had been 45 cm then there would have been a mere 50 years' grace). According to Brown (1978: 22) over much of the Earth's surface, topsoil is seldom more than 30 cm thick, so soil loss is likely to be a problem in the medium term.

There is little agreement in the literature: levels of 0.1 to 0.2 mm y^{-1} (4 to 20 t ha^{-1} y^{-1}) loss are often described as 'acceptable' (Kirkby and Morgan 1980: 5). (As a rough guide in the following section, 25 mm y^{-1} loss is roughly equivalent to 400 t ha^{-1} y^{-1}.) In the USA 11 t ha^{-1} y^{-1} has been seen as 'acceptable' and the average is said to have been 0.5 t ha^{-1} y^{-1} in the 1980s (Harrison 1992: 119); 180 t km^{-2} y^{-1} is seen by some as a cause for concern. However, Harlin and Berardi (1987: 219) warn that with soil seldom replaced at more than 2.5 cm per 200 to 1000 years the 'acceptable' level should be nearer 1 t ha^{-1} y^{-1}. The ideal 'acceptable' rate would be one that matches or is less than rate of soil formation. In the UK the natural replacement rate is 0.1 t ha^{-1} y^{-1}; compare this with losses of 20 to 50 t ha^{-1} y^{-1} which are not uncommon in the UK (*New Scientist* 113{1547}: 41). Zachar (1982: 86) felt that loss of over 0.5 mm y^{-1} (5 m^3 ha^{-1} y^{-1}) was a cause for deep concern, and classified soil erosion rates into:

- *serious* 40 to 130 years to lose the topsoil;
- *severe* 10 to 40 years to lose the topsoil;
- *catastrophic* loss of topsoil in one or a few storms.

The serious/severe/catastrophic classification may be the most useful type of approach, for it gives planners and people a clear idea of time to disaster unless action is taken, rather than what might be to them a meaningless numerical value.

Recognition of practical tolerance-levels or T-factors is not easy. Natural and human activity-related soil losses must be monitored before much can be usefully done. What planners call 'acceptable' now may not be if sustainable development become a goal. One millimetre per year loss may have little affect in a human lifetime but over several generations it might (especially if the soil is thin).

How bad are the rates of loss? Bearing in mind the warnings just given, there are typical values for virtually all parts of the world that are way in excess of 'acceptable'. For example, Iowa (presently the first-ranking maize producer, second-ranking soya producer in the USA and an important fodder-growing state) commonly has 35 $t\,ha^{-1}\,y^{-1}$ losses and in some areas 450 $t\,ha^{-1}\,y^{-1}$ (Canter 1986: 17). Some parts of the PRC with loess soils apparently lose 4000 to 10 000 $t\,ha^{-1}\,y^{-1}$. In the 'Dust Bowl' areas of the US midwest, records show wind erosion rates of around 10 $mm\,y^{-1}$ at the height of the problem in the 1930s (Kirkby and Morgan 1980: 3). In Haiti, things were so bad in 1989 that one-sixth of the total population had reportedly left the 'island', which once had good soils and a rich forest cover (Heathcote 1983; Boyle and Ardill 1989: iii). So far the world's farmed areas have been increasing in spite of soil degradation, but the expansion is often into poor quality marginal land whilst good land is degrading. Ultimately areas suitable for expansion will be scarce.

The impacts of soil erosion

The costs of soil erosion may be divided into direct (on-site) and indirect (off-site). The main direct cost is decline in crop productivity. Indirect costs include siltation of reservoirs, canals, streams (with associated loss of water supplies, power generation capacity, flooding, increased dredging costs, etc), land-slide damage to roads, infrastructure, housing and landholdings. The direct costs are vast; the indirect costs may also be and can sometimes exceed on-site impacts (many large and smaller hydropower and irrigation projects have been ruined by soil erosion). Soil contributes considerably to the value of land; once soil degradation begins, land values generally fall; ultimately the result may be virtually worthless, unsaleable wasteland. Soil loss or deterioration does not have to progress to the point where cropping or grazing is impossible before it has effect. Slight or moderate degradation may be sufficient to restrict what can be grown (due to insufficient depth of soil or poor quality soil); it may also, by reducing available soil moisture, reduce the land's resistance to drought.

The costs of halting and repairing soil erosion are not easy to calculate but often are considerable. Chisholm and Dumsday (1987: 91–5) examined the problems of estimating costs of soil erosion of control. In crude terms, soil erosion in Australia probably caused a decline in wheat yields of between 6 and 52 per cent in the period 1973–4 to 1983–4, equivalent to between A$13 and A$155 per ha over ten years. In the late 1970s, India probably spent about US$6 million a year on fertilizers just to match losses of soil.

Causes of soil degradation

The causes of soil degradation are often complex, cultural, institutional, social, economic and environmental factors play varying parts (Blaikie 1985; Blaikie and Brookfield 1987). However, there are vulnerable environments:

1 where soils are easily degraded
2 where conditions hinder maintenance of a good cover of vegetation (it may be too dry, too cold, too steep or ill-drained)
3 where rainfall is intense.

Soil degradation can be caused by the following factors:

1 shifting cultivation (land rotation agriculture)
2 livestock herding activity
3 arable cultivation
4 burning
5 human access, including tourism activity
6 commercial causes
7 agrochemical use
8 land drainage.

Shifting cultivation (land rotation agriculture)

Shifting cultivation is widely blamed for removal of vegetation, burning and soil exhaustion. Given the huge range of diversity of strategy and variation in degree to which each strategy has been disrupted by various development pressures, assessing the effects of shifting cultivation is difficult. If traditional shifting cultivation has not broken down, it is probable that most eroded material would be held within the cleared plot boundaries by surrounding weeds, etc. There would be damage and losses from the use of fire, but not all shifting cultivators use fire to clear their plots. Breakdowns may be caused by population increase or cash-cropping leading to reduction of fallow periods. Another problem is that people may revert to clumsy shifting cultivation when disadvantaged or disrupted (often these are not shifting but 'shifted' cultivators).

Livestock herding activity

Livestock have two main effects on soil: first, grazing animals remove vegetation exposing soil to wind and rain, and second, animals trample the surface and either dislodge soil particles or compact the surface reducing infiltration; the result of these actions is to increase soil removal. The rates of damage and pattern of damage varies according to type of livestock and factors such as water availability and whether particular trails are regularly followed. Erosion is likely to be worst around boreholes and stockades where livestock are kept for the night.

Arable cultivation

Ploughing is often a cause of land degradation. The degree of damage depends on the ploughing technique and upon the timing and vulnerability of the soil. Studies have shown that chisel-ploughing causes less damage than mouldboard-ploughing or disc-ploughing (Reganold 1989).

Burning

Not only do shifting cultivators use fire, but also it is a widespread technique for pasture management and land clearance and in many regions natural and accidental bush fires are common. Fire destroys above-ground vegetation and litter and, if the fire is sufficiently hot-burning (900°C or above), root and soil damage may occur. Most bushfires subject the soil surface to less than 300°C, hotter fires are possible if biomass has accumulated. When above-ground vegetation and litter burn there are losses of nutrients, directly in the smoke carried aloft (nitrogen in particular is lost), and because the ash left on the surface is rich in potassium and is very easily blown away or dissolved by rainwater.

Human access, including tourism activity

Walkers and vehicles may cause considerable damage: this tends to be mainly along specific routes. Tracks may collect rainwater and feed it to where it causes gully erosion. Off-road vehicles are a problem in some countries, and can cause widespread damage. Trail-motorcycles, 'dune-buggies' and jeep-type vehicles can start severe deflation in drylands or dunefields. In game parks off-road vehicles may also cause problems, especially in localities where animals congregate, such as around waterholes or salt-licks. In colder environments skidoos and skiers can damage the vegetation cover protecting delicate tundra or alpine soils, which then erode during spring and summer. In very arid environments tracks left by vehicles years ago may still show little sign of recovery. Trampling by walkers is a particular problem in peatlands, upland areas and in peri-urban areas where there is heavy usage. Some tourist authorities have had to fence off vulnerable areas like sand-dunes and provide erosion resistant pathways.

Commercial causes

In many tropical lowlands logging activities have been a major cause of soil erosion. Large areas of the world suffer soil degradation as a consequence of commercial agricultural demands, notably cotton, sugar and tobacco and groundnut production.

Agrochemical use

There has not been enough time for all the effects of using agrochemicals (chemical fertilizers, pesticides, herbicides, fungicides, etc) to become apparent. There may be slow, cumulative changes in soil chemistry, structure and populations of soil micro-organisms or a threshold may be reached and a problem may suddenly appear (chemical time bomb effect). There has been a vast increase in agrochemical use in DCs and LDCs, for example in the last 100 years the UK has increased chemical fertilizer use about 25-fold (Briggs and Courtney 1985: 101).

Land drainage

If land is drained there may be subsidence, acid-sulphate soil formation, oxidation of soil organic matter (even spontaneous combustion of peat deposits) and a risk that agrochemicals or heavy metals contaminated drainage water might damage floodlands and wetlands.

Controlling soil degradation

The steps in controlling soil degradation are:

- to assess the problem;
- to identify an appropriate landuse for the afflicted area (there may be a need for rehabilitation measures);
- to implement such a landuse;
- to manage and maintain such a landuse.

Land capability assessment is increasingly used to identify appropriate land uses. In practice, there are a plethora of problems such as:

- how to fund corrective efforts;
- how to overcome lack of interest or resistance by local people;

- how to establish suitable institutional arrangements to initiate and manage soil degradation control;
- there may be a need to alter legislation to support certain groups of society so that they have an incentive or chance to improve landuse.

There are situations where soil degradation can be easily controlled, possibly by selecting appropriate crops, but there are also situations where, because of the complexity of the problems, and because the relationship between agricultural productivity and soil degradation is non-linear, it is likely that computer mathematical modelling will be needed to predict the likely course of erosion and the best control strategy.

A problem with control of soil degradation is that the costs may begin immediately and be quite considerable, but the benefits do not appear for some time and may, at least in cash-terms, be slow to accrue. Erosion losses and problems are likely to affect particular groups of people especially poor people; the benefits from exploiting the soil are likely to be enjoyed by richer, more influential individuals. It is important that those involved in control efforts actually benefit: often they do not.

Technical measures for countering soil erosion

There are four groups of practices which help counter erosion caused by flowing water:

1. those which help maintain or improve soil infiltration;
2. those which help safe (non-erosive) disposal of excess runoff, should precipitation exceed the infiltration capacity of the land;
3. those which help bind the soil together;
4. those which reduce the impact of precipitation on the soil.

Measures for countering erosion caused by flowing water may be grouped into those which:

1. *Slow overland flow and reduce raindrop splash-effect* by maintaining vegetation cover, a mulch layer, contour cultivation, bund construction, grass strips, and so on.
2. *Reduce soil surface temperature* this may be achieved through shading and mulching.
3. *Improve infiltration* through tillage to open up the soil and construction of structures to retain runoff moisture so that it can infiltrate.
4. *Improve the organic matter content of the soil (and add plant nutrients)* through adding organic matter (manure, green manure, compost) and/or soil amendment compounds.

There is overlap between measures to control water erosion and those to control wind erosion. For example, techniques designed to maintain plant cover, bind the soil and minimize disturbance of the soil, such as 'zero'-tillage; direct drilling of seeds; rod-weeding (a technique whereby a subsurface bar is dragged through the soil to break weed roots with relatively little damage to soil surface); mulch; soil conditioners; contour planting of rows of plants. Measures that reduce water erosion by improving or maintaining soil organic matter also help reduce moisture loss and wind erosion.

Methods of reducing erosion caused by flowing water

Excavations, stone-lines and bunds To control erosion by water a major goal must be the slowing of runoff flows over the soil surface. There is a huge diversity of widely used measures: terraces, along-contour stone-lines, grass strips, trash-lines, drains, banks, ditches, etc. (see Canter 1986: 134–9). Few methods are as easy to apply as some would believe. There is a need to estimate likely runoff before measures can be safely selected and designed, but often this is not done. The result is that water is channelled to a point where it causes damage where concentrated flows escape. There may be problems if the erosion control measures hinder tillage (terraced land may require as much as 6 per cent more time to plough), restrict cropping patterns or take a significant amount of farm or grazing land out of production. Somehow the measures have to be financed, labour has to be found to install and maintain them, and farmers have to be convinced that the effort and loss of land is worthwhile (Little and Horowitz 1987: 32).

Funding and constructing soil erosion control/and moisture conservation measures (the two are often inseparable) can be difficult. Construction of erosion control measures is not so much a problem of lack of tried and tested methods as one of difficulty in getting labour in remote areas and motivating people to maintain and renew structures. In Ethiopia one estimate suggested that it takes 150 man/days work to dig 1 km of typical field terraces; to build 1 km of main terrace could take 350 man/days; 1 km of road may take 2000 man/days (Cross 1983). Moisture conservation terraces can last 20 years without much maintenance once built, but their construction may take 742 man/days per hectare protected. Terracing involves effort, but this need be no discouragement if it means that yields increase, erosion falls and security of harvest increases.

Some terracing and other soil conservation/and moisture conservation measures may not be so attractive. It may, for example, be better for people to allow their land to degrade and become migrant labour than to try and boost yields and market more or better produce (the roads and the market prices and demand for produce may make intensification unattractive). Indeed, where there is employment for migrant labour, rural workers may, at least seasonally, be scarce. In such circumstances, getting people to construct and maintain soil or moisture conservation structures may be difficult; there is a chance in some situations that by providing terraces and so on, the farm profits and security are boosted, the need for migrant labour earnings may be reduced, and labour may return. In Peru a USAID study showed that farmers accepted financial aid to construct terraces they then 'sub-contracted', paying labourers less than the aid they had received, made a profit and thereafter neglected the terraces. It seems that in the region there were sufficient opportunities for off-farm wage-earning to make terrace maintenance and better farming (with more effort involved) unattractive. The value of soil conservation in the eyes of locals may also depend on transport, labour, market and other factors. Clearly, to generalize can be dangerous, each situation must be carefully assessed before too large and costly a project of 'improvement' is embarked upon.

Soil conditioners There are a range of compounds that can be used to improve the fertility or resistance of soil to water and wind erosion. Some simply seal the soil surface, binding particles together so as to reduce dislodgement and removal, for example oil- or rubber-based emulsions. Other treatments simply cover the surface with a protective layer that holds and delays runoff, improves infiltration and reduces the effect of raindrop impact; there are many materials used as mulches with this in mind. Both surface treatment with emulsion sprays and mulching help conserve moisture and are useful as means of establishing vegetation where conditions are difficult. It is possible to include seeds with some of these treatments so that seeding and mulching are done in a single pass and there is good germination and survival. Some treatments and mulches are inert; others such as sewage sludge, sugar cane bagasse, or other nutrient-rich waste improve soil fertility as well as simply giving protection. There are risks: sewage sludge is probably best used where trees are to be grown or landscaping is the goal, rather than on cropland, unless it is clear that the treatment has no risks of disease transmission or pollution of streams and groundwater with excess nutrients, pathogenic organisms, heavy metals, etc.

There are soil conditioners which can be very efficient at conserving moisture (some can hold 1000-times their own weight of water) and stabilizing soil. These may be applied as a surface film or incorporated as an additive in the topsoil (Greenland and Lal 1977: 101; Zachar 1982: 385–6). However, such conditioners at present probably cost too much for general use in LDCs.

Green manures, organic fertilizers and chemical fertilizers There has been much interest in developing strategies like alley cropping, where shelter-belt shrubs or trees provide mulch or compost. There are plants which are particularly efficient at fixing nitrogen that might be mixed with or grown in rotation with crops or pasture, or which could be grown as hedges or farm woodland plots (and provide leaves and twigs for mulch).

There has been debate in horticultural circles as to the actual value of green manure for soil texture and nutrient content improvement. In coastal areas seaweed can be a valuable organic fertilizer which could be more widely adopted. Away from coasts aquatic weeds might prove to be a resource if collected and applied to cropland as a compost. The advantage of these organic treatments over chemical (artificial) fertilizers are:

- there is less need for precision in rate of application;
- there is much less handling risk;
- the release of nutrients is likely to be more gradual, so cutting pollution risks;
- timing of application is less critical;
- they add organic carbon to the soil.

There is debate, but it seems likely that, in the long run, maintenance of soil structure and fertility will require additions of organic material and not just artificial fertilizers. This conviction (added to worries about the damaging effects of agrochemical use on the land and the consumer) has led to increasing adoption of organic farming approaches, so far mainly in richer nations.

Groundcover plants The ideal groundcover plant provides protection from direct impact of raindrops, shed intercepted moisture in drops that are less damaging than raindrops, and has a dense mat of shallow roots which bind the soil. Things are more complicated if the groundcover is to be used beneath a tree or shrub crop (where light levels may be low) and if there are to be annual crops grown nearby (ideally the groundcover would fix nitrogen and improve the soil, rather than compete with crops for moisture and nutrients). The choice of appropriate groundcover plants, thus depends on a range of factors, including character of precipitation, the landuse, the slope, etc.

Methods of reducing erosion caused by wind

Soils with a particle size range of 0.002 to 0.100 mm are most prone to wind erosion and transport. The following gives only a brief overview of the many technical measures available for erosion control (for further information see Weber and Stoney 1989).

Shelter-belts and wind-breaks Wind-break is generally applied to non-living structures, such as fences of brushwood and wickerwork, and occasionally to small-scale plantings of, say, aloe. Shelter-belts are larger-scale plantings which give protection for at least 20 times their height downwind. Care is needed in siting wind-breaks or shelter-belts to ensure that they are at 90° to the most damaging winds (not necessarily the prevailing wind). Cross-sectional shape and the permeability of the barrier to airflow are also important; a shelter-belt or wind-break should not be too wind-proof, but should 'leak' a little or there will be damaging eddies in its lee (about 40 per cent porosity is generally considered desirable). Spacing and height also require attention (Zachar 1982: 380). Shelter-belts are not instant solutions; they take time to establish. Care must be taken to ensure that local people understand the need for and support the establishment of shelter-belts or erection of wind-breaks. If they do not, damage is likely. It is also important that the shrubs or trees used do not compete with nearby (generally shallower-rooted) crops or pasture for moisture.

Some trees or shrubs are particularly suitable for shelter-belts, for not only do they slow the wind, but also they supply fuelwood, fodder, compost, tan-bark or mulch that can be used for soil improvement. Shelter-belts have been used to conserve soil moisture in dryland regions to boost crop (possibly by as much as 30 per cent) or forage yields and help counter wind erosion (Barrow 1987: 157; Weber and Stoney 1989: 137).

Stabilizing moving sand or soil There are a great many ways (other than shelter-belts) of holding soil in place or catching that which has begun to move. Crops may be planted in a suitable pattern, usually a grid-pattern (*coulisses*). Sturdy, soil- or sand-trapping grasses or herbs may be planted to stabilize areas of soil erosion, for example marram grass (*Ammophila* spp.) has been found effective in Europe for sand-dune protection (Lenihan and Fletcher 1976). Simply laying a 'thatch' of tree branches may be helpful. It may also be worth spraying soil or sand with compounds to stabilize it; many have been tried, for example latex emulsion, oil, waste paper or fibres and water (with or without seeds incorporated in the mulch), shredded bark, to name but a few.

Reducing damage caused by agricultural activity

Mechanized cultivation can cause serious damage to soil, and there is a possibility that inappropriate tillage can also waste soil moisture (Barrow 1987: 156). Zero-tillage or minimal-tillage have been advocated to reduce degradation during arable cultivation. Zero-tillage may demand the use of herbicide or burning to control weed growth, although this can be avoided if special ploughs are used (a wide horizontal blade is drawn under the surface, cutting weed roots and disturbing the soil only around the supporting vertical blade). A promising development is the 'gantry': this consists of two sets of powered wheels several metres apart on a support from which tillage, spraying, seeding and harvesting, etc, can be carried out. The wheels follow narrow fixed-paths across the field so that there is little of the crop damage or soil compaction associated with conventional farm machinery. More efficient application of seed and agrochemicals should help to cut the cost of farming inputs, and more importantly, reduce the amounts of potentially polluting fertilizers, etc.

Soil degradation might be controlled by managing land use, e.g. set-aside, control of fertilizer use, recognition of phosphate sensitive areas, etc (see chapter 8). The USA started a Conservation Reserve Programme in 1985 to try and reduce soil erosion, apparently with promising results, although set-aside in the USA has received some criticism.

Encouraging, supporting and enforcing soil degradation control

Separation of administrative, extension and education efforts from consideration of soil protection techniques is not really practicable. Baker (1984) highlighted what he saw as a 'conventional' perception of land degradation which led to 'conventional' responses. Erosion is not just due to misuse of vulnerable physical environments. Erosion is often because vulnerable environment coincides with marginalized, powerless people who are forced to degrade land. Effective soil degradation control can utilize some of the following measures:

1 Educate people in the value of soil conservation.
2 Offer grants and loans to encourage people to practise soil conservation.
3 Tax or fine those who cause soil degradation, if it reaches a certain threshold.
4 Manipulate the market, hold down or drive up prices so as to aid or hinder sales to discourage cultivation of crops which lead to excessive soil damage and encourage growing of those that do less damage.
5 Discourage or prohibit practices or timing of practices that cause soil damage (e.g. discourage tillage of soil that is too wet or too dry, ban stubble burning, try to find alternatives to dung collection for fuel).
6 Reward those who manage to cut soil damage or improve soil quality.

Some of these options (e.g. 3 and 6) are potentially inequitable, if say a farmer takes over a farm with a poor soil degradation record, or a farm with especially difficult and vulnerable conditions.

Historically governments attend to land degradation only when public opinion forces them to, usually as a result of a perceived 'catastrophe'. The management of land development had usually been response-to-crisis management (Chisholm and Dumsday 1987: 47). There is a trend nowadays toward prevention of soil degradation, but it is not very well established, especially in LDCs where money is scarce. Another problem is that management often has to be according to the best available understanding (often only partially complete) of how the ecosystem and society functions.

Access to expertise and funds may be no guarantee of better land management. Chisholm and Dumsday (1987: 47) offer an explanation as to why Australia (which is relatively rich, has access to agricultural and soil conservation expertise, has a relatively light population and a short history of land development) has such a poor soil degradation record. It is, they suggest, because no conservative land ethic evolved in the short exploitative period since European settlement began; historically stewardship has been of little concern.

There are examples of governments offering tax incentives and grants to encourage soil conservation. Australian income tax offers a 100 per cent write-off of expenditure on structural works for soil conservation (Chisholm and Dumsday 1987: 239). Many irrigation management authorities regulate water supplies to growers to help avoid over-irrigation and the likely consequences of water-logging and salinization. Such approaches are likely to be much cheaper and more effective than remedial soil drainage or chemical treatment measures.

What is worrying is that, even in DCs, soil conservation appears to have a low priority. In spite of decades of work by bodies like the US Soil Conservation Service, US farmers were not managing soil as well in 1978 as they did 50 years earlier according to Brown (1978: 24). Hudson (1987a: 1987b) has discussed soil conservation strategies, and suggested that soil conservation might be more easily achieved if it were more closely integrated with agricultural extension. He advocated that soil conservation be renamed 'land husbandry' and that the emphasis be placed on:

- prevention of erosion rather than cure;
- stressing the loss of production (to keep farmers' attention);
- highlighting the potential of soil conservation for improving and sustaining production;
- promoting the use of local materials and practices;
- involving the local community at all stages;
- making programmes more attractive, so that external incentives and subsidies are less necessary.

References

Agarwal, B. (1986) *Cold hearths and barren slopes: the woodfuel crisis in the third world.* London, Zed

Ahmad, Y.J. and Kassas, M. (1987) *Desertification: financial support for the biosphere.* West Hartford, Conn., Kumarian; London, Hodder and Stoughton

Allan, R.J. (1988) El Niño Southern Oscillation influences in the Australasian region. *Progress in Physical Geography* 12(3): 313–48

Anon. (1987) Sea temperatures predict African drought. *New Scientist* 116(1580): 25
Arnon, I. (1981) *Modernizing agriculture in developing countries: resources, potentials and problems*. Chichester, Wiley
Aubréville, A.M. (1949) *Climats, forêts et desertification de L'Afrique tropicale*. Paris, Societé d'Edition Géographiques Maritimes et Coloniales
Baker, R. (1984) Protecting the environment against the poor: the historical roots of the soil erosion orthodoxy in the third world. *The Ecologist* 14(2): 53–60
Barrow, C.J. (1987) *Water resources and agricultural development in the tropics*. Harlow, Longman
Barrow, C.J. (1991) *Land degradation: development and breakdown of terrestrial environments*. Cambridge, Cambridge University Press
Beaumont, P. (1989) *Environmental management and development in drylands*. London, Routledge
Bell, R.H. (1987) Conservation with a human face: conflict and reconciliation in African land use planning. In D. Anderson and R. Grove (eds) *Conservation in Africa: people, policies and practice*. Cambridge, Cambridge University Press
Blaikie, P. (1985) *The political economy of soil erosion*. Harlow, Longman
Blaikie, P. and Brookfield, H. (1987) *Land degradation and society*. London, Methuen
Boels, D., Davies, D.B. and Johnston, A.E. (1982) *Soil degradation*. Rotterdam, A.A. Balkema
Bonnifield, P. (1979) *The Dust Bowl: men and dirt and the depression*. Albuquerque, N. Mex., University of New Mexico Press
Boyle, S. and Ardill, J. (1989) *The greenhouse effect: a practical guide to the world's changing climate*. London, Hodder and Stoughton
Briggs, D.J. and Courtney, F.M. (1985) *Agriculture and environment: the physical geography of temperate agriculture*. London, Longman
Brown, L.R. (1977) *Redefining national security* (Worldwatch Paper no. 24). Washington, DC, Worldwatch Institute
Brown, L.R. (1978) *The worldwide loss of cropland* (Worldwatch Paper no. 24). Washington, DC, Worldwatch Institute
Brown, L.R., Chandler, W., Florin, C., Postel, S., Starke, L. and Wolffe, E. (1984) *State of the world: a Worldwatch Institute report on progress toward a sustainable society*. New York, W.W. Norton
Bryson, R.A. (1967) Possibilities of major climatic modification and their implications in Northwest India. *Bulletin of the American Meteorological Society* 48: 136–42
Bryson, R.A. and Murray, T.J. (1977) *Climate of hunger: mankind and the world's changing weather*. Madison, Wis., University of Wisconsin Press
Canter, L.W. (1986) *Environmental impacts of agricultural production activities*. Chelsea, Mich., Lewis
Charney, J., Stone, P.H. and Quirk, W.J. (1975) Drought in the Sahara: a biophysical feedback mechanism. *Science* 187(4175): 434–5
Chisholm, A. and Dumsday, R. (eds) (1987) *Land degradation: problems and policies*. London, Cambridge University Press
Clarke, W.C. and Mann, R.E. (eds) (1987) *Sustainable development of the biosphere*. Cambridge, Cambridge University Press
Cross, M. (1983) Last chance to save Africa's topsoil. *New Scientist* 199(1368): 288–93
De Vos, A, (1975) *Africa, the devastated continent: man's impact on the ecology of Africa*. The Hague, Dr W. Junk
Dixon, J.A., James, D.G. and Sherman, P.B. (1989) *The economics of dryland management*. London, Earthscan
Dregne, H.E. (1983) *Desertification of arid lands*. Chur, Switzerland, Harwood Academic
Dunbar, R. (1984) Scapegoat for a thousand deserts. *New Scientist* 104(1430): 30
Eckholm, E.P. (1976) *Losing ground: environmental stress and world food prospects*. Oxford, Pergamon
Eckholm, E.P. and Brown, L.R. (1977) *Spreading deserts: the hand of man* (Worldwatch Paper no. 13). Washington, DC, Worldwatch Institute
El-Swaify, S.A., Moldenhauer, W.C. and Los, A. (eds) (1985) *Soil erosion and conservation*. Akeny, Soil Conservation Society of America
Fantechi, R. and Margaris, N.S. (eds) (1986) *Desertification in Europe*. Dordrecht, D. Reidel
Forestier, K. (1989) *The degreening of China*. *New Scientist* 123(1671): 52–8
Forse, B. (1989) The myth of the marching desert. *New Scientist* 121(1650): 31–2
Franke, R.W. and Chasin, B.H. (1980) *Seeds of famine: ecological destruction and the development dilemma in the West African Sahel*. Montclair, NJ., Allanheld, Osmun
Gee, H. (1988) Alarm bells in the forest fire front. *The Times* 22/7/88: 11
Glantz, M.H. and Katz, R.W. (1985) Drought as a constraint in sub-Saharan Africa. *Ambio* XIV(6): 334–9
Goudie, A. (1973) *Duricrusts in tropical and subtropical landscapes*. Oxford, Clarendon
Goudie, A.S. (ed.) (1990) *Techniques for desert reclamation*. Chichester, Wiley
Grainger, A. (1982) *Desertification: how people make deserts, how people can stop and why they don't*. London, Earthscan
Greenland, D.J. and Lal, R. (eds) (1977) *Soil conservation and management in the humid tropics*. Chichester, Wiley
Harlin, J.M. and Berardi, G.M. (eds) (1987) *Agricultural soil loss: processes, policies and prospects*. Boulder, Colo., Westview
Harrison, P. (1987) *The greening of Africa: breaking through in the battle for land and food*. London, Paladin Grafton
Harrison, P. (1992) *The third revolution: population, environment and a sustainable world*. Harmondsworth, Penguin
Heathcote, R.L. (1980) *Perception of desertification*. Tokyo, UN University
Heathcote, R.L. (1983) *The arid lands: their use and abuse*. London, Longman
Helldén, U. (1991) Desertification – time for an assessment? *Ambio* XX(8): 372–81
Higgins, G.M. (1988) *Soil degradation and its control in Africa* (Paper to the First All-African Soil Science Conference, Kampala, Uganda, 5–10 December 1988). Kampala, University of Kampala (mimeo)

Hudson, N. (1981) *Soil conservation.* London, Batsford
Hudson, N. (1987a) Limiting soil degradation caused by soil erosion. In M.G. Wolman and F.G. Fournier (eds) *Land transformation in agriculture.* Chichester, Wiley
Hudson, N. (1987b) The art of conservation. *SPLASH: Newsletter for SADCC Soil and Water Conservation and Land Utilization Programme (Lesotho)* 3(3): 22–3
IUCN (1986) *The IUCN Sahel report: a long-term strategy for environmental rehabilitation.* Gland, Switzerland, International Union for the Conservation of Nature and Natural Resources
Jackson, W., Berry, W. and Coleman, B. (eds) (1984) *Meeting the expectations of the land: essays in sustainable agriculture and stewardship.* San Francisco, Calif., North Point
Kassas, M. (1987) Drought and desertification. *Land Use Policy* 4(4): 389–400
Kirkby, M.J. and Morgan, R.P.C. (eds) (1980) *Soil erosion.* Chichester, Wiley
Lal, R., Hall, G.F. and Miller, F.P. (1989) Soil degradation: 1 basic processes. *Land Degradation & Rehabilitation* 1(1): 51–69
Land, T. (1987) El Niño holds the key to long range forecasting. *The Times* 26/11/87: 14
Lenihan, J. and Fletcher, W.W. (eds) (1976) *Reclamation* (environment and man – vol 9: the biological environment). Glasgow, Blackie
Lewis, L.A. and Berry, L. (1988) *African environments and resources.* London, Allen and Unwin
Little, P.D. and Horowitz M.M. (eds) (1987) *Lands at risk in the third world: Local perspectives.* Boulder, Colo., Westview
Lockwood, J.G. (1984) The Southern Oscillation and El Niño. *Progress in Physical Geography* 8(2): 102–10
Lockwood, J.G. (1986) The causes of drought with particular reference to the Sahel. *Progress in Physical Geography* 10(2): 25
Mabbutt, J.A. and Wilson, A.W. (eds) (1980) *Social and environmental aspects of desertification.* Tokyo, United Nations University
Mainguet, M. (1991) *Desertification: natural background and human mismanagement.* Berlin, Springer-Verlag
Middleton, N. and Thomas D.S.G. (1992) *World atlas of desertification* (for the UNEP). London, Edward Arnold
Moran, E.F. (1987) Monitoring fertility degradation of agricultural lands in the lowland tropics. In P.D. Little and M.M. Horowitz (eds) *Lands at risk in the third world: local-level perspectives.* Boulder, Colo., Westview
Mortimore, M.J. (1988) Desertification and reference in semi-arid West Africa. *Geography* 73(1): 61–4
Mortimore, M.J. (1989) *Adapting to drought: farmers, famines and desertification in West Africa.* Cambridge, Cambridge University Press
Nye, P.H. and Greenland, D.J. (1960) *Soil under shifting cultivation.* Harpenden, Commonwealth Agricultural Bureau
OECD (1977) *Strategy and programme for drought control and development in the Sahel.* Paris, Organization for Economic Cooperation and Development
Otterman, L. (1974) Baring high-albedo soils by overgrazing: a hypothesised desertification mechanism. *Science* 186(4163), 531–53
Pearce, F. (1992) Mirage of the shifting sands. *New Scientist* 136(1851): 38–42
Pimental, D. (ed.) (1993) *World soil erosion and conservation.* Cambridge, Cambridge University Press
Prior, J. and Tuohy, J. (1987) Fuel for Africa's fires. *New Scientist* 115(1571): 48–51
Reganold, J. (1989) Farming's organic future. *New Scientist* 122(1668): 49–52
Reining, P. (ed.) (1978) *Handbook on desertification indicators.* Washington, DC, American Association for the Advancement of Science
Ruddle, K. and Manshard, W. (1981) *Renewable natural resources and the environment: pressing problems in the developing world.* Dublin, Tycooly
Russel, N. (1988) Dirt gets in your eyes: comparisons between the American drought and the Dust Bowl of the 1930s are misleading. *New Scientist* 119(1625): 61
Schove, D.J. (1977) African droughts and the spectrum of time. In D. Dalby, R.J. Harrison Church and F. Bezzaz (eds) (1977) *Drought in Africa 2* (revised edn). London, International African Institute.
Scoging, H. (1993) The assessment of desertification. *Geography* 78(2): 190–3
Skujins, J. (ed.) (1991) *Semiarid lands and deserts: soil resources and reclamation.* New York, Marcel Dekker
Spooner, B. and Mann, H.S. (eds) (1982) *Desertification and development: dryland ecology in social perspective.* London, Academic Press
Thomas, D.S.G. (1993) Sandstorm in a teacup? Understanding desertification. *Geographical Journal* 159(3): 318–31
Thomas, D.S.G. and Middleton, N.J. (1994) *Desertification: exploding the myth.* Chichester, Wiley
Thornes, J. (1989) Solutions to soil erosion. *New Scientist* 122(1667): 45–9
Tiffen, M., Mortimore, M. and Gichuki, F. (1993) *More people, less erosion: environmental recovery in Kenya.* Chichester, Wiley
UN (1977a) *Desertification: its causes and consequences.* Oxford, Pergamon
UN (1977b) *World map of desertification.* New York, United Nations
UNEP (1987) *Environmental data report 1987.* Oxford, Basil Blackwell
UNEP and Commonwealth of Australia (1987) *Drylands dilemma: a solution to the problem* Canberra, Australian Government Printer
Weber, F.R. and Stoney, C. (1989) *Reforestation in arid lands.* Arlington, Va, Volunteers in Technical Assistance (1815 North Lynn Street, Arlington, Virginia 2209)
Whitney, J.B. (1987) Impact of fuelwood use on environmental degradation in the Sudan. In P.D. Little and M.M. Horowitz (eds) *Lands at risk in the third world: local perspectives.* Boulder, Colo., Westview
Wischmeier, W.H. (1976) Use and misuse of the universal soil loss equation. *Journal of Soil and Water Conservation* 31: 5–9
Wischmeier, W.H. and Smith, D.D. (1960) A universal soil loss equation to guide conservation farm planning. *Transaction of 7th International Congress of Soil Science* 1: 418–25
WMO (1986) *WMO report on drought and countries affected by drought.* Geneva, World Meteorological Organization

Wolman, M.G. and Fournier, F.G. (eds) (1987) *Land transformation in agriculture*. Chichester, Wiley

World Bank (1985) *Desertification in the Sahelian and Sudanian zones of West Africa*. Washington, DC, World Bank

World Commission on Environment and Development (1987) *Our common future: report of the WCED* (Brundtland Report). Oxford, Oxford University Press

Worster, D. (1979) *Dust Bowl: the southern prairies in the 1930s*. New York, Oxford University Press

Zachar, D. (1982) *Soil erosion* (Developments in Soil Science no. 10). Amsterdam, Elsevier Scientific

Further reading

Beaumont, P. (1989) *Environmental management and development in drylands*. London, Routledge. [Good introduction]

Grainger, A. (1990) *The threatening desert: controlling desertification*. London, Earthscan [Comprehensive introduction to nature, causes and scale, which also identifies key strategies for control]

Hulme, M. and Kelly, M. (1993) Exploring the links between desertification and climatic change. *Environment* 35(6): 4–11, 39–45 [Useful short overview]

Kassas, M. (1987) Drought and desertification. *Land Use Policy* 4(4): 389–400 [Overview]

Mainguet, M. (1991) *Desertification: natural background and human mismanagement*. Berlin, Springer-Verlag [Comprehensive overview]

CHAPTER 7 The degradation and loss of vegetation

Difficulties in recognizing and assessing vegetational change

There is a huge diversity of vegetation: sometimes floras are distinct and the boundaries are abrupt enough for easy mapping, sometimes the transition is gradual, perhaps imperceptible. Plant communities are dynamic systems, subject to a complex range of influences. Apparently similar formations in different localities may, on closer study, prove not to be, or are subject to dissimilar threats. It may be difficult to arrive at anything but the broadest of categories and generalized statistics.

Assessing whether there is degradation may not be easy because some damage may be necessary for a flora's long-term survival. Without periodic storm damage, occasional grazing, periodic fires or other disturbances, vegetation formations might become senile, cease to regenerate, and then give way to something different.

The terminology used to describe vegetation and what constitutes its degradation leaves a lot to be desired and this has often meant that attempts to compare successive measurements may in reality consider data that is different even if nothing has changed, i.e. observers are not using the same boundaries or categories. Structural similarities and parallel evolution may cause casual observers to overlook the considerable floristic and faunal differences between and within the grasslands, woodlands, forests, etc., of, say South and Central America, Africa and South East Asia and Australasia and, even within a limited area, vegetation may vary a lot because of differences in soil, drainage, aspect, altitude, history of disturbance, etc.

Vegetation terminology

Tropical rain forest (rainforest or tropical moist forest)

Once widespread in the humid tropics below roughly 1000 metres altitude where soils, terrain, drainage and rainfall were favourable (Parsons 1975; Caulfield 1982). Often there is tremendous diversity of tree and other species, indeed these can be some of the world's most complex ecosystems. To apply one label is generalization, but with caution it is useful to recognize predominantly evergreen, mainly broad-leaved forest with a canopy covering at least 20 per cent of the ground as closed tropical rain forests (see Figs 7.1 and 7.2).

Closed forest or woodland

That which covers more than 20 per cent of land with tree crowns (i.e. the canopy). Some forest in seasonally dry areas is closed.

Open forest or woodland

Implies the canopy covers 5–20 per cent of the ground. In general forests become more open with drier environment and disturbance.

Savanna

Generic term applied to land where tree cover is open enough for the herb and grass ground cover to be better developed. The reason for such conditions is debated.

Primary or virgin forest or woodland

Forest which is more or less unmodified by human activity, but may still include areas disturbed by limited numbers of indigenous hunter–gatherer people.

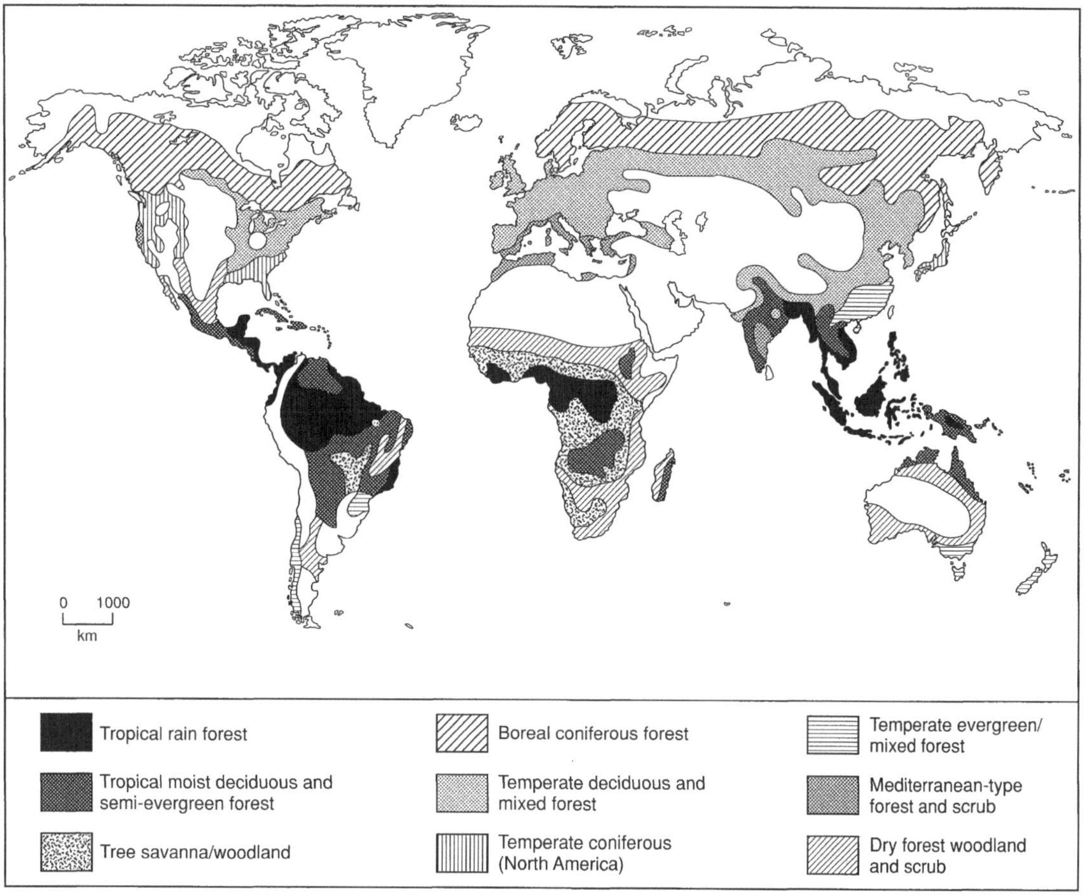

Fig. *7.1* World forests, woodland and scrub (actual extent has been reduced in many countries) ⟨*Source:* author from various sources⟩.

Secondary (regrowth or disturbed forest)

That which is subject to limited shifting cultivation or was once cultivated and has been abandoned, or in some other way disturbed. These areas have a shrub or tree regrowth, perhaps including some or all of the original species, often with opportunist species that have spread in from elsewhere, and whatever there is will be immature (Poore *et al.* 1989: 4; Mather 1990: 58).

Deforestation

Clearance of forest and conversion to some other use (some prefer to use the term 'conversion'). May be said to have been reached when woody vegetation has given way to mainly non-woody (but before that there could have been marked reduction of species diversity or other degradation). Rain forests generally comprise a 'mosaic' of ageing and younger trees, with some disturbance necessary to ensure a steady state. It is vital to recognize when disturbance is excessive and when it is not. Making an assessment is difficult when the land is rugged or remote, when tree canopies are high and the plant cover is dense or it is an aquatic or semi-aquatic environment. It may also be necessary to monitor some formations for decades to determine whether long-lived species are regenerating.

Conversion

Clearance and conversion to some other landuse.

Biodiversity

The variety and variability of all organisms on Earth, which may be viewed at three levels:

1 habitat diversity
2 species diversity
3 genetic diversity variability within species.

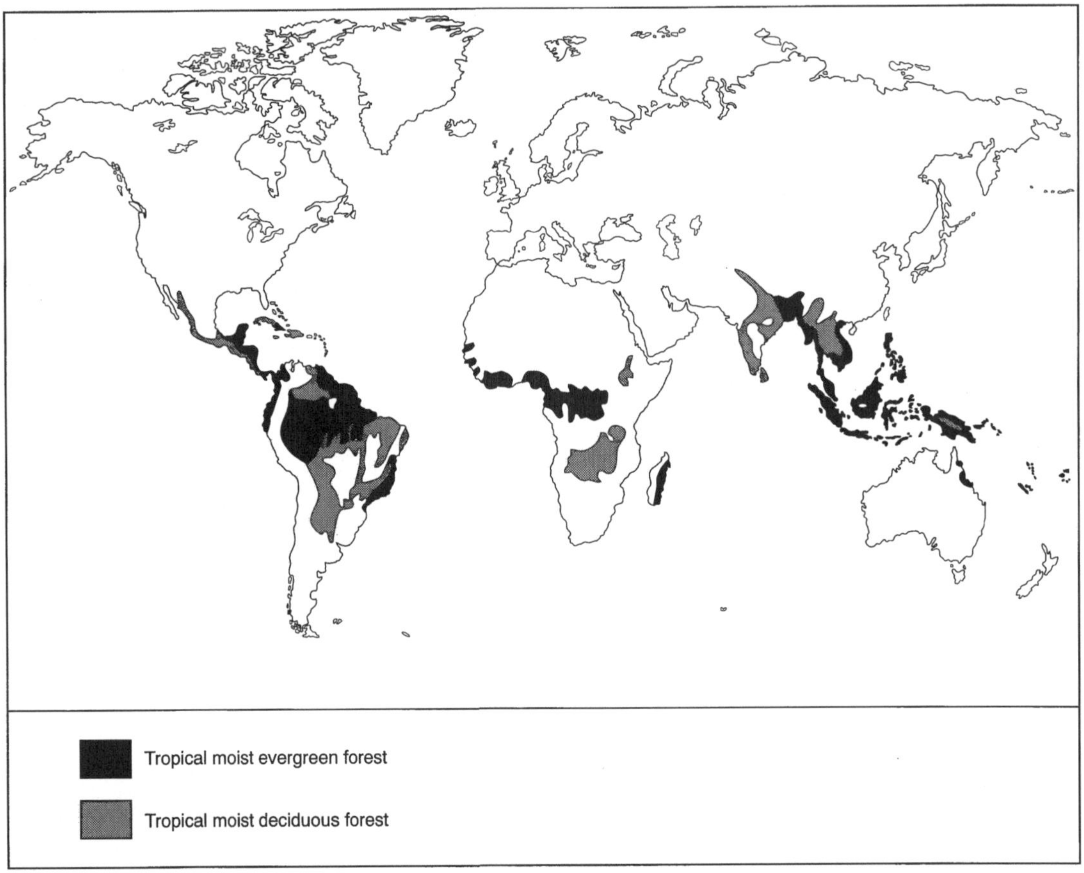

Fig. 7.2 Tropical moist evergreen (rain forest) and tropical moist deciduous forest (*Source:* based on Wyatt-Smith 1987: 20 (Fig. 1); Grainger 1993b: 31 (Fig. 1.1)).

Endemic species

An organism confined to a restricted distribution.

Terms like 'deforestation' must be treated with caution. Vegetation can degenerate through one or more of many possible causes before it is obvious, nor is it easy to distinguish between damage that will 'heal' and that which will not (Myers 1980: 7). Apparently unspoilt forest may have suffered reduction of plant or animal biodiversity; decrease in plant regeneration; thinning of cover; soil erosion; reduced plant vigour, etc. Considerable damage may take place before a grassland develops bare patches or a tree canopy becomes open or obviously changed (consequently successive surveys, often by different experts, may not accurately establish extent and rate of degradation. Remote sensing is improving assessment but ground verification tends to be inadequate and in some parts of the world where the flora and fauna are only partly known and skilled ecologists are in short supply monitoring will present problems for some time to come. An interesting reconnaissance-level inventory of wilderness (all natural land: forests, moorland, swamps, etc) remaining in the world was presented by McCloskey and Spalding (1989), together with an indication of what was *in theory* protected.

The scale and rate of forest and woodland degradation and loss

The world's forests, woodlands and savannas are potentially renewable resources provided they are not damaged beyond a certain point; unfortunately such damage has been commonplace. Human activity that destroys, damages or renders vegetation more

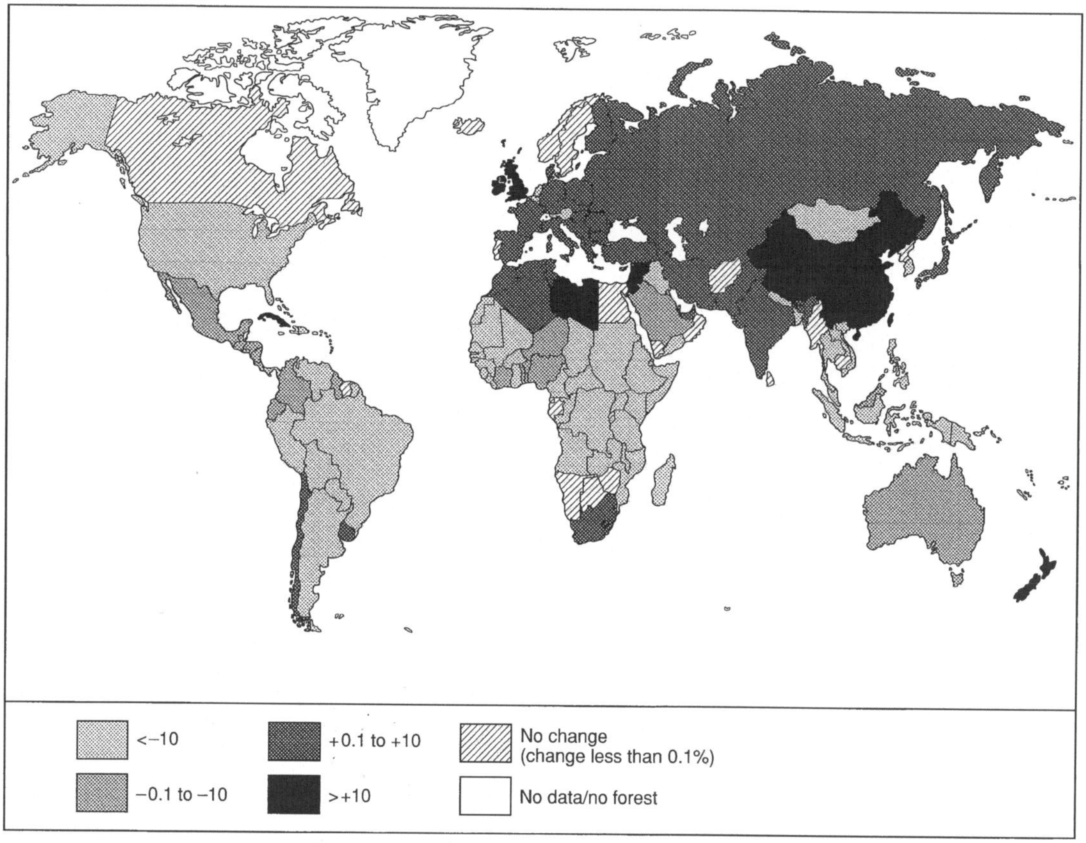

Fig. *7.3* Percentage change in area of forest and woodland 1975–85 ⟨*Source:* based on Mather 1990: 70 (Fig. 4.4)⟩.

vulnerable to various threats is nothing new. Clearance was under way in Papua New Guinean rain forests and upland forests by 10 000 BP (perhaps much earlier). Humans had decimated forests and woodlands around the Mediterranean by 3000 BP. Huge numbers of plant and animal species have become extinct or now have much restricted ranges thanks to humankind. Today the extent of damage and the rate at which such genetic erosion takes place threatens the extinction of an unacceptably high proportion of the world's plants and animals, some of which may well be needed to ensure humanity's long-term well-being (Myers 1988; Flint 1991). There are regions where it is too late, and others where there is little time left to take remedial action to conserve, protect or replant. There is no second chance: extinction 'is forever'.

Mather (1990: 58) noted that since 1970 estimates of global forest area have varied from less than 3000 to over 6000 million square kilometres. Mather (1990: 11) estimated that world-wide about 80 per cent of pre-agriculture forest survives, but its state of preservation varies greatly from locality to locality. Repetto and Gillis (1988: 2) suggested that humankind has probably destroyed (or 'converted') 32–35 per cent of Earth's temperate closed forests; 24–25 per cent of subtropical woody savanna and subtropical deciduous forests have been lost; and 4–6 per cent of moist tropical forests have been destroyed. There is still a lot of uncertainty over rates of tropical forest loss: Whitmore and Sayer (1992: 2–14, 58) attempted an objective assessment of rates of forest loss and of species extinction and concluded that many estimates have been too conservative. Those researching deforestation may sometimes make false assumptions about cause which does not help forecasting (Dove 1993). Grainger (1993a; 1993b: 124) reviewed estimates of tropical forest

Fig. *7.4* Deforestation and conversion. a) Former forest and savanna converted to cattle pasture, Roraima, Amazonian Brazil. Fire spreads into disturbed areas where there is dead, dry biomass (fire visible in background). *Note:* of the nine states in Brazilian Amazonia, four had over 98 per cent forest cover in 1992, three had over 80 per cent and two had over 40 per cent. The Brazilian Atlantic forests serve as a warning against complacency, as only about 12 per cent of original cover is left (Whitmore and Sayer 1992: 121). In Brazilian Amazonia smallholders have cleared forest in areas accessible by roads; since 1987 large landowners have had less tax advantages for clearance, and it is probable that the rate of loss peaked in 1987.

deforestation extent and rates, and provided a check list of criteria with which reliability of assessments may be tested (1993a: 35). There have been attempts to get an overall view of where the problem of forest degradation is worst; one recent discussion of 'hot-spots' and fronts of serious deforestation was provided by Myers (1993).

It is likely that forests and woods still cover roughly two-fifths of the world's land surface: about 66 per cent of that is in the tropics and of that roughly 43 per cent is closed tropical forest, with about 90 per cent of that being tropical rain forest (World Resources Institute *et al.* 1988: 70). That would mean that over roughly the last 7000 years humans have destroyed around one-fifth of the original tree cover (Repetto and Gillis 1988). Mather (1990: 59–67) estimated that in the mid-1980s forest and woodland covered roughly 31 per cent of the world's land surface, and that about 20 per cent of that was closed forest. Between 1975 and 1985, roughly 2 per cent of the world's total forest and woodland were destroyed (Fig. 7.3); at the local scale rates of loss vary a great deal (J.F. Richards and Tucker 1988; Mather 1990). If one relates global forest area to human population, then there has been a marked per capita decline in recent decades. Just over half of remaining forest and woodland (and the bulk of 'hardwoods') is in the LDCs.

Since 1945 LDCs with tropical forest appear to have lost it or have experienced accelerating forest degradation, for example between 1958 and 1968 Central America may have lost 38 per cent of its cover, and Africa may have lost 24 per cent of its cover. By the mid-1980s roughly 25 per cent of remaining rain forest was in Asia (mainly South East Asia), roughly 57 per cent was in the Americas (roughly 31 per cent in Brazil, which has the most extensive undisturbed blocks) and roughly 18 per cent was in Africa (Gradwohl and Greenberg 1988: 34).

Repetto and Gillis (1988: 6–9) provided a country-by-country list of deforestation between 1981 and 1985; Grainger (1933a: 37) reviewed estimates of rates and extent of deforestation; Collins *et al.* (1991) gave data on Asia and the Pacific; Sayer *et al.* (1992) presented information on tropical Africa; and Whitmore and Sayer (1992) outlined the situation for Brazil's Amazon and Atlantic forests.

Once well endowed with rain forest, Nigeria was 90 per cent deforested by 1992. By 1992 West Africa as a whole had lost 75 per cent of its treecover, and most of Africa's remaining rain forest was in Zaïre, Gabon and the Cameroon. Cameroon's rain forests probably have the highest species diversity in Africa, but in 1993, in the face of increasing logging and settler clearing, only about 4–5 per cent was protected or scheduled to be and funding to police and manage those areas was scarce (Alpert 1993). Grainger's (1993a) best estimate of the world total remaining tropical moist forest was 1081 million ha. Estimates of the (world) rate of loss of tropical rain forest vary between 2 per cent per annum (Myers 1988: 217) and 0.6 per cent per annum (Cross 1988) – even the lowest estimate is worrying.

Fig. 7.4 b) Logging trail in dipterocarp forest, West Malaysia.

Why are forests being degraded and lost?

Forest damage or loss is often due to a complex mix of reasons which vary from locality to locality (a list of causes may be found in Mather 1990: 245 or Aiken and Leigh 1992; a relevant bibliography is in Allen and Barnes 1985). In Central America, for example, deforestation caused by ranching is widely held to be 'driven' by the 'hamburger connection' – the supply of cheap beef in North America and Europe – a detailed examination reveals a complex of causes (Utting 1993: 14–33). But in Brazilian Amazonia although ranching is an important cause of forest loss, it is not driven by the export of cheap beef but by tax advantages, land speculation and mining, indeed widespread cattle disease hinders unprocessed meat sales to USA and Europe (Collins 1986; Lutzenberger 1987; Hecht 1989: 233; Mahar 1990; Biswanger 1991; Clearly 1992). Repetto (1988; Rietbergen 1993: 165) suggested that between 1965 and 1983 about 470 cattle ranches accounted for about 30 per cent of deforestation in Brazilian Amazonia (Fig. 7.4a).

In India the main cause of loss and damage in areas that still have a forest cover often appears to be smallholders cutting fuelwood and fodder (Nadkarni *et al.* 1989; R.S. Anderson 1988). In South East Asia and West Africa, the timber industry has probably played the major role in damaging forests; Porter and Brown (1991: 40) note that more or less all of Sarawak's timber exports are state-controlled (Figs 7.4b and 7.4c).

Martin (1991) suggested that 72 per cent of West Africa's rain forests have been lost because selective logging for timber export opens up the treecover, provides logging trails and roads, allowing in settlers who then finish the clearance. World-wide the

Fig. 7.4 c) Transporter with hardwood logs, West Malaysia (well above the altitude at which law allows logging, emphasizing the problems of enforcing conservation legislation).

greatest impact of logging is probably to ease access into forests so that other causes can come into play. Horta (1991) documents the shift of this process to the Cameroon and Wittle (1992) reports on Zaïre. At the time of writing Zaïre still had the world's third largest expanse of rain forest – about 12.5 per cent of the global total, but the rate of cutting was accelerating. On a world scale, Whitmore and Sayer (1992: 16) estimated that 55 per cent of tropical wood that is cut is used for fuel and 45 per cent for forest products (roughly two-thirds of which is pulp and one-third timber). World-wide forests have been felled for smallholder agriculture and to clear land for commercial plantations (Figs 7.5a–d and 7.6a–d).

Deforestation is not a problem restricted to poor nations. Few DCs have rain forest, but those that do have taken little care of it, for example in Australia, Queensland's rain forests have been decimated. DCs promote some of the forest damage by importing timber, woodchips, pulp and veneers; they often fund and encourage logging although LDC companies and joint ventures also share the blame (e.g. Malaysian companies are active in exploiting forests in Papua New Guinea). Many of the causes of forest damage listed on pp. 157–8 have roots beyond the country or region where its degradation is occurring. In South East Asia, Hurst (1990: 245–9) concluded that political, legal and economic causes of an 'external' nature were the cause of forest damage. These, Hurst suggested, were very much a twentieth-century phenomenon and to blame the local peasantry, which is common, was in his view like blaming the troops for warfare. In Amazonia the blame could be put on land speculation by the rich and upon in-migration of the poor who clear land for subsistence agriculture. These poor settlers have gained entry thanks to the roads built by nations with Amazon territory since the 1960s; it seems unlikely that road building will in the future match that of the 1970s and 1980s so it is to be hoped that the rates of forest loss of the last few decades will not be repeated (Fig. 7.4d). In many countries control of deforestation will not be

Fig. 7.4 d) Brazilian Amazonia, road through cattle pasture and savanna.

easy, given the levels of government corruption and the profits involved.

Improved technology has helped to speed forest clearing, notably the petrol-driven, hand-held chainsaw, also logging road construction and log extraction has been assisted by modern bulldozer and tractor developments. In the USA, Scandinavia and increasingly in the forests of Russia mechanized cutting and transportation is becoming highly efficient.

Causes of mangrove loss

1 *Clearing for building* in some regions real estate development is the reason, in others clearing is for industrial, port or airport facilities.
2 *Clearing for agriculture* often such clearance has quickly been followed by abandonment of land due to acid-sulphate soil problems.
3 *Clearing for fish or prawn ponds* aquaculture has been a major cause of mangrove loss in the Philippines, Malaysia, Thailand, Indonesia and Ecuador (Kundstadter *et al.* 1986; Burbridge 1988: 172). There is a growing market for the tiger prawn (*Penaeus monodon*) raised in ponds dug from mangrove swamps.
4 *Cutting for woodchips, timber, charcoal, tanbark* in South East Asia woodchip production for rayon manufacture or chipboard is a serious problem – much is exported to Japan (Ong 1982; Fortes 1988: 210; Vannucci 1988).
5 *Excessive siltation from rivers* development of catchments leads to silty runoff which damages mangroves.
6 *Industrial and sewage pollution* mangroves can withstand some pollution, and can help purify effluent, provided there are not excessive amounts.
7 *Decreased freshwater flows* river regulation and increased evapotranspiration resulting from development inland may reduce flows enough to allow excessive influx of undiluted seawater.
8 *Defoliation* large areas destroyed in Vietnam during war (Maltby 1986).

Fig. 7.5 Tropical treecrops. a) Oil palms, large areas of West Africa, Malaysia, Indonesia, and Latin America have been deforested to grow these treecrops to provide vegetable oil for export.

9 *Oil spillage* ocean tanker spillage, problem in areas with busy shipping lanes near shore (e.g. Straits of Malacca).

10 *Sea-level rise* not likely to be a problem unless development prevents mangroves from adapting by moving inland and upshore.

The problems caused by rain forest degradation and loss

Tropical lowland rain forests are important reservoirs of biodiversity: Prance (1990: 54) suggested that, with but 14 per cent of the world's land area, tropical forests had 50 per cent of all species. Estimates are inaccurate because only a proportion of species are recorded and named (Hadley and Lanly 1983), but the aforementioned suggestion is probably about the right magnitude. Already the degradation of these environments has led to the extinction of many thousands of species and, if anything, the rate of loss is accelerating (Wilson 1989; Shiva *et al.* 1991: 15). This loss of biodiversity is being monitored by a number of organizations including the IUCN, the Center for Plant Conservation (USA) and, more recently, the World Conservation Monitoring Centre (Cambridge, UK).

Once can list species known to have been lost (doubtless many more have disappeared without being recorded) and it is clear that the pace of extinction during the last few centuries is well above normal and increasing (Dayton 1991). Many important food, medicinal, timber and other crops originated in tropical forests and much more potential remains to be discovered, provided it is not lost first. Mooney (1979: 16) and Park (1992: 89) listed some of the important pharmaceutics derived from rain forests. Genetic material from the wild is needed to 'top-up' domesticated (crop) species so that they can withstand the constant challenges from disease, climate, pests and genetic drift away from vigour. Tropical rain forests are a particularly important source of raw material or ideas for synthesis of compounds for one of the growth industries of the late twentieth century: biotechnology.

Tropical lowland rain forests typically have over 50 species of tree per hectare: Prance (1990) recorded 283 tree species of over 10 cm girth in one hectare of Peruvian Amazonia; in Brazilian Amazonia there are seldom fewer than 87 and sometimes 300 different trees per hectare; in Indonesia diversity may be even higher (Mather 1990: 14). Brazil's remaining Atlantic forests (about 12 per cent of the original cover) also have a very high species diversity, some higher than much of the Amazonian forests. North American or Eurasian temperate forest diversity is unlikely to exceed ten tree species per hectare and three or four is more likely. The consequences of high diversity are that clearance of even a small area might cause many extinctions and that conservation may require a large reserve to avoid in-breeding over the long term because there are so few individuals of a given species per unit area. Rain forest vegetation is more dependent on animals for seed dispersal and pollination than other forests where wind pollination and seed dispersal is more common. Insects, birds and bats that pollinate and disperse seeds may be vulnerable if they nest, breed or roost in restricted areas, and can sometimes be lost without much forest

Fig. 7.5 b) Oil palm fruit bunches. Processing is generally at regional factories because the fruit ferments and spoils fast; unless carefully controlled the pollution can cause widespread damage to streams and rivers.

disturbance: in time the forest plants dependent on these creatures will suffer and perhaps become extinct. Sustained conservation may well require a considerable area of land to safeguard against disturbance and environmental change and ideally reserves should be duplicated.

Soils beneath rain forests are often poor and the vegetation has adapted to cope with this, so replacing rain forest with crops or pasture usually poses difficulties. A good deal of the nutrients in rain forest areas are in the biomass not the soil, if the trees are removed little is left in the soil and there may also be losses through burning and leaching: luxuriant vegetation is not a sure indicator of good soil. Rain forest clearance is likely to upset a complex web of interdependence (Mather 1990: 214).

Many trees have seeds that need a forest micro-environment to survive or germinate and are not viable after clearance; there is also a risk that important symbiotic micro-organisms (vital for nitrogen-fixing, etc) such as mycorrhizal fungi will be destroyed, hindering seedling growth. Forest disturbance can mean loss of seed dispersal or pollinating organisms.

Recovery of deforested land is less likely as the area cleared increases, for seeds and most pollinators are most likely to come from nearby patches of surviving forest cover. Once an area of forest is cleared there is a risk that there will be colonization by plant or animal species which hinder or prevent return of the original cover, either through competition or because the new cover supports frequent bushfires. Certain ant species have been blamed in South America for destroying seeds and seedlings, making forest regrowth impossible.

Animals and plants trapped in remnant areas of forest may have difficulty surviving in the long term, although Whitmore and Sayer (1992: 41) report quite good rates of species survival so far in Brazilian Atlantic forest patches. Large areas of South East Asia that were once luxuriant forest have become much less productive scrub, often dominated by

Fig. 7.5 c) The Jarí Project, Brazilian Amazonia. Extensive deforestation to plant exotic pine plantations in the foreground. Paper pulp factory in background has been fed with logs from forests as well as plantation. River pollution and acid deposition are other impacts.

Imperata spp. of grass with soils poor and subject to frequent bushfires. Island species have often had difficulty competing with introduced plants and animals. For example, in Hawaii and Mauritius invading weeds have seriously upset native floras, and in the latter hand-weeding of alien plants has made a major contribution to conservation.

There seems to have been large scale, catastrophic disturbance of forests in the past: extensive charcoal layers suggest there were widespread fires in various parts of the world including Amazonia, the Eurasian *taiga*, and elsewhere, before and since humans have had control of fire (Crutzen and Goldammer 1993). Present areas of high floral and faunal diversity, plus palaeoecological evidence suggest Amazonian forest was affected by drier and wetter periods in the past (Flenley 1979; Douglas and Spencer 1985: 35; Poore *et al.* 1989: 29; Whitmore and Sayer 1992: 125–30), and may have had considerable human settlement before the Conquest. Aiken and Leigh (1992: 29) wondered whether recent forest fires in South East Asia reflect increasing human activity opening up treecover or periodic drought due to ENSO events. Present conditions may thus be deceptive, the forest may not be stable and efforts to manage or conserve it may have to allow for this, select the most stable or species-rich areas, and plan for future environmental changes.

The pedalogical effects of deforestation on soil with sensitivity of soil, the location and the mode of clearance: vegetation is slow to recover in very cold or very arid environments, and exposed soil is more likely to suffer where there is intense rainfall, strong sunshine and steep slopes. Mechanized clearance does more harm than non-mechanized, mainly due to soil compaction; burning is also damaging. Selective logging is less damaging than many other forms of felling, but even if well managed to minimize canopy damage as trees fall, there will tend to be runoff concentration and erosion along

Fig. 7.5 d) The Jarí Project, *Gmelina arborea* plantation. As with the exotic pines, the plan was to clear natural forest plant gmelina and get rapid growth and sustained cropping. In practice there has been limited success.

Fig. 7.6 Deforestation and conversion. a) Clearance of upland rain forest for smallholder agriculture, Cameron Highlands, Malaysia. Remaining forest in background beyond steep terraced slopes. One problem has been heavy applications of pesticides leading to stream pollution.

skid-trails and logging roads and also people gain easier access.

Forest destruction releases fixed carbon from plant tissue and soil, it is claimed this has represented a major contribution to rising global carbon dioxide levels (Lal *et al.* 1986: 195–202). Mather (1990: 203) suggested that tropical forest destruction had contributed about 20 per cent as much carbon dioxide as fossil fuel combustion, another estimate (Mannion and Bowlby, 1992: 79) calculated that it had caused 25 per cent of total global forcing. If all tropical forest were destroyed, atmospheric carbon dioxide might rise between 10 and 20 per cent above 1991 levels (Foley 1991: 25). Deforestation-related carbon releases appear to have peaked in 1987 and are now decreasing. The importance of grasslands and savanna vegetation (and the world's soil) as carbon sinks should not be overlooked, grassland burning and desertification means more atmospheric carbon dioxide. Tropical grasslands probably fix as much carbon as the rain forests do (De Groot 1990).

Vegetative growth in the humid tropics can be very rapid: if it were decided to try to lock up atmospheric carbon dioxide through afforestation there are advantages in doing the planting in such environments. There have been cases where power companies have undertaken to plant forest, sometimes in a tropical LDC, to compensate for fossil fuel combustion (sink enhancement planting is discussed on p. 95).

Thousands of years of forest occupation have failed to guarantee most aboriginal peoples a legal claim to their surroundings on which they depend

Fig. 7.6 b) Conversion of lowland rain forest to cocoa plantation (left of picture), Malaysia.

and many have perished or survive as sorry remnants. Here and there in the 1990s a few groups have begun to protest and resist clearance (e.g. the Penan people of Sarawak) and even seek to control their forests and exercise a voice on national governments – like the Kayapo and Yanomami in Amazonia. Indigenous peoples with adequate tenure and whose voice is heeded by their state could play a key part in forest conservation, or they could sell timber rights to logging companies. Loss of forest and woodland deprives local people of game, fruit, staples, medicinal plants, construction materials, etc. Without such things they may lose subsidiary income, have difficulty in subsisting, and need to purchase things which once could have been gathered without cost.

Accurate quantification of the cost of forest loss is impossible. A guess can be made at the value of forest plants to the DC pharmaceutical industry: Hurst (1990: xii) felt that about 40 per cent of prescribed drugs in the USA originated from tropical forest plants, either leading to synthesis or as an ongoing source (many drugs cannot be easily synthesized) – which represents billions of dollars. It is likely that more often than not the economic gains from logging are much less than would have been any long-term gain from the land if left uncleared (Eckholm 1979). Some nations, for example Malaysia, have converted forest into export crop plantations important to their national economy; whether production from the plantations can be sustained indefinitely has yet to be proven, although some plantations have operated for 90 years or more.

The loss of forest and woodlands is an aesthetic loss: they have been and continue to be a source of inspiration for scientists, writers, artists and are attractive to tourists. In some situations such qualities may provide enough incentive for conservation.

There is little firm, undisputed evidence of regional climatic change as a consequence of deforestation (Salati and Nobre 1991; Aiken and Leigh 1993: 11), certainly when forests are cleared there may be alterations of water-tables, soil changes, altered microclimate, etc. Rain forests behave like large

Fig. 7.6 c) Conversion of lowland rain forest for rubber and oil palm production, recently cleared and terraced land, Malaysia.

waterbodies: they moderate surrounding climate, reducing diurnal temperature fluctuation, maintaining humidity, at least downwind, and reducing windspeeds in their lee. There does seem to be some evidence for reduction of rainfall following extensive forest clearance, perhaps due to one or all of the following: altered albedo, reduced evapotranspiration, altered 'roughness' to prevailing winds. It has been suggested that between 50 and 90 per cent of precipitation on rain forests is 'recycled', i.e. is derived from upwind evapotranspiration (Independent Commission on International Humanitarian Issues 1986b: 47; Anon 1988). As much as 54 per cent of Amazonian rainfall might be 'recycled' (Prance 1990: 58).

Removal of large areas of forest (say over 100 square kilometres) might bring about sufficient change in albedo to alter global heat flux affecting airmass movements (Salati *et al.* 1983; Molion 1989; Reynolds and Thompson 1989; Grainger 1993a: 163). Rain forests do seem to get rid of more heat (from solar radiation) through evapotranspiration than grasslands or shrublands. Clearance of forest does seem to have led to increased daytime temperatures and more diurnal range in Nigeria, the Gambia, Amazonia and Malaysia (Poore *et al.* 1989: 46; Molion 1991: 212) and, in southwest USA, daytime temperatures were found to rise where scrubland had been heavily grazed. There is a possibility that moisture rising from rain forests contains compounds that act as condensation nuclei and help in cloud formation; forest removal might therefore reduce cloud cover and change rainfall and airmass movements.

Many believe that there is likely to be increased peakflow and decreased lowflow after forest clearance (R. Freedman 1989: 243; Hurst 1990: xiii). Floods, high stream sediment-loads and landslips are commonly blamed on deforestation and there are many cases where dams or channels have silted-up after the surrounding region has suffered forest disturbance. However, blaming deforestation is often speculation (Independent Commission on International Humanitarian Issues 1986b: 35; Ives and Pitt 1988: 11).

Fig. 7.6 d) Shifting cultivation, Amazonia forest has been cleared by slash-and-burn and a cassava crop planted. Land will be cropped for one or two harvests and should be left fallow for at least 20 years.

Once saturated by rainfall, even undisturbed forest can produce floods and silty water. Drops falling from a canopy 30 metres above ground on to soil with little groundcover vegetation causes erosion but after deforestation thick weed growth may actually reduce it. Return to pre-clearance hydrological conditions can be rapid.

There are cases where forest clearance seems to have led to reduced and silty streamflow, for example in Panama (Prance 1990: 57). But in some cases forest removal increases streamflow and groundwater recharge because there is then less evapotranspiration; elsewhere the trees may have intercepted cloud and mist and their removal reduces streamflow and groundwater recharge (Mather 1990: 216). Molion (1991: 213) suggested that deforested land may differ little from forested with respect to mean runoff, but that peakflows might increase after clearance because of the reduction of surface litter. If leaf-litter is removed from below an undamaged tree canopy it is likely that there will be rapid runoff and erosion.

In South East Asia catchment experiments similar to that at Sungai Tekam Catchment (Pahang, West Malaysia) (see Fig. 7.7a) indicated that rain forest 'absorbed' 35 per cent of precipitation; after logging, absorption fell to less than 20 per cent and after conversion to rubber plantation, about 12 per cent (Shiva *et al.* 1991: 25).

Health impacts may follow forest disturbance as people and livestock come into contact with disease vectors, or existing vectors become infected from wild reservoirs of disease or the disturbance offers a favourable environment for some carrier of infection. In Brazil yellow-fever and a number of other mosquito, bug or tick borne infections are associated with deforestation (yellow-fever is harboured by forest monkeys). Elsewhere deforestation-related diseases include scrub typhus, Lhassa fever, Marburg green monkey disease, and there has been a suggestion that AIDS originally spread to humans bitten by forest monkeys.

Fig. 7.7 Deforestation. a) Sungei Tekam Experimental Catchment, Malaysia. Measuring flumes like this one (V-notch weir) monitor the flows from an area of rainforest part-converted to tree crops to try and assess the impacts.

The degradation and loss of upland tropical and subtropical forests and woodlands

At virtually all latitudes highland vegetation is likely to be different from that of the lowlands, often having less species diversity (although there may be interesting endemic species). In the tropics and subtropics there are many types of upland tree-cover, including:

1 *Cloud forest/woodland* Found on some windward slopes, between 1000 and 2500 metres in the tropics (lower outside the tropics), which receive mist and clouds. These are often rich in epiphytic species.
2 *Sub-montane broad-leaved evergreen forest* May be found between roughly 1000 and 3000 metres in the tropics, lower in the subtropics. Species diversity is likely to be less than that of lowland forest but there may still be many endemics.
3 *High altitude conifer forest* Possibly above 2500 metres in the tropics, lower in the subtropics.
4 *Stunted tree or shrub forest* The highest type of treecover. Sometimes called elfin forest/woodland.

At higher altitudes treecover is often under greater stress than at low, as a consequence of thin soils, wind exposure, low temperature, greater ultra-violet radiation, receipt of cloud-borne pollution and, at higher altitudes, temperature variation and even in the tropics frosts. Also at high altitudes, cold nights mean that people are likely to demand more fuelwood. Growth and regeneration may be slow so forest and woodland damage has catastrophic impacts.

In general tropical and subtropical uplands have had a marked increase of human population, for example, in Nepal, Andean countries and highland Ethiopia. Population and other pressures are not uniform and so some areas have suffered much more deforestation than others.

Fig. 7.7 Deforestation. b) Managed forest reserve, Malaysia. Selective logging and careful management are used to try and sustain timber production and to conserve some flora and fauna. In practice even careful removal of say 5 per cent of trees can damage up to 50 per cent tree cover.

Increasingly the upland forest is all that is left in a region because the accessible lowland forests have been cleared. Probably half the world's upland tropical forests are already severely degraded (Repetto 1987: 100; Ives and Pitt 1988: ix). In addition to the potential genetic value of endemics little is known of the role of these regions in the long-term survival of lowland flora and fauna, ideally lowland forest reserves should be linked to upland areas. Upland forests are also important for stabilizing and protecting steep slopes to prevent landslides and erosion and sometimes, by trapping mist and cloud, they recharge groundwater. For example, on Tenerife, mountain pine (*Pinus canariensis*) forests help groundwater recharge by intercepting mist and cloud and logging has adversely affected water supplies.

The degradation and loss of mangrove forests

Mangrove forests are restricted to shallow, brackish, saltwater environments, in regions mainly between

Fig. 7.7 Deforestation. c) Maracá Ecological Reserve, Roraima (Amazonian Brazil). The Maracá Reserve protects a forest area roughly equivalent to twice that of the Isle of Wight. Successful joint UK and Brazilian research and preparation of a reserve management plan involving SEMA, the Royal Geographical Society, and other research and conservation bodies has proved the potential for bilateral conservation efforts.

Fig. 7.8 Loss of mangroves. a) Straits of Malacca (Malaysia), mangroves have been disturbed or polluted by oil spills and stone-filled gabions have had to be emplaced to protect shore (foreground).

25°N and 25°S of the Equator where there are no frosts (locally mangroves may flourish to about 32°N) (Hutchings and Saenger 1987; Lidén and Jernelöv 1987). Mangroves probably comprise about 0.6 per cent of the world's total forest cover. The 60 or so common species of mangrove provide valuable coastal protection against waves and storms, are habitats for a rich wildlife and provide areas where many useful fish, shellfish, prawns and crabs feed and breed. Many of the plants in mangrove forests are salt-tolerant or salt-resistant and should have potential as sources of timber, fodder, and other products in saline environments (National Academy of Sciences 1990: 58; 119) (Figs 7.8a and 7.8b).

Table 7.1 lists the reasons for forest degradation and loss (not in order of importance). Between 1979 and 1990 perhaps 31 per cent of the world's mangrove forests were destroyed and by AD 2000 there may be only 46 per cent of what was present in 1979 (Kundstadter *et al.* 1986: 8). For Florida, Australia, parts of the Caribbean, the Philippines, Malaysia and Indonesia the situation is especially worrying.

Little effort has been made to replant the considerable areas lost (apart from in Vietnam and Bangladesh) or plant new mangrove forests or to improve mangrove species through breeding. If global warming does result in rising sea-levels, mangroves could offer a cheap, effective, 'low-tech', self-maintaining coastal defence (Lewis 1982; Hutchings and Saenger 1987: 309). There is also some potential for using mangrove forests as a means of sewage and organic effluent treatment. The polluted water could simply be led through an extensive network of channels through mangrove roots where micro-organisms would cleanse it. In some regions removal of mangroves has allowed silt and pollutants to reach the sea and this has contributed to coral reef damage, further exposing coasts to erosion and a loss of wildlife.

Fig. 7.8 b) Fuelwood and rattan probably from coastal forests, Kuala Trengganu (Malaysia).

The degradation and loss of seasonally dry tropical and subtropical forests, woodlands and scrublands

Wherever tropical and subtropical soils lose moisture for long enough to hinder tree growth, or if there are shallow or infertile or salty or poorly drained soils or the presence of hard layers or toxic compounds, forest and woodlands become thin. There are also likely to be fewer broad-leaved evergreen species than in tropical rain forest. The further away from the Equator these woody formations are, the more pronounced is likely to be the dry season during which treecover may lose its leaves; where rainfall is high all year, there may be subtropical or temperate rainforests, as in Chile, New Zealand, Tasmania, and so on. The harsher conditions become, particularly through drought and recurrent bushfires, generally the more open becomes the treecover. At some point the canopy thins enough for an understorey of grasses and herbs and finally the vegetation is likely to be regarded as savanna.

Tropical seasonal (or monsoon) forests can be found in northeast India, Indo-Malaysia, parts of South and North America and Africa. The extent of tropical and subtropical seasonally dry forests and woodlands, and their rate of change, is difficult to establish, not least because there has been little standardization on what constitutes specific formations (P. Harrison 1987b: 54). Open forests and woodlands probably comprise about 21 per cent of the world's total forest, woodland and scrub cover; scrublands comprise about 13 per cent and forest fallows (land recovering from clearance) about 1 per cent (Barrow 1991: 74). Steinlin (1982) suggested that open tree formations covered 734 million ha of the tropics, over half of this in Africa.

Some seasonal forests, like those of Burma, Thailand, parts of Brazil and some Caribbean islands have been important sources of teak and other hardwoods. Drier, more open woodlands in

Table 7.1 Reasons for forest degradation and loss[a]

Factor	Pressure exerted	Comment
Population increase	• demand for land • demand for fuelwood • demand for forest products • global or regional pollution	Not always a close link (Mather 1990: 74), pressure exerted in a range of ways
Improved communications roads railways canals powerlines pipelines	• settlers get access • easier to remove products • pest organisms gain access • forest divided up; animal or plant movement hindered • poachers get access	Caufield (1982) suggested 25 per cent of Brazilian Amazon loss due to roads (in East Amazonia and Gabon, due to railways) Waves of exploitation most profitable species first, later the less valuable
Large dams	• flooding of forest • influx of people • access roads • pollution/disturbance • downstream impacts on riverine forest/ mangroves • industry may be attracted and cause pollution	Reservoir may flood several thousand km^2 Downstream impacts often missed in planning Richest forest in region may well be that flooded
Demand for pulp/ woodchips/veneer	• mangroves badly hit • plantations may be established	Likely to be mechanized, fast and ruthless Clearcutting likely. Indonesia is active
Smallholder agriculture gov.-assisted spontaneous shifting cultivation[b]	• clearance • gathering forest products • hunting game • starting fires on land adjoining forest which 'nibbles away' remaining tree cover	Control may be a problem, settlers may relocate Can be a major cause of clearance (Steinlin 1982) Opened-up forest may be more prone to fires Tenure problems lead landuser to neglect land or to be dislocated, and fail to sustain landuse
Large-scale ranching	• conversion to pasture	Major cause of clearing in Latin America[c] and Botswana
Plantation crops	• conversion to crops • agrochemical use may affect nearby forest/streams • smallholders may get evicted for plantation land, move and clear forest, etc	May be associated demand for fuelwood/ pollution associated with processing Major cause in SE Asia and Caribbean Pressures will increase if fuel/industrial crops expand
Fuelwood	• local use • commercial collector • smelting	More damaging in drier forests Access enables commercial gathering Develops when forest already disturbed? Related to population increase?
Rattan and bamboo	• collection from wild damages trees and opens up forest	Little development of plantations yet Market for rattan expanding SE Asia problem
Failure to assist poor	• manifest through population increase; influx of settlers; poverty	May be 'driving force' but also forest is damaged where poverty is less of a problem
Warfare	• direct: defoliation/burning • demand for war materials • risk of nuclear war damage	Vietnam. In past 'scorched earth' in many countries Timber for warships denuded Mediterranean lands and Europe, later parts of USA and Caribbean
Demand for timber	• clearcutting • selective logging • genetic erosion • logging roads open up forests	Major cause. From 1950 to 1982 a 16-fold increase in LDC tropical timber exports. Late 1980s logs and other forest products exports from LDCs was worth over US$7500 million p/a (Wyatt-Smith 1987)

Table 7.1 (Cont.)

Factor	Pressure exerted	Comment
Mining/oil	• roads open up forests to settlers • may have associated air pollution • pipelines may open up access • mining may demand timber for props or fuel • mines may pollute and lead to railways	Problem on east slopes of Andes (Cleary 1992) and SE Asia
Administrative causes	• tax laws encourage land clearing • desire to establish sovereignty • government encourage settlement/clearing • poor conservation • poor logging control • failure to support alternatives to clearing	A cause in Nepal 1950s–60s (Little and Horrowitz 1987: 373) Wide range of administrative causes, often important
Fires	• More a problem in drier forests • disturbed rainforest may be vulnerable during drought • natural/set by humans	Serious fires in disturbed rainforest in East Kalimantan in early 1980s
Global warming	• altered climate, sea-level, altered storm incidence, new pests/diseases?	Serious threat to conservation areas
Acid deposition	• dieback • soil changes	Europe, Central Europe, parts of USA, South America, China, India, spreading Some tropical soils vulnerable
Spread of pest organisms	• kill or compete • remove pollinator or seed disperser	Islands vulnerable, Florida has problems

Note: [a] Mather (1990: 245) tabulates forest conversion causes by world region.
[b] Shifting cultivation usually starts to cause damage if fallows are shortened too much.
Disrupted people tend to resort to damaging shifting cultivation practices. Breakdown may be due to commercial causes or marginalization.
[c] 'Driving forces' vary: in Central America or Botswana it may be a 'hamburger connection', in Brazilian Amazon it is less likely to be beef production (due in part to cattle diseases hindering export) rather than land speculation, tax advantages (until 1987) and 'machismo' (Branford and Glock 1985; Myers 1986; 1987a; 1987b; Little and Horowitz 1987; Repetto 1987; Hecht 1989: 233; Pah 1992: 68; Cleary 1992)
Source: Compiled by author from various sources

Africa and South Asia yield gum arabic or pistachio nut, and are important to pastoralists and farmers for shade, fuelwood, fodder and soil protection. Growing human and livestock populations in many dry woodland and savanna areas has often meant the breakdown of traditional sylvipastoral strategies, increasing poverty and damage to the land through excessive demand for pasture and fuelwood. In seasonally dry lands, tree growth and regeneration can be slow and soil damage a serious risk if there is deforestation.

There has been debate over the term 'savannas' and on their origin, particularly on the role of fire in their formation and on whether they have been more or less extensive in the past (D.R. Harris 1980; Boutier 1983; Cole 1986: 11). Many questions are still unresolved; it does, however, seem likely that human activity has extended, and is extending, savannas in many parts of the world. Where extension of savannas has occurred it is not uncommon to find relict areas of once-extensive tropical or subtropical seasonal woodland on steep slopes, in ravines or on hilltops, where it has escaped fire, grazing or woodcutting or where conditions may be wetter. Savannas probably covered about 23 million square kilometres (roughly 20 per cent of the Earth's landsurface) in the mid-1980s (Cole 1986; Lal *et al.* 1986).

Seasonally dry forests and woodlands and open woodlands are being lost through grazing, bushfires, fuelwood collection and land clearance. Fire is probably more of a problem than it is to humid tropical treecover. These regions often have fertile, workable soils that are relatively easy to clear with modern stump removal techniques; pests are also likely to present fewer control problems than in the humid tropics (Joss *et al.* 1986: 243). In spite of the potential for agricultural development, less than 1 per cent had been cropped by the early 1970s (Garlick and Keay 1970: 60).

Fuelwood collection seems to be most damaging where treecover is already somewhat degraded, and there is easier access. Degraded areas are more likely to offer dead, dry, lighter-weight wood than unspoilt forest. Fuelwood may be in demand for industries like tobacco processing or tea-drying as well as for domestic consumers. In 1983 Brazil used charcoal for about 40 per cent of its steel production (much produced from eucalyptus plantations); there has been expansion of plantation area since. Often it seems that woodcutters and charcoal producers mainly 'finish off' the treecover of already disturbed areas (Bowander 1987; Little and Horowitz 1987: 120; Lewis and Berry 1988: 241).

Fuelwood and charcoal usage can be difficult to monitor and control. Usage of these fuels varies from family to family (in one Nepalese study by a factor of 67). Surveys are seldom year-round or thorough, and data seldom indicate whether wood is dead or live, good or poor quality: the measures are often simply vague 'bundles' or 'loads'. In all probability a household's economic status and culture affect fuel consumption as much as actual availability.

Charcoal and fuelwood supply is often a business activity with gangs shipping supplies to cities and able to bribe or intimidate forest conservation officers, so that even reserve areas may be damaged. Some developing country cities now have vast treeless zones around them as a result of fuelwood and charcoal demand (Agarwal 1986; Little and Horowitz 1987: 151). The problem might be countered through control of the price and amount of fuelwood and charcoal sold, encouragement of alternative energy, provision of fuelwood plantations and more efficient domestic stoves (Foley *et al.* 1984).

Global warming will probably alter rainfall and evapotranspiration, changing growing conditions and incidence of bushfires. Seasonally dry forests, woodlands and savannas may undergo considerable change. ENSO-related dry phases seem to lead to increased bushfires in Australasia, even in quite humid areas. Global climatic change, plus the ability of modern developers to easily clear land in the seasonal tropics and subtropics, probably means that there will be rapid loss of treecover in those lands in the coming decades unless more conservation and planting takes place.

It will be interesting to observe what happens to areas of African savanna presently infested with trypanosomiasis-carrying tsetse flies if these become available for cattle grazing as a result of improved insect control, disease treatment or global warming. Grazing might well spread and damage the land before pasture and woodland management can evolve to cope with the changes (Linear 1985). In Botswana, cattle ranching for production of beef for export is reported to be a threat to large areas of rangeland and its wildlife through overgrazing and erection of fences which restrict movements of game animals seeking pasture and water.

The degradation and loss of mediterranean-type forests, woodlands and scrublands

Today uncropped mediterranean-type environments, with hot, dry summers and cooler, moist winters, are likely to have open woodland or scrub cover. This is the result of grazing, woodcutting for construction or fuel and land clearance for agriculture. Bushfires have become more frequent in some regions, especially those where there is summer tourism. Present-day vegetation and, at least in the Old World, that of the last 3000 years or so, is unlikely to be a reasonable 'baseline' from which to assess true natural productivity.

In environments bordering the Mediterranean Sea humans have been damaging the vegetation for 9000 years or more, although most of the degradation will have been during roughly the last 2000 years. North Africa once had extensive forests of *Pinus halepensis*, *Juniperus phoenica* and *Tetreclinus articulata.* Wetter slopes of the Atlas, northern shores of the Mediterranean and parts of the Levant had forests of oak (especially *Quercus ilex*), sweet chestnut (*Castanea sativa*) and pines. Cyprus, as its name suggests, had extensive cyprus (*Cupressus* spp.), Aleppo pine (*Pinus halepensis*) and black pine (*P. nigra*) forests until quite recent times, and still has remnants in mountain areas (Fig. 7.9). Cooler uplands in North Africa and east of the Mediterranean once had extensive stands of cedar (*Cedrus* spp.) and Syria, Egypt and Morocco had much greater areas of acacia woodland before 2000 BC (De Vos 1975; Thirgood 1981: 211). The demands of metal smelting, shipbuilding, fuelwood

Fig. 7.9 Surviving upland pine forest, high on the Troodos Mountains, Cyprus.

and construction, plus grazing and agricultural clearances, have converted much of the original forest to various relatively species-poor, low-scrub formations within which a high proportion of plants are aromatic, drought- and fire-resistant. In many mediterranean-type environments terraced agriculture has declined and abandoned terraces have collapsed and become eroded. The cause of abandonment may be past episodes of rural population growth, followed by population decline; opportunities for off-farm employment which make terraced agriculture unattractive; mechanization of level areas, which has made sloping land uncompetitive; decline in demand for traditional crops like olives, raisins, almonds, perhaps due to new producers in say California, or the spread of consumption of frozen and canned foods (Barrow 1991: 215).

Attempts at conservation may be seeking to protect what has already been seriously degraded, but may prevent further decline to degraded grassland or even bare earth and rock.

In mediterranean-type environments of South Africa, Chile, California and parts of Australia degradation has mainly taken place in the last 200 years or less. Some mediterranean-type floras have a high species diversity, for example the *fynbos* bush of South Africa. This is now under threat from fires and competition from introduced plants from South America and Australia. The degree of damage, caused by fire or grazing, depends on many local factors which include: recurrence frequency, severity, timing of burn or grazing and vulnerability of the flora. Grazing or bushfire at brief critical times of year when plants or their seeds are vulnerable may have serious results yet for much of the time do little harm. If fires take place after a lot of dry matter has accumulated they may be particularly damaging. Light, frequent fires may be necessary to encourage or facilitate revegetation (De Booysen and Tainton 1984; Joss *et al.* 1986: 573; Vines 1987). Giant redwood (*Sequoiadenron giganteum*) and Douglas fir (*Pseudotsuga menziesii*) of North America cannot regenerate or survive competitor species without recurrent gentle brushfires so their conservation should not prevent all fire (Cooper 1974: 294; Kozlowski and Ahlgren 1974).

There is considerable debate about the role of livestock in the degradation of forests and woodlands, particularly the goat. In seasonal tropical, subtropical and mediterranean-type environments, dryland environments and islands, goats often get blamed for deforestation. Whether goats are wholly or part to blame along with woodcutters or shifting cultivators is still unclear; some hold that they simply administer the *coup-de-grâce* to damaged land (Thirgood 1981: 68; Dunbar 1984). Cattle and pigs also help degrade tropical and temperate, island and mediterranean-type environments, preventing seedling regeneration once forest has been opened up.

The degradation and loss of high-latitude forests and woodlands

Roughly half of the world's forest cover is in temperate and cool environments. Since the late 1930s and early 1940s these have not declined much in area (that may change as the CIS boreal forests are exploited) but in some countries the quality of the cover has altered as primary forest have been cut and replanted with plantations that are less species diverse. In the process of cutting and replanting much of the flora and fauna associated with primary forest is lost. Tree growth at high latitude can be very slow, so that regeneration takes centuries, there is also a risk where forest has permafrost beneath that there will be melting, waterlogging and no chance of forest re-establishment after clearing.

High-latitude, boreal forests form a huge belt from Alaska – western USA to eastern Siberia between the 13°C July isotherm in the north and the 18°C July isotherm in the south, which equates roughly to 45°N to 70°N (Gamlin 1988). The CIS (former USSR) has about 42 per cent of the world's temperate forests (roughly 25 per cent of all those trees are in Siberia). The boreal forests comprise about 29 per cent of the world's total forest (Kuusela 1992) and so are one of the largest carbon 'pools', or were until 1978, when net growth shifted to net loss (Sedjo 1992). These forests appear to be similar in Europe, Pacific northwest America and Siberia; however, they differ a good deal in species composition (Mather 1990: 18). Although they may have less species diversity than those in warmer climes, undisturbed temperature forest still harbour considerable biodiversity. As with tropical forests some of this biodiversity is already proving valuable, for example the Pacific yew of northwest America (*Taxus brevifolia*), now becoming rare because of logging, yields taxol from its bark which has proved an effective anti-cancer drug.

South of the boreal coniferous forests are broad-leaved deciduous or mixed broad-leaved and conifer forests. In Europe the latter have suffered over the last few thousand years and particularly since the 1940s. In the last few decades in Scandinavia, North America and the former USSR the mixed (pine, spruce and larch) forests have been increasingly felled and replanted with mainly single species conifer plantations (e.g. *Pinus contorta*, *P. sylvestris*, or *Picea abies*) (Barr and Braden 1988). In North America temperate coniferous forests of various pines and hemlocks (*Tsuga* spp.) occupied much of the land from Minnesota to New England until the nineteenth century; this has been heavily logged. From New Jersey to Florida and west to Texas a similar looking forest, but with a different species mix seems to have fared better (Mather 1990: 19). The felling which began in North America in the seventeenth century has destroyed 20–40 per cent of the original treecover, leaving little natural eastern lowland hardwood forest east of the Appalachians.

There has been much replanting in the last few decades but it lacks many of the qualities of the original treecover. Logging in northwest USA and Canada is mainly clearcutting; even when selective logging is practised; there is a tendency to remove the best trees and open up the cover enough to make it vulnerable to fires. The temperate forests of Eurasia and the Americas, especially in cooler regions, are slow growing and in colder environments very slow growing: a seedling may take centuries to reach maturity. Not only will it take a long time to remedy present timber use, but also the soils are very fragile and easily damaged during logging. Some forests may benefit from occasional partial logging, without which they develop stands of senile trees and suffer loss of species diversity and decline of soil fertility (Kuusela 1992: 13).

The logging of North America's old temperate forests has cleared large areas of trees, especially in British Columbia and Vancouver Island. Timber and pulp companies responsible for the conversion argue that selective logging, as opposed to the usual clearcutting, is too expensive. In regions where there is little employment other than forestry it might make long-term sense to subsidize more sustainable forms of forestry to maintain jobs and for the sake of wildlife conservation (Diem 1992). Logging is not the only threat to high-latitude forests, pollution, particularly acid deposition and increased tropospheric ozone, are affecting increasingly large areas of Scandinavia, Europe and the CIS (Nilsson and Sallnäs 1990; *The Times* 29/8/92: 7). Coniferous forests seem more susceptible to these threats, or have been the first to show signs of die-back, until recently, when broad-leaved species also began to show stress.

Forest clearance was under way in Europe by the Mesolithic and accelerated after arable farming arrived, anywhere between 7000 and 1000 BP, depending on locality. Forest cover remained on heavy clay lowland soils until after Roman times because available ploughs could not break those areas. By medieval times, after centuries of slash-and-burn clearance, swine and cattle herding, Europe's forests and woodlands were subjected to increasing demands from smelting, shipbuilding, and by agricultural clearances from about the sixth century AD as ploughs improved. By the nineteenth century there was little lowland forest left in western Europe outside Scandinavia, Austria and Germany. It has been estimated that by 1980 the UK had lost 90 per cent of its original forest cover; today, with about 8 per cent forest cover, the UK is one of Europe's least-forested countries. Some forest clearance in western Europe seems to have caused heath and moorland formation (Holdgate and Woodman 1976: 129; Hawkes 1978: 3).

Europe's own supply of oak and other straight timber ran short by the eighteenth century. Carpenters and shipbuilders increasingly imported teak and other substitutes and coal replaced wood for smelting, industry and domestic heating from the 1750s. Pulp and woodchips demand increased after the 1930s and now accounts for much of the logging. A natural catastrophe, Dutch Elm disease, struck the UK in the early 1970s, killing many of those broad-leaved trees. In spite of these demands and setbacks there has actually been expansion of forest in the UK, most of Europe and North America since late Victorian times (*c.*1860s). However, this has mainly been in the form of plantations of one or a few species of exotic conifers, sometimes poplars or, in warmer regions, eucalyptus.

In the UK the favourite plantation species in recent decades has been sitka spruce (*Picea sitchensis*), and planting has been quite sensitive to tax incentives. In the UK a lot of ancient, species-rich woodlands have been lost, mainly through clearance for agriculture or to replant with conifers which were seen to pay-back faster than native hardwoods. England lost about 45–50 per cent of its ancient treecover which remained in the eighteenth century between 1942 and 1992 (*The Observer* 17/5/92: 1; *The Times* 18/5/92: 3).

In recent years Scandinavia, North America and the former western USSR have expanded pulp production to meet the increasing demand for paper, cellulose, chipboard, rayon, nylon, etc. This has led, at least in Scandinavia, to clearance of native conifer forest (usually clearcutting) and replanting with a less diverse cover of exotics in the hope of sustaining production but, in practice, establishment rates are often poor. Finland has discovered that large-scale replanting has resulted in extensive draining of wetlands which may result in carbon dioxide emissions and wildlife disruption, and the pulp processing leads to serious water pollution (Isomäki 1991). In the CIS logging is accelerating at a worrying pace, especially in Siberia, with Japanese, North and South Korean and USA logging companies active as well as former Soviet timber interests. The Baikal–Amur Mainline Railway is likely to aid the export of timber to the Pacific Rim. If clearance of these CIS forests becomes extensive it might have an impact on global climatic change (Petrof 1992).

Species-rich temperate evergreen and mixed broad-leaved forests in Chile, especially the araucaria forests (*Araucaria araucana*), south Australia, Tasmania and New Zealand have suffered from logging, clearing for agriculture and from introductions of exotic plants and animals. In south Australia, New Zealand and Chile southern beech (*Nothofagus* spp.), Kauri pine (*Agathis australis*) and totara (*Podocarpus totara*) have suffered most and large areas have been cut, mainly for woodchips, and replaced with exotic faster-growing conifers, mainly sitka spruce in south Australia and New Zealand and Monterey pine (*Pinus radiata*) in Chile (where it now comprises about 20 per cent of forest cover). Japan and PRC still have extensive evergreen mixed broad-leaved forests, mainly in highland areas.

If enough global warming and carbon dioxide enrichment occurs, the effects on higher-latitude forest species composition and productivity could be considerable. Many tree species need cold winters to promote growth and there has been a tendency to plant conifers with such needs. If winters become too warm trees may not be physiologically suited and pest organisms and fungal diseases may flourish. In cool environments where tree growth is slow foresters are now likely to be planting trees that may have to face changed climate in thirty or forty years: the chances are that they will not be appropriate species or varieties.

Global warming, if it occurs, may result in greater fire risk for high-latitude forests (SCOPE 1983; Wolman and Fournier 1987: 59). Birch and conifer forest may spread into tundra areas and steppes and vegetation 'zones' will tend to move poleward in general and upslope in highlands (Mather 1990: 206). Some plants and animals will adapt and spread faster than others and it will probably take a long time for conditions to stabilize.

Recent EU agricultural policy changes and demands for recreational areas are likely to have considerable impact on forests and woodlands. In the UK several cities are planting or plan to plant large community forests for amenity and timber production while the Farm Woodland Scheme is designed to encourage smaller-scale on-farm planting. Removal of some tax incentives for commercial softwood plantations and increased benefits for those planting broad-leaved trees may help remedy the trend toward species-poor conifer plantations mainly in uplands (*The Times* 20/5/91: 2).

The loss and degradation of temperate hedgerows

In the UK, parts of Europe (notably Brittany, Normandy and Eire), New England (USA) and Tasmania hedgerows have become valuable conservation areas; they can also be important breeding habitats for predators of pest organisms. But in recent years many kilometres have been removed mainly to create larger fields that are more suited to mechanized cereal production and to obtain grant payments (Pollard *et al.* 1974; Briggs and Courtney 1985: 330–4). Estimates suggest that between 1945 and 1985 the UK lost 22–25 per cent of its hedgerows (Hooper and Holdgate 1970; Hooper 1979; *The Times* 30/8/91: 14; 18/11/93: 5); between 1984 and 1990 UK hedgerows may have declined from 341 000 to 266 000 miles, having been 500 000 miles in 1945 (*The Times* 17/8/93: 7) (1 mile = 1.61 km), with the losses being especially marked in eastern England. In 1877 Germany is believed to have had *c.*133 metres of hedge per hectare; in 1971 this had fallen to 60 and by 1979 it was only about 29 metres (OECD 1989: 41).

Few of the remaining UK hedgerows are now laid by hand in the traditional manner, rather they are close-trimmed mechanically, often several times a year, which reduces floral and faunal species diversity and prevents the development of tall emergent trees. Hedgerows are also being lost or are suffering loss of species diversity as a result of urban expansion, pesticide and herbicide use, traffic pollution, road widening and use of road de-icing salt in winter. As well as loss of species, hedge destruction can render an area more vulnerable to soil erosion, particularly by wind action.

Some of the UK loss has been caused by the availability of government grants for hedge removal; the practice has been discontinued and replaced by grants intended to encourage hedge planting. Hedge conservation can also be assisted by encouraging or enforcing trimming only at certain times of year and by restrictions on spraying the edges of fields.

It should be noted that the loss of dry-stone walls, scrubland, wetlands, ponds and saltmarshes, etc., are contributing to an accelerating world-wide loss of biodiversity: hedges are only one aspect of a multi-faceted destruction of nature.

Countering the degradation and loss of forests, woodlands and savannas

Ways in which degradation and loss of tropical forests could be reduced

Reduce demand for tropical timber

- use veneers instead of solid wood
- restrict trade in selected wood
- tax sales/exports of selected woods (invest revenue from duty in replanting/conservation)
- educate consumer/encourage use of alternatives
- recycle wood
- provide LDCs with alternative energy (reduce fuelwood use)
- reduce demand for pulp/woodchips.

Reduce volume of tropical timber traded

- strengthen/enforce logging licences
- improve profit to producers (sell as finished products, veneer, etc).

Reduce waste during extraction

- minimize damage during logging (restrict what is removed; minimize damage to surroundings by

choice of technique; control time of year cutting is done; better supervision; authority marks selected trees)
- promote sustainable timber production
- market a wider range of timber species.

Reduce forest clearing by locals

- encourage 'tolerant forest management'/extraction
- involve locals in forest management
- ensure locals benefit from forest management
- promote agroforestry, conservation, replanting
- identify why locals damage forests and counter
- control access to forest.

Encourage sustained yield management and reafforestation

- tax incentives/grants/aid/loans
- educate to promote social forestry.

Control extraction

- restrict areas where cutting allowed
- tighten up and enforce licences
- introduce quotas.

Reduce degradation

- control grazing
- alternative fodder
- control fires
- control acid deposition/pollution
- aid regeneration of damaged forest.

Support conservation

- establish reserves
- support multi-usage
- establish gene banks
- support on farm conservation.

Enrichment planting

- plantations
- plant part-cleared forest.

Countering the degradation and loss of non-tropical forests, woodlands and savannas

Many of the ways in which tropical forest loss and degradation might be reduced could be used to protect non-tropical forest, woodland and savannas. Savannas and open woodlands have attracted far less attention than rain forest; while these environments may lack the species diversity of the latter, they still contain valuable genetic resources (Burman 1991).

Attempts to reverse deforestation may fail because they address symptoms rather than causes (a familiar problem with efforts to control soil erosion and desertification) (Vanclay 1993). For example, settlers are responsible for much of the deforestation in Brazilian Amazonia; the root cause, however, lies in inequitable access to land: roughly 4.5 per cent of landowners in Brazil control roughly 81 per cent of all farmland. If all potential farmland outside Amazonia were redistributed, every Brazilian man, woman and child could have about 4 hectares, and there would instantly be no need to settle forest areas (A.B. Anderson *et al.* 1991: 2).

Attempts to restore natural forest vegetation

It may sometimes prove possible to regenerate something similar to natural forest where it has been lost, although the species diversity will probably remain much lower (Horta 1991). There has been some success at this, for example in Panama, Honduras, Cuba and Vietnam (Gradwohl and Greenberg 1988). The process may be slow: temperate and cold environment forest trees, in particular, may require many decades to mature and start regenerating.

Forest plantations

Increasingly the alternative to deforestation for fuel, timber, pulp and woodchips is likely to be tree plantations: typically single species or limited mixes, often eucalyptus species, caribbean pine (*Pinus caribae*) and gmelina (*Gmelina arborea* – see Fig. 7.5d) in tropical, subtropical and mediterranean-type environments, conifers or poplars in temperate areas, pines or other conifers in cooler environments or highlands. Efforts to establish plantations, in spite of widespread optimistic public statements, has led to planting well below rates of forest and woodland loss. Supporters of plantations claim that the timber produced is more uniform and the production more intensive than with mixed forests, consequently less land is needed for tree production.

Trinidad has had considerable success with replanting and has exported expertise (Poore *et al.* 1989: 76). Unfortunately, good forest management, not easy at the best of times, is hindered throughout the world because data on the existing forest and woodland are poor and because there are limited funds and a lack of skilled staff. As in other environmental protection fields, mobilizing people is easier than motivating them – those who plant trees will not necessarily care for them afterwards.

Planting is often marred by poor seedling survival rates; presumably this is the case in the PRC; which planted around 19 900 square kilometres between 1974 and 1984 but still has serious forest problems. It is not unknown for natural treecover to be cleared only to be replanted with single species plantation forests, for example in Zambia's Copper Belt or the Jarí Project of Amazonia (see Fig. 7.5c and d).

Case study: the Jarí Project, Amazonian Brazil

In the 1970s an American entrepreneur Daniel Ludwig purchased about 1.4 million hectares of land from the Brazilian Government on the Rio Jarí, eastern Amazonia (owned since 1982 by a consortium of 22 Brazilian companies). Ludwig calculated that there was a growing market for paper pulp, and that in a moist tropical environment, specially selected fast-growing exotic tree species could sustain profitable production. A pulp factory was built in Japan and towed on barges to Jarí. This factory was supposed to be supplied by about 200 000 ha of plantations of fast-growing (6- to 16-year cropping cycle) exotic trees, via a 200 km railway network, but only about 110 000 ha were planted by the early 1980s (with *Gmelina arborea*, *Pinus caribea* and *Eucalyptus deglupta*) (Fearnside and Rankin 1980). The *Gmelina* failed to prosper on the sandy-soils and growth rates of the other species have reputedly been only about one-fifth of plantations in Scandinavia, well below what had been expected (Hoppe 1992: 232). Consequently timber has also been cut from floodlands and *terra firme* (dryland) forests. It is too soon to be able to establish if treecropping is sustainable (Eden 1990: 97).

The pulp production results in considerable river pollution and some air pollution, there is also effluent from a china clay mine and disruption caused by hydroelectric generation (Halperin 1980). A township has been established at Monte Dourado (over 7000 inhabitants supported by Jarí in 1978), and a large squatter settlement has sprung up alongside the Rio Jarí. Jarí has helped to encourage the logging of an important wetland tree, virola (*Viriola surinamensis*); which is in demand for plywood, particle board, veneer, and manufacture of products like pencils; it is now Brazil's second most important timber export after mahogany (Macedo and Anderson 1993). Much of the virola is cut from floodland forests along the Rio Jarí.

Jarí has led to the deforestation of well over 110 000 hectares of lowland tropical forest and probably involved the expenditure of over US$1000 million; if that sort of money could have been spent on developing and promoting sustainable agriculture for Amazonian farmers there might have been considerably greater benefits.

The Jarí Project includes 12 000 to 14 000 hectares of high-yielding rice paddies sited on cleared floodlands subject to freshwater flooding when there are high tides in the Atlantic, irrigation is thus easy (Hall 1989: 8). The rice yields have been quite good – 10 to 12 tonne per hectare per year – but chemical fertilizers and pesticides are used and must inevitably get into the river system.

The record for large-scale agriculture in Amazonia remains poor. Jarí engaged teams of foresters, land capability surveyors and other specialists to research and establish feasibility; the message is that, even with study and resources, tropical development can be difficult.

Managed, sustained forestry

Attention needs to be given to the selection of more appropriate plantation species, to siting, tree mixtures and pest problems, wildlife conservation and the social and economic impacts of planting. Fast growth is little use if the trees provide poor quality wood for construction or fuel, damage the soil or groundwater, and cause unemployment or social unrest. Already a large literature is devoted to reported ill-effects of eucalyptus plantations: depletion of groundwater, reduced wildlife habitat and socioeconomic impacts (where traditional landuse is replaced by tree plantations there may be marked changes in employment opportunities) (Eckholm *et al.* 1984: 66–9; Kardell *et al.* 1986; *The Economist* 6/12/86: 93; Barrow 1991: 106–10; Saxena 1992).

There have been cases of commercial plantations competing with the interests of rural folk (Lohmann 1990), at worst acquiring common land and forcing poor people who used that land to make use of marginal environments or migrate in all probability to urban unemployment.

In cooler environments conifer plantations have been criticized for damaging wetlands and uplands, for causing loss of wildlife habitat, for hydrological impacts, possible release of 'greenhouse gases' from drained soils and increased trapping of atmospheric acid pollution. The latter plus altered soil processes under conifer forests may raise soil pH and give rise to high aluminium levels in groundwater and streamflow. In the UK and continental Europe there has been much criticism of block-planting conifers as unaesthetic; foresters now tend to try and landscape plantations and establish mixtures of trees.

Seldom have loggers paid for replanting or cut in a manner which ensures indefinitely sustainable forestry. Sometimes the intention has been to sustain logging but in practice it has proved an elusive goal. On a world scale, managed tropical forest area is negligible (Wyatt-Smith 1987; Kemf 1988; Mather 1990: 35). Many strategies of tropical silviculture have been developed (for an introduction see Grainger 1993b). For example:

1 *Shelterwood system* developed in Malaya in the 1950s: unwanted trees are removed after selecting marketable specimens, desirable seedlings are then encouraged using the remaining cover for shelter. This approach gives a greater and more uniform second harvest roughly 70 years after the first and maintains forest cover (see Fig. 7.7b).
2 *Enrichment planting* desired trees are planted in gaps or along trails after logging. With either of the aforementioned strategies local peoples may be engaged to 'nurse' the seedlings until established.

Whether in tropical or cooler environment forests, selective logging is often not as benign as its supporters claim. Selected trees are often the best specimens, which means in the long term genetic damage to the stock and, when cut, they tend to take down many surrounding trees and make further damage to ground vegetation when removed via skid-trails and logging roads. Selective logging can open up forest enough for people, grazing animals and fire to take hold and do further harm.

Reducing demand for forest cutting

There may be situations where non-forest biomass may be developed to reduce pressures on treecover for charcoal, fuelwood or paper pulp and feedstock for plastics like rayon or nylon. For example aquatic weeds, reeds (like *Miscanthus sinensis*) or willow may have potential. Biomass energy crops might be combined with sewage pollution control (the effluent would promote biomass production) (Dunn and Clery 1992). There are also possibilities for substituting fieldcrops like kenaf (*Hibiscus cannabinus*) or cereal straw for forest sources of woodpulp (Anon 1990). Denmark runs small- to medium-sized district heating and power plants using straw, livestock waste or refuse. Better use could also be made of hydroelectric, wind and solar power to cut demand for fuelwood.

Bamboo and rattan plantations could, by providing furniture construction materials, help to conserve natural forest areas. At present, however, much of the world's rattan is collected from the wild (the trade was worth *c.*UK£1500 million in 1988) and doing so damages forests; there would be huge benefits if rattan were grown as a plantation crop (Barrow 1991: 84). The Indonesian Government banned export of rattan stems in 1985 to try and get better returns from export of finished furniture (Grainger 1993b: 154). Bamboo is used to make paper in India and much comes from non-plantation sources, cultivation, perhaps in rotation with other crops, would take the pressure off forest sources and should be quite easy. Bamboo and rattan cultivation deserves much more support.

Wherever possible LDCs are well advised to export processed timber products (veneers, plywood or finished articles like doors or furniture) rather than uncut logs or rattan stems. By so doing, they obtain the maximum profit and employment and, assuming that timber processing is not wasteful and that the greed of those involved is controlled, it should not be necessary to cut so much forest (James 1989; Poore *et al.* 1989: 60). In practice attempts to acquire the know-how, equipment and funds for such processing can be difficult and the processed goods still frequently encounter trade barriers, tariffs, import quotas and other restrictions in 1993. Nevertheless, only Malaysia now exports a significant amount of raw timber (logs), most of that coming from Sarawak, timber producers having changed to processing before export.

In the late 1980s about 56 per cent of the world's industrial timber came from tropical forests (Wyatt-Smith 1987: 2). A significant proportion of tropical timber exported from LDCs is wasted (Cross 1988). More could be recycled and less well known species of trees could be used rather than left to rot, provided they could be identified and their qualities listed for buyers to see (unfortunately the timber trade is very conservative). Steel or plastic could be used for some applications where wood is used now, for example: to replace the wasteful use of plywood for concrete shuttering (building moulds) or softwood for disposable chopsticks. It should be stressed that many LDC logging companies are part or wholly owned by DC firms; the blame for deforestation is certainly not attributable only to LDCs.

There is scope for improving the efficiency of wood-burning stoves and for promoting alternative means of heating and lighting (Foley *et al.* 1984; P. Harrison 1987b; Munslow *et al.* 1988; Leach and Mearns 1989). Altering habits, providing access to alternatives to fuelwood and charcoal at affordable prices and controlling those who trade in fuelwood and charcoal will not be easy (Repetto and Gillis 1988: 82).

Demand for woodchips and pulp for manufacture of board, paper, rayon or nylon means both forest destruction and often pollution of the environment around the processing plant. Lake Baikal (CIS), the world's largest freshwater lake, has been seriously damaged by pulpmill waste and there are similar problems in Scandinavia, Chile, North America and other countries. More recycling or alternative feedstocks may help reduce the problem (plus better control of effluent).

Agroforestry, community forestry and social forestry

There is widespread and increasing interest in the promotion of community forestry and agrosilviculture, in some countries with considerable success. The former is the involvement of local people in tree planting and management, which if given proper encouragement, may ensure forest areas increase and are protected whilst benefiting local people rather than 'outsiders' (Arnold 1987; Westoby 1987). The latter is the deliberate association of trees, shrubs, crops and perhaps livestock, honey-production, latex or wax extraction, etc. The goals of both approaches are to help sustain supplies of wood and other forest commodities to local people, reduce the pressure to clear new areas and, if possible, afforest or reafforest (Foley and Bernard 1984: 128; Winterbottom and Hazelwood 1987; Weber 1989: appendix E). These ideas are not that new: attempts have been made to involve shifting cultivators in tree planting and aftercare since the 1870s.

Increasingly NGOs, women's organizations, schoolchildren, youth organizations, etc., play a part in tree planting and aftercare. Women's groups and children's groups are ideal routes to achieving 'grassroots' action as they are 'in touch' with the local environment and stand to be the first to gain from better availability of fuelwood and forest products, because they are the ones who generally have to spend hours walking great distances to get these things. With popular support for social forestry growing better funding is needed to assist tree planting and forest conservation.

More could be done to encourage poor people to invest in trees as a means of savings and security (Chambers *et al.* 1993). Many societies already do this, for example, by planting trees to use to pay for weddings, celebrations, funerals. This may be by selling the wood, the tree or rights to some or all of its wood or fruit, etc., may be given, or it may be used as security for loans or even mortgaged. Trees may be planted to provide their owners' funeral pyres or with material to repair their houses. Trees may be held as an investment and also provide ongoing support as dry season fodder or food (e.g. mangoes or locust *Parkia* spp.) or tan-bark to give regular cash income.

Tighter control of development funding

Aid and development loans may help promote deforestation. Even agencies like the World Bank, which profess a 'green' forest policy, have been accused of funding logging operations that threaten forests and which may be unsustainable, for example in the Congo (*The Times* 26/11/92: 4).

International and NGO controls on the timber trade

It has been claimed that, globally, the timber trade is the single greatest cause of forest degradation (Dudley *et al.* 1994); at the regional or even national scale that may not always be the case. Conservation and control of timber trade is promoted by a number of non-governmental bodies and by some nations (Pearce 1989). UNCTAD proposals in the 1970s ultimately led the establishment in 1985 of the International Tropical Timber Organization (ITTO) to try and follow up the 1983 International Tropical Timber Agreement (ITTA) (signatories of the ITTA are expected to harvest only sustainably managed forests by AD 2000; what 'sustainably' means has not been precisely defined). Designed to regulate trade in tropical timber, in 1992 the ITTO had been signed by 47 countries which produced or consumed tropical timber. The first commodity organization with a conservation mandate, the ITTO was given somewhat contradictory roles of encouraging sustainable use of forests and conservation plus the promotion of trade in tropical timber, not surprisingly it has come under repeated criticism for failing to protect forests (Rietbergen 1993: 185–207). For example, in 1993 WWF/UNEP-backed proposals to set up a research project to identify endangered species and try to curb trade in them were rejected by ITTO.

ITTO set up a Tropical Forestry Action Plan (drafted 1985) to try and provide a realistic, practical and flexible strategy to conserve tropical rain forests, promote sustainable development in those regions and feed in funds to assist (World Resources Institute *et al.* 1985; *IUCN Bulletin* 1989, vol. 20: p. 22; Sattaur

1990; Park 1992: 150). Although the Plan involved 73 LDCs in 1990 and received about US$1000 million, it appears to have done little to slow deforestation and there were soon calls for it to be restructured (Pearce 1989; Colchester 1990; Grainger 1993b: 253 is a little more optimistic). In 1987 the World Rainforest Movement, a coalition between DC and LDC environmental groups, proposed an Alternative Tropical Forestry Action Plan (Hunt 1990: 264).

A number of NGOs seek legislative or voluntary control of sales of endangered species, including certain timbers. Consumers may be educated to discourage buying practices that lead to deforestation; unions may monitor and perhaps refuse to handle imported goods which promote deforestation. Manufacturers can be encouraged to adopt codes of conduct and labelling of goods (e.g. an independent Forest Stewardship Council was established in 1993 to check and accredit timber that comes from sustainably managed forests). Friends of the Earth have tried to establish a Code of Conduct for Tropical Timber Traders. The latter aims to discourage sales of wood or wood products unless they originate from sustainable plantations or suitably managed forest (the World Wide Fund for Nature, the Rainforest Alliance and the Soil Association have promoted similar approaches). Attempts were being made in 1993 to write similar requirements into the ITTA; however, European nations had difficulty accepting the principle.

It is difficult to ensure that claims of sustainable forestry are true or even to establish the provenance of logs. To prevent fraud, logs might be tagged before cutting, possibly with a bar-code, recording species, location details, serial number, etc. The tag details could be transmitted when the tree is cut to the sales-point or importer to await arrival of the log. In 1989 100 UK companies, some retailers, signed an agreement to operate a UK Timber Trade Federation Code of Conduct which requires the sale of products made only from timber cut from well-managed, ecologically sound sources.

In 1993, in spite of efforts and agreements, less than 0.1 per cent of traded tropical timber came from sustainable forest (Brown *et al.* 1993: 162).

Landuse policy changes that favour forests and woodlands

In Europe, recent changes in agricultural policy have started to promote the conversion of cropland or pasture to alternative landuses including forestry. Within the EU it is likely that there will be increased plantation forestry, on-farm woodland, and multiple-use forests for timber and recreation or conservation. The UK Government announced plans in 1991 to compel farmers to preserve hedgerows with special landscape, conservation or historical value and to provide grants for the management of hedgerows to support wildlife (*The Times* 30/8/91: 14). Forest or woodland may be left untouched; it may be subjected to consumptive use (full or partial clearing, at infrequent, occasional or frequent intervals and extraction of natural products) or non-consumptive use (watershed protection, recreational and tourism use, conservation, etc).

Conservation

Conservation has a wide range of meanings. In general, though, it involves seeking one or both, of two main goals: to protect habitats, and to protect plants, animals or cultural features.

There are a number of ways of conserving plants and animals:

1 Areas may be protected as reserves or national parks (*in-situ* conservation) (Fig. 7.7c).
2 Plant germplasm can be banked as tissue cultures, dormant or deep-frozen seeds or bulbs, or as plants in botanic gardens (the Consultative Group on International Agricultural Research manages several collections of commercially useful species, also the FAO Plant Ecology Unit) (*ex-situ* conservation).
3 Animals can be conserved as zoo specimens, as domestic pets, and as frozen germplasm (provided there is a suitable surrogate mother) (*ex-situ* conservation).
4 It may be possible to conserve organisms in environments subject to other uses (e.g. farms, plantations, road verges, paddy-fields, reservoirs, city parks, etc), if need be by providing special conservation support features. As a general rule, the more intensive the land use, the less conservation opportunities it offers also, some organisms are much more sensitive to disturbance than others (*in-situ* conservation). Despite decades of experience in establishing watershed protection forests, there is still debate over their value for controlling erosion and streamflow (Smiet 1987).

Conservation which involves the selection of elements of flora or fauna (for gene banks or botanic gardens and zoos) risks the loss of what the collector(s) overlook or are forced by time or funding restrictions to miss. The dangers with such 'selective conservation' are, in addition to selectivity, that storage may break down or be destroyed by warfare, fire or some disaster; that stored material may have a 'shelf-life' and require periodic replenishment or re-breeding and that fails; that vested interests might wrest control of the conserved material and restrict access to it (in 1970 UNESCO promoted the Convention on the Means of Prohibiting and Preventing Illicit Import, Export and Transfer of Ownership of Cultural Property to counter these problems) (Juma 1989: 95). At the time of writing the legal status of *ex-situ* genetic material was still not settled.

Many collections of genetic resources could be made more secure and more accessible (see Juma 1989: 99), to try and overcome some of the threats the International Board for Plant Genetic Resources is seeking to establish a secure germplasm cold-store in the permafrost of Spitzbergen where there can be no breakdown of refrigeration, where the site is high enough to avoid flood or storm damage and is protected below ground. Access to the collection will be open, the contents will be published and the Board will encourage others to set up duplicate facilities elsewhere in the world. Even if collections are much improved, there is a need for some conservation to take place 'in the wild' where plant and animal associations can continue to evolve alongside each other (Smith *et al.* 1991: 8). Traditional crop or animal varieties might be kept by farmers paid a grand to do so to compensate for possibly less yield than with modern varieties.

Areas may be available for conservation because they are selected, more often because development has somehow bypassed them or they have been abandoned. Remote, steep, swampy and dangerous localities are likely to be available, also sites of deep cultural significance like sacred groves or burial grounds. However, reliance on historic accident to designate conservation sites is unsatisfactory and various bodies have become active in identifying what should be conserved. Having decided to conserve an area, it is desirable to make a taxonomic and environmental survey and determine a management plan. It is to be hoped that the chosen area will be large enough and secure enough from human interference and environmental changes or major disasters to ensure natural rates of extinction are not exceeded (Miller 1978: 194; Higgs 1981; Proctor 1984). In practice, reserve areas are likely to become 'islands' in a 'sea' of development: competitor species may then arrive, the surrounding environment may change and affect the conservation area and if there is change there may be no routes along which organisms can move to escape or gain opportunities to adapt (Grainger 1993b: 151). In an ideal world, reserve areas should be safely bigger than they need to be to ensure long-term conservation and withstand likely environmental change and human challenges.

Recognition of the best conservation areas may be difficult: often the best may have already been lost. There has been controversy over the existence and location of species-rich 'refugia' areas of Amazonia (and to a lesser extent West Africa) (Prance and Lovejoy 1985; Colinvaux 1989; B. Freedman 1989: 290; Aiken and Leigh 1992: 3; Whitmore and Sayer 1992: 125). It is widely accepted that those refugia (Fig. 7.10a and 7.10b) mark areas of greatest environmental stability: areas moist enough to sustain forest when, around 18 000 BP, Amazonia and other of the world's tropical forest regions probably became drier. Where the best areas have gone or are too vulnerable, then conservation may have to make do with others.

Great care is needed to ensure that local people are not dislocated or disadvantaged by creation of conservation areas. This has often been the case, and disadvantaged locals are likely to encroach on conservation areas and damage them. Wherever possible people should be involved from early planning stages in demarcating and managing such areas. If local people understand the value, participate in the management of and benefit from conservation areas, such sites have more chance of success (I. Anderson and Grove 1987: 85; Bunyard 1989; Redford and Padoch 1992).

There are also the moral and practical issues involved with ownership of genetic resources. Local people are seldom paid if a plant or animal is taken from their land and developed. It is not unusual for ethnobotanists to question rural folk on their uses of wild materials and then for pharmaceutical or seed companies to use that knowledge for profit. Ideally there should be some sort of royalty fee to assist conservation and improve the welfare of people who assisted commerce.

Conservation legislation is sometimes altered to allow damaging development, typically for 'strategic' reasons, for example Martin (1991) reports on an African nation in need of foreign exchange which reclassified a large part of a national park as a game reserve to enable logging to proceed (similar behaviour is not unknown in DCs). Sometimes there is a failure to monitor and enforce controls and sometimes they are unsuitable. In some countries

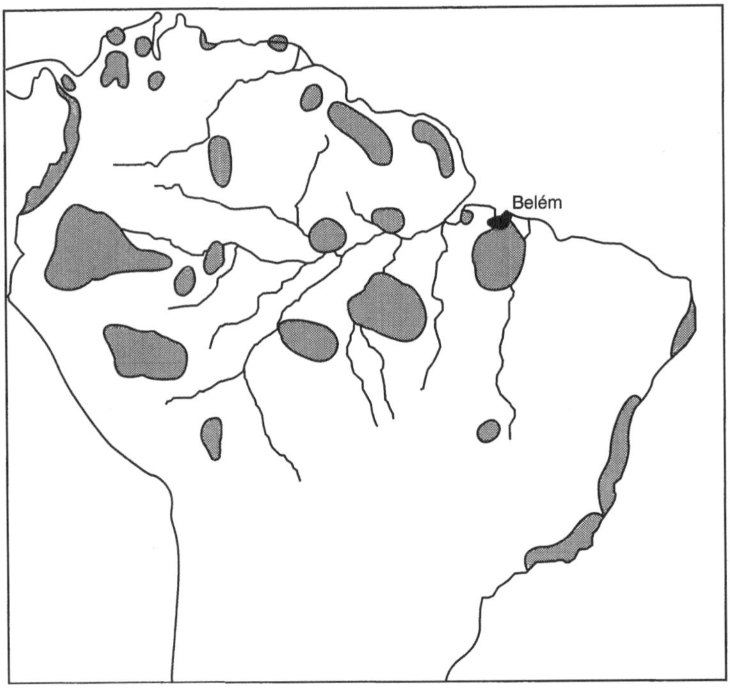

a)

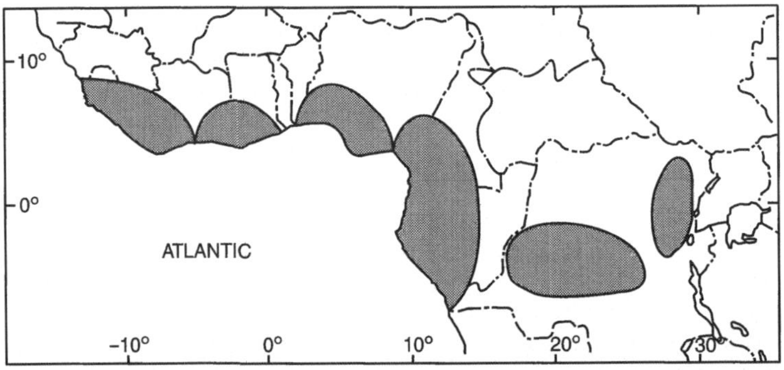

b)

Fig. *7.10* Lowland tropical rain forest 'refugia'. a) postulated for South America. b) areas of Africa which it has been suggested are refugia (*Source:* b) based on Meggars *et al.* 1973: 81).

forest laws date back centuries; in others they are recent, perhaps inappropriate applications of a colonial power's own legal system which have been ignored since independence. Conservation legislation may need modernization to meet needs adequately.

In 1972 the International Convention for the Protection of the World Cultural and Natural Heritage was drawn up by the UN. It came into force in 1975 and has been ratified by over 108 nations, providing an *international framework* for safeguarding humankind's heritage, including areas of scientific interest or outstanding natural beauty. The World Heritage List now contains over 315 World Heritage Sites, some of considerable extent.

Where trade is 'driving' environmental degradation or species loss, it may be possible to improve things by trade and sales restrictions, e.g. the Convention on International Trade in Endangered Species (CITES)

Fig. *7.11* Tolerant forest management. a) Typical homestead and modified forest cover, 5 km from Belém, Amazonian Brazil.

1975 or import bans and consumer education in the market countries.

Conservation must be paid for: people denied access to landuse may require compensation or relocation, and reserves may need policing and management. There has not been a lot of study of the value of conservation in economic terms, although one attempt found that in the late 1970s the contribution of plant and animal species to the US economy was about 4.5 per cent of GDP (roughly US$87 000 million) (McNeely *et al.* 1990: 18). If the value were better known, conservation might be given more support.

Debt-for-nature-swaps

Provided that they do not conflict with conservation, forest extraction, tourism or sport-use may help generate funds. After the mid-1980s debt-for-nature-swaps have been negotiated in several countries (see also chapter 11). These are felt by some, but not all, to provide a promising means for the recipient to 'pay-off' some foreign debt with less loss of face than defaulting. Recipients get aid for conservation (or other acceptable environmental management activities) and maintain control over the management of the conservation (or other activity) (Cartwright 1989: 124; McNeely 1989; Pearce 1989; Shiva *et al.* 1991: 63; Park 1992: 155).

A common form of debt-for-nature-swap works as follows: a bank decides that the chances that a loan debt will be repaid are slim; it sells the loan to another party at a discounted price – say an environmental group in the USA. A US$650 000 debt might be sold for US$100 000. The group then lets the debtor country pay them back the US$100 000 in local currency by doing conservation or environmental work to that value (US$100 000). Recent examples include the Korup Forest Reserve, Cameroon. (Hayter 1989: 258; Yearley 1991: 182; Park 1992: 135).

Where conservation benefits a particular producer or group of producers of a commmodity, say citrus or cocoa growers, it may be possible for them to provide funding. That would help remedy the anomaly, whereby 'gene-poor' developed countries dominate world food and commodity production. Bodies like the World Bank, NGOs and many national aid agencies have raised spending on conservation since

Fig. 7.11 Tolerant forest management. b) Fish trap used by riverfolk, lower Tocantins river, Pará, Brazil.

the mid-1980s and a 'revision' of the World Conservation Strategy has been released (IUCN, UNEP and WWF 1991).

Tolerant forest management

People in forests or woodland environments have long gathered natural products for their own use and sometimes for sale, for example in semiarid African savannas, acacias yield gum arabic (an important export crop for countries like the Sudan); in Canada's woodlands, maples yield syrup (worth in 1982 C$40 million per annum); virtually all of Brazil's rubber and Brazil nuts are still gathered from wild trees. Sometimes forest extraction can develop into what might be described as 'tolerant forest management': useful species are encouraged and the understorey may be cleared a little to improve access, but the treecover is left largely intact and many un-utilized species survive. Recent studies in Pará, Amazonia, reveal that tolerant forest managers can sustain a very good livelihood with the forest cover only slightly disturbed. If these gatherers can win higher profits by circumventing exploitative middlemen both their livelihoods and conservation may benefit and more may turn from destructive extraction to less damaging tolerant management. Considerable, sometimes controversial, publicity has been attracted to efforts by Anita Roddick, the owner of the 'Body Shop' stores, to support and encourage such extraction. In the long term extraction of forest, woodland or savanna products may well offer a far better return than careless removal of timber followed by land degradation.

Tolerant forest management (extraction of products leaving as much as possible of the forest cover intact) is a strategy that can provide people who might otherwise clear forest with a living and an incentive to protect and manage their resources. Useful species are encouraged and the understorey may be cleared a little to improve their growth and access for harvesting; a treecover is maintained and even in disturbed areas many non-utilized species survive; away from homesteads the forest is little altered. There have been a number of cases described in recent years of profitable and sustained tolerant forest management, e.g. Ilha Combu and Ilha das Onças, near Belém, Brazilian Amazonia.

Ilha Combu This island lies 2.5 km from Belém, a city at the mouth of the Amazon, and is inundated by seasonal and tidal flooding. About 15 square kilometres in area, the island supports about 600 riverfolk. Family groups cultivate gardens and tend a zone of managed forest surrounded by unmanaged forest, fishing provides protein and extra income. Management consists of opening paths to get to stands of useful trees and promoting the growth of those species (Nepstad and Schwartzmans 1992: 71). Açai (*Euterpe oleraceae* and *E. precatoria*) is the main source of income, but a diverse range of other products are extracted including other palms, rubber, cocoa, etc. In spite of a sharecropper situation, average incomes were US$1800 in 1992 (quite a reasonable livelihood). Whether açai (for fruit and palmhearts) could offer such returns further from a city market is uncertain, however, the diversity of actual and potential crops means there is a chance extractors could avoid swamping markets (for a discussion of potential crops see (Barrow 1990: 375; A.B. Anderson *et al.* 1991).

Ilha das Onças Studies on Ilha das Onças (also near Belém), reveal that tolerant forest managers there sustain a very good livelihood, with the forest cover only slightly disturbed. Açai fruit alone gave a net average income of US$109.83 per hectare per year (A.B. Anderson 1990: 81) (Figs 7.11a and 7.11b).

References

Adriawan, E. and Moniaga, S. (1986) The burning of a tropical forest: fire in East Kalimantan. What caused this catastrophy? *The Ecologist* 16(6): 269–70

Agarwal, B. (1986) *Cold hearths and barren slopes: the woodfuel crisis in the third world*, London, Zed

Aiken, S.R. and Leigh, C.H. (1985) On the declining fauna of Peninsular Malaysia in the post-colonial period. *Ambio* XIV(1): 15–22

Aiken, S.R. and Leigh, C.H. (1987) Queensland's Daintree Rainforest at risk. *Ambio* XVI(2–3): 131–41

Aiken, S.R. and Leigh, C.H. (1992) *Vanishing rain forests: the ecological transition in Malaysia*. Oxford, Clarendon

Allen, J.C. and Barnes, D.F. (1985) The causes of deforestation in developing countries. *Annals of the Association of American Geographers* 75(2): 163–84

Alpert, P. (1993) Conserving biodiversity in Cameroon. *Ambio* XXII(1): 44–9

Anderson, A.B. (1990) Extraction and forest management by rural inhabitants in the Amazon estuary. In A.B. Anderson (ed.) *Alternatives to deforestation: steps toward sustainable use of Amazon rainforest*. New York, Columbia University Press

Anderson, A.B., May, P.H. and Balick, M.T. (1991) *The subsidy from nature: palm forests, peasantry, and development on the Amazon frontier*. New York, Columbia University Press

Anderson, I. and Grove, R. (eds) (1987) *Conservation in Africa: people, policies and practice*. Cambridge, Cambridge University Press

Anderson, R.S. (1988) *The hour of the fox: tropical forests the World Bank and indigenous people in central India*. Seattle, Wash., University of Washington

Andrasko, K. (1990) Global warming and forests: an overview of current knowledge. *Unasylva* 41(163): 3–11

Angel, D.J.R., Comer, J.D. and Wilkinson, M.L.N. (eds) (1990) *Sustaining earth: responses to environmental threats*. London, Macmillan

Anon (1988) Chopping down rainfall. *New Scientist* 118(1612): 38

Anon (1990) Texas finds fresh fibres for tomorrows needs. *New Scientist* 125(1703): 35

Arnold, J.E.M. (1987) Community forestry. *Ambio* XVI(2–3): 122–8

Aubréville, A.M.A. (1985) The disappearance of the tropical forests of Africa. *Unasylva* 37(2): 18–27

Barr, B.M. and Branden, K. (1988) *The disappearing Russian forest: a dilemma in Soviet resource management*. Totowa, NJ., Rowman and Littlefield

Barrett, S.W. (1980) Conservation in Amazonia. *Biological Conservation* 18: 209–35

Barrow, C.J. (1990) Environmentally appropriate, sustainable small-farm strategies for Amazonia. In D. Goodman and A. Hall (eds) *The future of Amazonia: destruction or sustainable development*. London, Macmillan

Barrow, C.J. (1991) *Land degradation: development and the breakdown of terrestrial environments*. Cambridge, Cambridge University Press

Battistini, R. and Richard-Vindard, G. (eds) (1972). *Biogeography and ecology in Madagascar*. The Hague, Dr W. Junk

Biswanger, H.P. (1991) Brazilian policies that encourage deforestation in the Amazon. *World Development* 19(7): 821–9

Bourke, G. (1987) Forests in the Ivory Coast face extinction. *New Scientist* 114(1564): 22

Boutier, F. (1983) *Tropical savannas* (ecosystems of the world no. 13). Amsterdam, Elsevier

Bowander, B. (1987) Environmental problems in developing countries. *Progress in Physical Geography* 11(2): 246–59

Branford, S. (1988) Choking the Amazon. *New Scientist* 116(1581): 67–8

Branford, S. and Glock, O. (1985) *The last frontier: fighting for land in the Amazon*. London, Zed

Breunig, E.F. (1987) The forest ecosystem: tropical and boreal. *Ambio* XVI(2–3): 68–85

Briggs, D.J. and Courtney, F.M. (1985) *Agriculture and environment: the physical geography of temperate agricultural systems*. London, Longman

Brown, L.R. *et al.* (1993) *State of the world 1993*. New York, W.W. Norton

Brush, S.B. (1987) Diversity and change in Andean agriculture. In P.D. Little and M.M. Horowitz (eds) *Lands at risk in the third world: local-level perspectives*, Boulder, Colo., Westview

Buckman, R.E. (1987) Strengthening forestry institutions in the developing world. *Ambio* XVI(2–3): 120–1

Banyard, P. (1989) Guardians of the Amazon. *New Scientist* 124(1695): 38–41
Burbridge, P.R. (1988) Coastal and marine resource management in the Straits of Malacca. *Ambio* XVII(3): 170–7
Burman, A. (1991) Saving Brazil's savannas. *New Scientist* 129(1758): 30–4
Calabri, G. (1983) Fighting fires in Mediterranean forests. *Unasylva* 35(141): 14–21
Caldecott, J. (1988) Climbing towards extinction. *New Scientist* 118(1619): 62–6
Cartwright, J. (1989) Conserving nature, decreasing debt. *Third World Quarterly* 11(2): 114–27
Caufield, C. (1982) *Tropical moist forest: the resource, the people, the threat.* London, Earthscan
Caufield, C. (1985) *In the rainforest.* London, Pan
Chambers, R., Leach, M. and Conroy, C. (1993) Trees as savings and security for the rural poor. *IIED Gatekeeper Series*, no. 3. London, IIED, 15 pp.
Chapman, V.J. (1975) *Mangrove vegetation.* Vaduz, J. Cramer
Cherfas, J. (1986) How to grow a tropical forest. *New Scientist* 112(1531): 26–7
Chown, M. (1986) The red island is bleeding to death. *New Scientist* 111(1524): 57–8
Clark, S.S. and McLoughlan, L.C. (1986) Historical and biological evidence for fire regimes in the Sydney region prior to the arrival of Europeans: implications for future bushland management. *Australian Geographer* 17(2): 101–12
Cleary, D. (1992) *The Brazilian rainforest: politics, finance, mining and the environment* (The Economist Intelligence Unit Special Report no. 2100). London, The Economist
Colchester, M. (1990) The International Tropical Timber Trade Organization: kill or cure for the rainforest. *The Ecologist* 20(5): 166–73
Cole, M.M. (1986) *The savannas: biogeography and geobotany.* London, Academic Press
Colinvaux, P.A. (1989) The past and future Amazon. *Scientific American* 260(5): 68–74
Collins, J.L. (1986) Smallholder settlement of tropical South America: the social causes of ecological destruction. *Human Organization* 45(1): 1–10
Collins, N.M., Sayer, J.A. and Whitmore, T.C. (eds) (1991) *The conservation atlas of tropical forests: Asia and the Pacific.* London, Macmillan
Cooper, C.F. (1974) The ecology of fire. In Scientific American (ed.) *Ecology, evolution and population biology.* San Francisco, Calif, Freeman (Scientific American Publications)
Cross, M. (1988) Spare the tree and spoil the forest. *New Scientist* 120(1640): 24–5
Crutzen, P.J. and Goldammer, J.G. (eds) (1993) *Fire in the environment: the ecological, atmospheric, and climatic importance of vegetation fire.* Chichester, Wiley
Davidson, J., Thow Yow Pong, and Bijveld, M. (eds) (1985) *The future of the tropical rainforest areas in South East Asia* (Proceedings of a Symposium Organized by the Forest Research Institute, Kepong, Malaysia and IUCN Commission on Ecology, 1–2 September 1983, Kepong, Malaysia). Gland, Switzerland, IUCN
Davidson, J., Myers, D. and Chakraborty, M. (1992) *No time to waste: poverty and the global environment.* Oxford, OXFAM
Dayton, L. (1991) On the saving of the species. *New Scientist* 129(1752): 25–6
De Booysen, P. and Tainton, N.M. (eds) (1984) *Ecological effects of fire in S. African ecosystems.* Berlin, Springer–Verlag
De Groot, P. (1990) Are we missing the grass for the trees? *New Scientist* 125(1698): 29–30
De Laune, R.D. and Patrick, W.H. (1987) Forseeable flooding and death of coastal wetland forests. *Environmental Conservation* 14(2): 129–33
De Vos, A. (1975) *Africa the devastated continent: man's impact on the ecology of Africa.* The Hague: Dr W. Junk
Diem, A. (1992) Clear cutting British Columbia. *The Ecologist* 22(6): 261–6
Douglas, I. and Spencer, T. (eds) (1985) *Environmental change and tropical geomorphology.* London, Allen and Unwin
Dove, M.R. (1993) A revisionist view of tropical deforestation and development. *Environmental Conservation* 20(1): 17–24
Dudley, N., Jeanrenaud, J-P. and Sullivan, P. (1994) *Bad harvest: the timber trade and the degradation of global forests.* London, Earthscan
Dunbar, R. (1984) Scapegoat for a thousand deserts. *New Scientist* 104(1430): 30
Dunn, N. and Clery, D. (1992) Fast-growing reeds could fuel Europe's future. *New Scientist* 134(1820): 18
Eckholm, E.P. (1976) *Losing ground: environmental stress and world food prospects.* Oxford, Pergamon
Eckholm, E.P. (1979) *Planning for the future: forestry for human needs* (Worldwatch Paper no. 26). Washington, DC, Worldwatch Institute
Eckholm, E.P. (1982) *Down to earth.* London, Pluto
Eckholm, E.P., Foley, G., Bernard, G. and Timberlake, L. (1984) *Fuelwood: the energy crisis that won't go away.* London, Earthscan
Eden, M.J. (1990) *Ecology and management in Amazonia.* London, Belhaven
El-Ashry, M.T. (1985) Study shows mountain ecosystems may be especially sensitive to acid deposition. *Ambio* XIV(3): 179–80
Eyre, L.A. (1987) Jamaica: test case for tropical deforestation? *Ambio* XVI(6): 338–43
Ezealor, A.U. (1987) New threats to Nigeria's savanna woodlands. *Environmental Conservation* 14(3): 262–4
FAO (1981) *Eucalyptus for planting* (2nd edn). FAO Forestry Series 11, Rome, Food and Agriculture Organization
Fearnside, P.M. and Rankin, J.M. (1980) Jarí and development in the Brazilian Amazon. *Interciencia* 5(3): 146–56
Ferguson, J. (1988) Desert in the making. *New Scientist* 118(1613): 68
Fernando, V. (1989) The problem of shifting cultivation. *IUCN Bulletin* 20(1–3): 18
Flenley, J.R. (1979) *The equatorial rainforest: a geological history.* London, Butterworth
Flint, M. (1991) *Biological diversity and developing countries: issues and options.* London, Overseas Development Administration
Foley, G. (1991) *Global warming: who is taking the heat?* London, PANOS

Foley, G. and Bernard, G. (1984) *Farm and Community Forestry* (IIED Technical Report no. 3 – Energy Information Programme). London, Earthscan

Foley, G., Moss, P. and Timberlake, L. (1984) *Stoves and trees: how much wood would a woodstove save if a woodstove could save wood?* London, Earthscan

Fortes, M.D. (1988) Mangroves and seagrass beds of East Asia: habitats under stress. *Ambio* XVII(3): 207–13

Freedman, B. (1989) *Environmental ecology: the impacts of pollution and other stresses on ecosystem structure and function.* London, Academic Press

Freedman, R. (1989) Grazing cattle can change local climate. *New Scientist* 123(1680): 30

Gamlin, L. (1988) Sweden's factory forests. *New Scientist* 117(1597): 41–7

Garlick, J.P. and Keay, R.W. (eds) (1970) *Human ecology in the tropics.* Oxford, Pergamon

Gee, H. (1988) Alarm bells on the forest fire front. *The Times* 22/7/88: 11

Gradwohl, J. and Greenberg, R. (1988) *Saving the tropical forests.* London, Earthscan

Grainger, A. (1993a) Rates of deforestation in the humid tropics: estimates and measurements. *Geographical Journal* 159(1): 32–44

Grainger, A. (1993b) *Controlling tropical deforestation.* London, Earthscan

Hadley, M. and Lanly, J.P. (1983) Tropical forest ecosystems: identifying differences, seeking similarities. *Nature and Resources* XIX(1): 2–19

Hall, A. (1989) *Developing Amazonia: deforestation and conflict in Brazil's Carajás Programme.* Manchester, Manchester University Press

Halperin, D.T. (1980) The Jarí Project: large-scale land and labor utilization in the Amazon. *Geographical Survey* 9: 13–21

Harris, D.R. (ed.) *Human ecology in savanna environments.* London, Academic Press

Harris, L.D. (1984a) *The fragmented forest: island biogeographic theory and the preservation of biotic diversity.* Chicago, University of Chicago Press

Harris, L.D. (1984b) The fragmented forest. Island biogeography: fact or fiction? *Journal of Biogeography* 7(3): 209–35

Harrison, J., Miller, K. and McNeely, J. (1982) The world coverage of protected areas: development goals and environmental needs. *Ambio* XI(5): 238–45

Harrison, P. (1987a) A tale of two stoves. *New Scientist* 114(1562): 40–3

Harrison, P. (1987b) Trees for Africa. *New Scientist* 114(1560): 54–7

Hawkes, J.G. (ed.) (1978) *Conservation and agriculture.* London, Duckworth

Hayter, T. (1989) *Exploited earth: Britain's aid and the environment.* London, Earthscan

Hecht, S. (1985) Environment, development and politics in Eastern Amazonia. *World Development* 13(6): 663–84

Hecht, S.B. (1989) The sacred cow in the green hell: livestock and forest conservation in the Brazilian Amazon. *The Ecologist* 19(6): 229–34

Hemming, J. (ed.) (1985) *Change in the Amazon Basin, vol. 1: man's impact on forests and rivers.* Manchester, Manchester University Press

Higgs, A.J. (1981) Island biogeography theory and active research design. *Journal of Biogeography* 8(1): 17–24

Holdgate, M.W. and Woodman, M.J. (1976) *The breakdown and restoration of ecosystems.* New York, Plenum

Holland, P. and Olson, S. (1989) Introduced versus native plants in austral forests. *Progress in Physical Geography* 13(2): 260–94

Hooper, M.D. (1979) Hedges and small woodlands. In J. Davidson and R. Lloyd (eds) *Conservation and agriculture.* Chichester, Wiley

Hooper, M.D. and Holdgate, M.W. (eds) (1970) *Hedges and hedgerow trees* (Monks Wood Experimental Station Symposium no. 4). London, Nature Conservancy Council

Horta, K. (1991) The last big rush for the green gold: the plundering of Cameroon's rainforest. *The Ecologist* 21(3): 142–7

Hurst, P. (1990) *Rainforest politics: ecological destruction in South East Asia.* London, Zed

Hutchings, P. and Saenger, P. (1987) *Ecology of mangroves.* St Lucia, Australia, University of Queensland Press

Independent Commission on International Humanitarian Issues (1986) *The vanishing forest: the human consequences of deforestation.* London, Zed

Isomäki, R. (1991) Paper, pollution and global warming: unsustainable forestry in Finland. *The Ecologist* 21(1): 14–7

IUCN (1975a) *The use of ecological guidelines for development in the tropical forest areas of South East Asia* (Proceedings of a Meeting held at Bandung, Indonesia 29 May–1 June 1974, IUCN Publications new series no. 32). Morges, Switzerland, IUCN

IUCN (1975b) *Ecological guidelines for development in the American humid tropics* (Proceedings of an International Meeting, Caracas, Venezuela, 20–2 February 1974: IUCN Publication new series no. 31). Morges, Switzerland, IUCN

IUCN (1985) The future of tropical forests in South East Asia. *The Environmentalist* 5 (supplement no. 10)

IUCN, UNEP and WWF (1991) *Caring for the earth: a strategy for sustainable living.* London, Earthscan

Ives, J. and Pitt, D.C. (eds) (1988) *Deforestation: social dynamics in watersheds and mountain ecosystems.* London, Routledge

Jackson, W., Berry, W. and Colman, B. (eds) (1984) *Meeting the expectations of the land: essays in sustainable agriculture and stewardship.* San Francisco, Calif., North Point Press

James, B. (1989) A harvest from the rain forest. *The Times* 15/2/89: 12

Johnstone, B. (1987) Japan saps the world's rain forests. *New Scientist* 114(1554): 18

Joss, P.J., Lynch, P.W. and Williams, O.B. (eds) (1986) *Rangelands: a resource under siege*, Cambridge, Cambridge University Press

Joyce, C. (1988) The tree that caused a riot. *New Scientist* 117(1600): 454–9.

Juma, C. (1989) *The gene hunters: biotechnology and the scramble for seeds.* London, Zed

Kardell, L., Steen, E. and Fabaio, A. (1986) Eucalyptus in Portugal: a threat or a promise? *Ambio* XV(1): 6–13

Kartawinata, K., Adsoemarto, S., Riswan, S. and Vadya, A.P. (1981) The impact of man on tropical forest in Indonesia. *Ambio* X(2–3): 115–19

Kemf, E. (1988) The re-greening of Vietnam. *New Scientist* 118(1616): 52–7
Klein, R.M. and Perkins, T.D. (1987) Cascades of causes and effects in forest decline. *Ambio* XVI(2–3): 86–93
Kozlowski, T.T. and Ahlgren, C.E. (eds) (1974) *Fire and ecosystems*. New York, Academic Press
Kunstadter, P., Bird, E.C.F. and Sabhasri, S. (eds) (1986) *Man in the mangroves: the socioeconomic situation of human settlement in mangrove forests*. Tokyo, UN University (E.86.111.A7)
Kuusela, K. (1987) Forest products: world situation. *Ambio* XVI(2–3): 80–5
Kuusela, K. (1992) Boreal forests: an overview. *Unasylva* 43(170): 2–13
Lal, R., Sanchez, P.A. and Cummings, R.W. jnr (eds) (1986) *Land clearing and development in the tropics*. Rotterdam, A.A. Balkema
Leach, G. and Mearns, R. (1989) *Beyond the woodfuel crisis: people, land and trees in Africa*. London, Earthscan
Lewin, R. (1984) Parks: how big is big enough? *Science* 255: 611–12
Lewis, R.R. III (1982) *Creation and restoration of coastal plant communities*. Boca Raton, Fla., CRC Press
Lewis, L.A. and Berry, L. (1988) *African environments and resources*. London, Allen and Unwin
Lewis, L.A. and Coffey, W.J. (1985) The continuing deforestation of Haiti. *Ambio* XIV(3): 150–60
Linear, M. (1985) *Zapping the third world: the disaster of development aid*. London, Zed
Lindén, O. and Jernelöv, A. (1980) The mangrove swamps: an ecosystem in danger. *Ambio* IX(2): 81–8
Little, P.D. and Horowitz, M.M. (eds) (1987) *Lands of risk in the third world: local perspectives*. Boulder, Col., Westview
Lohmann, L. (1990) The scourge of eucalyptus in Thailand. *The Ecologist* 20(1): 9–17
Lovejoy, T., Bierregarrd, R.O., Rankin, J. and Schubart, H.O.R. (1983) Ecological dynamics of tropical forest fragments. In S.L. Sutton, T.C. Whitmore and A.C. Chadwick (eds) *Tropical rain forest: ecology and management*. Oxford, Basil Blackwell
Lutzenberger, J. (1987) Who is destroying the Amazon rainforest? *The Ecologist* 17(4): 155–60
Macedo, D.S. and Anderson, A.B. (1993) Early ecological changes associated with logging in the lower Amazon floodplain. *Biotropica* 25(2): 151–63
MacKenzie, D. (1988) Uphill battle to save Filipino trees. *New Scientist* 118(1619): 42–3
McCloskey, J.M. and Spalding, H. (1989) A reconnaissance-level inventory of the amount of wilderness remaining in the world. *Ambio* XVIII(4): 221–7
McNeely, J.A. (1989) How to pay for conserving biological diversity. *Ambio* XVIII(6): 303–18
McNeely, J.A., Miller, K.R., Reid, W.V., Mittermeier, R.A. and Werner, T.B. (1990) Strategies for conserving biodiversity. *Environment* 32(3): 16–20, 36–40
Mahar, D.J. (1990) Deforestation in Brazil's Amazon region: magnitude, role and causes. In G. Schramm and J.J. Warford (eds) *Environmental management and economic development*. Baltimore, Md., Johns Hopkins University Press
Malingreau, J.P. and Tucker, C.J. (1988) Large-scale deforestation in the southeastern Amazon basin of Brazil. *Ambio* XVII(1): 49–55
Malingreau, J.P., Stephens, G. and Fellows, C. (1985) Remote sensing of forest fires: Kalimantan and North Borneo in 1982–83. *Ambio* XIV(6): 314–21
Maltby, E. (1986) *Waterlogged wealth: why waste the world's wet places?* London, Earthscan
Mannion, A.M. and Bowlby, S.R. (eds) (1992) *Environmental issues in the 1990s*. Chichester, Wiley
Martin, C. (1991) *The rainforests of West Africa*. Berlin, Birkhäuser-Verlag
Mather, A.S. (1990) *Global forest resources*. London, Belhaven
Mather, A.S. (1992) Recent changes in afforestation rates in Britain. *Geography* 77(3): 270–4
Meggars, B.J., Ayensu, E.S. and Duckworth, W.D. (eds) (1973) *Tropical forest ecosystems in Africa and South America: a comparative review*. Washington, DC, Smithsonian Institution Press
Miller, R.I. (1978) Applying island biogeographic theory to an East African reserve. *Environmental Conservation* 5(3): 191–5
Molion, L.C.B. (1989) The Amazonian forests and climatic stability. *The Ecologist* 19(6): 211–13
Mooney, P.R. (1979) *Seeds of the earth*. London, International Coalition for Development Action
Munslow, B., Katerere, Y., Ferf, A. and O'Keefe, P. (1988) *The fuelwood trap: a study of the SADCC region*. London, Earthscan
Myers, N. (1980) *Conservation of tropical moist forest* (Report for Committee on Research Priorities in Tropical Biology of the National Research Council). Washington, DC, National Academy of Sciences
Myers, N. (1985) *The primary source: tropical forests and our future*. London, Norton
Myers, N. (1986) Economics and ecology in the international arena; the phenomenon of 'linked linkages'. *Ambio* XV(5): 296–300
Myers, N. (1987a) Book review: Tropical Forest Agriculture. *Forest Ecology and Management* 20: 321–4
Myers, N. (1987b) Seeing the wood for the trees. *Nature* 330(6145): 286
Myers, N. (1988) Tropical deforestation and remote sensing. *Forest Ecology and Management* 23: 215–25
Myers, N. (1993) Tropical forests: the main deforestation fronts. *Environmental Conservation* 20(1): 9–16
Nadkarni, M.V., Pasha, S.A. and Prabhakar, L.S. (1989) *The political economy of forest use and management*. New Delhi, Sage
National Academy of Sciences (1990) *Saline agriculture: salt-tolerant plants for developing countries*. Washington, DC, National Academy Press
Nepstad, D.C. and Schwartzmans, S. (eds) (1992) Non-timber products from tropical forests: evolution of a conservation and development strategy. *Advances in Economic Botany 9*. New York, New York Botanic Gardens
Nilsson, S. and Sallnäs, O. (1990) Air problems and European forests: policy implications based on simulation models. *Unasylva* 41(163): 34–41
OECD (1989) *Agricultural and environmental policies: opportunities for integration*. Paris, OECD

Olson, S. (1987) Red destinies: the landscape of environmental risk in Madagascar. *Human Ecology* 15(1): 67–89
Ong, Jin Eong (1982) Mangroves and aquaculture in Malaysia. *Ambio* XI(5): 252–7
Pádua, M.T. and Quintão, A.T.B. (1982) Parks and biological reserves in the Brazilian Amazon. *Ambio* XI(5): 309–14
Park, C. (1992) *Tropical rainforests.* London, Routledge
Parsons, J.P. (1975) The changing nature of New World tropical forests since European colonization. In IUCN (ed.) *Ecological guidelines for development in the American humid tropics.* Morges, Switzerland, IUCN
Pearce, F. (1989) Kill or cure? Remedies for the rainforest. *New Scientist* 123(1682): 40–3
Pearce, F. (1993) Botswana: enclosing for beef. *The Ecologist* 23(1): 25–9
Petrof, D. (1992) Siberian forests under threat. *The Ecologist* 22(6): 267–70
Pollard, E., Hooper, M.D. and Moore, N.W. (1974) *Hedges.* London, Collins
Poore, D., Burgess, P., Palmer, J., Rietbergen, S. and Synnott, T. (1989) *No timber without trees: sustainability in the tropical forest.* London, Earthscan
Porter, G. and Brown, J.W. (1991) *Global environmental issues.* Boulder, Colo., Westview
Prance, G.T. (1990) Deforestation: a botanist's view. In D.J.R. Angel *et al.* (eds) *Sustaining earth: response to environmental threats.* Macmillan, London
Prance, G.T. and Lovejoy, T.E. (eds) (1985) *Key environments: Amazonia.* Oxford, Pergamon
Price, M.F. (1987) Tourism and forestry in the Swiss Alps: parasitism or symbiosis? *Mountain Research and Development* 7(1): 1–12
Prior, J. and Tuohy, J. (1987) Fuel for Africa's fires. *New Scientist* 115(1571): 48–51
Proctor, J. (1984) Tropical forest conservation. *Progress in Physical Geography* 8(3): 443–9
Redford, K.H. and Padoch, C. (eds) (1992) *Conservation of neotropical forests: working from traditional resource use.* New York, Columbia University Press
Repetto, R. (1987) Creating incentives for sustainable forest development. *Ambio* XVI(2–3): 94–9
Repetto, R. (1988) *The forest for the trees? Government plans and the misuse of forest resources.* Washington, DC, World Resources Institute
Repetto, R. and Gillis, M. (1988) *Public policies and the misuse of forest resources.* Cambridge, Cambridge University Press
Reynolds, E.R. and Thompson, F.B. (eds) (1989) *Forests, climate and hydrology. Regional impacts.* Tokyo, UN University
Rietbergen, S. (ed.) (1993) *The Earthscan reader in tropical forestry.* London, Earthscan
Richards, J.F. and Tucker, R.P. (eds) *World deforestation in the twentieth century.* Durham, N.C., Duke University Press
Richards, P.W. (1973) The tropical rain forest. *Scientific American* 229(6): 58–67
Salati, E. and Nobre, C.E. (1993) Possible climatic impacts of tropical deforestation. *Climatic Change* 19(1): 177–96
Salati, E., Lovejoy, T.E. and Vose, P.B. (1983) Precipitation and water cycling in tropical rain forests with special reference to the Amazon Basin. *The Environmentalist* 39(1): 67–72
Sanders, D. and Hobbs, R. (1989) Corridors for conservation. *New Scientist* 121(1649): 63–7
Sattaur, O. (1987) Trees for the people. *New Scientist* 115(1577): 58–62
Sattaur, O. (1990) Last chance to save the rainforest plan? *New Scientist* 129(1758): 20–1
Saxena, N.C. (1992) Eucalyptus in farmlands in India: what went wrong? *Unasylva* 43(170): 53–8
Sayer, J. and NcNeely, J. (1984) IUCN, WWF and wetlands. *IUCN Bulletin* 15(4–6): 46
Sayer, J.A., Harcourt, C.S. and Collins, N.M. (1992) *The conservation atlas of tropical forests: Africa.* London, Macmillan (South American volume in preparation)
SCOPE (1983) *The role of fire in northern circumpolar ecosystems* (SCOPE Report no. 18). Chichester, Wiley
Sedjo, R.A. (1992) Temperate forest ecosystems in the global carbon cycle. *Ambio* XXI(4): 274–7
Shiva, V. and Bandyopadhyay, J. (1988) The Chipko Movement. In J. Ives and D.C. Pitt (eds) *Deforestation social dynamics in watershed and mountain ecosystems.* London, Routledge
Shiva, V., Anderson, P., Schücking, H., Gray, A., Lohmann, L. and Cooper, D. (1991) *Biodiversity: social and ecological perspectives.* London, Zed
Shyamsunder, S. and Parameswarappa, S. (1987) Forestry in India: the forester's view. *Ambio* XVI(6): 332–7
Simons, P. (1988) Costa Rica's forests are reborn. *New Scientist* 120(1635): 43–7
Sinclair, L. (1985) International Task Force plans to reverse tropical deforestation. *Ambio* XIV(6): 352–3
Smiet, F. (1987) Tropical watershed forestry under attack. *Ambio* XVI(2–3): 156–8
Smith, N.J.H., Williams, J.T. and Plunknett, D.L. (1991) Conserving the tropical cornucopia. *Environment* 33(6): 6–9, 30–2
Steinlin, H.J. (1982) Monitoring the world's tropical forests. *Unasylva* 34(137): 2–8
The Environmentalist (1985) vol. 5(10) supplement issue on tropical forests
Thirgood, J.V. (1981) *Man and the Mediterranean forest: a history of resource depletion.* London, Academic Press
Utting, P. (1993) *Trees, people and power: social dimensions of deforestation and forest protection in Central America.* London, Earthscan
Vanclay, J.K. (1993) Saving the tropical forest: needs and prognosis. *Ambio* XXII(4): 225–31
Van Der Hammen, T. (1972) Changes in vegetation and climate in the Amazon Basin and surrounding areas during the Pleistocene. *Geologie en Mijnbouw* 51: 641–3
Vannucci, M. (1988) The UNDP/UNESCO Mangrove Programme in Asia and the Pacific. *Ambio* XVII(3): 214–17
Vietmeyer, N.D. (1979) Tropical tree legumes: a front line against deforestation. *Ceres* 12(5): 38–41
Vines, B. (1987) Fire in the bush. *New Scientist* 113(1549): 49–52
Watts, D. (1966) *Man's influence on the vegetation of Barbados 1627 to 1800* (Occasional Papers in Geography no. 4). Hull, University of Hull
Weber, F.R. (with Stoney, C.) (1989) *Reforestation in arid lands* (2nd edn). Arlington, (1815 North Lynn Street, Arlington, VA, USA), Volunteers in Technical Assistance

Westoby, J. (1987) *The purpose of forests: follies of development*. Oxford, Basil Blackwell

Whitmore, T.C. and Sayer, J.A. (eds) (1992) *Tropical deforestation and species extinction*. London, Chapman and Hall

Wilson, E.O. (1989) Threats to biodiversity. *Scientific American* 261(3): 60–6

Winterbottom, R. and Hazlewood, P.T. (1987) Agroforestry and sustainable development: making the connection. *Ambio* XVI(2–3): 100–10

Wittle, J. (1992) Deforestation in Zaïre: logging and landlessness. *The Ecologist* 22(2): 58–64

Wolman, M.G. and Fournier, F.G. (eds) (1987) *Land transformation in agriculture*. Chichester, Wiley

World Resources Institute, World Bank and UNDP (1985) *Tropical forests: a call for action* (3 parts: *I The Plan*, 47 pp., *II Case Studies* 55 pp., *III Country Investment Profiles*, 22 pp.). Washington, DC, World Resources Institute (1735 New York Avenue, N.W. Washington DC 20006)

World Resources Institute, International Institute for Environment and Development and UNEP (1988) *World Resources 1988–89: an assessment of the resource base that supports the global economy*. New York, Basic Books

Wyatt-Smith, J. (1987) *The management of tropical moist forest for the sustained production of timber: some issues* (IUCN/IIED Tropical Forest Policy Paper no. 4). London, IUCN, IIED, WWF (3 Endsleigh Street, London WC1H 0DD)

Yearley, S. (1991) *The green case: a sociology of environmental issues, arguments and politics*. London, Harper Collins

Further reading

Caufield, C. (1985) *In the rainforest*. London, Pan [Introduction to rain forests and their problems]

Cleary, D. (1992) *The Brazilian rainforest: politics, finance, mining and the environment* (Special report no. 2100 of The Economist Intelligence Unit). London, The Economist [Recent study of forces driving Brazilian deforestation]

Collins, M. (1990) *The last rain forest*. London, Mitchell Beazley [Addresses the problem of forest loss]

Gradwohl, J. and Greenberg, R. (1988) *Saving the tropical forests*. London, Earthscan [Review of deforestation with several case studies, sections on sustainable agriculture, forests, natural forest management and restoration of tropical forests]

Hurst, P. (1989) *Rainforest politics: the destruction of forests in South East Asia*. London, Zed [Examines forces behind forest loss]

Mather, A.S. (1990) *Global forest resources*. London, Belhaven [Good introduction]

Whitmore, T. (1990) *An introduction to rainforests*. Oxford, Oxford University Press [Good introduction]

CHAPTER 8

Environmental problems associated with agriculture

Introduction

Once everyone had to accept what nature provided, but people have increasingly come to depend on agriculture to try and modify the environment to favour plant growth and livestock production. Agriculture may be seen as the 'stretching' of nature. Opportunities may be lost if there is no 'stretching' but risks are involved: 'careless stretching' or 'over-stretching' may cause unwanted environmental impacts or might result in a 'snap' and partial or total, temporary or permanent damage to food and commodity production. Maintaining 'optimum stretch' may be difficult, given the complexity of many agricultural processes and the external and internal pressures that may assail them.

Agriculture has been described as an 'open system' (Briggs and Courtney 1985: 19). The system receives natural inputs such as solar radiation; weathering products which provide nutrients; precipitation, genetic resources and human inputs which may include irrigation, agrochemicals, hydrocarbon fuel-based energy, human labour and know-how (Hobbelink 1991). The system loses energy and matter (outputs) as harvested crops; soil losses; moisture losses; material lost through leaching or oxidation; livestock waste. The world is finite, so provision of inputs and assimilative capacity (the ability of the environment to render output of pollutants harmless) are limited, yet growing populations, some of which seek increasingly better lifestyles, continue to increase demands made of agriculture.

The extent to which agriculture causes environmental problems and the maintenance of production depends, according to James (1978: 938), upon:

- The absorptive capacity of the environment.
- Usage and rate of depletion of non-renewable inputs.
- Convertibility of inputs to outputs, i.e. how wasteful is the process?
- Durability of the goods produced.
- Rate of population growth.
- Rate of economic growth.
- The nature of institutions and the legal system.

To which should be added:

- Technological innovation (including new crops).

Simple history of world agriculture

Shifting cultivation

In some areas cultivation may have been underway before 10 000 BP. The transition to sedentary agriculture in many, but not all, parts of the world occurred between roughly 5000 BP and the present. In some places transition still presents a problem for agricultural development.

Pastoralism

New livestock was introduced to various parts of the world from roughly AD 1500; improved breeding and fencing made an impact from the sixteenth century; refrigeration and steam shipping were introduced in the late nineteenth century; the spread of modern veterinary care occurred in the second half of the twentieth century.

Sedentary cultivation

This had started in a number of places by 7000 BP, and earlier in some parts of the world. Probably sedentary cultivation was developed independently by different groups of people, at least in Eurasia and the Americas.

By the thirteenth century Medieval farming systems in Europe had begun to change: those areas with the open-field system began to convert to enclosed fields or other systems with shorter fallows.

Roughly AD 1500 onward Transfer of crops from New World to Old and Old to New World was established, and from the seventeenth century; botanic gardens assisted.

From roughly AD 1600 Plantation agriculture in Brazil, Caribbean, North America, East Indies, South Asia, Malaya, Kenya and other parts of Africa lead to clearance and often land degradation.

Roughly AD 1650 – 1850 'Agricultural revolution' in northwestern Europe. European population seems to have increased rapidly in the ninth, fourteenth and eighteenth centuries AD. The response to increasing population was to move from long fallow to shorter fallow and ultimately to annual cropping with improvements in draft-animal harnessing, ploughs and range of crops and introduction of better crop rotations. Mouldboard-plough and horse harness improvements allowed cultivation of heavy clay soils to help feed growing populations.

From 1830s to 1914 There was increased use of horsepower on the land and improved sea transport of produce.

From 1870 to 1915 First chemical fertilizers were developed.

From 1914 to present Pace of mechanization of agriculture increases in DCs from 1930s to 1960s; LDCs follow more slowly outside the large-scale producer sector.

From 1945 to present Chemical pesticides, chemical fertilizers and modern crop varieties spread.

From 1950s to present Commercialization of agriculture increases.

From 1950s to present Traditional shifting and sedentary cultivation and pastoralism degenerate, due to a range of 'development pressures'; land degradation often results.

From 1960s to 1970s 'Green revolution' emphasizes maize, wheat and rice.

From 1960s to present Use of irrigation increases.

Mid-1970s OPEC oil 'crisis' raises cost of inputs to green revolution agriculture.

From 1975 to present 'Green revolution' deals with wider range of crops and trend toward more appropriate 'grass roots' approach. There is increasing multinational corporation control of seeds and other inputs.

From 1980s onwards The beginnings of biotechnology and the start of a 'gene revolution'. There is less certainty that DCs will continue to produce large cereal surpluses.

Agriculture in 'crisis'?

The sheer diversity of the world's agriculture and the different circumstances in which it is practised make reliable generalizations difficult. Challenges to agriculture are likely as a consequence of rising human population; global climatic change; growing poverty; global pollution; spread of new technology; social changes resulting from civil unrest; employment opportunities that compete with agriculture; unfavourable terms of trade; and health problems like AIDS.

Agriculture in the west has faced, and so far overcome, many environmental constraints and challenges largely through crop improvements and, since the 1920s, inputs based on fossil hydrocarbon resources. But these developments, especially since the 1940s, have led to new forms of environmental damage which may directly or indirectly affect crop production. Some question whether harvests can continue to increase or even be maintained at present levels (Canter 1986: 10). Porrit (1984: 102) observed that the world was paying 'a high price for cheap food'. It is too early to say that the hopes that biotechnology will permit continued agricultural improvement or even a huge breakthrough are well founded or whether it will lead to yet more unwanted impacts.

Problems may not be on-farm or even in-country: they may be global or they may be indirect, cumulative (synergistic) and insidious. Agricultural impacts may propagate via physical then social or economic processes and back again to physical, or vice versa, making the advance recognition of critical thresholds difficult.

The unwanted impacts of agriculture must be weighed against the world demand for food, commodities and livelihoods. Between 1990 and 2030 world population increase will probably roughly double food demand. Most of that increased demand (about 90 per cent) will be to feed LDC populations (World Bank 1992: 7). Between 500 million and 800 million people could not get or could not afford adequate food in 1980; the situation has generally deteriorated and will probably get worse. Agriculture struggles to keep up but in many places in so doing destroys the capability to sustain production. Overall world agricultural production has increased in recent decades, but not relative to population increase. In the future, agriculture is likely to be called upon to meet new demands like biofuel, industrial feedstock and carbon sink plantations.

Meeting the challenges

Agriculture may improve productivity through expansion of landuse or intensification or a combination of both. Usable land is often said to be increasingly scarce, even allowing for the European tendency to overlook potential, so the expansion option is assumed by many to be becoming more difficult to follow. Biotechnology may assist by facilitating the use of presently unusable land, and by enabling intensification through new crops and crop-enhancement measures and could make possible more food and commodity production through fermentation using virtually no land.

There are many opportunities to intensify 'conventional' production: Boserüp (1990: 97) noted the frequent amazement of Chinese advisors in Africa that abundant suitable land was not used for paddy rice production. There are clearly a lot of opportunities for shortening long-fallow cultivation to annual cropping or even more than one crop a year. In some regions caution will be needed to ensure that nutrient inputs do not exceed output in produce or there could be phosphate and nitrate pollution of surface and groundwaters.

Agricultural modernization requires caution: from the 1960s many countries have seen expansion of crop production for export as a route to 'development'. In following such a route they have become vulnerable to market forces beyond their control, locked into production of what they do not consume and consumption of what they do not produce (e.g. export of cocoa and import of food grains, fertilizer and fuel). The emphasis has tended to be on large-scale production of food or commodities for export or cheap grain mainly to feed urban populations. There will be little intensification of agriculture in many countries if there are no reasonable returns for farmer's labour. Low food prices have tended to be of little help to the bulk of the LDCs, largely subsistence, smallfarmers ('smallfarmer' or 'smallholder' encompasses tremendous diversity of agricultural strategies).

One can recognize several goals for agriculture: yield maximization; security against harvest failure; sustainability of production; equity (ensuring the widest section of the population benefit); reduction of environmental impacts; provision of employment. If at least the basic needs of poor people can be met and they are given enough security of livelihood, there is more chance that population growth will slow. So far there has been too much emphasis on yield maximization and the other goals just listed have been neglected. The improvement of smallholder agriculture, especially in remote and difficult environments, has received less attention than large-scale commercial production, yet it is the means for millions to feed themselves or to find farm employment.

The character of agricultural development

For at least 7000 years (in some areas perhaps over 20 000 years), vegetation has been cleared for agriculture or altered by grazing and some of this has been rendered unproductive and derelict. By 3000 BP parts of the Middle East were affected by salinization as a consequence of agriculture. Agriculture responds to demand and the ability to satisfy demand: New Zealand and Australian pastures were developed to provide the UK with cheap meat and dairy produce; *laissez-faire* development helped cause the 1930s 'Dust Bowl' of the US midwest; medieval plagues, changes in legislation like the modification and repeal of the UK Corn Laws in 1846 and 1869, the present spread of AIDS, a myriad of social, economic or environmental factors determine the success of agriculture and its impact on the environment.

One day biotechnology might reduce the need for agriculture but, for the foreseeable future, agricultural impacts on the environment will probably continue to increase.

Environmental impacts of agriculture

Clearing vegetation

- Regional hydrology and probably local climate altered.
- Habitat damage leading to wildlife losses.

- Loss of biodiversity.
- Significant contribution to greenhouse effect.
- Air pollution.

Tillage

- Erosion (greater than stream silt loads).
- Altered groundwater (saline seeps).
- Dust.
- Soil structural changes.
- Loss of soil organic carbon.
- Altered soil flora and fauna.
- Altered albedo (weather changes?).

Irrigation

- Polluted return flows (stream pollution).
- Raised groundwater (waterlogging and salinity).
- Soil structural changes.
- Altered soil flora and fauna.
- Possibility of reduced soil erosion.
- Disease vector problems.
- Reduced streamflow and groundwater due to extraction and evapotranspiration.
- Methane generation from paddy-fields, etc. (greenhouse effect).

Altered drainage

- Pollution from land drains.
- Polluted groundwater (saline seeps).
- Groundwater raised (waterlogging and salinity).
- Groundwater lowered (loss of soil organic carbon and subsidence).
- Genetic losses as wetlands destroyed.

Agrochemicals

- Pesticides: local, regional and global pollution possible.
- Pest resistance or loss of predators.
- Fertilizers: eutrophication of surface waters.
- Nitrate/phosphate problems with groundwaters.
- Alteration of soil structure.
- Alteration of soil flora/fauna.
- Greenhouse gases generated.
- Agricultural lime: alteration of soil flora/fauna and changes to aquatic flora/fauna (agrochemicals may affect marine life).
- Genetic losses as wildlife poisoned.

Agricultural 'wastes'

- Straw/sugar cane burning: air pollution and loss of soil nutrients, especially nitrogen – much depends on when and under what environmental conditions it is burnt.
- Processing crops: stream/groundwater pollution.
- Livestock wastes: water pollution (genetic losses); smell; methane generation (greenhouse effect).
- Silage effluent: water/air pollution.
- Noise: aesthetic and wildlife damage.

Pest control/hunting

- Reduces wildlife (including malaria-vector control impacts).

Pasture damage

- Over/undergrazing.
- Burning to encourage sward.
- Reseeding/improving pasture.
- Genetic losses.

Marine and freshwater aquaculture

- There were 460 coastal salmon farms and 400 onshore trout farms in UK alone in 1992.
- Pesticides used to control fish-lice, also antibiotics, fungicides and additives to control growth (formaldehyde, nuran, dichlorovos pesticide and malachite green used).
- Waste nutrients may be a problem.
- Genetic erosion of wild stock when fish escape.

Biotechnology

- Risk of uncontrolled escape, terrorism or military use of organisms or genetic material.
- Risk of increased poverty for those who cannot adopt biotechnology leading to poor farming or use of marginal land, both leading to land degradation.
- Promotion of new crops may discouraging growing of traditional varieties – loss of genetic diversity.
- Possibility that biotechnology might help reduce pressures on agricultural land.

Agricultural modernization

In the DCs from roughly the 1930s agricultural modernization (that has given huge improvements in yields) has usually meant more capital-intensive production, greater mechanization, introduction of agrochemicals and high yielding seeds. There has been a trend for farms in these countries to increase in size and become more specialized. One consequence of the specialization is that arable and livestock farming become separated. In the past manure was a resource spread on the land to improve the soil and sustain production; now intensive livestock rearing away from arable farms means that animal wastes have often become difficult to dispose of and a potential or actual pollutant. For the DC arable farmer, use of livestock manure has fallen and artificial fertilizer use has increased. The Netherlands and Denmark are already installing facilities to process livestock waste to reduce pollution and either obtain energy or provide compost that can replace some of the artificial fertilizer used; it will take time for this to become widespread.

More food or commodities are produced nowadays but the technology is often 'dirty' and wasteful (Farvar and Milton 1972); also, far fewer people find employment on the land: in late-nineteenth-century USA an 'advanced' farmer probably took 300 manhours to produce about half a hectare of maize; today it probably requires less than two manhours (and a lot of petroleum, which was not used in the past). Without transport, processing, marketing and storage, increased yields may be of little value.

Agricultural development has often meant commercialization, yet over 60 per cent of the world's farmers market less than half of what they produce, and are little involved in commerce (Harwood 1979: 13). There has been a neglect of subsistence agriculture and the role of women. The latter is ridiculous as many of the world's farmers are female: extension services, funding, education, land ownership and aid more often than not by-passes them.

The breakdown of long-established agriculture is increasingly common: the causes are varied and often complex. Traditional pastoralism and shifting agriculture (non-sedentary or land-rotation cultivation) pose particular problems, the cause of which are often population increase in difficult or marginal environments or the commercialization of production. Pastoralists and shifting cultivators can also be hindered by developments such as restrictions on movement or land allocation to sedentary cultivators or large-scale producers. Practitioners of traditional agriculture have often been seen to be 'backward' and undesirable. In reality there may be little better with which to replace traditional methods of using difficult environments, agricultural modernization having concentrated on commercial crops and more favourable land (Harwood 1979).

Degeneration of pastoralism or shifting and traditional sedentary agriculture often leads to land degradation; there is a need for alternatives to be developed and disseminated, some will be improvements of proven traditional methods, some will be new innovations (Greenland 1975; Okigbo 1981).

There is abundant evidence of productive agriculture in the past, often in harsh conditions, where there is none today. For example, the skills of the Nabateans of the Negev Desert (Israel), who practised farming between 1000 BC and AD 600 where rainfalls were as low as $50\,\mathrm{mm\,y^{-1}}$, have contributed to modern rainfall harvesting agriculture (Barrow 1987: 174). Farmers of the Altiplano (Andean highlands) grow *tarwi*, a lupin (*Lupinus mutabilis*), in rotation with potatoes. The *tarwi* fixes nitrogen, enriching the soil and exudes a compound that controls nematode worm pests of potatoes. By rotating with *tarwi*, potato cropping can be sustained with little or no inputs of fertilizer or pesticides – saving costs and reducing pollution. Today there is growing interest in the use of lupins in Europe as a nitrogen-fixing crop which also produces an oil like soya oil. Many other old techniques are awaiting discovery (in some cases through archaeology) and perhaps modernization, and hold great potential for both poor smallholders and commercial agriculture.

Alteration of natural soil

Agriculture may result in the loss of soil through erosion and it may be rendered less productive as a consequence of compaction, pollution, salinization, waterlogging, and many other processes. Soil degradation is a major problem world-wide: for example, in the USA roughly one-third of all cropland is badly affected (Hobbelink 1991). Some of the damage due to natural or human causes is difficult to avoid, but much is a result of inappropriate methods, the farming of unsuitable environments or the attempt to squeeze too much out of the land to meet the demands of growing populations or commerce.

Arable cultivation has generally involved tillage, the overturning and rearrangement of soil to improve physical conditions for crops (Briggs and Courtney 1985: 43). Before the Middle Ages simple ploughs and hand-tools were used but from the eighteenth century there was a move toward more horse-ploughing in Europe and lands under European influence; this meant deeper disturbances of a greater variety of soils. Much of the credit for this move has been given to Jethro Tull (who published in London in 1731 *The New Horse-Ploughing Husbandry: or an essay on the principle of tillage and vegetation, wherein is shown a method of introducing a sort of vinyardculture into the cornfield*). Since the nineteenth century the introduction of mouldboard-ploughs has led to even deeper disturbance and more overturning of the soil, which has affected breakdown of organic matter in the soil, rates of erosion, and may lead to compaction.

Tillage, provided it is carried out at the right time when soil has the correct moisture content, can bury weeds, incorporate organic matter into the soil, expose pests to predators, and may improve rooting and drainage conditions. If done in an appropriate pattern tillage may improve infiltration and, in colder climates, hold snowcover to improve soil moisture. Unfortunately, less careful tillage often results in soil degradation including wind or water erosion, and the passage of plough and other machinery can lead to compaction. The latter is more of a risk where soils have a high clay content and ploughing is done when the land is too moist or too dry.

Soil compaction may occur after very few passes of machinery. Generally the soil is affected to the depth of previous tillage but, especially in drier soils, it can be a half-metre or deeper (Holloway and Dexter 1990). Once compacted, it may take several years or more for the land to recover, provided compaction has ceased. Wheelmarks may concentrate runoff and cause gullying. While it is desirable to minimize soil compaction due to mechanization, it may be a price to pay for much improved yields. For example in France during the period 1850 to 1980, tillage changed from horse-plough to 1 400 000 tractors and 150 000 self-propelled machines: with a slight decrease in area farmed, production doubled. How much this was due to mechanization, new seeds or fertilizer is not easy to quantify, but mechanization played a part (Spoor *et al.* 1987: 134).

Compaction and water movements in tilled soil may result in the formation of impermeable plough-pans typically at several centimetres depth. Compaction often leads to increased runoff causing greater sheet erosion and gullying. There have been reports that ploughing in temperate environments probably leads to a reduction of soil organic carbon and fewer earthworms (Briggs and Courtney 1985: 59).

The alternatives to tillage include direct drilling and subsurface tillage. With direct drilling seeds or crops are inserted into soil, often cleared of cover by herbicide, burning or harrowing, without tillage (zero-tillage). This is generally considered to have the potential to reduce erosion and to better conserve soil organic matter and soil moisture than normal tillage. Some soils in temperate and cold environments may warm faster with direct drilling. Not all soils are suitable and there do not seem to be increased crop yields when it is practised. Subsurface tillage involves dragging a horizontal blade through the soil which disturbs a swathe of soil beneath a protective surface crust. Various forms of mulching and raised-bed cultivation also involve minimal tillage but require soil and compost moving from time to time.

Tillage, grazing and the passage of vehicles along unpaved roads may raise dust. At the local to regional level this may damage vegetation and human health, dust-related respiratory problems were common in the US midwest 1930s 'Dust Bowl'. Dust may also be sufficient nuisance to prevent certain types of industry, tourism or commodity processing. Wildlife impacts include damage to epiphytic plants, the biota of lakes and even quite distant coral reefs. Sometimes dust may have a more benign effect, for example, alkaline dust from seasonally dry north China blows into regions suffering acid deposition and helps to counter it.

Alteration of natural vegetation

Agriculture has followed two broad routes: first, *non-sedentary*, i.e. shifting cultivation (swidden, slash-and-burn and other local names) and nomadic pastoralism; and second, *sedentary* farming or livestock keeping. Thus agriculturalists and herdsmen have either moved to fresh land when that in use becomes tired or have tried to sustain use of a plot of land. Whether non-sedentary or sedentary, agriculture may seek to adapt to the environment or may try to adapt the environment to its needs; Geertz (1963: 16) noted 'Any form of agriculture represents the flow of energy to man: but a wet-rice farming terrace accomplishes this through a *bold reworking of the natural landscape*; a swidden plot through *a canny imitation* of it' (my italics). At these extremes – mimicry or marked change of vegetation cover – agriculture seems sustainable and environmental impacts may be acceptable, moderate change may lead to problems. Much of the agricultural development of the last hundred years has been a moderate reworking of nature, usually seeking to adapt the environment to fit the crop or suit the livestock; in the long run that may be difficult to sustain, costly in inputs and likely to generate unwanted impacts.

Throughout history clearance for cultivation or pasture has usually depended on fire. Forest, grassland and scrub burning has been a significant source of greenhouse gases and atmospheric soot particles, a cause of loss of soil carbon and nutrients and damage to wildlife. Modification of natural vegetation and subsequent alterations of crops or agricultural practices is likely to change: soil infiltration, groundwater recharge, evapotranspiration losses, albedo of the land, roughness in relation to windflow over the land and wildlife. The impacts depend on what is done and how it is done (e.g. mechanical clearing and burning will have more impact than hand-clearing without fire). Not all impacts may be negative and the degree of impact depends on the vulnerability of the environment and to some degree on the extent of the area modified.

Keepers of livestock modify pasture both consciously and by accident. Pastoralists often rely on burning to 'improve' pasture, ranchers may plough and re-seed, adding alien grass and herb species, may spread artificial or organic fertilizers, remove scrub, use irrigation, fence and rotate grazing to allow forage to recover.

Overgrazing, undergrazing (imprecise and relative terms) and trampling damage vegetation and initiate erosion, weed invasion and, on steep slopes, terracettes (trampling may result from passage of people, not just grazing herds). Trampling damage depends on time of exposure, type of beast, a great deal on weather conditions, vegetation characteristics, soil type, slope and factors leading to concentrated numbers in a restricted area. Undergrazing, overgrazing and trampling do not exert a uniform pressure on the land: damage is greatest where there is some attraction like shelter, water, gateways, etc. Similarly livestock are likely to deposit dung in a non-uniform manner, leading to more organic matter and greater earthworm numbers and diversity in some areas and less where animals have not concentrated.

A given level of grazing may be necessary to maintain a desirable vegetation cover. In Europe's Alpine environments pasture tends to become too long if undergrazed. The consequences are a reduction of attractive flowering plants, which are a major tourist attraction, and long grass, which fails to hold snow, leading to increased avalanche risk. In the UK semi-natural downland pastures became more scrub covered and lost flowering herbs when the rabbit population was struck by myxomatosis in the 1950s.

Where livestock damage develops, the problem may be countered by rotation of grazing, tethering, controlled grazing (e.g. with an electric fence) or stall-feeding. The latter has the advantage that cattle can be better monitored for disease and, by using less energy grazing, put on weight faster.

Alteration of natural drainage

Agriculture may involve alteration of soil moisture, either increasing it through irrigation or moisture conservation or removing it by some form of drainage. Moisture removal may be by simple ditching, buried tile-drains, mole-drains or similar subsurface structures. The main impacts may include the need to dispose of drainage water which may contain pollutants, soil micro-organism changes, acid-sulphate conditions (see chapter 6). Drainage sometimes promotes oxidation of soil organic matter leading to subsidence, in some peatlands of several metres, and sometimes underground peat-fires.

Where agriculture has caused waterlogging through the formation of subsurface impermeable layers (pans or plough soles), subsoiling – the passage of a heavy plough or ripper-blade deep enough to break up layers – may restore drainage and improve growing conditions. Subsoiling is expensive, and in common with most other forms of drainage, may not be a permanent solution and so requires periodic renewal. Some soils shrink and crack when drained, disturbing roots and infrastructure. Certain clay soils are especially prone to this problem, for which there is so far little solution.

The green revolution

Green revolution is a catch-phrase coined in the mid-1960s (and has no relation to the recent use of 'green' to mean environmentally friendly) to describe a process or movement whereby modern science is applied to LDC agriculture ostensibly to try to reduce hunger and poverty (Glaeser 1987: 15). The phrase implies technical breakthrough; bloodless, promotion of crop production and reduction of agrarian conflict perhaps – 'science in the service of counter-revolution'. In the 1950s and 1960s amidst widespread fear of peasant unrest, some may have seen the green revolution as a way to avoid or reduce social and economic reform (R. Andersen and Morrison 1982: 262). Many more saw limited opportunities for agriculture to expand into new land of reasonable quality, yet growing populations had to be fed; technological advance – green revolution – might allow intensification of production; it offered hope of side-stepping natural limits, of converting 'Malthusian to Boserüpian victory' (Lipton and Longhurst 1989: 346); it was a 'techno-political strategy for peace'.

Table 8.1 International agricultural research centres supported by the Consultative Group on International Agricultural Research (CGIAR).

Centre	Location	Date founded	Coverage	Crops/ activities
Asian Vegetable Research and Development Centre (AVDRC)	Taiwan	1971	Humid tropics, especially Asia	Beans, tomatoes, cabbage, etc., insect-resistant crops
Centro Internacional de Agrocultura Tropical (CIAT)	Colombia	1967	World-wide, tropical lowlands, especially Latin America and Caribbean	Cassava, field beans, rice, pasture. Cattle, pigs. Modernizing shifting cultivation
Centro Internacional de Mejoramiento de Maiz y Trigo (CIMMYT)	Mexico	1966	World-wide	Maize, wheat, barley, triticale
Centro Internacional de la Papa (CIP)	Peru	1971	World-wide	Potatoes
International Board for Plant Genetic Resources (IBPGR)	Italy	1972	World-wide	Collection, documentation and conservation of genetic material of important crop species
International Centre for Agricultural Research in Dry Areas (ICARDA)	Lebanon, Syria, Iran	1976	World-wide, semiarid/seasonal rainfall regions	Dryland farming systems for subtropics. Wheat, barley, pulses and irrigation research
International Crops Research Institute for the Semi-Arid Tropics (ICRISAT)	India	1972	World-wide	Sorghum, millet, food legumes and farming systems for the semiarid tropics
International Food Policy Research Institute (IFPRI)	USA	1979	World-wide	Food policy research
International Institute for Tropical Agriculture (IITA)	Nigeria	1968	World-wide, lowland tropics, especially Africa	Farming systems for the humid tropics. Root crop food legumes, maize, rice. Alternatives to shifting cultivation
International Livestock Centre for Africa (ILCA)	Ethiopia	1974	Africa	Livestock production
International Laboratory for Research on Animal Diseases (ILRAD)	Kenya	1973	Africa	Livestock disease research (trypanosomiasis and theileriasis in particular)
International Rice Research Institute (IRRI)	Philippines	1960	World-wide, special emphasis on Asia	Rice and rice-based farming systems
International Service for National Agricultural Research (ISNAR)	The Netherlands	1980	Developing countries	Assistance to national agricultural research
West Africa Rice Development Association (WARDA)	Liberia	1971	West Africa	Rice research

Note: CGIAR is an informal association of countries, private donors, funding bodies, international aid organizations, national and international aid and research groups. Chaired by the World Bank, it was set up in 1971 with its main objective – to help LDCs grow food. CGIAR is independent of the UN-FAO. Shiva (1991a: 38–9) provides a list of membership for 1983. P.R. Mooney (1979: 38–9) claimed that the Rockefeller, Ford and Kellog Foundations still retained a strong influence on these bodies in the late 1970s.
Source: P.R. Mooney 1979: 39; 87–8; Barrow 1987: 48; Glaeser 1987; Juma 1989: 87

Although there have been claims it was promoted by agrochemical companies seeking profits (P.R. Mooney 1979: 38), or the CIA, and as an alternative to meeting demands for land reform and increasing crops without redistributing estates, most accept that the green revolution was founded on less sinister reasons, but was sometimes misguided or mismanaged. A good deal of what has been written about the green revolution has depended more on the stance of the writers than fact and too often it is forgotten that 'revolution' can be diffuse, varying from one situation to another and evolving with time. Some argue that this evolution reached a dead-end about 1974 and was a failure; others recognize diversification and change after 1974 and that, far from failing, the green revolution is still making progress.

By the 1930s it was clear in a number of DCs that they had considerably improved harvests through the use of improved seeds, mechanization, fertilizer and other agrochemicals. In Japan the peasantry had achieved a four- or five-fold improvement of production in a few decades without abandoning traditional ways and free of serious strife, largely through use of improved seeds, fertilizer and better crop mixes. Efforts to improve major grain crops were underway in the USA and Europe by the 1930s; significant advances came after 1941 when the Rockefeller Foundation began funding agricultural modernization through improved seeds and other inputs in Mexico, the Philippines and then South and South East Asia. Improved maize and wheat hybrids from the USA and Rhodesia were adapted for Central America and East Africa. From the early 1960s the Ford Foundation began to fund IRRI (International Rice Research Institute) to improve rice production and the Kellog Foundation then became involved in funding. IRRI and CIMMYT (Centro Internacional de Mejoramiento de Maiz y Trigo) played a pioneering role in developing and promoting the new seeds and techniques (see Table 8.1). Normal Borlaug, a driving force behind CIMMYT, was hailed as 'father' of the green revolution and in 1971 was awarded the Nobel Peace Prize for his work on high yielding varieties of crops.

The 'components' of the green revolution had started to evolve before the mid-1960s (see Table 8.2), but it was then that there were breakthroughs in breeding improved seeds and the 'package' of the green revolution was put together and achieved results. The new seeds were the crucial component of the revolution.

Table 8.2 Developments in agricultural science, some of which contributed to the green revolution

Date	Development
1930s	HYV maize
1940s	DDT
1945–7	Advances in fertilizers
1951–3	New herbicides
1961	HYV wheat
1965	HYV rice
1966	HYV 'miracle rice', e.g. IR-8 released 1966
1969	HYV barley

Note: dates approximate.
Source: author.

Modern varieties (MVs) or high yielding varieties (HYVs) of maize spread through Mexico and other parts of Latin America; HYV (mainly semi-dwarf) rice spread to East Asia, especially Bangladesh, North Mexico, India (1964) and Pakistan, and South East Asia (including the Philippines); HYV wheat spread in Mexico (after 1954), the Middle East and India (1960–5), especially the Punjab (Wolf 1986).

The HYVs promoted from the 1960s and 1970s (e.g. IR-8, IR-20 and IR-26 rice varieties) allow improved harvests by:

1 forming grain rather than leaf in response to fertilizer (improved grain to foliage ratio)
2 giving better yield
3 being less prone to lodging (falling over before harvesting)
4 having a reduced growing period so more crops may be had over time
5 having a reduced sensitivity to daylength, allowing crops at virtually any time of year and possibly in areas once outside the traditional range.

From its foundation in 1970 the Consultative Group on International Agricultural Research (CGIAR) has coordinated the world's 'international' agricultural research and directs funding of the Agricultural Research Centres which have played, and still play, the lead role in the green revolution.

Impacts of the green revolution

Between 1961 and 1980 India became self-sufficient in wheat and Indonesia, once an importer of rice, became an exporter by 1986, largely thanks to HYVs. India's food grain production was about

Fig. 8.1 The Muda Dam and new high yielding rice varieties have considerably improved the livelihoods of rice farmers in Kedah State, Malaysia. Forest still clothes the hills in the background.

50 million t y^{-1} in 1947; in 1989 it was about 170 million t y^{-1} – roughly a 240 per cent increase (whilst population rose roughly 140 per cent in that time) (Davidson *et al.* 1992: 81). Other countries converted from importers to exporters or achieved self-sufficiency in grain or at least raised production. In 20 or 30 years HYVs seem to have considerably improved food production in many countries (Fig. 8.1). The green revolution may have helped minimize the impact of droughts in the 1960s and 1970s: without it many poor people might have starved (Lipton and Longhurst 1989: 9). Asian and Latin American crop yields have increased by more in the last 25 years than the previous 250 years, some claim largely through HYVs.

Increasingly since the 1950s LDCs have imported cheap DC-produced grain (in the early 1970s the trade was around 40 million tonnes, by 1980 it had grown to about 96 million tonnes, by AD 2030 it could be 333 million tonnes, according to Boserüp, 1990: 39), this may well discourage LDC production. In the future it is uncertain whether large DC grain surpluses will be available. There are those who would claim that the green revolution has helped reduce LDC dependency on imported cereals, critics would counter that this is offset by dependency on agrochemical and seed imports and overshadowed by socioeconomic and environmental side-effects. Some of the problems associated with the green revolution are as follows:

1 *Monoculture problems* If there are drought, frost, pests or diseases, these are likely to affect the bulk of the crop. Traditional varieties may offer more resilience and thus security. HYV monocultures have often suffered from pests or environmental problems: in 1968–9 the HYV rice IR-8 suffered bacterial blight throughout South East Asia in 1968–9 and 1970–1; in 1975 a virus struck (some sources also report virus problems in 1971); a HYV rice developed for better disease resistance IR-36 was soon hit by viruses; HYV maize in the USA has suffered serious blight, and in one year 15 per cent of the crop was lost. In 1991 about 80 per cent of the rice grown in the Punjab, India's 'rice-bowl', was of one HYV (PR-106). For a table of outbreaks in the Punjab (India) see Shiva (1991a: 92).
2 *Adoption of HYVs tends to displace genetic diversity on two levels* First, it leads to loss of traditional varieties and landraces; for millennia farmers have increased the diversity of the crops

they grew, now HYVs are causing the opposite. Second, HYV seeds generally cannot be saved by a farmer and must be purchased each year; the way they are supplied and legal protection afforded to suppliers discourage genetic diversity.

3 *Increased demand for water* Waterlogging or drawdown of groundwater may occur as a consequence of irrigation.
4 *Mechanization may lead to soil compaction.*
5 *Maintaining vigour of HYVs* This depends on ongoing breeding and availability of genetic material for such breeding, with dependency on commercial or DC research and development. Problems are created as wild 'reservoirs' of genetic diversity are lost.
6 *Increased pollution from agrochemicals* Pollution means risk to people, soil under cultivation, wildlife, fisheries and coral reefs. Groundwater is contaminated and the long-term effect on soil is unknown.
7 *Pesticides may kill weeds or wildlife that people use as a resource.*
8 *Pesticide use may lead to insect resistance* As a consequence there are problems in controlling crop pests and mosquitoes responsible for malaria transmission. In many areas which have adopted HYVs the locals joke that 'miracle seeds have led to miracle insects'.
9 *Dependency on seed companies* (often DC-owned or MNCs) HYVs are hybrids that cannot be successfully regrown from farmers' own seed (changes of law also tend to discourage this).
10 *Some farmers incur debt or are driven off land as green revolution technology is acquired by larger landowners* Mechanization may reduce rural employment; other problems are the drift to city slums, rural poverty and land degradation of marginal land.
11 *Other crops reduced as grain expands* In India and elsewhere this has meant less production of nutritious pulses, and thereby reduction of a crucial protein-rich foodsource for local consumption. Pulses have probably played an important role in crop rotations that have helped maintain soil fertility; repeated growth of HYV grain may lead to soil fertility problems.
12 *Dependency on inputs* People are vulnerable to cost increase or supply difficulties related to oil or fertilizer, pesticides, etc., produced from it.
13 *Reduced stalk* Less straw and fewer residues for craft industry, fodder and thatch mean that poor people may lose free or cheap resources. In parts of South Asia weeds are an important source of vitamins for children: HYVs and herbicide reduce this resource and lead to more risk of malnutrition and blindness caused by vitamin A deficiency.
14 *Poor people may help produce HYV crops but there is no guarantee that they can obtain or purchase grain* In Brazil export crops increased 13 per cent between 1967 and 1972 but over that period food crops increased by only 1.7 per cent (Hobbelink 1991: 127). Efforts were concentrated on large-scale production of soya, etc.
15 *Diversion of funding from more appropriate things?*
16 *HYV often have greater water demand than traditional varieties: grain to water yield may not be that favourable compared with non-HYV crops* (Shiva 1991a: 128).
17 *Greenhouse effect* Expansion of paddy rice, fertilizer use and irrigation lead to methane production by soil micro-organisms.
18 *Larger landowners may make more profit.*
19 *Production of cheap grain* Poor people may not be able to buy it if they are unemployed or lacking cash; it may reduce market prices and remove incentives for farmers to intensify.

Boserüp (1990: 114) argued that about 75 per cent of the cereals that have fed growing LDC populations between 1930s and 1960s came from *expansion of sown area*; it is less clear whether post-1960s grain production improvements resulted from similar causes or from HYVs and fertilizer use. Certainly Asia was intensifying production before the green revolution and socioeconomic problems blamed on the green revolution may have arisen without the spread of HYVs (Boserüp 1990: 116).

The green revolution, at least prior to the 1973/74 OPEC 'oil crisis', focused on grain – maize, wheat and rice. It was petroleum-based: in the 1960s oil was cheap and the idea of technical miracles was attractive. The revolution was 'top-down' and breeder-led; local peoples were seldom consulted and had to adapt to the package on offer. The HYV package gave better results than traditional farming only if all its components worked: there had to be a functioning chain of inputs supply and farmers had to obtain and correctly use inputs; farming conditions had to be suitable; marketing and credit were also important. Disparities thus arose between those that could and those who could not successfully adopt, and commercialization of social relations has commonly been associated with adoption of green revolution agriculture, sometimes leading to social breakdowns.

With less genetic diversity than a field of traditional varieties, HYV monocultures tend to be more vulnerable to environmental change, pests and disease and successful cultivation has come to depend on pesticides and fertilizers – resulting in pollution.

Critics of the green revolution abound. A common argument is that the increased grain production mainly benefits richer farmers and merchants, i.e. it is not 'scale-neutral' (Farmer 1977). Shiva (1991b) argued that even in the Punjab, hailed as a green revolution success, relatively few really gained. Another criticism is that the revolution has 'bypassed' many: according to Lipton and Longhurst (1989: 109) less than one-third of LDC country people live in areas that have adopted HYVs, so the overall benefit to farmers has been relatively limited. Often adoption of green revolution techniques have led to a reduction of employment opportunities for landless labourers and the buying-out of smaller or less competitive farmers. In some regions adoption of HYVs has discouraged cultivation of traditional crops, including pulses that are of considerable nutritional value and accessible to poorer people.

It might have been more sensible to have called HYVs potentially high yielding varieties or high response varieties, for their yields depend on inputs and conditions and if these are not favourable, they can give less than traditional varieties; even if successful, they often give less than research station trials suggest. Generalization is risky: there are situations where HYVs without fertilizer have not failed and situations where HYVs did better than traditional varieties, even with poor inputs (Pereira 1989: 181). Shiva (1991a: 76) suggested that some traditional varieties could claim the title HYV as they equal or exceed actual HYV yields, sometimes without expensive, polluting inputs.

Although many HYVs can be grown by rainfed farming, production concentrated on the land with best soils and reliable, adequate rainfall or irrigation; consequently other areas were largely bypassed, for example subSaharan Africa and East India. In Africa many cultivators grow a mix of crops; if a 1960s–70s HYV were added to the mix it would probably give little improvement. The development and spread of MV millet, sorghum and cassava (suitable for more marginal, harsher environments) came after 1974. Africa depends more on rootcrops than other continents, which is another reason why it got less benefit from the 1960s–70s 'grain' green revolution; for the most part Africa has been slow to convert from extensive to intensive agriculture (these may not be the only reasons why the continent benefited poorly from the green revolution).

MV rootcrops and robust cereals like sorghum or millets have been developed since the mid-1970s but it is still too early to judge the benefit from improvements; there is a risk crops like cassava could be grown on a large scale for export as feedstuff (as is the case in parts of Thailand) or raw material for industry, rather than food for poor people (Lipton and Longhurst 1989: 230).

Crop breeders have increasingly tried to give HYVs improved resistance to environmental problems, pests and disease and to reduce dependence on fertilizer.

To summarize, the 'first wave' of the green revolution (1950s–74) was characterized by: production of a few cereal types (wheat, maize, rice); oil-based inputs; command agricultural extension (top-down); need for better land and access to water; neglect of ecological and social issues. The 'second wave' (1974 onwards) is characterized by a movement away from oil-based inputs toward appropriate crops, nitrogen-fixation and bred-in pest resistance; diversity of crops, not just grain; some consideration of marginal environments and poorer farmers; more awareness of the need for bottom-up community development, careful extension and farming systems research; and more concern for environmental and social issues. While a more appropriate green revolution is now preached there is still displacement of local knowledge and increasing use of MNC-controlled methods and inputs.

In the 1960s and 1970s 'first wave' green revolution, it has been argued: 'The west tried to apply its own concept of rural development to the Third World, working through local élites, pretending the benefits showered on those élites would "trickle-down" to the less fortunate'. Shiva (1991a) suggested that it might be better for more LDCs to take an 'agricultural involution' path (see Geertz 1963), rather than a green revolution path, to agricultural improvement, i.e. improve traditional methods through intensification and re-investment of profits back into agriculture. Such a route would provide rural employment and avoid many of the ills associated with the way HYVs have been adopted since the 1960s.

Glaeser (1987) attempted an objective summing up, concluding that it was not a success, but it was progress and that there is still potential. Today the task is to develop and promote the right package for a given people and environment. Suitable components for the packages are already available and the range is increasing.

Modernization of agriculture: problems with modern crop varieties

Agriculture increasingly relies on MVs: carefully bred seeds or propagated plant material. These MVs offer tremendous potential – some talk of a 'seed revolution' or 'gene revolution' – but these innovations also raise problems, notably access to seeds that can only be commercially produced, genetic erosion, and sustainability and security of cropping.

Loss of biodiversity: genetic erosion

Before the green revolution most farmers grew a crop mix of more than one traditional variety or landrace and therefore maintained considerable genetic diversity (Box 8.1). Farmers kept their own seeds for sowing the following year or relied on many small family-run seed producers, most offering many different varieties. Since the 1960s that has changed, the spread of genetically uniform crops ('monocultures') have replaced fields of genetically diverse traditional varieties and landraces. Commercial forces, political pressures and legal developments are reinforcing these trends. Twenty years ago one could have found over 3000 varieties of potatoes in the Andes; today many have been lost and the trend is toward fewer. Fowler and Mooney (1990: 63) estimated that of the apple and pear varieties available in the USA between 1804 and 1904, about 86 per cent had been 'lost' or were extinct. The pattern of loss is repeated around the world for many crops.

Box 8.1 Variety and landrace: definitions

Variety a distinct, rather uniform creation (or cultivar). There is some variation of meaning depending on whether it is used by a botanist or a crop breeder.

Landrace (more-or-less synonymous with traditional variety) something usually more variable than a variety, less distinct, and less uniform. Within one field landraces vary in height, growth time required, resistance to disease, drought, and so on.

Meanwhile habitat damage has destroyed wild genetic resources and efforts to conserve material in gene banks, nature reserves or on-farm are inadequate.

The Irish 'potato blight' serves as a warning of the risks run when people come to depend upon genetically near-uniform crops. By the early nineteenth century potatoes were the staple of Ireland's growing population; unfortunately these had inherited little genetic diversity and lacked resistance to the fungus 'blight' *Phytophthora infestans*. Between 1845 and 1850 possibly as many as 2 million people died and many emigrated as a consequence of potato crop failure (Fowler and Mooney 1990).

Crops often require 'injections' of new genetic material, via traditional crop or livestock breeding or biotechnology, to cope with pest adaption, environmental change, altered consumer taste and other new challenges. The destruction of natural habitats, as well as the loss of traditional crop varieties are eroding genetic resources that must be drawn upon to maintain existing and develop new crops, Fowler and Mooney (1990: 53) likened this to the loss of colours from an artist's palette. In recent years crops that have been saved from serious disease or pest risks by incorporation of wild or landrace genetic material include maize, sunflower, sugar cane, tomatoes, potatoes, cocoa and strawberries.

In order to retain desirable qualities, like high yield, and to speed up release of a new variety HYVs/MVs are often hybrids (inbred with homozygosity or outbred with heterozygosity – they fail or deteriorate if regrown from seeds saved by farmers) and may in future be genetically engineered. Some crops are propagated by cloning (cuttings, etc., that produce genetically identical copies of parent plant) which also acts to reduce genetic diversity. Popular HYVs/MVs are widely adopted and tend to drive out other varieties. The monoculture risks would be more acceptable if there was free, uninterrupted access for farmers to the seeds, and if genetic material were well conserved and available for 'crop maintenance' or new crop development by anyone: seldom is that so (Lipton and Longhurst 1989: 36–43).

Crop and livestock breeding and biotechnology is now heavily controlled by DCs and MNCs, making it difficult for others to get the 'know-how' and raw materials (germplasm) and there are growing legal constraints. In general the DCs are poor in genetic resources, the LDCs are often 'rich'. Wild material and traditional varieties are taken from LDCs, with little or nothing paid for this 'raw germplasm', it is then bred or developed, turned into corporate property and then often sold back to the areas that provided the raw material. Shiva *et al.* (1991: 56) provided an example of this 'genetic imperialism': a wild tomato variety (*Lycopersicon chomrelewskii*) was obtained by North American commercial growers (at no cost) from Peru in 1962; by 1991 it had helped earn US$8 million for the US canning industry. In 1993 Indian protesters accused a US chemicals company of 'intellectual piracy' in patenting a natural pesticide derived from the neem tree (*Azadirachta indica*); similar processes and compounds have not been patented in India. This is nothing new: China had a monopoly on silk production until the worms were smuggled out in the sixth century; in the early twentieth century the British broke Brazil's monopoly in natural rubber in a similar manner.

It has been suggested that royalties on the benefits derived from the development of LDC genetic material would pay off a very large portion of the international debts acquired by the latter nations. Bodies like the World Intellectual Property Organization and UNESCO have tried to improve compensation for germplasm but legislation is still inadequate (Goldsmith 1991). The Shuar people of Ecuador and Peru are by no means unique in that they have never profited from informing outsiders of the value of quinine and curare. An estimated US$43 million a year is made in the USA from forest-derived drugs with little or no payment made to the countries where the compounds were first found. One Brazilian tribe who provided pharmaceutical company ethnobotanists with details of a medicinal plant that has been developed as a profitable drug sought 'royalties' with little success (Shiva *et al.* 1991: 67).

As the, often quite restricted, areas that are sources of useful germplasm become disturbed, some companies and institutions have established conservation collections (gene banks); most are in DCs and of the remainder the bulk are controlled by DC or MNC companies. Commercial producers may be reluctant to share and duplicate their banks as it could aid a competitor or reduce their control of the market, so for some crops one or a few companies hold most of the remaining genetic resources. There may also be problems funding and organizing conservation of material presently of little commercial interest.

The seed trade has become a rapidly growing, profitable industry which offers opportunities for those who control it to manage food and commodity supplies as well as inputs, and increasingly major petrochemical, pharmaceutical, vehicle and food businesses are in charge. Genetic material is lost in careless collection, shipment and storage – a tragedy if it is also lost in the wild.

At first glance there may appear to be quite a substantial network of commercial and non-commercial genetic conservation, a number of authorities are, however, concerned that collections are inadequate, insecure, are poorly documented (making it difficult to find out what is held where), and have restricted access. Some gene banks have been called 'gene morgues'. P.R. Mooney (1979: 28) noted that the US National Seed Storage Laboratory (at Fort Collins, Colorado, close to nuclear power and other hazardous sites) had a poor budget, weak security and high rates of storage failure, i.e. much of the collection, supposedly in suspended animation, was actually dead plant material. Hobbelink (1991: 5) claimed that only 28 per cent of Fort Collins's stock was viable. A list of gene banks and a ranking of their standard may be found in Juma (1989: 90–1 and 99). In 1973–4 the CGIAR set up the International Board of Plant Genetic Resources (IBPGR) to facilitate germplasm preservation, oversee the genetic collections and try and improve conservation (but with limited funding).

Seed companies, commercial livestock breeders and biotechnology companies argue that they invest funds, time and effort in developing new varieties or biotechnology, and therefore deserve legal protection – patents, patent-like rights or plant breeders' rights (PBR). A patent is a legal right which confers exclusive rights for an invention over a limited period. PBR are property rights extended to breeders to afford them exclusive rights over new plant varieties (Juma 1989: 252–4).

The application of patent legislation and PBR to agricultural development, and of intellectual property laws and patents to biotechnology is a complex field, one largely initiated in 1930 by the US Plant Patent Act. The legal status of 'wild' genetic resources or traditional varieties 'common property' is not settled, and there is debate as to whether germplasm is common heritage or patentable private property. One criticism of the move toward patent-like controls on seeds and the products of biotechnology is that breeders and genetic engineers do not create life (i.e. something 'new'): they merely relocate genetic material often in the form of imprecise, difficult to map, groups of genes or just parts of such genetic groupings.

Companies breeding and distributing new seeds have sought restrictive varietal legislation – 'seed laws'. In 1961 an international convention: the Union for Protection of New Varieties of Plants gave private breeders patent-like rights in the USA. Patent-like rights were further strengthened in 1970 by the US Plant Variety Protection Act which updated the 1930 Plant Patent Act. Pressure has grown for further strengthening of plant breeders' rights which include patent-like restrictions, like those of the US Plant Variety Protection Act (P.R. Mooney 1979: 51; Juma 1989). The consequence of PBR is a trend toward the following:

1 Reduced diversity of crop varieties for administrative simplicity.
2 Reduced diversity of crop varieties for ease of marketing and control of marketing.
3 Enforcement discourages diversity of seeds. Breeders generally try to avoid diversity in their crops in case it generates legal problems over 'patenting'.
4 Dependence on dangerously few varieties, e.g. in 1979 75 per cent of the Canadian wheat crop came from four varieties (Anon 1979); by 1991 about 40 per cent of the US potato crop came from one variety ('Russet Burbank'), largely because it was favoured by one fast-food company (according to Shiva *et al.* 1991: 202).
5 Farmers forced to become dependent on commerce for seeds (in LDCs this may mean dependence on imported seeds or a seed supply controlled by a MNC). There are growing demands for 'farmers' exemption' – the right to plant seeds they harvest from their own land without paying the original supplier of seeds (who may have had to foot development costs and thus seek profit).
6 Commerce may take genetic material (wild or traditional varieties) from a LDC, breed from it, then sell it back at a profit, paying no royalty to the source; this is seen by many as 'theft' of genetic resources. There can sometimes be a dilemma, for if the germplasm had not been collected it might well have been destroyed in the wild and lost (restrictions on commerce must be married to support for genetic conservation).
7 Companies may develop a crop variety that requires a chemical input ('dovetailing'); the user then has to buy the company's fertilizer or herbicide.
8 Proliferation of brand names may disguise the fact that many are the same variety produced under licence.
9 Smaller seed companies have increasingly been taken over by a few large, often MNCs (for a list of powerful seed interests see Juma 1989: 82).
10 Whoever controls seeds controls the world's food and commodity trade.
11 PBR will probably restrict free exchange of genetic material.

In Europe and elsewhere, governments have enforced rules that favour new varieties at the expense of old. The European Common Catalogue lists 'patented' varieties; other 'un-patented' varieties are difficult to sell, and may even be illegal to grow (for fear they might cross-pollinate with the legal ones). The pressure is to grow what companies want: new, distinct, uniform varieties, not necessarily what may be best in the long term or for all types of growers (Belcher and Hawtin 1991: 7). Seeds may be promoted because they give a crop that suits the canning industry or demands of supermarkets for uniformity of size and looks, not for flavour, nutritional qualities, resilience to environmental change and so on.

To counter genetic erosion, some people advocate that at least some farmers should be 'curators' of traditional varieties; this may prove cheaper and safer for genetic conservation than reliance on gene banks alone. PBR so far discourages this. In 1983 the FAO published an Undertaking on Plant Genetic Resources which declared that these resources should be available without restriction (but not necessarily free of charge); many DCs did not ratify it. To assist farmers, UNESCO and the FAO have supported the concept of farmers' rights (FR), something similar to copyright, whereby breeders pay royalty on varieties they create from traditional varieties or wild genetic material. So far a workable system has not been agreed.

Modernization of agriculture: problems with pollution

Changes in agricultural technology, population increase or demand generated by affluence and greed may each or in combination cause pollution. Intensification of agriculture has often led to pollution from agrochemicals use, livestock manure, livestock feedstuffs, wastes like straw or crop processing effluent. Sometimes pollution may result from subsidy, ignorance, marketing pressures, grower's fears of crop damage and consumer's concern for produce appearance. Those responsible for such problems often do not perceive it or bear the costs and this can hinder remedial measures.

Environmental pollution caused by agriculture

Pesticides

Pesticides can cause pollution of surface waters; pollution of food (local concentrations and global background); and death of wildlife on land, in streams and lakes and in oceans. Pesticides may weaken organisms leading to disease, immune system and other problems including carcinogenic effects and pest resistance (see Conway and Pretty 1991: 3).

Heavy metals in livestock feed

They contaminate soil, crops, livestock, surface water and groundwater, and may contaminate wildlife.

Environmental contaminations caused by agriculture

Fertilizers

N-Fertilizer Groundwater contamination above 25 mg l^{-1} may cause human gut bacteria to produce nitrite (NO_2) which can cause 'blue baby' syndrome (methaemoglobinaemia) in young humans and may also affect livestock. (First noted in Minnesota, USA, only 14 cases have been recorded in the UK up to December 1989: Hornsby 1989.) Excess nitrates can lead to nitrosamines in the gut of organisms, possibly resulting in carcinogenic effects: gastric and oesophagal cancer in humans is suspected. Excessive nitrates may damage flora, e.g. reeds get too leggy and suffer wind damage (Saull 1990). The effect of excessive nitrates in food is not fully known. Release of nitrous oxide (N_2O) derived from breakdown of N-fertilizers damages stratospheric ozone and is a greenhouse gas contributing to global warming. Pollution is associated with nitrate fertilizer manufacture.

N- and K-fertilizer Contamination of groundwater is already a problem in many areas and likely to get much worse. Surface water contamination includes eutrophication of streams (the process of nutrient enrichment such that a waterbody may support so much aquatic plant life that oxygen is depleted, wildlife dies and rotting material may then cause stagnation), waterbodies and marine environments (algal blooms 'red-tides'). The Baltic Sea is badly affected (over 10 000 square kilometres) and the North Sea, Aral Sea (CIS) and Chesapeake Bay (USA) are also badly affected. Algal blooms can poison people and livestock through drinking water or contaminated shellfish and may choke water purification and drip-irrigation equipment. Algal blooms can occur when there are very low levels of nutrient enrichment (even as low as 0.03 ppmv nitrate, less for phosphorus). Eutrophication may alter species mix in streams and waterbodies, and may cause excessive aquatic weed and reed growth. Nutrient-poor lakes and streams may have their flora and fauna completely altered through eutrophication. Phosphates (K-fertilizer) have been causing river and wetland problems in Europe for the last 30 years or so. Norfolk Broads (UK) waterways had about 80 mg l^{-1} phosphates before fertilizer use; today it is usually 150–300 mg l^{-1} and sometimes 2000 mg l^{-1}, causing turbid waters and fish kills. Pollution is associated with fertilizer manufacture.

Livestock waste (manure slurry and dung)

Apart from smell, ammonia gas contributes to acid deposition and can cause plant damage downwind and enhanced greenhouse effect. Stream contamination and eutrophication leads to high biological oxygen demand (BOD) and fish kills. Heavy metals from feedstuffs (mainly pigfeed) may be concentrated in slurry, which may accumulate in soil and cause human or livestock poisoning if crops or meat are consumed. In the UK and some other countries there have been problems with contamination of drinking water by the micro-organism *Cryptosporidium* often derived from the dung of cattle or sheep; the organism is highly resistant to chlorine treatment and causes gastro-enteritis if ingested. Some EU countries are already investing in incineration facilities to deal with excess livestock waste.

Agricultural lime

Lime may alter natural vegetation and fauna if it drifts as dust or leaches away from farmland. Streams and surface waterbodies may be affected. Naturally acid aquatic environments may be damaged; however, with acid deposition increasingly a problem lime may be one means of counteracting. In some situations lime use may allow snails to breed in acid water environments and raise the risk of disease transmission (sheep flukes in temperate environments and shistosomiasis in warmer regions); this may be the case in parts of eastern Amazonia (field observations by author 1985 and 1987).

Silage effluent

Not only smell, but also stream, soil and groundwater contamination, which leads to excessive BOD and fish kills.

Processing wastes

Wastes such as oil palm, rubber, sugar waste and effluent contaminate streams and sometimes groundwater, resulting in high BOD and kills of aquatic life. Fuelwood used for processing may cause smoke pollution and deforestation, leading to land degradation.

Paddy-fields/irrigation

Moist soil conditions lead to anaerobic decomposition and methane emissions, which increasingly contribute to the greenhouse effect and may prove more of a challenge than carbon dioxide. Since 1970 Asian paddy-fields alone have increased by roughly 40 per cent and emissions from paddy-fields in 1990 may represent 45 per cent of total human-made and natural global emissions of methane (Conway and Pretty 1991: 321; 382).

Silt and dust

Silt lost from agricultural land may choke stream channels causing flooding, hindering navigation and filling storage reservoirs. Irrigation canals and equipment may be choked or damaged by abrasion. Coral reefs may be damaged. If alkaline, silt and dust may counter acid deposition.

Plant growth regulating compounds

Compounds used to stimulate growth, flowering, fruiting, ripening, latex flow, etc, may cause problems to other crops, terrestrial and aquatic flora and fauna, and pose possible public health risks.

Antibiotics and animal growth stimulants

Prophylactic use of antibiotics for livestock (sometimes added to feed) may cause problems for wildlife. There are public health risks as diseases become resistant to antibiotics and are thus difficult to treat. Growth regulating hormones and steroids or chemicals may affect wildlife (problem with fish farming). There is controversy over BST (bovine somatotrophin), a growth hormone used to increase dairy cattle milk yield.

Factors which must be considered when using fertilizers if problems are to be reduced

The right type of fertilizer must be used for crop, soil and environmental conditions

Some fertilizers stimulate leaf growth and some flowering/fruiting; misuse may delay or cause a crop to be premature or damage soil or soil microorganisms. Different fertilizers may lead to different crop moisture demands. Fertilizer must dissolve at a suitable rate to ensure that the crop gets neither too much nor too little nutrient during as much of the growing period as possible; excess may be leached into streams and groundwater causing pollution. Work is in progress developing controlled-release fertilizers.

Application of the right amount

Soil analysis, linked to knowledge of fertilizer qualities, should help ensure appropriate application rates so the maximum amount of fertilizer is taken up by crops. Simplified methods and instruments may help. Few farmers do anything other than guess what should be applied, indeed may be urged by agrochemical sales staff and advertising to apply too much (sometimes of the wrong type of fertilizer). Careful guesswork may sometimes suffice, but plants vary in their uptake according to variety and environmental conditions so even careful guessing is often inaccurate.

Appropriate placement

Placement can be 'contact' as a seed dressing or added to the soil at or after sowing; 'broadcast' scattered before seeds emerge; 'ploughed in' at tillage before or after planting; 'side dressing' added alongside row crops; 'top dressing' applied after crop appears or spread on pasture; added to irrigation water or sprayed on leaves (foliar feed). The closer the placement to crop roots, the less is the chance of leaching and waste.

Application at the right time

Ideally all fertilizer would be used by the crop or excess would alter to become harmless. In cooler environments it has become popular to spread fertilizers in autumn (e.g. in the UK for winter-sown wheat); this may allow considerable 'escapes' to contaminate groundwater and surface water and denitrification (greenhouse gases may be generated by the latter). Spring application may mean less wastage and less escapes but may demand very careful timing if crop yield is not to suffer.

Chemical (artificial) fertilizers

Before the late eighteenth century, agriculture relied on livestock and human manure, compost, bonemeal, dried blood, green manures, marl, agricultural lime and crop rotations. To supplement those, the UK imported guano (seabird dung) from South America in the nineteenth century. Work on plant nutrition (see Box 8.2) in the UK by Lawes, Gilbert and others, and in Germany by von Liebig, led by the 1840s to the development of superphosphate artificial fertilizer. Between 1885 and 1985 the UK, one of the earliest countries to widely adopt artificial fertilizer, increased applications 25-fold. From 1935 to 1985 nitrogen or nitrogenous (N) fertilizer use in the UK rose by over 16 times, potassium (P) fertilizer use by over 5 times and phosphorus or phosphate (K) fertilizer use tripled (Briggs and Courtney 1985: 34, 101). In practice crops are more likely to receive a combined NPK fertilizer (N plus P_2O_5 plus K_2O typically mixed in the ratio 20 : 10 : 10). Application may be as dry pellets or powder scattered on to and sometimes dug into the soil or as concentrated or dilute solution with water: the latter may be applied directly to leaves as foliar feed.

After 1945 fertilizer use in the UK showed a marked increase; the same period also saw changing agricultural practices leading to reduced use of animal manure and agricultural lime, more mechanization and less input of manual labour. In the 1930s UK agriculture could not feed the population; between 1952 and 1972 UK agricultural outputs rose by about 60 per cent and nowadays it more than feeds a larger population; this progress is largely thanks to artificial fertilizers, although changes in the crops grown make it difficult to be specific (Briggs and Courtney 1985: 34).

On a world scale, fertilizers, particularly N-fertilizers, have probably played a key role in increasing crop production in DCs (Pinstrup-Andersen 1982: 148). However, because of the diversity of factors involved in cropping, it is difficult to be sure just how much benefit has been gained from fertilizers and how much from improved crop varieties. In 1950 the world used about 14 million tonnes of N-fertilizer; by 1985 this had risen to about 125 million tonnes (Saull 1990). In the late 1970s on average the LDCs used 28 kg ha^{-1} and the DCs 107 kg ha^{-1}. Most of the fertilizer used in the LDCs is for large-scale grain and export crop production. Boserüp (1990: 40) noted that in 1970 80 per cent of India's chemical fertilizer was used by only 15 per cent of districts; it is probably reasonable to say that most LDC farmers use little or no chemical fertilizers. The Far East tends to use more N-fertilizer and Latin America more P- and K-fertilizer. Japan's success in modernizing its agriculture and the roughly one-third increase in food production in the PRC between 1970 and 1985 are attributed largely to fertilizer use (G.R. Allen 1977; Wolf 1986: 12). An estimate by IIASA (1980 *IIASA Options* no. 3: 5) was that at least 40 per cent of the agricultural yield of the USA could be due to N-fertilizer use. The world has clearly become dependent on what is a costly input and often a serious source of pollution.

There is still inconclusive evidence about the relative merits of artificial and natural fertilizers,

Box 8.2 N and N-fertilizer

Plants cannot use nitrogen (N_2) directly from the atmosphere, which is about 78 per cent composed of that gas, nor can they use nitrites; they have to take in, via their roots, ammonium ions (NH_4+) and if available, nitrates (NO_3−). The two compounds are water soluble and tend to leach away. A little nitrate is added to the environment as a consequence of lightning and some is produced by micro-organisms which 'fix' N_2 from the atmosphere as ammonium ions.

Rhizobium spp. bacteria can fix N_2 and a relatively small number of plants (particularly legumes – plants of the pea and bean families and clover family) have them living in symbiosis in root nodules, in effect acting as fertilizer sources.

Levels of nitrates usually cited as mean nitrate concentration (mg N l^{-1}) or as nitrate-N, the nitrate to nitrogen concentration (kg N ha^{-1} y^{-1}).

NPK-fertilizer farmers usually apply a combined nitrogenous, phosphate, potassium mix (ratio of components varied to suit specific need) – NPK-fertilizer.

Eutrophication an excess of plant nutrients, especially phosphates, in water may lead to excessive aquatic and semi-aquatic weed growth or a phytoplankton 'bloom'. There may then be such a demand for oxygen (high BOD) that conditions become stagnant. If certain types of plankton 'bloom', shellfish and fish which consume them may become toxic (a particular problem in some shallow seas and lakes).

although some are sufficiently opposed to the former to reject it and practise organic farming. There is some indication that where year-round use of HYVs has replaced crop rotation and use of livestock manure, fertility problems arise, in particular a net loss of organic matter from the soil and in some areas zinc or sulphur deficiency. In DCs there are problems caused by use of artificial fertilizer and problems caused by livestock manure that no longer gets returned to the land. In the future it is likely that livestock manure and wastes like straw, domestic refuse and sewage will be composted and used to reduce artificial fertilizer use and solve the problem of their disposal; agriculture will have come full circle, back to recycling of wastes.

Artificial fertilizers do offer the following advantages:

1 They can be easier to store, handle and transport than most natural fertilizers presently used.
2 There is less smell, lower risk of pathogenic contamination (although well-composted organic material is virtually pasteurized).
3 Land spread with manure cannot be properly grazed for some time due to the risk of disease transmission, and because cattle dislike unclean pasture, artificial fertilizers allow intensive use of grazing land more rapidly after treatment.

Today fertilizers are one of the main costs involved in intensive farming and are often a cause of environmental contamination. If not applied with caution, artificial fertilizer may cause contamination (fertilizers are contaminants rather than pollutants) and may also fail to give its full potential (Mellanby 1970; Gunn and Stevens 1976). Both organic manures and artificial fertilizers (both N- and P-fertilizers) are blamed for stream and lake eutrophication (Box 8.2) and increased nitrates in groundwater. Successful fertilizer use depends upon a range of factors. Even when care is taken, presently used fertilizers may cause problems, one of which is the accumulation of phosphates and nitrates in the soil; gradual or sudden environmental change could lead to conditions under which enhancement of leaching or some other associated difficulty suddenly develops to crisis proportions (what Stigliani and others termed a chemical time bomb).

Phosphate pollution

Phosphates have been accumulating in soils, river and lake sediments for decades as a consequence of the use of phosphatic fertilizers, spreading of livestock manure, disposal of sewage and leaching of poorly sealed refuse dumps (landfill sites). The accumulation poses a serious threat, particularly in Europe and North America, possibly in other parts of the world. Studies in the EU suggest that, even if application of phosphates is controlled, steady leaching and possibly more rapid mobilization if there is soil acidification as a consequence of air pollution or global warming will lead to a 6- to 10-fold increase of contamination. Such levels of contamination would raise problems for domestic water supply and for the ecology of rivers, lakes, the Baltic and the North Sea (Behrendt and Boekhold 1994).

Nitrate pollution of surface and groundwater

Excessive levels of nitrates (NO_3) in groundwater and surface water are increasingly a problem in the UK, mainland Europe, the USA and other parts of the world (Fig. 8.2 shows the UK situation). Although livestock wastes, sewage and altered landuse may sometimes be to blame, the indications are that it is N-fertilizers which are responsible for a good deal of groundwater contamination. In areas where the problem arises in Europe and the USA, application rates for N-fertilizer are often between 120 and 550 $kg\,ha^{-1}\,y^{-1}$ (for a table of application rates by country see Conway and Pretty 1991: 159). In the UK the River Stour showed no significant rise in nitrate between 1948 and 1964, but between 1965 and 1985 it rose to three times the pre-1951 average (Briggs and Courtney 1985: 280). What percentage of this rise reflects increased N-fertilizer use, more deep ploughing, sewage pollution, conversion of pasture to arable or land drainage is difficult to establish.

Generally less nitrate is lost from grassland than cropped land. Sewage probably contributes more to river pollution and less to groundwater contamination in DCs. Clay soils can hinder nitrate contamination of groundwater, whereas sandy soils tend to leach easily. Tropical soils may be less prone to nitrate leaching although the use of irrigation in such environments may cancel out any advantages (Saull 1990). Where nitrate leaching to groundwater is a problem it may be difficult to judge how much is due to fertilizer and how much to other landuse activities. In parts of the USA irrigation using N-fertilizers seems to be a major cause of groundwater nitrates. In the UK borehole studies show correlations between conversion of pasture to arable and N-fertilizer use and high groundwater nitrate levels (Conway and Pretty 1991: 186).

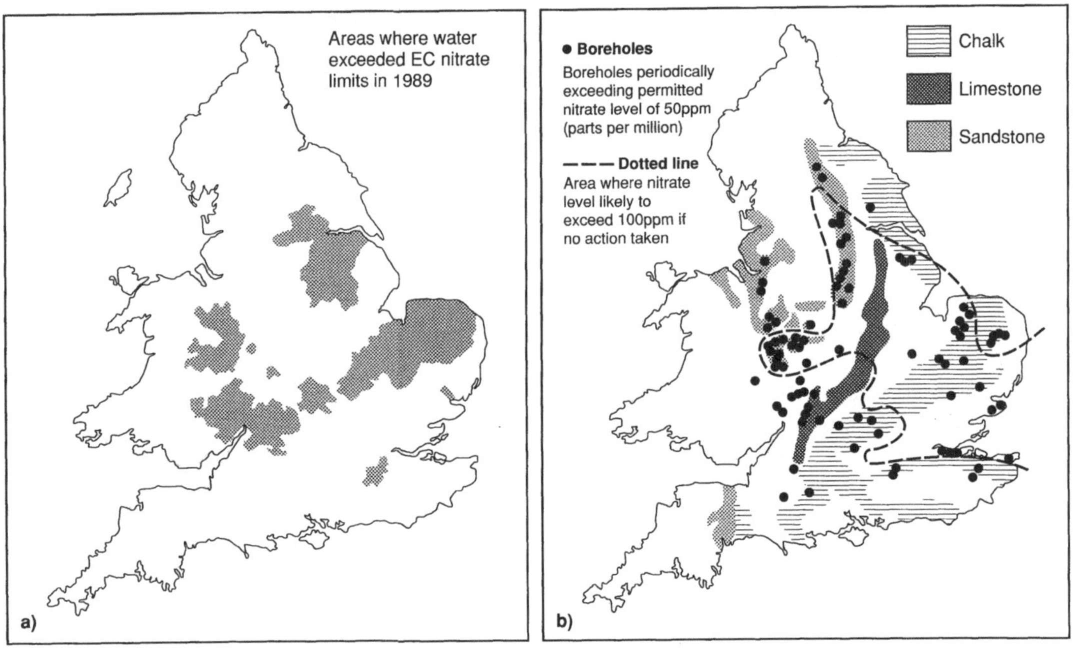

Fig. 8.2 Nitrate pollution of UK water supplies. a) Nitrate pollution of groundwater 1992 (*Source: The Times* 24/2/92). b) Areas where water exceeded EC nitrate limits in 1989.

There is no full agreement on how much nitrate in water or food constitutes a hazard: the World Health Organization (WHO) established a 'safe' limit for groundwater of $11.3\,mg\,N\,l^{-1}$, the EC set a drinking water limit at $500\,mg\,N\,l^{-1}$ in 1975 (Conway and Pretty 1991: 183) and the latter has been enforced in the EU since 1985. Nevertheless, in 1989 75 UK water supplies, mostly groundwater sources, failed that limit (Saull 1990). In 1991 between 30 and 35 per cent of the UK population depended on groundwater; in some areas contamination was high and rising, notably in Lincolnshire and East Anglia. Things are as bad or worse in The Netherlands, Denmark, parts of Germany, Poland, other parts of Europe and elsewhere. Acid deposition may aggravate the problem by speeding up nitrate release.

Landuse changes and nitrate in groundwater

Ploughing buries organic matter which is broken down by micro-organisms to form nitrates; tillage of grassland may thus supply nitrate to groundwater (Burt *et al*, 1993). Certain crops seem to allow more nitrates to escape. In Europe, potatoes leave a lot of nitrates after harvest to leach away. Winter-sown or autumn-sown cereals in the UK have N-fertilizer applied before spring growth can lock it up; much of it seems to get leached, especially if there is heavy rainfall after a period of drought. In areas of the UK suffering groundwater nitrate contamination, ploughing of pasture is a probably significant cause. After tillage, nitrate leaching will continue for several years as organic matter decays. Nitrates moving through the soil and underlying rocks may take 50 years to reach groundwater in chalk, up to 20 years for sandstone aquifers, and less for limestone aquifers (Hornsby 1989). Restriction of fertilizer use will therefore not have anything like an immediate effect. Present-day nitrate pollution began in many instances three or four decades ago (Conford 1992) and will continue to pose problems well into the future even if accumulation is controlled (the chemical time bomb effect).

Reducing nitrate pollution

The major sources of nitrates are animal wastes, fertilizers and sewage. There are two possible routes to cutting nitrate fertilizer use: first, measures like quotas, set-aside, reduction of price supports for crops (i.e. abolition of a minimum guaranteed price) or regulation of crops grown (quotas are permits which seek to limit expansion of an activity; set-aside is the withdrawal of land from production) (Fig. 8.3). The second method involves taxation of nitrate fertilizers.

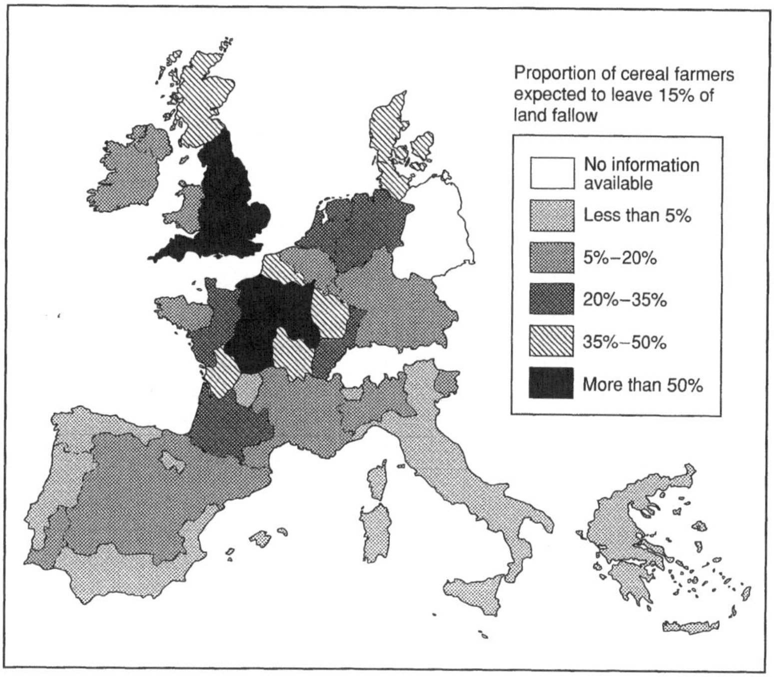

Fig. 8.3 Expected impacts of EC set-aside under rules in force in 1992 (*Source:* based on *The Times* 3/11/92: 6).

A risk with set-aside is that remaining lands will be used more intensively and so pollute more (Clunies-Ross 1993). Restricting nitrogen fertilizer use in Europe, say a 20 per cent cut, could decrease pollution and cut over-production leading to 'produce mountains' whilst maintaining some livelihood for farmers. Conversion from autumn- or winter-sown, to spring-sown cereals could offer some assistance in reaching the same goals. Nitrate pollution control via restrictions on fertilizer use and changed landuse will be slow to have effect, pollution will probably get worse, perhaps for decades, so it seems likely that authorities will be forced to treat domestic water to remove nitrates, blend contaminated and pure supplies or store water in surface reservoirs for long enough to reduce nitrate content. Water supply agencies in the UK, like Thames Water, are installing carbon and sand layer filter-beds to help reduce nitrates, pesticides and other contamination.

The 1989 Water Act allows the UK Government to designate 'nitrate sensitive areas' where mandatory requirements could be insisted upon to reduce nitrate leaching. In late 1993 the EU required control of nitrate at source, i.e. designation of 'vulnerable zones' where limits are or might be exceeded within which landuse will be restricted to reduce contamination. This affects parts of the UK, France, Belgium, Germany (where problems are mainly due to cereals production) and The Netherlands (livestock waste is probably to blame). To protect aquifers it may require the conversion of huge areas of arable land to woodland or lightly grazed pasture to meet the EU Directive on nitrate in groundwater (Hornsby 1989). The UK Ministry of Agriculture Fisheries and Food (MAFF) designated 'nitrate sensitive' areas in 1989 and offered compensation to farmers who undertook to grow covercrops to reduce winter leaching and make payments to encourage conversion of arable to unfertilized land. The UK Countryside Commission suggested (in 1991) that it was practical to take as much as 15 per cent of cereal land gradually out of production and use it for conservation and recreation (*The Times* 2/10/91: 7). Unfortunately in the UK the compensation has so far been inadequate to give much incentive (Hornsby 1989). In the UK the National Rivers Authority were expected to designate source protection zones for England and Wales by late 1993.

It should be noted that conversion from farmland to some other use (set-aside) could, because agricultural liming ceases, lead to increasing soil pH and greater releases of nitrates, phosphates and heavy metals (this has been the case with land abandoned over 100 years ago at Rothampstead Experimental Station, UK). Leaching control may require slow-release liming treatment of set-aside and other lands.

Another possibility might be to encourage rotational set-aside, whereby farmers are paid to leave a different part of their farms each year under something like white clover. Planting winter wheat with white clover might help to reduce nitrate leaching, would cut costs of fertilizer inputs and may help discourage pests.

Pesticides

Pesticides are compounds used to kill or disable unwanted pests (see Table 8.3); the reasons for their use are:

- to maximize crop or livestock yields;
- to improve quality and looks of crops or livestock;
- for disease control (human health and veterinary use);
- for preservation and maintenance of buildings, clothing, boats (anti-foul paint), furniture;
- to control weeds which hinder transport and access (road and railway use, weed control on boat hulls);
- for aesthetic or leisure reasons, lawn-care, garden flowers, golf courses.

Pesticides may be dispersed on to pests, laid as barriers, or used in conjunction with attractants, such as chemicals, baits, noise, hormones, sticky compounds or light sources. Nature produced pesticides and pest deterrents long before humans evolved; plants may have toxic or unpleasant sap, wax or hairy coverings, and in some cases natural pesticides have been adopted by humans or have provided ideas for synthesis of artificial compounds. Natural compounds or 'organics' should not be seen as always harmless alternatives: some are very toxic or carcinogenic.

Agriculturalists have long tried to kill or deter pests; Shakespeare referred to the use of ratsbane (a plant-derived poison) for controlling rodents in *Othello*, the Sumerians dusted vines with sulphur as an insecticide before 2500 BC and Amerindians have long used rotenone for fishing. From the late seventeenth century European and American growers employed compounds like nicotine, derris, pyrethrum ('botanical' insecticides) or Bordeaux mixture (fungicide for vines and fruit trees made from lime and copper sulphate), Paris green (a copper and arsenic fungicide, probably the first synthetic pesticide, which came into use in the 1860s) and soap solutions or coal-tar derivatives against insects. Although some of these were toxic, the usage was in general restricted and the effects short lived and localized. It should be noted that many natural plant

Table 8.3 Pesticides

Types by role	*Intended target*
Insecticides[a]	kill insects
Herbicides[†]	kill weeds or scrub
Molluscicides	kill snails/slugs
Vermicides	kill worms[b]
Acaricides	kill ticks/mites[c]
Algicides*	kill algae
Fungicides*	kill fungus[d]
Rodenticides	kill mice/rats, etc

Types by chemistry

Organochlorines e.g. DDT, aldrin, dieldrin, heptachlor (most developed in 1950s). Residues persist for years and accumulate in fatty tissues of higher organisms, most now banned or restricted in DCs.

Organophosphates e.g. parathion, malathion, ronnel, dimethoate, bromophos, dichlorovos (first appeared in 1930s). Not as persistent as organochlorines, but in general more toxic to animals. Linked to incidence of aplastic anaemia in those exposed to pesticide.

Carbomates e.g. aldicarb, methiocarb (developed from organophosphates). Moderate to highly toxic, but breakdown fast.

Synthetic pyrethroids (synthetic versions of natural pyrethrum compounds). Not especially toxic to livestock and humans, but kill fish.

Note: Organisms may be exposed to more than one pesticide. For many of the targets listed there are artificial and 'natural' (biotic) compounds.

* May be added to paint, anti-foul paint (until recently tributyltin oxide (TBTO) or copper-based and caused considerable damage to shellfish in shallow marine environments), building materials, wood preservatives, plastics.

† Accidental/deliberate poisonings common using stored herbicides or contaminated containers. On the whole herbicides tend to do less damage to wildlife than insecticides, but large-scale use in warfare (as in Vietnam) has caused damage to flora, fauna and humans.

[a] Crops, stored crops, household, disease vectors control.

[b] Control in environment or internal control in body of infested organism (tablets, drench, etc).

[c] Sheep or cattle-dip or spray.

[d] Includes 'rusts' on crops or mould on buildings or stored foods, etc.

Fig. *8.4* Pesticide use. a) Spraying a riverside settlement against mosquito-transmitted diseases with persistent insecticides, Pará, Brazil.

compounds are toxic or carcinogenic; agrochemical risk must be weighed rationally.

Modern pesticides were developed as human populations grew and demand for food increased after 1945. DDT was one of the first synthetic organic compounds (organochlorines or chlorinated hydrocarbons), synthesized in 1874, rediscovered in 1939 by a Swiss chemist, Paul Muller, who received the Nobel Peace Prize for his work in 1948; and patented in 1940; it was adopted for louse and mosquito control during the Second World War.

The second main group of synthetic pesticides, the organophosphates, were discovered in Germany in the 1930s, and include parathion, malathion and ronnel. After the 1940s other synthetic organic compounds were developed and widely used for agriculture and public health measures along with DDT (Fig. 8.4). Other insecticides to become widespread by the 1960s were: gamma-lindane (Γ-HCH) and dieldrin. Phenoxy herbicides (e.g. 2,4-D; 2,4,5-T and MCPA) and compounds like diquat and paraquat were in use for weed control by the 1960s. There has been a tendency to replace many of these pesticides with 'safer' organophosphate (e.g. methyparathion insecticide), carbomate and pyrethroid insecticides and synthetic herbicides and fungicides. Organophosphates can be more toxic than organochlorines but are less persistent (for details see Conway and Pretty 1991).

It should be noted that the benefits from pesticide are considerable in terms of improved harvest, fewer storage losses, human and livestock disease control. Successful pest control might well reduce crop and produce losses by 20 to 30 per cent, perhaps more, and improves security of harvest and storage. But it is difficult to quantify the benefits and the risks of pesticide use; for LDCs it may do more harm than good as a large proportion of what is used is applied to luxury export crops, not staples (and labourers and people off-farm suffer the side-effects). In the UK the Pesticide Exposure Group has been established to fight for better controls on use, monitoring and compensation and care of exposure victims.

Realization that pesticides were causing serious environmental problems came by the early 1960s, the public being alerted by Carson (1962). DDT attracted

Fig. 8.4 b) Amazon fish, already there is widespread mercury pollution, and increasing risk of contamination and kills due to pesticide and other agricultural chemicals. Some species are also over-fished. Apart from moral conservation issues, Amazonia has a very diverse fish fauna which pollution might reduce before it can be assessed for pisciculture, etc.

much attention: although reputed to be less toxic than aspirin, it was found to concentrate in the fat of higher organisms through 'biological magnification' and by 1972 use in the USA was banned (but not manufacture and export). The problems associated with pesticide use can be summarized as:

- poor selectivity of compounds (not narrow-spectrum, i.e. not very specific in terms of what is killed or injured);
- overuse;
- toxicity and slow breakdown (persistence);
- tendency to be concentrated by food-web;
- misuse or unsafe methods of application;
- the effects of long-term usage of pesticides on soil fertility is little known;
- the impact of cumulative effects on the global environment is not known.

Ideally a pesticide should be specific, i.e. kill or disable the pest and affect nothing else. Unfortunately most compounds are far from specific. Non-pest organisms may be directly or indirectly affected: on-farm (workers, livestock, crops, soil, wildlife and groundwater); off-farm (nearby woods, hedges, housing, streams) or globally. The impacts may be short term or long term and may have cumulative (synergistic) effects. Releases may be point sources or area sources, rare, intermittent or virtually constant. Some pesticides are more mobile than others (e.g. 2,3,6-TBA; TCA; delapon), some less (e.g. lindane; parathion; benomyl; paraquat).

Tracing lines of pollution (and proving liability) from pesticide use back to point of application from impact can be difficult as the links may be indirect, insidious and occur away from site or use and well after the usage. Use in certain environmental conditions may have far less impact than it would under different circumstances, for example, spraying ten minutes earlier in the morning may avoid killing bees; spraying in warm conditions may do more damage than in cold; when it is windy pesticide is likely to 'drift' to affect organisms well away from the target. Weir and Shapiro (1982: 4) publicized the way in which the export of pesticides banned in the USA had an impact on DC people through food imports with excessive levels of contamination.

Some pesticides break down after use into less harmful and some into more harmful compounds; occasionally there is contamination with toxic compounds in manufacture or through decay during storage. How long conversion to safer compounds takes after pesticide has been used and to what extent depends partly on environmental conditions (generally the time taken is stated as the half-life – the time taken for half the toxicity to be lost). Toxicity is widely measured as lethal dose-50 (LD-50 – the amount of the compound delivered orally or through dermal exposure required to kill half the exposed organisms under test conditions; the lower this value, the more toxic is the compound. The relationship between mortality and dose of pesticide is complex, to try and better estimate it some use effective dose-50 (ED-50), which indicates how much affect 50 per cent of exposed organisms.

Much pesticide is used as a pre-emptive measure and may not be necessary: world-wide overuse is a major problem. Usage increased rapidly from roughly 1950, partly reflecting the green revolution and the spread of HYV seeds. World sales in 1991 were over US$20 000 million a year and rising, with 12 DC-based MNCs contributing about 70 per cent of that trade (Hurst *et al.* 1991: 19–21). Overuse implies that in some situations pesticide may be underpriced, possibly due to subsidies. About 50 per cent of all pesticide is applied to wheat, maize, cotton, rice and soya. Most used in LDCs goes on to plantation crops like cocoa, coffee and oil palm. Japan is probably the most intensive user; the largest user is the USA followed by western Europe (Conway and Pretty 1991: 19; 29). More and more pesticides, it seems, are being used to achieve less and less: in spite of increased pesticide use the world's crop losses to pests may have increased over the last 30 years.

Practices that lead to problems include farmers' mixing of 'cocktails' of pesticides; careless application that allows wind drifting; disposal of containers or mixing of compounds where spillage can get into surface or groundwater; and harvesting of crops too soon after spraying, leading to contaminated food. Fish, amphibia and crustaceans are vulnerable to pesticide contamination and in many LDCs a side-effect of the green revolution has been damage to or destruction of fisheries on which people depend for protein or income. Paddy-field fishery contamination with pesticides and fertilizers has occurred in India (notably the infamous Malnad case), in Thailand, Malaysia and many other countries (Bull 1982: 63; Boardman 1986: 157). Roughly half of all known pesticide poisonings and at least 80 per cent of deaths so caused have occurred in LDCs, yet these nations use only 15–20 per cent of world total pesticides (usage is growing fast) (Pimbert 1991: 3). Deaths are common: one estimate is that 80 000 people die each year through pesticide poisoning, much through poor handling (Hurst *et al.* 1991: 174); another is that 3 million people were poisoned every year, of whom 1 in 14 died (WHO estimate, 1989). Pesticides can remain in soils, river, lake and marine deposits; worries have been voiced that soil acidification or global warming might reach a threshold, unbind the compounds from clays and allow a sudden leaching of many years accumulated pollution (a chemical time-bomb effect).

Pesticides are usually slow and very costly to develop. Companies, even those acting with the best of motives and care, may therefore be unable to test them fully; may be reluctant to withdraw the compound if there is some fault; may resist developing specific pesticides (i.e. those that act on just one or a few types of pest) because it restricts sales; may resist giving safety advice that could cut sales; may promote use (or the middlemen may do so) in inappropriate situations to maximize profit. Side-effects may become apparent only after extensive use and can be difficult to foresee. Pesticide development may neglect important pests if there seem to be limited profits to be made: witchweed (*Striga* spp.) and quela birds, pests of Africa's sorghum and millet crops, have had little attention, perhaps because subsistence farmers are unlikely to be able to afford to pay much for control.

Demand for pesticides may not be due to rational reasons: the public or supermarkets may demand uniform size, colourful and blemish free produce; a canning company may do the same even if the finished product is pulped. Growers may use chemicals to ensure ripening of the bulk of a crop at a given moment to assist gathering and processing. Citrus and apple growers commonly use growth-regulating compounds and treat fruit to reduce spoilage when transported and stored. There has been much publicity for a recent US ban on use of alar, widely used to regulate apple ripening, which some suspect of causing cancer. Development planners may promote unwise pesticide use. For example tsetse flies spread sleeping sickness to humans and livestock so their presence discourages pastoralism, use of pesticides would mean environmental pollution and the risk that land would be used and probably degraded before sustainable strategies have been developed (De Vos 1975: 149; Linear 1982; 1985: 233).

Pesticide impacts

Acute toxicity This leads to sudden injury or rapid death.

Chronic toxicity Long-term exposure causes injury or death directly or indirectly.

Non-lethal (chronic) but significant effects Birds eggs may become thin and survival of offspring be reduced (first noted in the USA in the 1940s, now widespread); organisms may become sterile, abort their young, become debilitated, suffer behavioural changes (e.g. increased vulnerability to predation or upset breeding habits) or cancer. Possible reduction of fertility of animals, including humans (in western Europe human sperm-counts have been falling in recent decades; pesticide pollution, as well as other industrial pollutants and pharmaceutics, might be a contributing factor).

Contamination of groundwater Central Valley (California, USA) has been badly affected, particularly by dibromochloropropane (DBCP), a pesticide used against nematodes and fungus by irrigators. Although DBCP was banned in 1978, by 1991 1473 wells had been closed as unsafe. In Europe and some other countries the use of atrazine (a herbicide) has been banned. In Germany atrazine was used on maize crops, in spite of cessation of use it is increasingly leaching from soil to contaminate water supplies.

Effect on organisms well removed from site of use in space and time Successful use of rodenticide may poison birds of prey and other predators: owls have been especially vulnerable, and rodents then multiply.

'Safe' levels of contamination may cause problems if the food chain concentrates compounds For example in the 1960s Lake Michigan (Canada) had 0.0085 ppmv DDT in bottom mud, primary consumer organisms had 0.41, fish 3.0–8.0 and gulls 3177. There have been reports elsewhere of pollutant magnification (bioaccumulation) by 100 000-times.

Species vary in sensitivity to pesticides Factors like state of health, weather, time of day, and many others may be involved. Within human populations social differences may mean different exposure levels.

Pest resistance Pests behave in a way that reduces exposure or have physiological resistance. Within a few generations resistance can increase.

Predators, feeding at the top of foodwebs, can easily be damaged Target organism may escape or recover while the predators do not, consequently with less predation and with pests possibly becoming resistant or avoiding pesticides or new pests appearing, problems increase. In the 1970s some South East Asian rice monocultures suffered from brown planthoppers (*Nilaparvata lugens*) as a consequence of predator loss and pest resistance. Pimbert (1991: 3) noted that roughly 450 insect and mite species, 100 plant pathogen species and 48 weed species were resistant to one or more pesticide.

Accumulation in bodyfat or other tissues DDT and other compounds are found in wildlife even in remote regions like Antarctica. In some regions wildlife accumulates dangerous levels of pesticide (e.g., irrigated areas of Mississippi, USA have 'mosquito fish', minnows poisonous enough to kill a predator which eats one or a few). Some pesticides are known or are suspected to cause cancer (particularly organochlorine and organophosphate pesticides and those containing PCBs or similar), damage nerve cells or are teratogenic (cause birth defects). Recently whales, dolphins and seals have suffered immune system damage, nervous system damage or have been found dead, perhaps related to pesticide pollution. A claim has been made that UK bovine spongiform encephalopathy (BSE or 'mad cow disease', first noted in 1985 in the USA and in 1986 in the UK and attributed to a virus or protein fragments similar to that causing 'scrapie', in sheep, and thought to be transmitted in high-protein feed made from dead livestock) might be related to pesticide pollution (organophosphorus use against warble fly) (Purdey 1992). BSE had led to the slaughter of over 87 000 cattle in the UK by mid-1993: post-mortem examination confirmed the disease in most cases.

Coral reefs seem to be damaged by pesticides.

Fears that the poor, especially women, may lose employment weeding if herbicides used.

Factory accidents Leakage from Bhopal killed and injured many people; there is pollution and hazard from a number of pesticide plants around the world.

Dependency LDC and DC farmers become 'locked in' to purchase of pesticides.

Factors affecting pesticides released into the environment

Leaching Some pesticides are less soluble than others, some bind to clays, etc. Soils differ in capacity for leaching; moisture for leaching varies in availability; moisture that is acidic may leach more.

Adsorption Soil colloids may actively hold pesticide. Character and quantity of colloids (clay and organic matter) varies and certain pesticides are more likely to be held. Acid deposition will speed release of pesticides.

Volatilization Some pesticides are less volatile than others (DDT is not very); temperature also helps determine rate of volatilization, i.e. decomposition to gases.

Chemical decomposition Hydrolysis and photodecomposition are the main routes. Soil pH, clay content, temperature, sunlight and moisture all affect decomposition and movement of pesticides and are not constant.

Biological decomposition Environmental conditions and type of pesticide affect biological decomposition. Organophosphates tend to be broken down more readily than organochlorines.

Reducing pesticide problems

Pesticide problems might be controlled by:

- Banning dangerous compounds: there have been some recoveries by bird of prey populations, which fell in the 1960s in the UK and elsewhere, since bans on DDT and other organochlorines.
- Developing alternatives like biological control or integrated pest management (IPM).
- Restrict trade movements and sales of pesticide contaminated produce.
- Controlling pesticide usage by monitoring, inspection and licensing to ensure sensible procedures.
- Developing less dangerous pesticides.
- Controlling prices of pesticides to discourage use of too much. Education to discourage the use of the wrong types or unsound strategies.
- Rotation of crops to upset pest breeding and access to food.
- Hand or mechanical weeding, which may also provide employment.
- Encourage funding agencies to cut funds for pesticides or to try to control use.
- Treat drinking water to remove pesticides (e.g. with ozone and activated charcoal filters).

Between 1972 and 1975 the US Environmental Protection Agency cancelled the sale and prohibited the use of some of the worst (in terms of toxicity and persistence) insecticides, including DDT, aldrin, dieldrin, chlordane and heptachlor; many other countries followed. Unfortunately, such bans may not stop manufacture and sale abroad: in 1979 at least 25 per cent of pesticides exported from the USA were banned for use in the USA (Weir and Shapiro 1981: 4). Some pesticides, the use or sale of which is banned in DCs, are still manufactured there or in LDCs and are widely used in the latter countries.

Most countries have established departments responsible for reviewing pesticide use with the powers to initiate controls, but there are still problems in disseminating information about pesticides and their effects, in monitoring, and with political and economic aspects of control (Ghatak and Turner 1978; Boardman 1986). Various databases and networks are now established to assist with monitoring and control. The FAO and WHO have set up the Codex Alimentarius Commission ('Codex System') to establish food standards and one of its tasks is to check on pesticide residues in produce. Another of its tasks is to try and get governments to agree on international standards to control pesticides manufacture, export and usage (each year it publishes a Codex of information to assist in this). Under GATT proposals (see chapter 11) the Codex will have increased influence over the way countries set their food and agriculture standards; it has been argued, Codex meetings are dominated by DC and MNC representatives (Avery *et al.* 1993). Any nation setting standards too high compared with the Codex might have to justify doing so or suffer trade sanctions.

In 1986 the FAO issued an International Code of Conduct on the Distribution and Use of Pesticides and in 1990 got 100 countries to sign a code of conduct on pesticides. The UNEP, WHO, OECD, ILO, EU and the Pesticides Action Network (PAN) and other international bodies and NGOs make efforts to improve pesticide use and controls, but in practice there is a long way to go before controls are anything like satisfactory.

Alternatives to indiscriminate use of pesticides

Focus use of pesticide Strategies include reducing quantities through ultra-low volume spraying; minimizing spray-drift; using pesticide-treated bait; improving pest forecasting for treatment at optimum moment; and using less harmful pesticides like pyrethrum.

Direct control Methods include weeding to remove cover; erection of barriers to prevent pest access; use of masking scents to hide crop from pest, or deterrent compounds that prevent movement or feeding; hunting and removal of pest; and noise or visual scarers.

Trapping Pests may be lured by attractants (light, pheromones, crop or prey scent or bait) to traps (with or without pesticide compounds) to be killed or rendered sterile or given a transmissible genetic defect and released to breed.

Biological pest control Natural enemies of pests should be conserved and supported; if need be, augment natural predators with exotics or genetic engineering; breed and release predatory species, parasites of pest, diseases of pest; provide nestboxes or cover for predators (e.g. 'beetlebanks' at intervals across fields to encourage predators); release sterile pests or pests with genetic faults to breed with wild population. There are various examples of biological control: a number of large Dutch and UK glasshouse tomato and cucumber growers now use biological methods for pollination (bumblebees) and pest control (whitefly and spider mite control), including release of parasitic wasps (e.g. *Encarsia formosa* for whitefly). The use of insects for pest control and pollination has proved effective and means a crop free of pesticides which enhances sales. Stour Valley, Dorset (UK), suffers from blackfly (*Simulium post-icatum*), which breeds on aquatic weeds that cannot be safely sprayed with herbicide. In 1989 a species-specific bacteria (variety of *Bacillus thuringiensis*) was tested, sprayed on to weeds infected with blackfly (Fullick and Fullick 1991; Hurst *et al.* 1991: 271). The same bacteria is used for control of cabbage white butterfly caterpillars on cabbage. Also in the UK, bracken (*Pteridium aquilinum*) has become a nuisance; the South African moth *Conservula cinisigna* seems to have promise for biological control. The Argentine moth *Cactoblastis cactorium* has been used in Australia to successfully destroy 25 million ha of prickly-pear cactus (*Opuntia* spp.). In Scotland there has been trial use of a wrasse (*Ctenolabrus rupestrus*) to control sea lice on farmed salmon rather than use pesticides.

Tractor-drawn or knapsack sprayer delivering boiling water/steam Trials show great success in controlling weeds, insects and fungal infections. It has been tried in Florida on cotton, citrus and potatoes. It is cheap, easy, effective and can be used by those who are organic farming. The disadvantage is treatment of field crops when in growth, rather than before or after or between crop rows.

Breed resistant crops Seed merchants now market carrot-fly resistant varieties of carrot (e.g. 'Fly Away'–F1), powdery mildew-resistant peas and numerous vegetables that withstand viruses.

Cultural methods These include planting and harvesting at times or in ways to avoid pests; crop rotation to reduce pest build-up; growing deterrent plants near crop, for example 'bugles' (*Ajuga* spp.), aloes (e.g. *Aloe vera*) and neem tree (*Azadirachta indica*) may be used to discourage insect pests when planted close to crops. Further methods include spraying crop with a scent or other compound that hides, discourages or hinders feeding; tillage, mulch or netting to exclude pests; removal of pest habitats or breeding sites near crops.

Hormones or pheromones Attract pests to traps; dose with juvenile hormone analog to ensure pest does not mature or breed.

Vacuum-clean Certain crops may be treated to rid them of pests (successful use of 'bug-vacs on soft fruit in California).

Barriers Nets, plastic tents, rows of certain plants or belts of deterrent material are placed around crops on ground surface.

Biotechnology including genetic engineering This will probably open up new possibilities for biological control, crop resistance, synthesis of compounds like hormones, and so on.

Integrated pest management

Integrated pest management (IPM) is likely to include a mix of the aforementioned measures appropriate to given pests and situations. As we have seen, alternatives to present pesticides and their patterns of use need to be more appropriate as well as safer. IPM may prove an effective substitute for the imprecise chemical pesticides in general use today. IPM involves study of the pest(s) and its or their context, which may necessitate approaches like participatory rural appraisal to diagnose the situation by learning by and with agriculturalists. A range of cultural, breeding and pest control techniques may be used, perhaps backed by legal controls like quarantine on plant movements, enforcement of anti-pest practices, bans on certain activities, etc. To be successful over the long term IPM will need to be coordinated with wildlife conservation, land and water management, social and economic development, public health, etc. IPM uses pesticides as a last resort but judicious use is possible.

Biological controls

As with chemical pesticides, there is a need for caution over biological controls. History has taught that alien organisms may attack non-pest species or somehow become a problem, for example the mongoose introduced to control rats which instead attacks an island's wildlife. At present Australia and New Zealand are seeking a means of controlling rabbits, release of the myxoma virus (which causes myxomatosis) having become ineffective or unacceptably cruel. Viral rabbit haemorrhagic disease might be the solution: there are proposals to test it on remote Macquarie Island where unforeseen problems might be better contained. Genetic engineering could lead to serious problems if an unwanted trait escaped or a dangerous organism were released. Biological pest management and biotechnology demand careful advance studies.

Agricultural wastes

Agriculture increasingly produces 'wastes'; ironically these may once have been valuable resources and might still be if managed in a sensible way (the PRC makes full use of urban nightsoil and livestock manure). These wastes include animal dung and urine slurry; silage effluent; cereal straw; crop residue (husks, stalks, etc). The impacts of agricultural waste are global: an ice core from the CIS's Vostock Station (Antarctica) shows an increase of nitrous oxide from 1750 to 1990, probably due to use of fire for brush clearance for agriculture and to increased N-fertilizer use since the 1930s (Conway and Pretty 1991: 326).

There are three main routes to reducing agricultural waste problems:

1 *Quotas* limits on quantities that a farm may produce.
2 *Fertilizer tax* to discourage excessive use and pollution, which satisfies the 'polluter pays principle' (Clunies-Ross 1993).
3 *Set-aside* withdrawal of land from production (Fig. 8.3); the risk is that remaining agricultural land will be more intensively used and thus there would be little overall reduction of pollution.

Livestock wastes

In Europe and the USA livestock waste has become a major problem as farms increased in size and arable and livestock farming have tended to become separated. In The Netherlands livestock producers have increasingly been locating near ports so as to reduce costs of transport of imported feed. Livestock waste is a major cause of nitrate pollution of ground and surface waters in Europe, North America and other parts of the world. The dependence of DC livestock producers on imported feed should be noted; forest clearance and land degradation in some LDCs may be driven by the DC livestock feed market for crops like soya (e.g. in the Brazilian *cerrado* savannas) or cassava (e.g. in North Thailand).

Manure once a resource used to sustain cropping and pasture, is now often a problem to livestock owners, while arable farms without animals have to buy in fertilizers. A dairy cow may produce as much waste as 16 humans and two pigs produce as much as 5 humans: a UK farm of 40 ha with just 50 cows and 50 pigs can present a waste disposal problem equivalent to a town of nearly 1000 people. For the UK as a whole livestock produce 2.5 times the total human sewage, in the USA about 10 times (Conway and Pretty 1991: 276). Already Denmark and The Netherlands have begun to build disposal and processing plants and the manure may one day be returned to the land again or sold, safely composted. In 1993 The Netherlands had about 40 million tonne 'manure mountain' – more than twice that which could be safely disposed of on to all the farmland. The authorities introduced dung quotas for pig farmers in 1993, and a few enterprising landowners even resorted to arranged marriages to boost their quotas (*The Times* 4/3/93/: 13)!

Stored in pits or lagoons, livestock waste generates methane, ammonia and hydrogen sulphide, which cause nuisance smells, damage vegetation downwind (the ammonia) and act as greenhouse gases. If slurry escapes, as it often does, it can cause serious stream, lake or groundwater pollution, through chemical oxygen demand (COD), biological oxygen demand (BOD), harmful bacteria and parasites, excreted antibiotics or growth promoting hormones and steroids and sometimes heavy metals (especially copper and zinc added as growth accelerators to pig feed). With large quantities of slurry, disposal by spraying on to farmland is impractical as it may transmit diseases; heavy metals from feedstuffs tends to concentrate in the soil and nitrates and phosphates can leach to contaminate surface and groundwaters (Fig. 8.2). The alternatives are to use livestock waste for on-farm generation of biogas, to pool material from several farms to provide biogas or district heating and power for villages and towns (this practice is now widely adopted in Denmark) or to mix it with materials like straw or refuse for conversion to compost. Residues left after biogas generation may also be recovered as a safe compost. Dutch biotechnologists have developed ways of treating pig manure to convert the ammonia that it releases into harmless protein which can be fed back to the pigs; there are plans that this will be in commercial operation by 1994.

Silage effluent

Silage has become popular in Europe since the late 1960s as a feedstuff for livestock. When it is made moisture is released, the amount depending upon how dry the grass or other crop was on collection and on the mode of production. It is not uncommon for 330 litres of effluent to be formed for each tonne of silage made, which means a farmer must store and dispose of large quantities of highly acidic (often pH 3.4) liquid. If the effluent escapes, it may damage soil and cause havoc with aquatic life and groundwater. Stored in lagoons or pits, the effluent gives off ammonia and hydrogen sulphide, both active as greenhouse gases.

Straw and crop residue burning

Cereal straw burning has been a problem in Europe for some years until recently when legislation began to curb it. In Brazil, Mauritius, parts of Australia and the Caribbean, sugar fields are burnt before harvesting. Crop residue burning helps control weeds and pests but may also destroy harmless or useful wildlife, damage soil, cause visibility accidents, and generate soot and greenhouse gases. Cottage industries which use straw, such as weaving and thatching, may be deprived of raw material. Modern cereal straw may not be as long or strong as in the past (often the latter is because of rapid growth due to fertilizer leached from cropland), and thus less useful for thatching, but there is still potential for manufacturing strawboard, paper, cardboard, or for on-farm or district heating. The problem in recent years in DCs has been the cost of collection and transport. In the UK the Ministry of Agriculture, Fisheries and Food (MAFF) announced a complete ban on straw burning to take effect from 1992.

Brush clearance

Clearing land for agriculture is often done with fire which each year generates vast amounts of soot and greenhouse gases, promotes soil degradation and kills wildlife. It is a worldwide problem; large areas of savanna and Amazon forest areas have been burnt since the 1960s.

Wastes from agricultural products processing

A wide range of agricultural products can cause problems when processed because of toxic by-products or because effluent gets into streams and causes eutrophication. The processing of palm oil, rubber, pineapples, sugar, meat, fish, and many other products can cause difficulties. In West Malaysia and Sabah, oil palm fruits are processed in local factories, partly to ensure that treatment takes place without delay to ensure minimal fermentation and high quality palm oil: consequently few major streams and rivers have escaped pollution and considerable damage has been done to middle and lower reaches in spite of tougher legislation since the late 1970s. In Brazil, the processing of sugar, cassava and yams to produce alcohol for automobiles results in about 13 litres of high-BOD effluent for each litre of fuel. With the expansion of the alcohol programme many rivers, especially in the northeast, have suffered.

Crop processing often demands fuelwood and large areas can be deforested to provide for the needs of tobacco curing, tea drying, and preparing many other locally consumed or export crops. Where hides are turned into leather, there is a risk of river pollution. Tanning with oak bark, wattle bark or other natural compounds produces acidic and high-BOD effluent, considerable smell and nuisance from flies. Recently toxic chemicals have been adopted for tanning; some contain heavy metals like chromium or mercury, which can cause tremendous damage to stream ecology and groundwater.

Mechanization problems

A major problem associated with mechanization is compaction of soil. Exhaust emissions from tractors contribute to vehicle-related pollution. In the USA catalysers have been required on farm machinery since 1974. Disposal of lubricating oil and hydraulic fluid may cause local problems, especially if it gets into streams or groundwater. Machinery noise may also cause a nuisance and disturb wildlife.

On-farm conservation

A whole range of measures could enhance conservation when adopted by farmers, foresters or other landusers (e.g. golf course managers, those who maintain burial grounds or road and rail verges or armed forces training areas): areas of land at the ends of fields could be left undisturbed and unsprayed; forest fire-breaks could incorporate ditches and scrub; ponds could be dug and small blocks of on-farm woodland could be planted. To get such measures adopted requires farmer awareness, and incentives like tax allowances or grants.

In 1988 the EC agreed set-aside rules and payments to farmers; this has also been tried in the USA. The UK and other parts of western Europe had a high take-up of set-aside by 1993, with those setting aside over 15 per cent of their land qualifying for area payments based on estimated 'average' yields for regions. Coupled with a gradual reduction of cereal subsidies, set-aside may prove a good way to encourage reduced inputs of agrochemicals. Unfortunately set-aside can be unsatisfactory if not carefully controlled, e.g. by allowing grass-cutting at times that could kill nesting birds and in some cases has reputedly led landowners to intensify use of their non set-aside land, leading to more agrochemical pollution and soil degradation. In 1993 when the rules were due to be renegotiated there were fears that landusers would plough up set-aside land if their payments were delayed or ceased. Some feel that quotas on certain crops might be a better route to a reduction of environmental problems. Rotational set-aside (i.e. the block of land taken out of production changes over time) might be used to ensure that farmers do not simply intensify use of their best land and set-aside the rest.

Beyond the green revolution

Challenges and possibilities

The 1973/4 OPEC oil crisis marked a turning-point for the green revolution. Since the mid-1970s demands upon agriculture have increased and the problems and mistakes of the pre-1974 'first-wave' of the green revolution have led to a reappraisal of agricultural improvement approaches with more emphasis on learning from farmers (Richards 1985; Wolf 1986). Rather than just seeking to boost yields, interest now focuses on better resource management, sustainable production, resilience to environmental challenges and equitability (better access to crops or profit from crops).

Raising and maintaining agricultural yields in DCs and LDCs will be a challenge, and temporary boosts of production are easier than sustained. Environmental problems appear to be increasing and agriculture is often degrading the land it uses. Three recent agricultural 'revolutions' might be recognized: 1920s–50s – mechanical revolution (mainly in DCs); 1960s–80s – agrochemical and HYVs-based 'green revolution'; the 1990s onward – a biotechnology 'gene revolution'.

There is a limit to how much new land can be found, so in the future efforts will concentrate more on intensification and sustaining production. Better management of moisture and nutrients generally leads to reduction of soil erosion and better crops. There could be considerable reduction in wastage of inputs and harvested produce. Methods like inter-cropping, agroforestry and agrosilviculture 'mimic' natural conditions and may have potential for sustainable, appropriate agricultural production, including the replacement of degraded shifting cultivation. Some of these improvements offer little or no profit for agroindustry, and may actually reduce the need for inputs; their promotion may therefore rely upon NGOs, bodies like the UN, FAO and CGIAR, and enlightened industries. There are hopes that biotechnology will be a route to better yields and sustained crops, less pollution, and crops capable of growing in difficult conditions with few inputs. However, biotechnology may be a 'double-edged sword', agroindustry may control it and it could cause environmental problems and allow industrial production of some of the crops LDCs depend upon selling for foreign exchange.

More use could be made of green manures and other means of biological nitrogen fixation to reduce usage of chemical fertilizers. Rice production has been improved as much as 25 per cent (and following crops also benefit) by adding to paddy-fields the aquatic fern (*Azolla* spp.) which, in symbiosis with blue-green algae (*Anabaena azollea*) and various bacteria, fixes nitrogen and suppresses weeds. These and other green manures could be used for other crops, as animal feed or for biogas production. *Azolla* requires water and occasional dressings with phosphate fertilizer (Gould and Giller 1990).

It is now accepted that traditional methods can be rediscovered and improved to be environmentally friendly and require low-inputs so they can be adopted in areas with poor communications and limited resources. There are many cases where in the past people lacking modern materials and engineering cropped lands now regarded as unusable (Barrow 1987: 174: Pearce 1991). Such approaches would reduce agriculturalists' dependency on outside sources of inputs and skills.

Since the 1950s consumption of maize and wheat bread has increased in the LDCs with an associated decrease in cassava and sorghum or millet consumption (wheat consumption in tropical countries increased by about 3 per cent per annum between 1961 and 1981). The rising market has encouraged overproduction in DCs, which has kept prices low and discourages LDC producers.

Since the 1960s increasing numbers of people in LDCs have abandoned consumption of traditional foods in favour of bread. This may be due to new tastes stimulated by marketing or the consequence of urban women having jobs and less time to cook or because of cooking fuel shortages. Local crops are often more nutritious, their supply more secure and, above all, they do not have to be imported. More could be done to improve production of cassava and traditional grains like sorghum and it is worth encouraging people to consume them; one route may be to find ways to use these staples to mimic wheat bread.

A MV sorghum has been developed which resists witchweed (*Striga* spp.), a pest that often greatly reduces harvests in Africa. Other sorghum and millet MVs are being bred to resist moisture stress (drought) and to avoid various pests through rapid growth which means the plants have reduced vulnerable stages. MVs of barley, wheat and other cereals are being developed which resist saline conditions; this should allow production to be continued in areas suffering natural or anthropogenic salinization and possibly allow use of poor quality irrigation water, even seawater. Some of these MVs are being developed with 'traditional' breeding methods, some will be the product of biotechnology.

Markets and controls

Farmers do not operate in a political or economic vacuum. DC and LDC producers can easily become trapped on a 'modernization treadmill' where they struggle to produce for a market they do not control. In Europe many depend on subsidies and are controlled by quotas; the Common Agricultural Policy concentrated in the 1970s and 1980s on promoting cereal production which increase fertilizer and pesticide use; recent Common Agricultural Policy reform has tried to encourage less intensive landuse to reduce overproduction of some crops, to cut pollution and to reafforest. After decades of support for intensification, farmers are now encouraged to set-aside land to encourage wildlife conservation, reduction of pollution, forestry, tourism, etc (Clunies-Ross and Hildyard 1992).

Organic farming

Until roughly the early 1940s all farming was organic. Organic farming relies on alternatives to agrochemicals: crop rotation, green manures, animal dung and other organic waste for soil fertility maintenance; cultural methods, biological pest control and other measures to avoid toxic chemicals for pest control (Conford 1992) (Box 8.3). Modern methods are used, provided these do not entail agrochemicals. Interest in organic farming was established in the 1930s, Lady Eve Balfour, who spread the idea through her book *The Living Soil* (1943) and by founding a NGO, the Soil Association, in 1946. The Soil Association runs a standard for organic farming in the UK and actively campaigns for its adoption. Similar bodies are actively promoting organic agriculture elsewhere in the world. In order to sell produce as 'organically farmed', a grower may need to practise agrochemical-free production for some years, thus conversion can be a slow process, during which growers may need assistance.

In North America and Europe some consumers are willing to pay more for organically produced crops, while others prefer the attractive appearance of chemically treated crops. Organic farming NGOs

Box 8.3 Organic farming

Characteristics

- Avoidance of chemical pesticides and soluble N-fertilizers.
- Crop rotation used.
- Nutrients recycled.
- Non-chemical pest and weed control.
- Mixed farming or horticulture.
- Care for environment and livestock well-being.

Benefits claimed

- Less pollution.
- Sustainable production.
- Reduced requirement for energy, especially lower use of hydrocarbons, reducing pollution.
- Generally reduction of soil erosion and other land degradation and environmental damage.

Critics question

- Whether levels of soil potassium and phosphates can be maintained in the long term.
- Whether by generally being less productive organic agriculture may lead to farming or grazing of greater areas of land.
- Whether most organic farming is really self-sufficient as feedstuffs are often brought in.

Critiques of organic agriculture can be found in Canter (1986: 83); Adams (1990).

and growers have established marketing networks for organic produce; further expansion now depends mainly on fashion and attitudes.

The claim is made that organic farming prevents land degradation and so sustains production; a counter-claim is made that it is less productive than established methods and so could well lead to more land being farmed for the same yield. Reganold (1989) compared two similar Washington State (USA) farms, one organic and one conventional; the indications were that, compared with the average for the region, the organic farm did give better cereal yields and its soil retained more moisture and kept in better condition. Hurst and colleagues (1991: 284) argued that for the UK organic farming could be a viable alternative: there may be no significant fall in output and although crop mixes would alter, there would be benefits from reduction of pollution.

Permaculture ('permanent + agriculture') is a sustainable form of agriculture related to organic farming originally developed by Bill Mollinson, a Tasmanian whose works include *Permaculture* (Mollinson and Holmgren 1978) and *Permaculture Two* (Mollinson and Holmgren 1979). Permaculture seeks independence of external inputs and has a central concept the goal of working with nature (managing rather than modifying environment to fit agriculture) (Bell 1992). Permaculture is promoted in the USA by the Permaculture Association (Durham, North Carolina).

Energy crops

Bioenergy has the advantages of virtually no net pollution, because as plants grow they lock-up carbon released by the batch burnt; lower sulphur content than many fossil hydrocarbon fuels; probably less NO_x emissions than fossil hydrocarbon fuels; sustainability; the potential to reduce dependence on fuel imports; some possibilities for combination with pollution control and disposal (sewage may be fed to willow beds or combustion may be with refuse). A number of European countries are interested in developing fast-growing willow (*Salix* spp.) or reedbeds for energy cropping. Several countries, notably Brazil, produce alcohol by fermentation of sugar cane, cassava, yams or maize (usually ethanol is produced). When petroleum prices are high, this makes economic sense and reduces harmful vehicle exhaust emissions in cities, although pollution of rivers from fermentation effluent can be a problem (it need not be because the phosphate-rich effluent could be recovered for use as fertilizer).

Rudolf Diesel tested vegetable oils when developing his engine in the 1890s; however, cheap petroleum has since the 1920s reduced interest in biofuels. There are many suitable vegetable oils: some need little or no processing and for a few the cultivation and oil production is already familiar to many LDCs and DCs (e.g. palm oil, groundnut oil, sunflower oil, coconut oil and soybean oil). Europe has begun to grow oilseed rape for diesel fuel substitution; the town of Reading (UK) experimented with this fuel for public bus transport in 1992–3 but ran into cost problems due to unsupportive UK fuel duty and the lack of UK rapeseed oil producers. Rapeseed oil needs little processing and emits little sulphur.

Energy crops have great potential, provided that they do not force out small farmers and cause them to resettle on marginal land where they can cause land damage (which is what alcohol production has done in parts of Brazil) or reduce food surpluses to a dangerously low level. Production of vegetable oil for diesel fuel offers LDCs several advantages: the technology is straightforward and thus accessible; crop culture is familiar, and in the case of the babassu (*babaçu*) palm (*Orbignya* spp.) in Brazil something that could be done on poor soil with little skill or outlay (A. B. Anderson *et al.* 1991); it may save a great deal of foreign exchange; productivity is greater in the tropics than in temperate (generally DC environments) so LDCs have a competitive advantage; it can offer self-sufficiency; crops like coconut and palms can be combined with other land uses and do not demand frequent tillage or input of fertilizers – unlike groundnut, sunflower, soya or rapeseed. It is also possible to use biotechnology and produce single-cell oils from cultures of micro-algae, bacteria or fungi. (For a review of the potential of a number of oil crops see Shay 1993.)

Sustainable agriculture

Often the assumption has been that sustainable agriculture meant 'low-input', 'low-output', i.e. reduced returns to the farmer. However, there are indications from the UK, USA and LDCs to suggest that 'there exists an economically and environmentally viable alternative to both industrialised and organic agriculture' (Pretty and Howes 1993: i). Such an agriculture seeks a more thorough integration of natural processes into the agricultural production process, combined with a reduction in the use of off-farm inputs, to achieve profitable and efficient production. It does not constitute a return to low technology or low output farming. It is a spectrum of technologies and practices which 'usually involve the substitution of labour, knowledge and management skills' for high use of external inputs (Pretty and Howes 1993: i). According to Pretty and Howes (1993) this sort of integrated farming can match or better the gross margins of conventional farming in the UK, although there may be a yield per hectare reduction of some 5–10 per cent for crops and 10–20 per cent for livestock. In addition to this level of production there are likely to be benefits for farmers, rural communities (more employment, less pollution), consumers and the environment. There is then an outlook of promise not unmitigated gloom.

Biotechnology

So far, agriculture has drawn upon, and increasingly degrades, soil, water, genetic resources and air. Biotechnology may reduce the pressure on some of those resources or might cause further damage, especially loss of genetic diversity (synonymous with loss of biodiversity or genetic erosion). Biotechnology (a subdiscipline of bioengineering) involves the utilization of biological processes and manipulation of living material without genetic change or transfer – using culture techniques like micro-propagation (cloning many plants from one seedling) and tissue culture (cloning from a callus grown from a portion of a parent plant), fermentation and selective breeding; it also includes genetic engineering – genetic manipulation, i.e. manipulation of DNA to produce recombinant DNA (ρDNA), to modify or transfer genetic material; at its most extreme a transgenic organism may be produced (Forster and Wase 1987; Mannion 1991: 291; PANOS 1991).

Biotechnology offers opportunities and poses threats hardly dreamed of before the 1980s. If the potential is realized it could ensure that agriculture does not decline or remain on a plateau and might cut pollution. The economic, social and environmental impacts could be huge: biotechnology and genetic engineering may be the weapon needed to combat the food problem, avoid fertilizer use, avoid pesticide pollution, save energy, reduce need for land, reduce greenhouse gas emissions and stabilize production in spite of environmental changes (Ministry of Environment 1990; Mannion 1992; Bud 1993). Alternatively, biotechnology could cause major problems for the environment and human well-being. The 'gene revolution' is more commercialized and more difficult to oversee and control than the green revolution.

Biotechnology: actual and potential innovations

1 *Speed up and reduce costs of developing new crops and livestock.*
2 *Enhanced photosynthesis* through breeding or gene transfer.
3 *Biological nitrogen-fixation* either by gene transfer from a N-fixer plant to others which are not, or inoculation of crop roots or seed with a culture of natural or 'engineered' bacteria or fungus which can or which stimulates symbiosis. This may be difficult as many genes are involved in most crops.
4 *Pest deterrence, pest avoidance, pest resistance* breeding or gene transfer to produce organisms that kill selected pests or give plants ability to avoid, kill or resist pests. *Bacillus thuringiensis* is already widely used for insect control. Mycoherbicides are cultures of a fungus that attacks weeds, e.g. 'Casst'® or 'DeVine'®. Whitefly and mosquito control with fungi (e.g. *Verticillium lecanii* or bacteria). The advantages of biotechnology over agrochemicals for pest control are less risk of pests becoming resistant, no chemical pollution, better kill, cheaper, better focus (i.e. kills more specifically).

5 *Enhanced drought tolerance, salt tolerance, etc* by breeding or gene transfer.
6 *New 'transgenic' crops and livestock* i.e. human-made species from 'parents' that cannot naturally breed.
7 *Enhanced protein content* crop engineered or bred to have improved protein (or some other quality) enhanced.
8 *Culture of material to control pests* through production of bacteria or other organisms or extracts like toxin from cultures.
9 *Substitution of 'field' agriculture with industrial production* single-cell protein (SCP) has been used since the 1940s to produce cattle feed (and to lesser extent human food) from hydrocarbons. A major producer is the CIS (about 430 000 tonnes per year in the 1980s) (Rimmington 1985). SCP is produced using suitable algae, fungi, yeasts or bacteria cultured in vats. At present cheap soya, fishmeal and cassava makes SCP less attractive; cheaper production might cut demands for the aforementioned produce, perhaps releasing more food for humans, perhaps causing havoc to established trade and affecting LDC exports (SCP trade names in UK include 'Mycoprotein'® and 'Pruteen'®). Fermentation could produce commodities like cocoa butter from cheap vegetable oils (cocoa is the second most important tropical commodity in international trade), new 'super sugars', gum arabic, single cell oils to substitute for presently used vegetable oils, alcohol for fuel, vanilla, etc (these being identical to natural product but not grown on farmland, production may require only inputs like air, water, sunlight and wastes or petroleum). If these commodities were to be produced in bulk in DCs, the impact on the export earnings of a number of LDCs would be catastrophic. A West African plant with sap reputedly over 1000 times stronger than cane sugar (*Thaumatococcus donielli*) might well be cultured to provide cheap sweetener, which would threaten traditional sugar producers already suffering from substitutes like corn-syrup and synthetic sweeteners (Hobbelink 1991: 79). In 1991 Brazil got 28 per cent of its auto fuel from fermentation of biomass (sugar, cassava or yams); to do this there had been a growth of cane fields from 1.5 million ha in 1972 to 3.8 million ha in 1985 and this had displaced many farmers who became unemployed or moved to degrade marginal environments. There might be ways to dispense with biomass grown on farmland, use biotechnology and reduce pressure on the land and perhaps agrochemical pollution or methane emissions. Factory production of commodities or food might be linked to waste disposal or sewage treatment and may also give by-products that add to profits. The possibilities are enormous, and include the following already in commercial use.
10 *Frost-countering* bacteria culture provides freezing nuclei to promote freezing, provide an ice layer and so reduce frost damage, e.g. 'Ice-minus'® or 'Frostban'®. Fears have been voiced that an 'escape' could alter cloud, snow and rain incidence, i.e. change climate.
11 *Pharmaceutics* vaccines, human insulin, human interferon, antibiotics, growth hormones, etc.
12 *Growth hormones* BGH (which stimulates lactation) and bovine somatotropin BST (which increases milk yield). Other hormones and steroids may be used to stimulate production or improve quality of meat.
13 *Diagnostics* (compounds for identifying things) chemicals; food/commodity production; energy production; pollution control.
14 *Bio-control of pollution* micro-organisms (bacteria and yeasts) produced to attack pollutants, perhaps including some radioactive spills, and render them harmless, easier to leach from soil or to bind them to the soil to reduce leaching. Oil spills may be countered with petroleum consuming bacteria. Bioreactors are now available to treat pollutants, including VOCs.
15 *Waste disposal* micro-organisms or larger creatures used to compost refuse or waste or convert it to biogas.
16 *Bio-mining* micro-organisms (e.g. bacterium *Thiobacillus ferrooxidans*) used to release required mineral from rock or water. Used on copper-bearing sulphide ores in USA to aid leaching-out metal (Mannion 1991: 298; Mannion and Bowlby 1992: 151). For example, Ghana's huge Ashanti Goldmines are adopting 'Biox' (bio-oxidation) to extract gold from pyritic ores: this not only is effective, but also allows the discontinuation of use of toxic arsenic.
17 *Bio-desulphurization* use of micro-organisms to leach sulphur out of coal or oil before it is burnt, thus reducing air pollution.
18 *Enhanced oil recovery* micro-organisms injected down oil or gas wells to improve release and flow of oil/gas.

Tools of biotechnology

There are various techniques in biotechnology (see Lipton and Longhurst 1989; George 1990; Hobbelink 1991). Several tools are now discussed briefly.

Tissue culture Growing organisms or parts of organisms in cultures (nutrient-rich mediums held under appropriate environmental conditions). Ensures more rapid production of material, more uniform product, possibility of disease-free (especially virus-free) crops, replication of organisms that may be difficult to achieve naturally, exact copies of original. A well-established approach is the meristem culture – formative tissue that can still grow to maturity (such as a plant growing-tip or apical bud, anther or a seed embryo) is dissected out from the parent plant (possibility of application to animals in future?), surface sterilized, placed in or on nutrient and kept under warm, sterile conditions. The cells are stimulated to reproduce disorganized tissue (callus) and ultimately plantlets are taken from the mass and are grown on until they can be treated as normal seedlings. Cloned material is 100 per cent identical to the parent, which means performance should be predictable but in time genetic diversity will be eroded (it also offers a chance to replicate rare plant species quickly to aid conservation or new crop development). Widely used for tropical root crops, cocoa, oil palm, bananas, rubber, orchids, cassava. Freedom from disease often means better yields and new variety is faster and cheaper to produce. Predictable qualities may offer legal advantages for producer. Tissue cultures may be chemically treated, temperature stressed or irradiated to produce mutations in the hope of creating new varieties.

Embryo transfer Embryo conceived naturally or *in vitro* is transferred to a surrogate mother. Already used for livestock (especially cattle) and possible for humans. Reduces genetic diversity but economically attractive and predictable results.

Somatic embryo genesis Artificial seeds – 100 per cent genetic uniformity, like clones produced in cultures, but ease of use as seeds. Same advantages and risks as cloned material.

Enzyme and fermentation technology or cell culture These techniques have tremendous potential for production of food and commodities in 'factory' vats. This could be in the market country, could use cheap, accessible raw materials, reduce need for land, pesticide pollution and greenhouse gas emissions. Success would ruin many LDC exporters, and cut their employment.

Gene transfer, cell fusion, gene splicing, recombinant DNA (ρ-DNA) – 'Genetic engineering' Can 'map' DNA and identify which bits do what; these can be cut out and transferred, deleted or replaced with material from another organism. Recipient (host) cells have to be induced to accept transferred genetic material; this may be done by infecting the host cells with a bacteria carrying required genetic material or by blasting the host cell with minute particles from a 'gene gun'. Suitable infective bacteria include those which cause plant tumours (e.g. *Agrobacterium tumefaciens*), the tumour genes are removed and required genetic material is substituted. These techniques have been used to impart herbicide tolerance to crops and to create bacterial cultures capable of producing human insulin, interferon and growth hormones.

Biotechnology: problems

Problems will arise if biotechnology is used carelessly or if DCs and commerce control it (Hindmarsh 1991). The potential benefits are so great that such risks will have to be faced: there is already talk of a 'biotechnology revolution' or 'gene revolution' having begun (Goodman and Redclift 1991: 167; Marx 1989). Biohazard assessment, avoidance and licensing of release of biological materials is vital (Gustafsson and Jansson 1993; Molin and Kjelleberg 1993). The risks are that an uncontrollable escape of material may occur and have unforeseen impacts or, techniques could be misused for commercial gain or warfare and terrorism (H. A. Mooney and Berardi 1990). For example, if an organism that was engineered to carry something like spider venom for use as a pest control passed on resistance to insects, natural predation would be upset. There have been risky experiments already (Hatchwell 1989; Juma 1989: 132). In 1986 trials of recombinant DNA vaccine for bovine rabies in Argentina infected two workers and potentially could have had wider impacts. In the USA genes from genetically altered radish were found to have spread to other varieties nearby, carried by pollinating insects. In the UK rules have been drawn up to control release of human-made lifeforms (the GENHAZ risk assessment system); other countries are also legislating. It should be noted that countries like the USA argue that caution should not be restricted to organisms created by biotechnology, transfer of naturally bred lifeforms from one environment to another can cause problems so *any* release is a risk.

Recombinant DNA techniques could be used to develop biological weapons. Stockpiling, research and production of offensive biological weapons has been banned since 1972 by the Biological and Toxic Weapons Convention signed by 105 nations, and in theory by the 1925 Geneva Protocol. Nevertheless, in recent years at least 10 countries have been proven or are suspected of having developed or used biological weapons.

Primitive biotechnology, e.g. aerobic treatment of sewage, fermentation and brewing are long established (the latter dating back to 700 BC or even earlier in Egypt). The first 'modern' bioengineering company was established in California in 1980; already transnational corporations (TNCs) control a major part of enzyme and fermentation technology (for a list of companies see Hobbelink 1991: 37). It is possible that DCs will control biotechnology so that much research will be orientated toward DC needs and that which is not will possibly lead to LDC dependency. Those who invest will do so only if there are returns; consequently there has been interest, shared with seed, pharmaceutic and agrochemicals companies, in extending patent rights from inanimate to animate things. This could deny LDCs access to biotechnology, a sort of 'bio-imperialism' (Belcher and Hawtin 1991; Goodman and Redclift 1991: 180).

Some progress has been made with the development of crops with herbicide resistance ('dovetailing'). For example, Hoechst have genetically engineered crops that resist their herbicide 'Basta'®. Such crops could be more effectively weeded and might lead to less herbicide use if farmers were able to wait and see if weeds grew enough to need treating. But they would offer agroindustry the chance for 'double profit': farmers would have to buy seeds and appropriate herbicides. *Bacillus thuringiensis* has been engineered to contain genes that kill insects, the cultured bacteria is now available to the public as an effective means of non-chemical pest control. However the release of bacteria to control pests, consume pollutants like oil spilt on soil, assist rotting of landfill refuse tips, or provide freezing nuclei to reduce frost damage to crops like strawberry and potatoes (e.g. 'Ice-minus'® based on *Pseudomonas flurescens* engineered from *P. syringae*) has raised worries and controversy.

Problems associated with irrigation development

Provided techniques and management work well, irrigation has huge potential to improve yield, quality, security and diversity of crops. There is also a chance that irrigation may enable production to be sustained and may even reduce land degradation, particularly erosion associated with rainfed production. It could also reduce demand for land because yields should be better. This potential has encouraged worldwide expansion in less humid and humid environments, but in practice, in both LDCs and DCs, full potential is often not realized, and serious problems can arise: irrigation may be a 'panacea' or a 'Pandora's box' (Heathcote 1983: 186).

Irrigation may be broadly defined as the practice of applying water to the soil to supplement natural rainfall or flooding, and thereby provide more moisture for plant growth. While there is a huge diversity of irrigation techniques and management strategies, a rough subdivision is possible, into *complete irrigation*, where the bulk or all plant moisture needs are met; *supplemental irrigation*, where enough extra moisture is provided to improve the quantity and/or quality and/or allow diversification of the crop; *protective irrigation*, i.e. provision of water when drought threatens. The first of the aforementioned tends to cause more problems because levels of moisture are maintained above natural levels and there is likely to be standing water.

Irrigation is the dominant form of agriculture in a number of regions and countries. Sometimes it is in the hands of numerous independent small farmers seeking subsistence or semi-subsistence livelihood, sometimes it is practised by tenants on government or commercial schemes; increasingly it is run as large-scale commercial projects producing crops for export or grain for city populations. Roughly 50–60 per cent of the increase in agricultural output achieved in the LDCs since the early 1970s has come from irrigated land. Only about 11 per cent of the world's total arable cropland is irrigated but it includes roughly two-thirds of all LDC farmers. About 20 per cent of LDC cropland is irrigated and provides over 40 per cent of annual crops, especially foodstuffs (Barrow 1987: 200). Much of the agricultural development effort and expenditure of the last few decades has been focused on irrigation and a high percentage of LDC fertilizer and HYV use is on such land, the success of irrigation must be judged against the share of inputs it enjoys.

Many of the best sites and most suitable water sources have already been exploited for irrigation. Increasingly there will be competition between agriculturalists, and with industry, urban areas and wildlife conservation for water supplies, necessitating more use of less than optimal land and reliance on less satisfactory water supplies. There may be improvements in techniques and crop breeding which allow the use of presently unsuitable water, particularly saline water, but many parts of the world will still remain unsuitable, being too remote or having unsuitable terrain or soil. In AD 2000 around 84 per cent of the world's farmers will still be rainfed cultivators and will produce about 51 per cent of total crops (Barrow 1987: 311). There is a lot of scope for improvement of rainfed agriculture and it may well be that with little expenditure and research reasonable advances will be made. In an ideal world the improvement of rainfed agriculture might attract as much funding as irrigation development, in reality that is unlikely.

A worrying proportion of once-productive irrigated land has been ruined or is being ruined and often ends up wholly useless even for rainfed production. The most common cause of degradation of irrigated land is salinization – saline or alkaline accumulation. Perhaps half of the world's farmland is badly salinized (Barrow 1987: 297; Pereira 1989: 178). It is a problem of both LDCs and DCs; the challenge for the coming decades will be to try and prevent further damage, rehabilitate as much as possible of what has been damaged and find ways of using that which cannot be easily de-salinized. There are some hopes for the latter, based on either salt-tolerant versions of existing crops, the development of new saline-tolerant crops as food, fodder, wood or fuel sources, etc., or cost-effective cultivation methods that allow the use of saline water for normal crops (Barrow 1987: 217–18; National Academy of Sciences 1990).

Irrigation problems may be caused by numerous things, including careless exploitation of water supplies, the development of unsuitable land, over-application of water, under-application of water, misapplication of water, construction of channels which harbour or allow the movement of pests, impacts associated with the use of agrochemicals, HYVs and other intensive farming practices. Often the suppliers and users of water are very wasteful: few large irrigation projects actually get as much as 20 per cent of the flow they divert or abstract to the crops. The losses are due to design faults, construction or maintenance problems, poor management, theft of water and overuse by some recipients. Wastage of water means that less land can be properly irrigated; there may be crop failure for some or all irrigators if moisture supplies falter, and there may be areas of waterlogging and associated salinization.

Irrigation development is often carried out by bodies most concerned with engineering, costs or political gain: the environmental and social aspects are then ignored or missed. Adams (1990: 156) attributes considerable blame for irrigation problems to 'irrigationism' – the preoccupation with large, capital-intensive projects which, once started; are difficult to modify to deal with any difficulties that arise.

Leaking canals waste water and can raise groundwater to cause waterlogging and salinization. If funds are available, canal leakage can be virtually eliminated by lining or replacement with pipelines; the latter also controls many pest and disease problems. Surface application of irrigation water, by flooding fields or releasing water down furrows, is one of the most common approaches, but it wastes a considerable amount of water through evaporation and seepage. If funds are available more precise methods of application may be employed: various sprinklers, emitters, buried perforated-pipes, dribblers, etc. These waste less water and may help avoid waterlogging and salinization (although methods that water only a small area, like dribblers, tend to form a fringe of salts some distance from the water release point and in time these may be spread by rainstorms or wind).

Salinization (and alkalinization) may arise because there is waterlogging or because there is inadequate application of water (either the wrong quantity and/or poor quality – i.e. containing too much salt/alkali compounds) or poor drainage which prevents the maintenance of a salt (or alkali) balance. Such a balance is a satisfactory level of salinity or alkalinity sustained in the soil where crops root because enough water has been left after meeting crop needs to wash away the contaminants (meet the leaching requirement). A common cause of waterlogging and salinization is the inadequate provision of drainage for irrigated land. Waterlogging and salinity control depend on satisfactory design, installation and management of irrigation. The former two are difficult to rectify after construction, the latter is often the cause of problems and might be rectified.

Some irrigation impacts are beneficial, wildlife and migrant birds may find suitable feeding, nesting and resting sites and local people may obtain fish from channels and paddy-fields. These advantages are too often lost because of agrochemical pollution.

Irrigation-related problems

Water supply problems

River diversion Diversion of river water may cause problems by reducing downstream flow leading to saltwater intrusion in estuaries, siltation, reduced river absorptive capacity leading to increased pollution problems, less water available for downstream agricultural, industrial, power generation and domestic supply users. Barrages cause less flow disruption and do not form large impoundments. Dams constructed to store water impound, often large, reservoirs which may cause resettlement, local disease problems, seismic disturbance, and may severely disrupt downstream flows. Whether from dam or barrage, transfer of water, especially in open channels, may lead to dispersal of pollutants and nuisance organisms. Canals may disrupt communications and, if breached, damage nearby land. Unlined canals may leak and damage nearby land through waterlogging and salinization.

Groundwater abstraction Use of groundwater may depress watertables causing social disparity and conflict over water access, may cause springs to run dry and streams to have reduced baseflows. Wetlands may dry out with a loss of flora and fauna if groundwater falls. Drawdown of watertable may lead to aquifer collapse, and saltwater intrusion, or subsidence. The problems afflict LDCs and DCs.

Inadequate drainage

Irrigation is often provided with inadequate or no drainage, waterlogging and salinity problems result. During rainstorms erosion may occur if there is no drainage provision.

Groundwater or surrounding land contamination

Surplus water may leach through soil or via poorly sited drains to pollute groundwater or surrounding land. Water is likely to contain salts and agrochemicals.

Return flow problems

Waste drainage water (return flow) discharged into streams, surface water bodies or on to low-lying land may cause difficulties due to salts and agrochemicals content. Problems can reach regional or national dimensions affecting areas well removed from the causative irrigation if a large area of irrigation is involved (e.g. catastrophic shrinkage and contamination of the Aral Sea by pesticide contaminated return flows from CIS Central Asia and Kazakhstan irrigated cotton production). Alkaline return flows may encourage snail breeding and increase risk of schistosomiasis, fertilizer contaminated return flows may promote weed growth raising maintenance costs and providing sites for insect breeding and increased transmission of malaria or river blindness (oncocerciasis). Return flow problems generally grow as irrigated areas expand or agricultural use intensifies.

Waterlogging

Unlined or poorly lined supply canals and channels, excess water applied to cultivated areas and poor drainage and poorly levelled irrigated land lead to waterlogging. It may occur when groundwater is exploited by farmers using tubewells or in areas supplied by canal. Waterlogging may arise after a period of no difficulty if agriculturalists modify their practices. Modification may be change to a crop that needs less water or result from fear of interruption of supply ('apply as much as you can while you have it').

Soil structural changes

Irrigation plus agrochemicals, tillage and compaction through use of machinery may lead to suffusion, reduction of soil organic matter, hard pans, compacted soil.

Biological impacts

Difficult to predict in advance, soil flora and fauna likely to alter. Micro-organisms likely to be reduced in diversity but those left tend to increase in numbers (White 1978: 43). Nematode worms sometimes become a crop pest (Kovda *et al.* 1973: 413). Some changes due to disturbance and tillage, some to increased moisture, some to agrochemicals.

Health problems

Schistosomiasis, malaria and mosquito-borne fillarial and viral infections are often introduced to an area or increase following irrigation development. Wetter conditions favour vector organisms and people may have altered their agricultural practices bringing them into increased contact with them. There are likely to be a number of actual or potential mosquito disease-vectors, which makes control and advance assessment of problems difficult. Shistosomiasis incidence may be reduced by careful design of the channels to discourage snail breeding. In many cases these problems will be solved by integrated measures: prophylactic and remedial drug treatment, improved hygiene, weed removal, introduction of mosquito or snail eating fish or other animals, disease vector control (chemical or organic pesticides or molluscicides, insect screening of buildings, etc) and encouragement of preventative behaviour through health education. The lack of funds, logistical difficulties and expertise are unlikely to be available to counter irrigation-related health problems in many countries for the foreseeable future.

Resettlement

Local people may be moved off land to develop it for irrigation.

Social impacts

One community may benefit from irrigation development, others may not leading to social and economic disparity. People may become more dependent on outsiders – water supply authority, marketing network, agrochemical and seed suppliers.

Irrigation may divert money from other things

Roads/conservation/healthcare/rainfed farming, etc. Might an equivalent or better improvement of crop yields be obtained through existing agriculture with better crop storage, pest control, subsidy on fertilizers?

Risk climatic change will cause difficulties for irrigation

If climate alters and causes crops to demand more moisture irrigation supply systems may not be able to meet this or prevent salinization. If rainfall increases or becomes more intense there may be waterlogging or soil erosion as surplus water escapes. May mean more moisture is available and scheme can expand to become more productive.

Possibility irrigation will cause climatic change

Large areas of irrigation may sufficiently alter albedo to change winds, cloud formation and airmass movements. Downwind atmosphere may be moister due to evapotranspiration. Methane and N_2O emissions from irrigated land, especially flooded rice paddy-fields and from livestock used to work them likely to be increasingly important as a cause of global warming.

References

Adams, N. (1990) The case against organic farming. *New Scientist* 128(1734): 68

Allen, G.R. (1977) The world fertilizer situation. *World Development* 5(5–7): 525–36

Allen, J. (1983) In search of the perfect cocoa bean. *New Scientist* 97(1343): 293–6

Anderson, A.B., May, P.H. and Balick, M.J. (1991) *The subsidy from nature: palm forests, peasantry and development on the Amazon frontier*. New York, Columbia University Press.

Anderson, R. and Morrison, B. (eds) (1982) *Science, politics and the agricultural revolution in Asia*. Boulder, Colo., Westview

Anon (1979) The seeds of disaster. *New Internationalist* 81: 18–20

Avery, N., Drake, M. and Lang, T. (1993) Codex Alimentarius: who is allowed in? Who is left out? *The Ecologist* 23(3): 110–12

Barrow, C.J. (1987) *Water resources and agricultural development in the tropics*. Harlow, Longman

Behrendt, H. and Boekhold, A. (1994) Phosphorus saturation in soils and groundwaters. *Land Degradation & Rehabilitation* 4(4): 233–44

Belcher, B. and Hawtin, G. (1991) *A patent on life: ownership of plant and animal research*. Ottawa, IDRC, 40 pp. (PO Box 8500 Ottawa K1G 3H9)

Bell, G. (1992) *The permaculture way: practical steps to a self-sustaining world*. London, Thornsons

Boardman, R. (1986) *Pesticides in world agriculture: the politics of international regulation*. London, Macmillan

Boserüp, E. (1990) *Economic and demographic relationships in development* (*essays selected and introduced by T.P. Schultz*). Baltimore, Md., Johns Hopkins University Press

Briggs, D.J. and Courtney, F.M. (1985) *Agriculture and environment: the physical geography of temperate agricultural systems*. London, Longman

Bud, R. (1993) *The uses of life: a history of biotechnology*. Cambridge, Cambridge University Press

Bull, D. (1982) *A growing problem: pesticides and the third world poor*. Oxford, OXFAM

Burt, T.P., Heathwaite, A.L. and Trudgill, S.T. (eds) (1993) *Nitrate: process, patterns and management*. Chichester, Wiley

Buttel, F.H. (1990) Biotechnology and agricultural development in the third world. In H. Bernstein, B. Crow, M. Mackintosh and C. Martin (eds) *The food question: profit versus people?* London, Earthscan

Canter, L.W. (1986) *Environmental impacts of agricultural production activities*. Chelsea, Mich., Lewis

Carson, R. (1962) *Silent spring*. New York, Houghton Mifflin

Clunies-Ross, T. (1993) Taxing nitrogen fertilizers. *The Ecologist* 23(1): 11–17
Clunies-Ross, T. and Hildyard, N. (1992) The politics of industrial agriculture. *The Ecologist* 22(2): 65–71
Commoner, B. (1971) *The closing circle*. London, Jonathan Cape
Conford, P. (ed.) (1992) *A future for the land: organic practice from a global perspective*. Bideford, Green Books
Conway, G. and Barbier, E. (1991) *After the green revolution: sustainable agriculture for development*. London, Earthscan
Conway, G.R. and Pretty, J.N. (1991) *Unwelcome harvest: agriculture and pollution*. London, Earthscan
Davidson, J., Myers, D. and Chakraborty, M. (1992) *No time to waste: poverty and the global environment*. Oxford, OXFAM
De Vos, A. (1975) *Africa the devastated continent: man's impact on the ecology of Africa*. The Hague, Dr W. Junk
Dudley, N. (1990) *Nitrates: threat to food and water*. London, Green Print
Farmer, B.H. (ed.) (1977) *Green revolution? Technology and change in rice-growing areas of Tamil Nadu and Sri Lanka*. London, Macmillan
Farvar, M.T. and Milton, J.P. (eds) (1972) *The careless technology: ecology and international development*. Garden City (New York), Natural History Press
Forster, C.F. and Wase, D.A.J. (eds) (1987) *Environmental biotechnology*. Chichester, Ellis Harwood
Fowler, C. and Mooney, P. (1990) *The threatened gene: food politics, and the loss of genetic diversity*. Cambridge, Lutterworth
Fullick, A. and Fullick, P. (1991) Biological pest control (Inside Science Supplement no. 43). *New Scientist* 129(1759), 4 pp.
Geertz, C. (1963) *Agricultural involution: the process of agricultural change in Indonesia*. Berkeley, Calif., University of California Press
George, S. (1990) *Ill fares the land*. Harmondsworth, Penguin
Ghatak, S. and Turner, R.K. (1978) Pesticide use in less developed countries: economic and environmental considerations. *Food Policy* 5(2): 134–46
Glaeser, B. (ed.) (1987) *The green revolution revisited: critique and alternatives*. London, Allen and Unwin
Goldsmith, A. (1991) The gene hunters. *The Geographical Magazine* LXIII(1): 36–9
Goodman, D. and Redclift, M. (1991) *Refashioning nature: food, ecology and culture*. London, Routledge
Gould, J. and Giller, K. (1990) A green solution to the green revolution. *New Scientist* 128(1731): 57
Greenland, D.J. (1975) Bringing the green revolution to the shifting cultivator. *Science* 190(4217): 841–4
Gunn, D.L. and Stevens, J.G.R. (1976) *Pesticides and human welfare*. Oxford, Oxford University Press
Gustafsson, K. and Jansson, J.K. (1993) Ecological risk assessment of the deliberate release of genetically modified microorganisms. *Ambio* XXII(4): 236–42
Harwood, R.R. (1979) *Small farm development: understanding and improving farming systems in the humid tropics*. Boulder, Colo., Westview
Hatchwell, P. (1989) Opening Pandora's box: the risks of releasing genetically engineered organisms. *The Ecologist* 19(4): 130–6
Haugerud, A. and Collinson, M.P. (1991) *Plants, genes and people: improving the relevance of plant breeding* (IIED Gatekeeper Series no. 30). London, IIED
Heathcote, R.L. (1983) *The arid lands: their use and abuse*. London, Longman
Hindmarsh, R. (1991) The flawed 'sustainable' promise of genetic engineering. *The Ecologist* 21(5): 196–205
Hobbelink, H. (1991) *Biotechnology and the future of world agriculture: the fourth resource*. London, Zed
Holloway, R.E. and Dexter, A.R. (1990) Compaction of virgin soil by mechanized agriculture in a semiarid environment. *Land Degradation & Rehabilitation* 2(2): 107–17
Hornsby, M. (1989) Farming's nitrate nightmare. *The Times* 11/12/89: 11
Hornsby, M. (1992) EC nitrate targets 'will harm farms'. *The Times* 24/2/92: 5
Hurst, P., Hay, A. and Dudley, N. (1991) *The pesticide handbook*. London, Journeyman
James, J. (1978) Growth, technology and the environment in less developed countries: a survey. *World Development* 6: 937–65
Juma, C. (1989) *The gene hunters: biotechnology and the scramble for seeds*. London, Zed
Kloppenburg, J. (1988) *First the seed: the political economy of biotechnology, 1492–2000*. Cambridge, Cambridge University Press
Kovda, V.A., Van Den Berg, C. and Hagan, R.M. (eds) (1973) *Irrigation, drainage and salinity: an international sourcebook* (published for FAO/UNESCO). London, Hutchinson
Levintanus, A. (1992) Saving the Aral Sea. *The Environmentalist* 12(2): 85–91
Linear, M. (1982) Gift of poison: the unacceptable face of development. *Ambio* XI(1): 2–8
Linear, M. (1985) *Zapping the third world: the disaster of development aid*. London, Zed
Lipton, M. and Longhurst, R. (1989) *New seeds and poor people*. London, Unwin Hyman
Mannion, A.M. (1991) *Global environmental change: a natural and cultural environmental history*. Harlow, Longman
Mannion, A.M. (1992) Biotechnology and genetic engineering: new environmental issues. In A.M. Mannion and S.R. Bowlby (eds) *Environmental issues in the 1990s*. Chichester, Wiley
Mannion, A.M. and Bowlby, S.R. (eds) (1992) *Environmental issues in the 1990s*. Chichester, Wiley
Marx, J.L. (ed.) (1989) *Revolution in biotechnology*. Cambridge, Cambridge University Press
Mellanby, K. (1970) *Pesticides and pollution*. London, Fontana
Ministry of Environment (1990) *Developments in biotechnology*. London, HMSO
Molin, S. and Kjelleberg, S. (1993) Release of engineered microorganisms: biological containment and improved predictability for risk assessment. *Ambio* XXII(4): 242–5
Mollinson, B. and Holmgren, D. (1978) *Permaculture: a designer's manual*. Hobart, Tasmania, Tagari Publications
Mollinson, B. and Holmgren, D. (1979) *Permaculture two: practical design for town and country in permanent agriculture*. Hobart, Tasmania, Tagari Publications

Mooney, H.A. and Berardi, G. (1991) *Introduction of genetically-modified organisms into the environment* (SCOPE 44). Chichester, Wiley

Mooney, P.R. (1979) *Seeds of the earth: a private or public resource?* London, Coalition for Development Action (Bedford Chambers WC2)

Mooney, P.R. (1985) The law of the seed revisited. *Development Dialogue* 1985(1): 139–52

National Academy of Sciences (1990) *Saline agriculture: salt-tolerant plants for developing countries*. Washington, DC, National Academy of Sciences

Okigbo, B.N. (1981) Alternatives to shifting cultivation. *Ceres* 14(6) (no. 84): 41–5

PANOS (1991) *Miracle or menace? Biotechnology and the third world* (PANOS Dossier no. 3). London, PANOS

Pearce, F. (1991) Ancient lessons from arid lands. *New Scientist* 132(1798): 42–8

Pearse, A. (1980) *Seeds of plenty, seeds of want: social and economic impacts of the green revolution*. Oxford, Oxford University Press

Pereira, H.C. (1989) *Policy and practice in the management of tropical watersheds*. London, Belhaven

Pimbert, M.P. (1991) *Designing integrated pest management for sustainable and productive futures* (IIED Gatekeeper Series no. 29). London, IIED

Pinstrup-Andersen, P. (1982) *Agricultural research and technology in economic development*. Harlow, Longman

Porritt, J. (1984) *Seeing green: the politics of ecology explained*. Oxford, Basil Blackwell

Pratt, C.J. (1965) Chemical fertilizers. *Scientific American* 212(6): 62–72

Pretty, J.N. (1990) Agricultural pollution: from costs and causes to sustainable practices. In D.J.R. Angell, J.D. Comer and M.L.N. Wilkinson (eds) *Sustaining earth: response to environmental threats*. London, Macmillan

Pretty, J.N. and Howes, R. (1993) *Sustainable agriculture in Britain: recent achievements and new policy changes*. IIED Research Series vol. 2(1). London, IIED

Purdey, M. (1992) Mad cows and warble flies: a link between BSE and organophosphates? *The Ecologist* 22(2): 52–7

Reganold, J. (1989) Farming's organic future. *New Scientist* 122(1668): 49–52

Richards, P. (1985) *Indigenous agricultural revolution: ecology and food production in West Africa*. London, Hutchinson

Rigg, J. (1989) Green revolution and equity: who adopts the new rice and why? *Geography* 74(2): 144–50

Rimmington, A. (1985) Single-cell protein: the Soviet revolution? *New Scientist* 106(1462): 12–15

Samways, M.J. (1980) *Biological control of pests and weeds*. London, Edward Arnold

Saull, M. (1990) Nitrates in soil and water (Inside Science Supplement no. 37). *New Scientist* 127(1734): 4

Shay, E.G. (1993) Diesel fuel from vegetable oils: status and opportunities. *Biomass and Bioenergy* 4(4): 227–42

Shiva, V. (1991a) *The violence of the green revolution: third world agriculture, ecology and politics*. London, Zed

Shiva, V. (1991b) The failure of the green revolution: a case study of the Punjab. *The Ecologist* 21(2): 57–60

Shiva, V., Anderson, P., Schücking, H., Gray, A., Lohmann, L. and Cooper, D. (1991) *Biodiversity: social and ecological perspectives*. London, Zed; Penang, World Rainforest Movement

Spoor, G., Carillon, R., Bournas, L. and Brown, E.H. (1987) The impact of mechanization. In P. Wheale and R. McNally (eds) *The bio-revolution: cornucopia or Pandora's box?* London, Pluto

Stigliani, W.M. (1991) *Chemical time bombs: definition, concepts and examples* (IIASA Executive Report no. 16). Laxenburg, Austria, International Institute of Applied Systems Analysis

Weir, D. and Shapiro, M. (1981) *Circle of poison: pesticides and people in a hungry world*. San Francisco, Calif., Institute for Food and Development Policy

White, G.F. (ed.) (1978) *Environmental effects of arid land irrigation in developing countries* (MAB Technical Notes no. 8). Paris, UNESCO

Wold, S. (1992) Biotechnology, agriculture and food. *The OECD Observer* 177 (August/September): 4–8

Wolf, E.C. (1986) *Beyond the green revolution: new approaches to third world agriculture* (Worldwatch Paper no. 73). Washington, DC, Worldwatch Institute

Wolman, and Fournier, F.G.A. (eds) *Land transformation in agriculture* (SCOPE 32). Chichester, Wiley

Wood, R.K.S. and Way, M.J. (1988) *Biological control of pests, pathogens and weeds: development and prospects*. London, Royal Society

World Bank (1992) *World development report 1992: development and the environment*. Oxford, Oxford University Press

Further reading

Bull, D. (1982) *A growing problem: pesticides and the third world poor*. Oxford, OXFAM [Looks at pesticide problems, especially impacts on LDC poor, and how usage might be improved]

Conway, G. and Barbier, E. (1991) *After the green revolution: sustainable agriculture for development*. London, Earthscan [Argues for a new approach to agriculture which considers sustainability and LDC farmer circumstances]

Conway, G.R. and Pretty, J.N. (1991) *Unwelcome harvest: agriculture and pollution*. London, Earthscan [Introduction to pesticide, fertilizer and agricultural waste problems, comprehensive and readable]

Dudley, N. (1990) *Nitrates: threat to food and water*. London, Green Print [Useful review]

Glaeser, B. (ed.) (1987) *The green revolution revisited: critique and alternatives*. London, Allen and Unwin [Objective examination of green revolution from an 'ecodevelopment' standpoint and with regard for LDC poor]

Hobbelink, H. (1991) *Biotechnology and the future of world agriculture: the fourth resource*. London, Zed [Good introduction]

CHAPTER 9 Problems with river development

Introduction

Large parts of the world, periodically, seasonally or permanently, do not have enough water of the quality required. This may always have been the case or it may have become so because population has increased and/or living standards have improved raising demand for domestic supplies (drinking, cooking and washing needs), or it is due to development of industry or agriculture, or because new demands have developed. Water supply problems may also arise because there has been alteration of available rainfall, streamflow or groundwater. New demands include the spread of water-based sewage and industrial effluent disposal (sewerage); hydroelectric power generation; new industrial demands; recreational, navigational or conservation demands. Change of precipitation, streamflow or groundwater may be due to water exploitation, climatic change or subtle alteration of landuse. Even parts of the humid tropics, where rainfall may exceed 10 000 mm y^{-1}, can have periods when evapotranspiration exceeds precipitation – enough to check or kill crops, especially on free-draining and thin soils.

Water shortage may be countered by adapting activities to reduce moisture use and, if that is insufficient, impossible or undesirable, natural moisture supplies can sometimes be augmented. The chief ways of augmenting natural moisture supplies are to intercept runoff, to divert or store streamflow, to exploit groundwater, to recycle waste water, to desalinate seawater or other saline sources (expensive) and – so far on a small scale – intercept and collect mist cloud or dew.

Apart from very arid and extremely cold regions most parts of the Earth's surface can be divided into river basins. Sometimes gentle topography makes it difficult to recognize a precise boundary between river basins, but there is seldom significant overlap or gaps between. River basins have attracted the attention of those involved in planning and managing development because they offer relatively stable, discrete biogeophysical units within which water offers an integrative 'theme' between different facets of resource use and human activity. Much of human history has been acted out and has involved the exploitation of runoff, streamflow and groundwater resources within river basins. Development continues to take place within basins, often necessitating management of water resources which may be in private, common or state ownership, and often, shared by more than one group of people or countries. For the aforementioned reasons river basins are often regarded as a practical unit for planning and management. Watershed is widely used to indicate the boundary between basins, sometimes it used to refer to the area contributing water to a stream or river (sometimes called catchment). In the USA 'divide' is indicative of the basin boundary (Pereira 1989: 7).

Streamflow is a potentially renewable resource but poor management may convert it to a temporary or permanent finite resource. The same may hold for a groundwater that it sufficiently well recharged to meet human demands: poor management, particularly excessive use, can convert a renewable to a finite resource (some groundwaters are not being recharged enough, or at all, to be considered potentially renewable). Once polluted a groundwater may take a very long time to recover and may be effectively ruined even if pollution ceases.

Within river basins terrestrial and aquatic ecosystems are coupled: drought and flood are interrelated (parched land generally has reduced infiltration capacity and less vegetation and when rains come there is likely to be sudden runoff and flooding); alterations of vegetation cover can change groundwater or streamflow (by varying infiltration, evapotranspiration and ground albedo); urban development feeds more water rapidly to rivers. Measures which improve moisture availability for plants generally also reduce runoff and soil erosion and hence the amount of silt in streams, so it makes sense to coordinate watershed management (including groundwater management) and river basin management (Barrow 1987a: 37).

River basin development

Secure, productive agriculture, industrial production or urban development are often impossible relying on rainfall, natural runoff or streamflow. Solutions may be to exploit groundwater (if available), control runoff within a watershed, practise low moisture demand agriculture or streamflow regulation. When using groundwater that is renewed relatively fast, it is desirable that watershed management should ensure integration of surface and underground water resources so that when there are unused surface flows these can be used to recharge groundwater so that when surface water is scarce groundwater can be exploited. Storms seldom swell all of a river's tributaries at the same time; with a network of telecommunications and suitable dams and barrages, it may be possible to delay storm runoff and to hold back some flows and allow stormwater to take their place, avoiding flooding in mid and lower basin.

There is a tendency for the literature on river basins to deal with watershed development or with river development. The former focuses on surface runoff (overland flow – water flowing between interception as precipitation and a recognizable channel) plus ephemeral – streamflow and flow in minor channels; the latter, on more regular streamflow in channels. However, they interlink and it is in many respects undesirable to deal with them separately.

Watershed management

Careful watershed landuse (with appropriate tillage, groundcover and possible use of runoff control structures) can help stabilize and often increase groundwater recharge and streamflow, reduce erosion and stream silt content, and improve crops or forestry. The wrong sort of changes in vegetation cover, soil compaction, inappropriate land drainage and use of agrochemicals can each alter surface flows (increasing streamflow during and shortly after storms and reducing baseflow between precipitation), cut groundwater recharge and alter the quality of both. In colder environments, if the right type of planting or retention structures are used to prevent snow blowing off the watershed this can give more spring and early summer moisture. Attempts to control runoff and increase moisture availability for crops, to improve groundwater recharge, to hold adequate soil on sloping land and to reduce erosion, rely on various forms of earthen-banks, on structures of stone, wood or patterns of crop or shrub planting (for an introduction to watershed management see Pereira 1973; 1989).

When good watershed management is maintained, streamflow management is easier with less risk of reservoirs, irrigation channels, etc., running dry or silting-up. Clearer rivers are less likely to choke and flood so structures like bridges will be safer. However, there is sometimes increased channel erosion after silt reduction, reduced food for detritus-feeding fish and other organisms, loss of the deposits on floodlands which helped maintain fertility, and increased delta and coastal erosion.

Streamflow management

Streamflows are often inconstant, sometimes silt-laden and often located away from areas in need of water. The solution to these shortcomings has generally been to store water or divert water (or a combination of both). Engineering has long been used to modify streamflow to improve water supply, drive machinery, and sometimes to deprive an enemy of water or flood land they occupy or wish to traverse (Smith 1972).

Using streamflow

Micro-hydroelectricity (less than 300 kW generation) and mini-hydroelectricity (typically less than 6 MW per generation unit and 12 MW total for a plant) and direct mechanical power can be obtained from even quite small streams. The power of moving water has been harnessed since Roman times or earlier for flour milling, and so on. In the UK and western Europe flowing or falling water powered flour mills in many villages by medieval times and some textile mills well before and during the early part of the industrial revolution. Between the First World War and the 1950s there were many mini- and micro-hydropower plants generating electricity in the UK, mainland

Europe and other countries, but from roughly the 1930s until recently few western manufacturers continued producing the equipment, and skills and interest have only recently been revived (Fraenkel *et al.* 1992).

The PRC reportedly obtains about 40 per cent of its hydroelectricity from small dams on streams (Agnew and Anderson 1992: 203), and there is increasing interest in this proven, flexible form of power generation in DCs and LDC (e.g. Nepal). With small-scale electricity generation or directly harnessing mechanical power there is little or no need for reservoirs; such 'run-of-the-river' exploitation does virtually no environmental damage (a fall of 1.5 to 2.0 metres can generate 100 kilowatts, using a simple rotor and generator). In countries like the UK there are promising opportunities for a renaissance of clean, sustainable local power generation. In the 1800s there were probably more than 20 000 water mills in England and Wales; some of these sites and others might be developed; in the future interest in such technology may also help promote development of tidal generation. Ultimately a country like the UK might generate over 25 per cent of its power needs, sustainably and pollution-free, from tidal and small-scale hydroelectricity (*The Times* 30/10/93: 18). Numerous recent studies suggest that many existing river weirs in the UK, if they could be developed, might give as much power as a moderate size nuclear power station.

Check-dams

As already stressed, it is difficult and unwise to separate watershed and streamflow management; check-dams on streams (a diversity of semi-permeable stone or brushwood structures that let some flow pass but which divert or spread enough to usefully benefit agriculture though increased moisture retention, accumulation of cultivable soil, control of gully erosion, reduction of flooding and improved groundwater recharge) may divert water on to watershed pastures or crops. Alternatively flows may be diverted into near-contour canals or across floodplains. Check-dams made from local materials and labour can be a cheap, effective way of controlling soil erosion and improving landuse and water resources, although in practice mobilization of people to build and maintain them may be difficult (Vlaar and Brasser 1990; Vlaar 1992).

Small-scale surface storage

Small reservoirs, tanks and cisterns may be constructed to store surface runoff from a watershed or from streamflows. These take many forms but, in general, unless equipped with silt-traps they are vulnerable to rapid siltation, and can also suffer leakage and high levels of evaporation loss. This may not matter if labour rates are low enough to make it worthwhile to remove the silt from time to time for application to farmland. Some tanks are on-farm, constructed and managed by a single family or group, some are village-built and under common ownership. Small reservoirs are common in DCs and LDCs: Southern Asia has a long history of their construction for irrigation, although this seems to have peaked in 1958–9. When linked by canal or stream, small reservoirs can collect flow over a period of time (perhaps night-time) and meet short-term demands that would be too much for the streamflow or reservoir alone to meet (Barrow 1987a: 187).

Where appropriate construction methods are used and local people manage and maintain small reservoirs they can be valuable and effective as sources for irrigation, domestic water supply, micro- and mini-hydroelectricity and for fisheries.

Barrages

Streamflow can be modified with weirs or barrages. Like check-dams, these structures raise enough head of water to feed a canal, maintain upstream depth for navigation, control floods, prevent upstream salt-water intrusion or generate low-head (i.e. less than about 2 metres) hydroelectricity. No large reservoir is impounded, there is thus little or no storage and much less disruption of downstream flow than is the case with dams, and fish and other aquatic organisms are more able to move upstream and downstream. Barrages cause fewer difficulties than large dams, but they store less water and provide less head so where streamflow is erratic dams may be required for hydroelectric, irrigation and flood control purposes.

Large-scale surface storage

There is no clear distinction between small and large dams and reservoirs; however, one could reasonably describe the latter as impoundments of over 100 square kilometres surface area. Quite large dams were built by the Romans or even earlier. During Victorian times large brickwork and earthfill dams and barrages were built in the UK, mainland Europe, the USA and other countries to provide cities with water, to supply irrigation or control floods. During the 1930s 'New Deal' era in the USA and 'Stalinist' era in the USSR reinforced-concrete dams were developed – an American example being the Hoover Dam (Smith

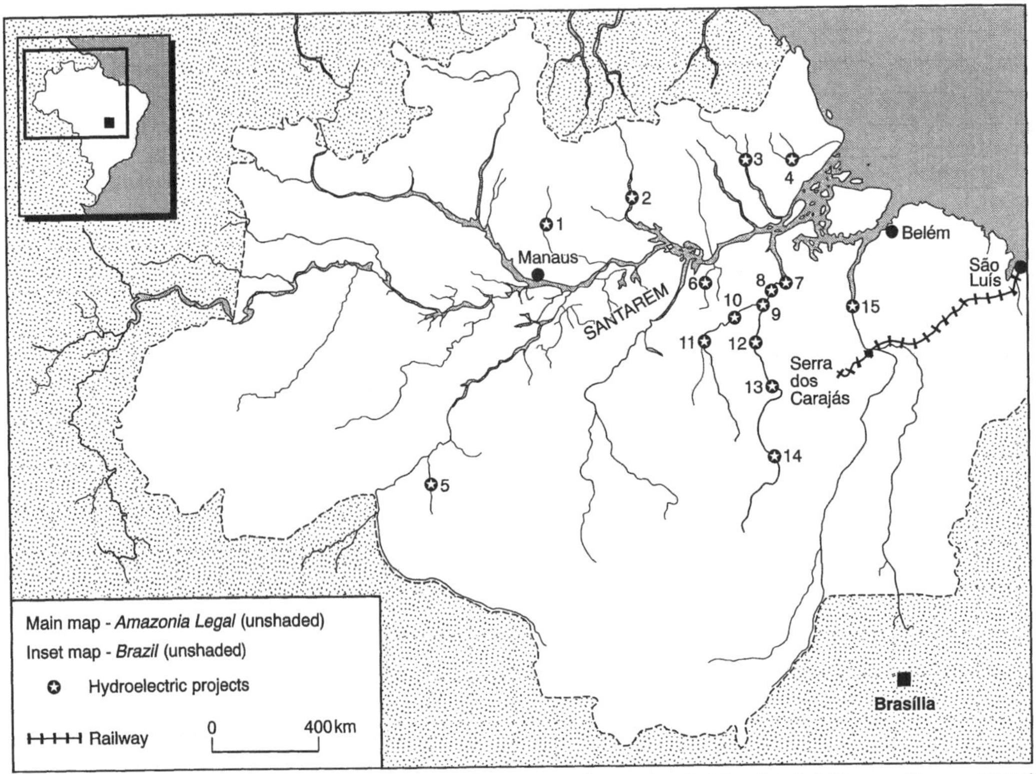

Fig. 9.1 Brazilian Amazonia: large dams completed and proposed (see Table 9.1 for details) ⟨*Source:* Barrow 1988⟩.

1972). Large dams started to spread after 1945 as economies recovered from the Second World War and as concrete and earthmoving technology improved. An increasing number of impoundments since the late 1950s are a thousand or more square kilometres (Goodland 1989: 14).

Large dams either store water in high flow months for use in low flow months or accumulate supplies for periodic shortage years (intra-year storage); they may also be necessary where power generation requires a constant head of water, impossible to achieve with a barrage where a river has variable flow (an alternative might be to develop turbines that need less head or alternative ways of converting streamflow to electricity).

Numerous large dams now support irrigation, generate power, supply industrial and domestic users, provide flood control and improve navigation. It has been estimated that about 13 per cent of the world's riverflow is now channelled into reservoirs, and that in Europe and North America over 40 per cent of stable discharge is regulated by dams (about 15 and 21 per cent of riverflow respectively) (McDonald and Kay 1988: 90–4). The trend has in general been toward multipurpose projects, i.e. dams are intended to provide more than one benefit. Since the early 1960s huge sums of money have been spent on large dams, many in the LDCs: Ponting (1991: 342) estimated that nearly half the spending of the World Bank in recent years has gone on such projects and may have supported over 3 per cent of the world's dams, and Thukral (1992: 8) suggested that India spent 15 per cent of its total national expenditure on dams in the 1980s.

Large dams have attracted a lot of criticism; this has focused on the opportunity costs which result from tying up large sums on these projects (many are funded by loans and the interest payments burden the recipient state) and on the environmental and socioeconomic impacts. There is still debate over the degree of success of a number of large impoundments completed in the 1970s or earlier (Fahim 1981; Hart 1980; Abu-Zeid 1987), and much criticism of many more recent projects.

Table 9.1 Brazilian Amazonia: large dams completed and proposed

No.†	Name of dam	River	Approx. size reservoir (km^2)	Initial installed power capacity (MW)	Status	Reported impacts (see notes below)
1	Balbina	Uatumã	2100	250	C	I, O, F, R
2	Porteira	Trombetas	1400	1191	UC	N
3	Itapeuara	Jarí	1800	252	UC	
4	Coaracy-Nunes	Araguari	100	40	C	W, F, O
5	Samuel (Cachoeira Samuel)	Jamari	645	216	C	
6	Curuá-Una	Curuá-Una	90 to 100	20	C	W, F, O, S
7	Cararão	Xingú	1200		P	
8	Babaquara	Xingú	6100		P	
9	Juruá	Xingú			P	
10	Carajari	Iriri			P	
11	Iriri	Iriri			P	
12	Carajás	Xingú			P	
13	Kaiapo	Xingú			P	
14	Gorotire	Xingú			P	
15	Tucuruí	Tocantinus	2400	4000 (increased to 8000)	C	F, DF, R, I

Status: C = completed; UC = under construction; P = proposed. Data are approximate throughout this table.
Key to impacts: F = fish; I = indians; O = deoxygenation; W = weed problems; DF = use of defoliants; R = resettlement problems; S = schistosomiasis; N = flooding of nature reserve.
† Numbers refer to position of dams on Figure 9.1.
Note: Interest in Amazonian dams is for generation and transmission of power to northeast and centre-south population and industry [see inset map of transmission routes] and to support mineral extraction and processing industry in Amazonia. By AD 2000 Brazil plans to obtain 40% of its electrical power from Amazonia from as many as 40 large dams.
Although there are large flows in most of the Amazon Basin topography slopes gently, to get sufficient head to run presently used turbines requires dam construction. The gentle topography also means reservoirs are large and shallow and their extent varies a lot as water levels rise and fall.
Source: Barrow 1988.

There have been and will continue to be catastrophic dam failures; nevertheless many see the technology as less threatening than nuclear power generation. Hydroelectricity is a potentially sustainable form of energy with little or no carbon dioxide emission provided it is well planned and engineered; it can also help reduce use of hydrocarbon fuel, which reduces that source of pollution, and for some states can cut costly oil or coal imports. The proportion of total world energy use met by hydropower has declined since the early 1940s, in spite of completion of many large projects, but with revived interest this could change. There is still huge potential for irrigation, flood control and hydroelectric dams in South East Asia, Africa, Amazonia (Fig. 9.1 and Table 9.1), Australia and elsewhere.

Dam impacts may result from the following:

1 Disturbances linked to construction and related communications improvements and availability of power for industrial and urban development.
2 Interruption of flow: silt is trapped behind the dam, fish and other creatures may not be able to pass upstream or downstream, water released downstream may be altered in quality (reduced suspended solids, often reduced oxygen, possibly contaminated with products of vegetation and soil decomposition and of lower temperature than normal river flow in the tropics) and in all probability diminished in flow.
3 Creation of a large body of standing water: raised local groundwater, problems with insect breeding, resettlement of people, loss of wildlife and farmland to flooding.
4 Seismic disturbances, which can threaten to breach an impoundment, caused by pressure and weight of stored water.

While positive impacts (benefits) tend to be understated and most projects meet technical and often economic goals, there have been considerable negative impacts and a worrying tendency for these to be repeated in project after project in spite of hindsight experience. There is now a large literature on dam and reservoir impacts (see Barrow 1981; 1987a and b; 1988: 270; Goldsmith and Hildyard

Box 9.1 Case study: Bangladesh flood control

Three-quarters of the Bangladeshi population live in rural areas of the Ganges-Brahmaputra-Meghna Delta and other floodlands. Seasonal flooding is normal but, if floods come early, late or are higher than usual (as is often the case if there are cyclones) there are disastrous losses of crops, livestock and human life. Areas of 'green revolution' farmland and urban areas like Dhaka City are also vulnerable to flooding. Flood disasters are not a new threat, but population in vulnerable areas is growing. The 1987 and 1988 floods prompted the UNEP to recommend a National Flood Master Plan involving extensive embankment and creation of polders. USAID objected to the costs and the risks involved in relying on embankments and in 1989 a compromise was agreed: the Flood Action Plan (1990–5). Funded by multilateral aid and loans, this could cost US$500 million or more.

The Flood Action Plan is widely seen as a 'technological fix' with emphasis on embankment and engineering, rather than early-warning, storm and flood shelters, and provision of relief. There are fears that the 'fix' will exacerbate the problem: there has been remarkably little collection of baseline data or environmental assessment before work started. There have also been warnings that, by embanking Dhaka, floods will be shifted to hit the millions of rural poor living beyond the banks, the very people who suffer most in floods (perhaps a manifestation of urban bias). More study might have suggested a better investment, perhaps in a system of lower embankments maintained by local people, perhaps in storm shelters, but overall more equitable, more relevant and likely to last longer (Brammer 1990a; 1990b; Dalal-Clayton 1990; Custers 1992; Pearce 1992).

The Bangladesh flood question also illustrates how environmental myths become established: it is common to see reports that river sediment loads and flooding are getting worse and are due to misuse of land in the Himalayas. There is, however, little evidence that these assertions are true (Ives 1991).

1984; 1986a; 1986b; Adams 1990: 130), this *should* provide sufficient guidelines for effective problem avoidance, however, as Dasmann *et al.* (1973: 183) noted, there was 'good reason to think that [large dam] development projects are spreading faster than efforts to anticipate their full consequences' (Box 9.1).

Caution should be exercised before condemning large dams as harmful technology: like other technology it may be 'hijacked' for motives that are political, by those seeking prestige and by those seeking profit, and by those who are naïve or just careless. Successful developments often have less successful components for critics to focus on: the Mahawelli Programme (Sri Lanka) has brought great benefits for example, the transfer of water from the Victoria Dam to drier regions – but siltation and wildlife impacts get noticed.

Dam and reservoir impacts may be local, regional, national or international and affect environments, societies and economies upstream and downstream. The suite of impacts alters from pre-construction to construction to post-construction phases and may continue in tropical and subtropical environments for 20 or 30 years before a steady-state is reached, and even then alter if dam management is modified, local people modify their habits or environmental change occurs. Only the more important and common impacts are discussed in the following section.

Impacts of large dams

Reservoir siltation

Reservoir siltation often results from poor basin management or unwise siting of a dam (see Fig. 9.2). No cost-effective, widely applicable way to unsilt a large reservoir has yet been developed, so the useful life of dams is often reduced and some projects fail to pay for themselves. Even reservoirs of more than 1000 square kilometres are affected.

Examples of serious reservoir siltation

1 *Anchicaya Project (Colombian Amazonia)* lost at least 80 per cent of its storage within 12 years (Cummings 1990: 22).
2 *Nizamsagar Dam (Andra Pradesh, India)* lost two-thirds of its storage in a short period.
3 *Sriram Sagar Dam (Andra Pradesh, India)* completed 1990, one-third of storage lost within two years (Pearce 1991a).
4 *Sanmenxia Dam (Yellow River, PRC)* completed 1960, abandoned 1964.
5 *Laoying Project (Yellow River, PRC)* abandoned before completion due to siltation, having necessitated relocation of about 300 000 people; following extensive re-engineering, it has now got one-third of its originally planned storage; re-engineering meant loss of turbines and installation of sluices (Pearce 1991b).

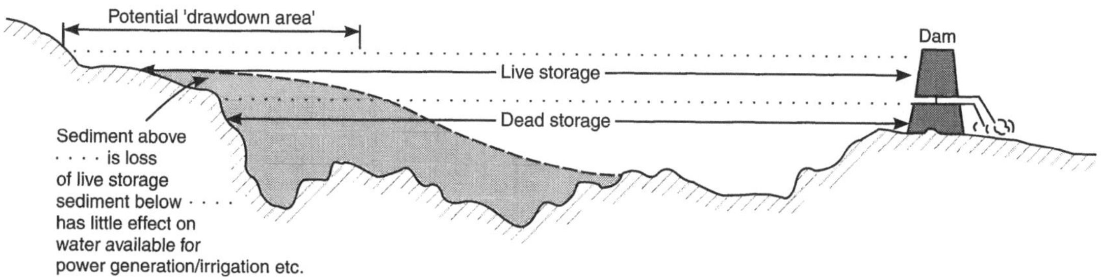

Fig. *9.2* Reservoir live and dead storage and likely pattern of siltation.
Note: Dead storage contributes little to turbines or irrigation – siltation of this is less crucial. In practice siltation may affect portion similar to shaded – significant loss of storage.

6 *Proposed Three Gorges Project (Yangtze River, PRC)* intended to protect lower and middle Yangtze from flooding, to provide power for industry, to improve river navigation. Many people would need to be resettled as a consequence of reservoir formation. There are risks of damage to river fisheries, coastal erosion and Shanzia Reservoir sedimentation (Gu Hengyue and Douglas 1989; Long Li 1990). There are also fears that it would endanger millions of people if subject to engineering failure, earthquake, terrorism or warfare.
7 *Khashm el Girba Reservoir (Atbara River, Sudan)* severe siltation and loss of storage.

Downstream impacts

Downstream impacts often receive inadequate attention from planners, yet may be felt for hundreds of kilometres, perhaps affecting millions of people. Where a river is shared by more than one state, alteration of flow may cause conflict (see p. 232), those below a dam tend to suffer from: loss of floods, risk of dam failure, and altered water quality (Interim Committee for Coordination of Investigations of the Lower Mekong Basin 1982). Aid and compensation tends to focus on those flooded and forced to relocate from the reservoir area, not on those downstream. By blocking riverflow a dam causes silt to settle, and the impounded water can become stratified (i.e. unmixed so that deeper layers may be cold and/or stagnant). Water released downstream may be colder, stagnant and more clear than normal river flows. These water quality changes can seriously affect fisheries and wildlife. Not only may flows be chemically or thermally injurious or lack oxygen, the loss of suspended detritus deprives aquatic organisms of nutrients. Clear water may make it difficult for animals to conceal themselves and may render fishing nets and lines sufficiently visible that they no longer work which may mean many people lose their livelihoods (Barrow 1987b: 88). The loss of silt can also lead to serious channel scouring and erosion of delta and coastal lands.

Flow regulation can upset the natural pattern of floods and low flows with catastrophic impacts on floodland ecology (in particular through loss of floodwater and the silt it deposited) and on those farming, grazing or fishing such environments (Barrow 1981; Adams 1989). There have also been problems with post-impoundment (or post-barrage construction) salt intrusion into lower reaches of rivers and loss or reduction of seasonal flow changes that trigger or enable aquatic life to breed. The solution is to release enough water to mimic natural floods. In practice this has seldom been done, usually because reservoir management demands are seen to conflict with it (Adams 1992: 201–3). When controlled flooding is possible careful timing is vital if agriculture and wildlife are not to be damaged, if done well it should be more predictable than nature. Controlled flooding has been tried with the Manantali Dam (Senegal River), the Pongolapoort Dam (South Africa), the Itezhitezhi Dam (Zambia) but surprisingly few others (Scudder 1989a).

Examples of downstream impacts

1 *Tucuruí Dam (Pará, Brazilian Amazonia)* (Fig. 9.3) Fishing and downstream flood recession agriculture has been affected. There is a possibility that agricultural development (in particular use of agrochemicals) prompted by the dam will cause reduction of the region's surface water acidity leading to more widespread snail breeding and schistosomiasis: carrier snails are present locally, as are people who have contracted the disease in the northeast of Brazil, but there is little or no

Fig. 9.3 Tucuruí Dam on the Toncantins River, Pará, Brazil (about midway through construction). Completed in 1984, it impounds a reservoir of *c.*2100 km^2 and has led to settlement, industrial and agricultural development in the region.

transmission at present because of acid conditions (Barrow 1988).

2 *Turkwell Gorge Dam (Kenya)* Reduction of flooding by Turkwell River downstream of the dam has damaged floodplain forest affecting wildlife and the economy of local peoples. There are hopes for improvement through controlled releases (Adams 1989).

3 *Floodlands of the Tana River (Kenya)* Dam regulation has reduced seasonal flooding affecting forests, agriculture and fishing.

4 *Aswan High Dam (Egypt)* Nile Delta, and coastlands eroding, reduced detritus is affecting marine organisms and fisheries in the southeastern Mediterranean. Farmlands downstream of Aswan have largely lost annual deposition of fertile silt (Fahim 1981).

5 *Akosombo Dam (the Volta Project, Ghana)* Coastal erosion, salt intrusion 30–50 km upstream, fisheries problems.

6 *James Bay Hydroelectric Projects (Quebec, Canada)* Three large dams built by Hydro-Québec have already flooded over 10 000 square kilometres (the 'James Bay I–Le Grand Project' which has diverted the Estmain, Opinaca and Caniapiscan Rivers). A second phase of development is proposed (the 'James Bay II–Great Whale River Project and the Nottaway, Broadback and Rupert Rivers Project). There are fears that the waters acidified by drowned vegetation will become methyl mercury contaminated (from flooded rocks rich in the normally insoluble substance). Downstream fisheries will suffer from the mercury and the pollution will in turn affect wildlife and humans. Being relatively warm, the flows released could alter ice cover and the salinity of James and Hudson Bays (Raphals 1992). The local Cree and Inuit peoples, who have lived in the region at least 5000 years, were not consulted before the start of

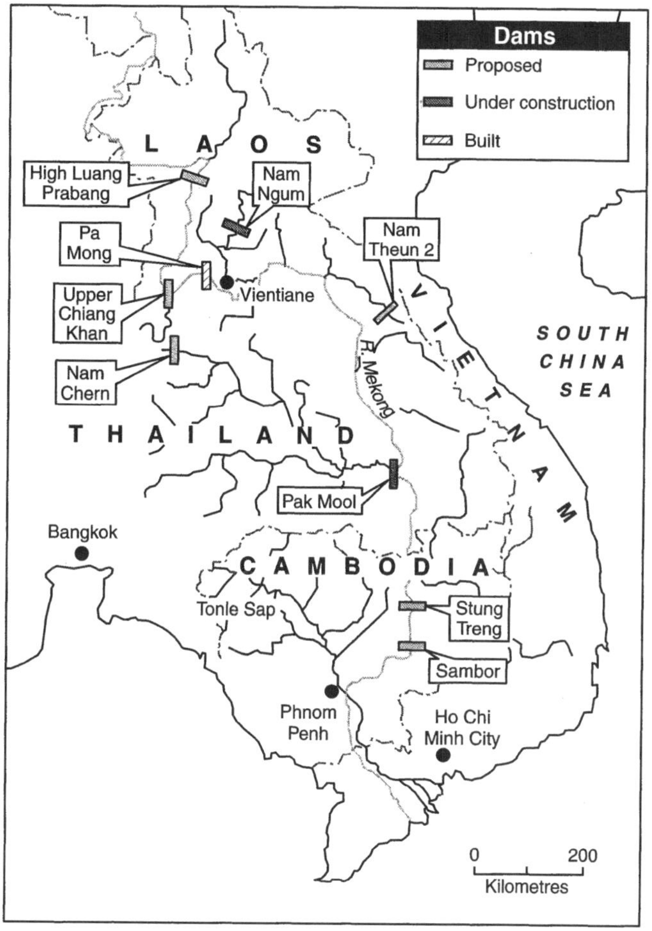

Fig. *9.4* Projects completed, under construction and proposed for the Mekong Basin (*Source:* Lohmann 1991 and other sources).

the Project and went to court in 1973 to try to stop James Bay II.

7 *Mekong Basin* (see Fig. 9.4) Plans, well advanced in the 1960s to develop up to 100 dams, were abandoned because of unrest, including the Vietnam War. Successful development will require cooperation and careful integrated basin development by several nations. There are increasing signs of interest in Mekong development (Lohmann 1990; 1991). Worries have been voiced that there will be serious impacts, including problems for those downstream from dams. Hydroelectric projects may disrupt fisheries and flood recession farming. So far only one dam (Nam Ngum, Laos) has been completed (Lohmann 1990 lists projects likely to proceed).

Catastrophic failure

Risk of catastrophic failure is significant: both developed and developing countries have had problems. All recently-built dams should meet the safety standards (including earthquake resistance) of the International Commission on Large Dams. India has probably built more large dams than any other nation, and the PRC has considerable experience, yet both still have occasional difficulties. Sometimes engineering has been faulty: a particular risk is earthfill dams, which are liable to failure if any leakage occurs below or through their structure (they are, however, cheap and quick to construct). Reinforced concrete dams sometimes fail; there have been problems in warmer climates keeping concrete

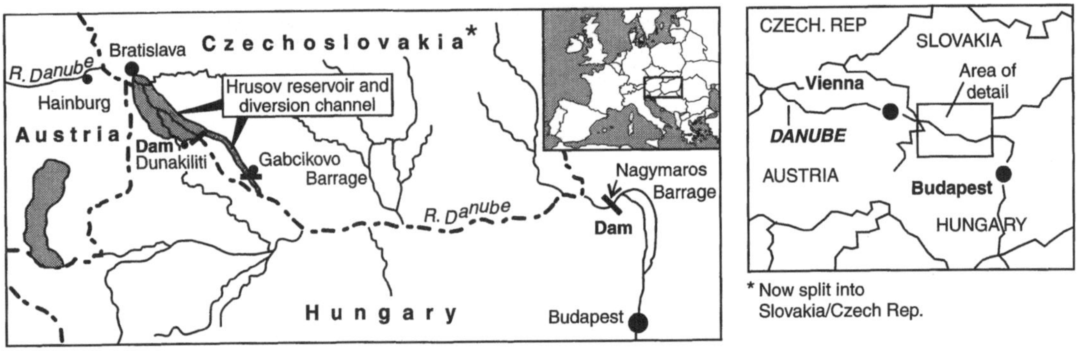

Fig. 9.5 Gabcikovo Barrage and Danube River diversion ⟨*Source:* compiled from various sources⟩.

cool while it cures to prevent cracking and earthquakes may cause damage. Large dams may themselves cause seismic disturbance, apparently the weight of water in the reservoir stresses rocks and may open and lubricate faults if the impoundment is deep (Rothé 1968). Construction errors seem to have led to collapse of the Malpasset Dam (Southern France, 1959) and the Morvi Dam (Gujarat, India, 1972) and at least seven others in various countries since the late 1950s.

Construction errors *should* be less of a threat nowadays but seismic and landslide threats remain (landslides into a reservoir can cause overtopping resulting in downstream floods even if the dam holds together), plus the problem of terrorism and warfare.

Examples of catastrophic failure or risk of catastrophic failure

1 *Vaoint Dam (Italy, 1963)* This disaster claimed 2000 lives and was caused by a landslide that spilled water over the dam (but which did not break it) (Ward 1978: 54).
2 *Dams which appear to have triggered seismic disturbances with the potential to damage their structure* Boulder Dam (USA, 1930s); Koyna Dam (near Bombay, India, 1967) which broke: although it impounded quite a small reservoir 177 people were killed; Kariba Dam (Zambia); Mangla Dam (Pakistan); Marathon Dam (Greece) (Dasmann *et al.* 1973: 217; Rothé 1968). At both the Koyna Dam and the Kariba Dam, an earthquake of Richter 6 or more was recorded.
3 *Huayuankow Dyke (bordering the lower Yellow River, PRC)* Deliberately breached to counter the Japanese invasion in 1938, killing between 250 000 and 1 million Chinese people.
4 *Tehri Dam (Bhagirathi River, Andra Pradesh, India)* Work was underway by 1991 in spite of strongly adverse environmental impact assessments. The river is silt-laden, the region earthquake prone, the risk of landslides is high and a downstream population of at least 200 000 people are vulnerable. This project, if completed (in 1993 funding seemed uncertain), will also require resettlement of perhaps 80 000 people (Pearce 1991a).
5 *Gabcikovo/Nagymaros River Barrage System Barrage and embanked canal on Danube River* Probably Europe's largest scheme (Fig. 9.5): originally a joint Hungarian (with considerable Austrian investment) and Czechoslovak project, by 1988 many environmental groups and the Hungarians disapproved of completion on cost, environmental impact and human risk grounds. A revised version of the original scheme was under construction in late 1992, under Slovak control (the Nagymaros Barrage will probably not be built), at a cost reported to have risen to over UK£1000 million, in spite of adverse environmental impact assessments. The Gabcikovo Barrage will probably flood large areas of land that are agriculturally productive, will damage ecologically important wetlands and upset the Danube's flow regime. A 25 km long embanked canal carries water above the land surface and might be vulnerable to seismic disturbance, putting many people at risk (*The Times* 24/7/91: 9; *The Sunday Times Review*, 6/6/92: 4–6; Rich 1992). (Recently the WWF has reviewed its opposition.)
6 *Tarbela Dam (Indus River, Pakistan, 1974)* This project has suffered rapid loss of storage through

siltation; it was also built at a site felt to be at risk from landslides and seismic disturbance.

7 *Basha Dam (Pakistan)* Seismic activity in region yet dam still built.

8 *Peruca Dam (Cetina River, former Yugoslavia)* Completed in 1990, it was seriously damaged by sabotage in 1993: the breach threatened 20 000 lives downstream.

Loss of wildlife

Wildlife loss associated with reservoir inundation has attracted attention since the Kariba Dam in the late 1950s. As already noted, downstream impacts have had rather less attention yet are common and often considerable.

Examples of loss of wildlife

1 *Silent Valley Hydroelectric and Irrigation supply Project (Kunthipuzha River, Kerala, India)* Opponents feared that a significant part of India's remaining tropical forest would be destroyed and put up a fierce opposition that was finally successful in halting construction in the early 1980s (Adams 1990: 135).

2 *Amazonian dams* Some large dams have already been completed, including Balbina, Curuá-Una and Tucuruí (Pinto Paiva 1977; Barrow 1987b; Cummings 1990: 35). More are planned – perhaps 30-odd will be built by AD 2010. Considerable areas of forest have already been flooded and downstream impacts are apparent; migration of fish and riverine mammals has been disrupted.

Resettlement

Relocation and resettlement has often been a serious impact of large dams' construction. It has not been uncommon for authorities to neglect relocation and resettlement of those displaced and where efforts have been made, these are too often unsatisfactory and delayed (Barrow 1981). The flooding of several hundred or even thousands of square kilometres will generate large numbers of people needing resettlement, even if the population is relatively sparse, and in well-populated regions or where townships are involved a million or more people may be affected.

Resettlement efforts are hampered by rapid construction schedules that give little time for preparation, by speculators who move in to try and profit from compensation schemes, by lack of suitable land or livelihood for those to be resettled and by the trauma that those uprooted suffer. Compensation of those without written proof of land ownership and those who own no house or land may be difficult and tends to be neglected. Conflict between resettled people and the population in the area they relocate to is not uncommon. A significant proportion of those resettled fail to regain a satisfactory stable livelihood (Thukral 1992).

Examples of resettlement problems

1 *Volta Project (Ghana)* Large numbers of people were resettled but, in spite of considerable efforts, difficulty and conflicts arose.

2 *Three Gorges Project (PRC)* There are huge potential benefits from planned projects like the Shanxia Reservoir (flood control, power, irrigation, navigation) but 1.4 million or more may need to relocate (Long Li 1990), and there may be considerable ecological impacts (Barber and Ryder 1993).

3 *Narmada River Valley Development Project (India)* (see Fig. 9.6) This has become one of the world's most controversial development projects. It consists of the Sardar Sarovar Projects (Gujarat, Maharashtra and Madhya Pradesh) (completion is expected in the mid-1990s) and the Narmada Sagar Projects. The proposal is, over the next 50 or so years, to exploit the Narmada and its 41 tributaries, using 31 large dams, 135 medium dams, perhaps 3000 small dams and large irrigation supply systems (also water and electricity will go to Delhi). The hope is that 2.2 million ha of Madhya Pradesh and Gujarat will be irrigated, plus 70 000 ha in Rajasthan. There will also be flood control, hydroelectricity and domestic water supply benefits to perhaps 30 million people. Gujarat has had poor monsoon rain in recent years (droughts one year in three) and together with Rajathan State hopes to benefit from the development. The projects may be subject to sedimentation, will submerge some forest lands and might displace 100 000 to 1 million people (most estimates are over 750 000); in 1992 they were expected to cost over US$11.4 billion (Sattaur 1991). Following scathing reports on the probable impacts of the projects and the failure to carry out satisfactory environmental impact assessments, the World Bank (having committed *c.*US$450 million by 1993) insisted on improved resettlement and environmental management before releasing the final portion of funding; in 1993 India decided to try and proceed without the final US$170 million of the Bank's loan. In late

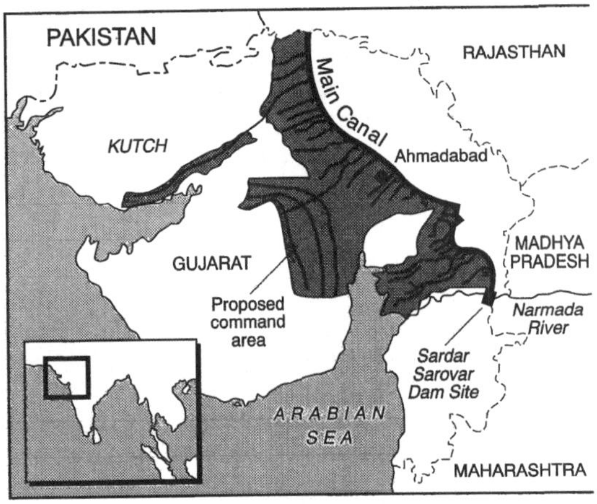

Fig. 9.6 Sardar Sarovar Dam, Narmada Project (India) and planned irrigation command area (land to be irrigated) (*Source:* compiled from various sources).

1993 there had still been no adequate arrangement for compensation of relocatees and fears were being voiced that the full canal network would not be finished, the likelihood being that it would reach only as far as big cotton farmers. Local resistance to the projects quickly developed, especially after August 1993, when the first settlements along the Narmada River were flooded.

Inter-state conflicts

Inter-state conflicts may arise when dams divert a large portion or all of the flow of a river. The likelihood that such projects will cause difficulties in the future is increasing as states compete for less resources. Not all diversions involve dams, some make use of canals and barrages, potential conflict situations are therefore also discussed later when dealing with large scale diversion and interbasin transfer of water.

Examples of interstate conflicts

1 *Ataturk Dam (River Euphrates, Turkey)* From 1990 the flow of the Euphrates has been reduced to the concern of Syria and Iraq (Agnew and Anderson 1992).
2 *Farraka Diversion Barrage (River Ganges, India)* This barrage, sited *c.*25 km upstream of the Bangladesh border, diverts water to the Hooghly River, consequently Bangladesh has suffered reduction of river flows leading to seawater intrusion into the distributaries of the Ganges Delta.
3 *Yarmuk Dam (River Yarmuk, a tributary of the River Jordan, Jordan)* The half-built dam was demolished by the Israelis in 1967. Israel had redirected much of the flow of the River Jordan into its National Water Carrier distribution system in 1964.

Reducing problems associated with large dams

Large dams were spread rapidly between 1958 and 1974 by planners and engineers largely trained and experienced in temperate or semiarid continental DC environments. Attention generally focused on engineering and economic goals and experience of environmental and social conditions in LDC was often limited. To describe this as 'careless transfer of technology' is perhaps a little harsh. Since the 1970s there has been more preventive planning, and those involved in dam projects could draw on over 30 years of hindsight and the guidelines that have been published (e.g. Garzon 1984; Marchand and Toornstra 1986; World Bank notes on avoidance of unwanted impacts of dam construction, in Goodland 1989). In the late 1970s the Organization of American States (OAS) and the Government of Argentina focused on the Rio Bermejo Basin (northern Argentina) to try and develop a basin development model that included an economic development account and environmental quality account long before talk of environmental accounting in the 1990s (OAS 1978).

Unfortunately many factors can conspire to prevent satisfactory project planning and implementation for virtually any human enterprise, including large dams. Large dams are plagued by poor databases, political pressures that overrule sound planning, and unforeseen environmental or economic changes that upset progress. For example, a dam may be built by a MNC for aluminium processing or an irrigation authority with little consideration of other factors. In LDCs, particularly since the late 1970s, shortage of funds and debts have encouraged rapid completion of projects and may mean that once construction is finished there is no recurrent funding to cope with unforeseen problems and maintenance. Many of the experts involved in development projects are consultants who will have moved on when problems start to appear, and the lessons learnt and information gathered tends to become hidden in 'grey literature' (material returned to DCs or held by consultants and difficult to access). There is an understandable reluctance for governments and consultants to broadcast news of failures, so problems may not be well studied and, when there is a study, it has often been 'compartmentalized', i.e. carried out by too narrow a group of specialists who may not communicate well, and who may have predetermined ideas.

It is possible to plan and engineer to avoid river development problems but this mainly has to be done at the outset and it is not easy to effectively add measures at a later stage. For example:

- Spillways can be designed to discourage blackfly breeding and thus reduce river blindness transmission.
- Fish-ladders or bypass channels can be installed during construction but are difficult to add later or the fish have died out.
- Irrigation channels can be designed to discourage snails and be fitted with traps and screens to cut schistosomiasis transmission.
- Gates can be installed to let out floodwater and reduce siltation.
- Silt-traps can help reduce reservoir siltation.
- Outlets from a dam can be designed to draw water from different depths allowing the mixing of stratified water to reduce downstream impacts.
- Controlled releases of water can maintain river floodplain ecology and livelihoods. If well managed, controlled releases are predictable compared with natural flooding (Adams 1992: 201–3).

A possible means of reducing dam impacts, or at least ensuring that problems are better dealt with, is the Build-Own-Operate-Transfer (BOOT) approach. Under BOOT a company builds a dam, retains ownership, operating perhaps for decades, before transferring control to a suitable local body. The World Bank has shown interest in BOOT since about 1989 as a means of reducing unwanted problems and coping with finding recurrent funds and expertise after project completion. BOOT does pose problems of resources control and dependency.

Channelization

To improve flow or navigation or reduce the risk of flooding stream channels may be cleared, embanked, have their cross-section altered and perhaps be straightened. The impacts depend on the actual measures taken but in general because water no longer spills on to floodplains or fills side channels, cutoffs, etc (overbank storage) there is less water to maintain streamflow, there may be less recharge of groundwater and the impact on floodplain and aquatic wildlife is likely to be significant. Floods cannot spread on to floodplains if a river is embanked and consequently flood peaks may increase. There is also a risk that embankment will give people false confidence to settle areas that could be flooded suddenly if waters overtop the defences. Sometimes channel improvements that prevent a stream from altering its course lead to the formation of a levee, which in time may lead to increased flood risk to the surrounding land, seepage problems and the need to pump drainage water up to the channel.

Channel dredging and bank-protection engineering like concrete sides or camp-sheathing will discourage some forms of wildlife, as will the more 'flashy' streamflow and the faster current or increased depth (if weirs or barrages are used to maintain water levels). There may be some situations where channel improvement reduces pollution and health risks. Fig. 9.7 illustrates a commonly adopted improvement for streams which vary a lot in a flow: a flood relief channel reduces flood peak hazard and a low-flow channel ensures that sediment and pollution is kept moving and flow is fast enough to discourage insects and other pests.

At its extreme streamflow may be contained within a culvert, as has happened with a number of streams in London, perhaps the most famous being the River Fleet tributary of the Thames. When this is done there is drastic alteration of aquatic and river margin wildlife, and of groundwater recharge.

Large-scale diversion: in-basin, interbasin and inter-regional transfer

The Romans were building impressive aqueducts nearly 100 km long and able to supply cities of many thousand people before 200 BC. With modern technology and engineering some of the most difficult and rugged terrain can be crossed, provided it is cost-effective. Terrain may be more of a limitation where 'low-tech' and manual labour are used, for example in India and the PRC. In the past there were problems in diverting water from turbid flood-prone rivers, the silt deposition soon filled channels and the headworks got damaged. Nowadays engineering solutions and better siting of headworks (structures that divert water from a stream) reduces these problems.

Water transfers are often judged by their direct costs with little consideration of maintenance or

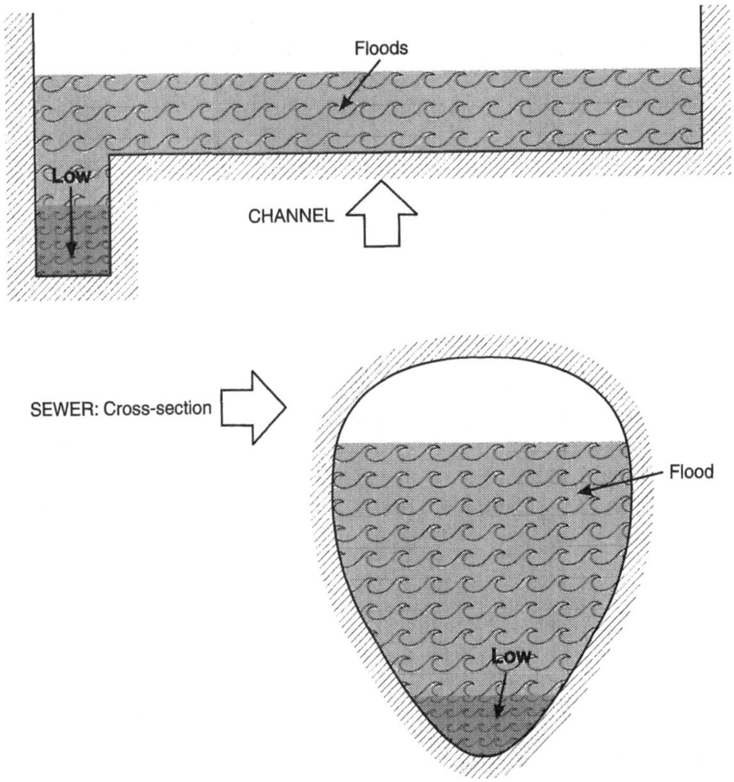

Fig. 9.7 Channel or sewer (cross-section) which can cope with considerable variation of flow without overflow at high flow or silting and stagnant water at low flow (for sewers an egg-shaped cross-section, narrow part at the bottom serves a similar purpose).

repair costs, opportunity costs (whether the costs would have been better spent on something else) or the external costs affecting those who do not directly benefit, if at all, from the project.

The system in receipt of transferred water will have increased flow and possibly altered water quality after transfer. The 'donor' system suffers reduced downstream flows that may lead to water shortage, salt intrusion in estuaries and deltas, reduced dilution and carriage of pollutants. In-basin transfers from river to irrigation project or to supply a city raise relatively few administration problems and probably have slightly less serious environmental impacts than interbasin transfers. Interbasin transfer is more likely to transport organisms over land barriers, for example: fears have been expressed, but now appear to be unfounded, that the South–North Water Transfer (PRC) might spread aquatic snails carrying schistosomiasis northward. Long-distance transfer, even if not involving supplies shared by different states, may raise development problems: towns develop more than natural water supplies would have allowed and other environmental factors may become stressed.

Water transfers may generate problems along their course. Floods, earthquakes, landslides, structural failure, terrorism, even burrowing animals may cause leakage or a serious breach. Major leakage can drown those living near transfers, leakage can waterlog land or wash away soil. Appropriate lining can reduce almost all of these threats and ensure wastage is reduced. Canals, aqueducts and pipelines can upset communications and animal migration across their path but may facilitate movements along their route if roads or tracks are provided. Weed growth in canals reduces flow and may harbour snails and other disease-carrying organisms, weed growth along banks provide a valuable conservation area but might also allow movement of pest organisms.

The reduction of impacts associated with large-scale water transfer depends largely on better pre-project appraisal of environmental, social and economic impacts and on those impacts being heeded by planners and administrators (Howe and Easter 1971; Herren *et al.* 1982).

Where groundwater use, surface water storage or transfer changes (or threatens to change) the availability of water for other states, issues of sovereignty and water sharing arise and must be addressed if conflict is to be avoided. International (shared basin) laws are far from perfect: there is no fully accepted principle yet and agreements depend on mutual goodwill between co-riparians and shared groundwater users (Barrow 1987a: 70; Agnew and Andersen 1992: 228).

Examples of large-scale diversions

Countries with large-scale diversions finished, under construction or planned include India, PRC, Sudan, Israel, Mexico, Peru, Chile, Libya, Turkey, USA, CIS (former USSR). Shared international basins, with the potential for conflict if cooperative use is not achieved and maintained, include the Nile (Egypt, Sudan, Ethiopia, Uganda and Kenya): the Tigris–Euphrates (Turkey, Syria, Iraq); the Jordan (Israel, Jordan, Syria, Lebanon) (Agnew and Andersen 1992: 231; Thomas and Howlett 1993).

CIS (former USSR) There have been proposals since at least the 1940s, initially to improve transport and generate electricity, nowadays for irrigation and to counter environmental problems with southern river systems. Studies have been carried out into the diversion of northward flowing rivers into the Aral and Caspian seas (Voropaiev 1984). This 'Siberian Scheme' involves the possible diversion of the Ob, Irtysh and Yenesei rivers via the Syrdar'ya and Amudar'ya rivers to the Aral Sea. The water would be used en route for irrigation in Kazakhstan and Central Asia. This might help counter the catastrophic fall in the level of the Aral Sea. By 1988 work had begun on transfers from the Pechora River (which flows into the Arctic Ocean) to the Volga; progress has now slowed or halted.

Proposals have been put forward for the southward diversion of the Ob and Yenesei rivers (which flow into the Arctic Ocean) (McDonald and Kay 1988: 98) and similar plans have been drawn up for the Volga and Kama rivers (Agnew and Anderson 1992: 205). There has been much disagreement about the impact of reducing freshwater flows into the Arctic Ocean: this might alter summer or winter ice conditions and in turn affect climate in northern Eurasia.

The Aral Sea (CIS) The Aral Sea has been seriously affected by water transfers and is regarded as a crisis area, the consequence of faulty centralized planning (Kotlyakov 1991; Levintanus 1992). Once the world's fourth largest lake, the sea now covers less than 50 per cent of its extent in the 1950s and has lost over 70 per cent of its water volume, the impact on wildlife has been harsh: for example, 23 of the sea's 24 fish species have been lost. Dust laden with salts, pesticides, radio-nucleides and industrial pollutants blows across the land, ruining vegetation, regional economies and affecting the health and livelihoods of over 50 million people (World Bank 1992: 38). The reduced size of the waterbody seems to be responsible for recent colder and longer winters and hotter, drier summers in central Eurasia and the frequency of dust storms has markedly increased. The population of the Aral region is increasing rapidly, and this compounds existing problems. Recently the authorities have started to install expensive de-ionization facilities to provide safe drinking water that is free of pollutants (Perera 1993). The fall is due mainly to abstraction of water from the Amudar'ya and Syrdar'ya rivers since the 1950s for cotton and rice irrigation in the central Asian republics (Middleton and Thomas 1992: 68; Glantz *et al.* 1993).

California State Water Transfer Diversion of water from northern California (the northern third of the state has about 70 per cent of the state's water) supplies the southern two-thirds of the state, where cities, industry and agriculture use roughly 80 per cent of the state's water. Nine major aqueducts were built between 1913 and the 1970s, causing environmental, political and developmental impacts. 'Donor' regions have suffered flow reduction and hindrance of andronomous fish (notably salmon spp.) migration in rivers. Dependence on aqueducts and pipelines where there is a high risk of earthquake may be unwise. Some have suggested that the northern surplus may have been exaggerated, and that transfer has caused planners to ignore better alternatives when water shortage might have been a welcome means of limiting excessive urban development in southern California (Seckler 1971; Robie and Reynolds 1978).

Dams in northwestern USA and British Columbia (Canada) Since the 1900s these have enabled transfer south and east to supply irrigation and reduce flood dangers in the source systems. The North American Water and Power Alliance proposed in 1964, if implemented, would divert water from several Canadian rivers to central Canada, southwestern USA and northern Mexico.

Muda Scheme (West Malaysia) A successful inter-basin transfer, whereby water from one basin is supplied via a tunnel to a neighbouring valley to enable multiple-cropping of rice and significant agricultural improvements.

Snowy Mountain Scheme (Australia) Completed in the early 1970s, water is transferred from the Snowy River westward over and through the Great Dividing Range to augment the Murray and Murrumbidgee Rivers. The scheme includes nine hydroelectric generating stations.

India National Water Grid Planned to transfer flows from the Ganges and Brahmaputra to water shortage regions (Golubev and Biswas 1979; Zaman 1983; Barrow 1987a: 287).

PRC Plans for a South-to-North Water Transfer Southern regions of the PRC have abundant water but northern regions need supplies (some work started in 1987). Flows from the Yangtze River will be diverted up to 1000 km to water shortage areas around Beijing (Biswas *et al.* 1983: 127; Biswas and Qu Gepeng 1987: 112–26; Barrow 1987a: 288). Of three proposed routes (east, west and mid) the western has been rejected.

Israel National Water Carrier (built in the 1960s) redistributes water between the Sea of Galilee and aquifers in western Israel and supplies cities and agriculture Open channels and conduits feed water between the upper Jordan River and aquifers to the west and supply cities and agriculture as far south as the Negev. The bulk of Israel's rainfall is in the north and the carrier helps redistribute supplies (see McDonald and Kay 1988: 46–9).

Jonglei Canal (from White Nile at Bor, Sudan to Malakal, Sudan) Joint Egyptian–Sudanese project, intended to bypass the Sudd Swamps in southern Sudan. Construction started 1976, but was suspended when three-quarters finished by the Sudanese in 1983 due to local opposition and guerrilla warfare. Had the scheme been completed; the impacts on the ecology and the pastoralists of the Sudd region would have been considerable: between 200 000 and 400 000 people and their livestock depend on annual flooding which the scheme would have probably terminated (Barrow 1987a: 286–90, 292; Adams 1992: 114, 141–3).

Libya: the Great Man-made Rivers Project Transfer of (what many argue is non-renewable) groundwater from Kufra in the Sahara Desert *c.*1900 km northward via a very large diameter pipeline to supply irrigation and industry may reduce groundwater available to Egypt (Agnew and Anderson 1992: 227). By 1992 the project had cost roughly UK£2.4 billion and was expected to run to UK£13.5 billion to complete (*The Times* 20/6/92: 13).

Proposals for a pipeline from Turkey to Syria, Jordan, Saudi Arabia, Bahrain, Qatar and the UAE This seems technically and economically feasible but is unlikely until there are stable political conditions in the Middle East (Agnew and Anderson 1992: 220).

Texas Water System Proposals in 1968 to divert water from the Mississippi into Texas have not yet started.

Mahaweli Ganga and Gal Oya Basins (Sri Lanka) Rivers dammed and linked for interbasin transfer by canal.

UK has recently proposed to use navigation canals of British Waterways Board Constructed in the eighteenth and nineteenth centuries for transportation of goods, these could help provide a 5000 km transfer network for domestic supply, industry and stream recharge. North of a diagonal line from the Humber to Bristol the UK has above average rainfall, southeast of that line there is below average, added to which there is a greater population in the southeast – hence the interest in the transfer.

River basin planning and management

The generally clearly defined, biogeophysical unit functioning as a system – the river basin – has attracted researchers, planners and administrators as something which can be monitored and effectively managed (Laconte and Haimes 1982). Basins or sub-basins (i.e. the catchments of tributaries) are often used for measuring modelling streamflow and sediment discharge. Precipitation and flows can be monitored for periods before and after landuse changes so that the effect of different basin management on streamflow and groundwater can be judged.

The use of the river basin as a planning and administrative unit was suggested as early as 1752 (Chorley 1969: 20). The river basin has often been adopted as an operational landscape unit (some would use the term ecosystem) with functional unity, relative homogeneity (in practice there may be marked differences between lower, mid and upper basin) and non-ephemeral nature, compared with many planning or administrative units (Friedmann and Weaver 1979: 67; Saha and Barrow 1981). There may be some movements of labour, electricity and funding into and out of a basin and settlement may not always be confined to a basin; even so, few units coincide as well with human activities and natural processes.

At its simplest, river basin planning aims to coordinate the different demands made on water within a basin. The earliest proposals for basin planning approach were probably those of Sir William Wilcocks for the Nile in the 1890s. Modern river basin planning and management in large part evolved from the Tennessee Valley Authority (TVA) in the US midwest; the TVA has been described as the USA's first practical example of regional development. The TVA was part of President F.D. Roosevelt's 'New Deal' response to the Depression through which the Federal Government became involved in investment in natural resources development. Using the Tennessee, a sub-basin of the Ohio River Basin, the US Government led by F.D. Roosevelt moved away from its established ways of *laissez-faire* development inducing and guiding association between state and private resource developers (see also coverage of the 'Dust Bowl' in chapter 6; Creese 1990).

Between its foundation in 1933 and about 1936 there was considerable progress with basin-wide comprehensive development (after that the TVA lost its regional development personnel and drive to become more of the power generation authority it is today). For a relatively backward region of over 200 000 square kilometres with a population of more than 3 million people, mostly living in poverty in rural areas, the TVA constructed 27 multipurpose dams, improved flood control and navigation, generated and distributed electricity to stimulate industrial development, reduced health problems (notably malaria), illiteracy and unemployment, improved agricultural production (roughly six-fold) and reduced land degradation, created employment, stimulated social development and led to new town construction (Lilienthal 1944; Selznick 1949; Martin 1956).

The TVA was a seminal idea – though the US Corps of Engineers in 1914 and the Niagara Frontier Planning Board in the 1920s had proposed similar development authorities – and marked a turning-point in river development from simple resource development to resource development integrated, and as far as possible in harmony, with other aspects of development (Le Marquand 1989; Mitchel 1989; Rowntree 1990; Adams 1992: 116).

River basin development has gone through four broad stages of evolution:

1 *Mainly specified goal* ('single-purpose') flood control or hydroelectricity or irrigation supply.
2 *Multipurpose planning approach* multiple use but water simply divided between users. In practice there is a possibility of conflicting uses which simplistic division of the resource fails to resolve (e.g. industrial use may pollute and make allocation to fisheries pointless).
3 *Integrated river basin planning approach* attempt to coordinate and develop use of water in basin 'in harmony with other development processes both within and outside the basin' (OAS 1978: 5).
4 *Comprehensive river basin planning and management* extension of integrated river basin planning which goes beyond water development to involve other resources. It is a multi-objective approach in which water is seen as a 'tool' for development. Both (3) and (4) share three features: multipurpose water development; a basin-wide programme; comprehensive regional development as goals.

The TVA generated great interest but there were few 'copies' until the 1950s. The Second World War, post-war lack of funding and a shift in fashions toward 'economic growth' models of development probably diverted attention away from river basin planning and management (Krutilla and Eckstein 1970). US river basin commissions, rather than authorities, were established after the Second World War (e.g. Colorado River Basin Commission – six US states and

Table 9.2 Some river basin planning and management authorities and commissions established between 1945 and 1975

Country	Date	Body (roles)
Mexico	1945	Integrated basin authorities in Rio Tepalcalepec, Rio Palaloapan, Rio Grijalva, Rio Fuerte
India	1948	Damodar Valley Corporation (regional development/flood control)
Philippines	1973	Bicol River Basin Development Project: USAID/Gov. Philippines pilot project to assess basin management as a development 'tool'
Philippines		Pampagna Basin (irrigation/flood control/community development)
Nigeria	1973	Eleven River Basin Development Authorities overseen by a National River Basin Development Committee (regional development/administrative)
Kenya		Tana Basin Authority (irrigation/power/regional development)
Colombia		Corporacion Autonomica del Valle del Cauca (irrigation/pollution control/regional development/power)
NE Brazil		CODEVASF – Rio São Francisco (power/irrigation/regional development)
Brazil		PRODIAT – Rios Araguaia/Tocantins (power/navigation/regional development)

Mexico; New England River Basin Commission – a number of US states; Delaware River Basin Commission – a number of US states; Potomac Basin Commission – a number of US states); they seem to have had more of a planning and advisory than implementational role.

After the Second World War some countries began to establish river basin planning and management authorities or commissions (see Table 9.2). Some of these have been more development corporations than true river basin planning and management bodies, and the results have been mixed (for reviews of the basin approach, see Faniran 1981; Ayoade 1988: 177). In a number of cases they have been used by governments in an attempt to attract development away from cities to more remote areas and perhaps to 'side-step' stagnant planning. Provided the leadership is good and has the power, a basin authority may be able to plan and manage in situations where established bodies fail to cooperate; therefore a central government may establish a river basin planning authority to bypass local administration or national ministries. To succeed, it is likely that a basin authority will need a multidisciplinary planning and management team and authority (not just an advisory role) over the *whole* catchment.

Nigeria and Mexico have used river basin planning and management: Nigeria has named these River Basin and Rural Development Authorities and they function as instruments of the Federal Government (Ayoade 1988: 182). The Senegal River Basin Commission (established 1964 between four nations) and the Gambia River Basin Commission (involving three nations) are shared international basin authorities. In South Asia shared international basin authorities operate for the Ganges (India, Nepal, Bangladesh), Brahmaputra (India, China, Bangladesh, Bhutan) and Indus (Pakistan, Kashmir, India). In 1993 worldwide there were in excess of 280 such shared international basin bodies (Rogers 1993).

Two of the Mexican efforts seem to have been successful (see Table 9.2: Rios Fuerte and Tepalcalepec), perhaps because they had the stronger management (Barkin and King 1970). African experiences with river basin development have been reviewed by Scudder (1989b) and Adams (1992: 113, 116–27). Adams, while acknowledging the value of river basin planning in principle, was sceptical of the practical value so far, especially when used in Africa.

The UK has altered its water planning and management units from Catchment Boards (in the 1930s) to River Boards (from 1948) to River Authorities (1963) and Regional Water Authorities (1974). At each change there has been an increase in the scale of operation, and the Regional Water Authorities are now large enough to practise effective integrated water management (Agnew and Anderson 1992: 245).

According to the World Bank (1992: 156), 40 per cent of the world's population live in river basins shared by more than one state. Increasingly, river development will demand coordination and cooperation between interested parties; increasing population and development activities are putting ever greater demand on water resources increasing the risk of conflict. Since the 1930s a number of countries have been trying to use the river basin planning and management approach to improve water-sharing and flow regulation. There is still a long way to go: international laws are not yet finalized and rivers like the Nile are still not covered by satisfactory agreements. Rogers (1993) noted the need for a strong international legal framework to control basins shared by more than one nation and suggested ways of organizing cooperation.

Successful river basin planning and management requires more than an effective administrative structure, without a database on water resources status and behaviour and ongoing monitoring it is difficult or impossible to predict flows, seasonal water quality changes and so on. When establishing a database and monitoring system it is important not to overlook infrequent but significant events. Floods, earthquakes or landslides may have a 50-year or longer recurrence interval but should still be allowed for. The 1952 Lynmouth Flood Disaster (UK) was unexpected but might well have been predicted as a rare event from palaeoecological or geomorphological evidence; building restrictions could then have been enforced to minimize vulnerable settlement. With the possibility of climatic change in coming decades it is important that river basin planners try to predict and develop mitigation strategies where necessary.

References

Abu-Zeid, M. (1987) Environmental impact assessment for the Aswan High Dam. In A.K. Biswas and Qu Gepeng (eds) *Environmental impact assessment for developing countries.* London, Tycooly

Ackerman, W.C. *et al.* (1973) *Man made lakes: their problems and environmental effects.* London, Academic Press

Adams, W.M. (1985a) River basin planning in Nigeria. *Applied Geography* 5(4): 297–308

Adams, W.M. (1985b) The downstream impacts of dam construction: a case study from Nigeria. *Transactions of Institute of British Geographers* (new series) 1093: 292–302

Adams, W.M. (1989) Dam construction and the degradation of floodplain forest on the Turkwell River, Kenya. *Land Degradation & Rehabilitation* 1(3): 189–98

Adams, W.M. (1990) *Green development: environment and sustainability in the third world.* London, Routledge

Adams, W.M. (1992) *Wasting the rain: rivers, people and planning in Africa.* London, Earthscan

Agnew, C. and Anderson, E. (1992) *Water resources in the arid realm.* London, Routledge

Ayoade, J.O. (1988) *Tropical hydrology and water resources.* London, Macmillan

Barber, M. and Ryder, G. (eds) (1993) *Damming the Three Gorges: what dam builders don't want you to know.* London, Earthscan

Barkin, D. and King, T. (1970) *Regional planning and economic development: the river basin planning approach in Mexico.* London, Cambridge University Press

Barrow, C.J. (1981) Health and resettlement consequences and opportunities created as a result of river impoundment in developing countries. *Water Supply and Management* 5(2): 135–50.

Barrow, C.J. (1983) The environmental consequences of water resources development in the tropics. In Ooi Jin Bee (ed.) *Tropical resources development.* Singapore, Singapore University Press

Barrow, C.J. (1987a) *Water resources and agricultural development in the tropics.* Harlow, Longman

Barrow, C.J. (1987b) The environmental impacts of the Tucuruí Dam on the middle and lower Tocantins River basin, Brazil. *Regulated Rivers* 1(1): 49–60.

Barrow, C.J. (1988) The impact of hydroelectric development on the Amazonian environment: with particular reference to the Tucuruí Project. *Journal of Biogeography* 15(1): 67–78

Biswas, A.K. and Qu Gepeng (eds) (1987) *Environmental impact assessment for developing countries.* London, Tycooly

Biswas, A.K., Zuo Dakang, Nickum, J.E. and Liu Changming (eds) (1983) *Long distance water transfer: a Chinese case study and international experiences.* Dublin, Tycooly

Brammer, H. (1990a) Floods in Bangladesh 1: geographical background to 1987 and 1988 floods. *The Geographical Journal* 156(1): 12–22

Brammer, H. (1990b) Floods in Bangladesh 2: flood mitigation and environmental aspects. *The Geographical Journal* 156(2): 158–65

Canter, L.W. (ed.) 1985 *The environmental impact of water resource projects.* Chelsea, Mich., Lewis

Chorley, R.J. (ed.) (1969) *Introduction to geographical hydrology: spatial aspects of the interactions between water occurrence and human activity.* London, Methuen

Cohen, M (1982) River basin planning: observations from international and Canada–United States experience. *Natural Resources Forum* 6(3): 247–61

Creese, W.L. (1990) *TVA's public planning: the vision, the reality.* Knoxville, University of Tennessee Press

Cummings, B.J. (1990) *Dam the rivers, damn the people: development and resistance in Amazonian Brazil.* Harmondsworth, Penguin

Custers, P. (1992) Banking on a flood-free future? Flood mismanagement in Bangladesh. *The Ecologist* 22(5): 241–7

Dalal-Clayton, B. (1990) *Environmental aspects of the Bangladesh Flood Action Plan.* Issue series no. 1. London, IIED

Dasmann, R.F., Milton, J.P. and Freeman, P.H. (1973) *Ecological principles for economic development.* London, Wiley

Fahim, H.M. (1981) *Dams, people and development: the Aswan High Dam case.* Oxford, Pergamon

Faniran, A. (1981) On the definition of planning regions: the case for river basins in developing countries. *Singapore Journal of Tropical Geography* 1(1): 9–15

Fearnside, P. (1991) Brazil's Balbina Dam: environment versus the legacy of the Pharos in Amazonia. *Environmental Management* 13(4): 401–23

Fraenkel, P., Paish, O., Harvey, A., Brown, A., Evans, R. and Bokalders, V. (1992) *Micro-hydropower: a guide for development workers.* Rugby, Intermediate Technology Publications

Friedmann, J. and Weaver, C. (1979) *Territory and function: the evolution of regional planning*. London, Arnold

Garzon, C.E. (1984) *Water quality in hydroelectric projects: considerations for planning in tropical forest regions* (World Bank Technical Paper no. 20). Washington, DC, World Bank

Glantz, M.H., Rubinstein, A.Z. and Zonn, I. (1993) The tragedy of the Aral Sea basin. *Global Environmental Change* 3(2): 174–98

Goldsmith, E. and Hildyard, N. (eds) (1984) *The social and environmental effects of large dams: vol. 1: overview*. Camelford, Wadebridge Ecological Centre

Goldsmith, E. and Hildyard, N. (eds) (1986a) *The social and environmental effects of large dams: vol 2: case studies*. Camelford, Wadebridge Ecological Centre

Goldsmith, E. and Hildyard, N. (eds) (1986b) *The social and environmental effects of large dams: vol 3: bibliography*. Camelford, Wadebridge Ecological Centre

Golubev, G.N. and Biswas, A.K. (1979) *Interregional water transfer*. Oxford, Pergamon

Goodland, R. (1989) *The World Bank's new policy on the environmental aspects of dam and reservoir projects*. Paper presented to Congress on Research Needs and Strategies for the Self-Sustaining Development of the Amazon, Manaus, 28–31 August 1989. Washington, DC, World Bank

Goodland, R. (1990) World Bank's new environmental policy for dams and reservoirs. *Water Resources Development* 6(4): 226–39

Gu Hengyue and Douglas, I. (1989) Spatial and temporal dynamics of fluvial erosion in the middle and lower Yangtze Basin, China. *Land Degradation & Rehabilitation* 1(3): 217–36

Hart, D. (1980) *The Volta River Project: a case study in politics and technology*. Edinburgh, Edinburgh University Press

Herren, G.B., Hansen, B.K. and Wandesforde-Smith, G. (1982) *Environmental impact assessment in the tropics: guidelines for application to river basin development*. Davis, Calif., Center for Environmental and Energy Policy Research, University of California

Howe, C.W. and Easter, K.W. (1971) *Interbasin transfers of water: economic issues and impacts*. Baltimore, Md., Johns Hopkins University Press

Interim Committee for Coordination of Investigations of the Lower Mekong Basin (1982) *Environmental impact assessment: guidelines for application to tropical river basin development*. Bangkok, UN-ESCAP

Ives, J.D. (1991) Floods in Bangladesh: who is to blame? *New Scientist* 130(1764): 34–7

Koppel, B. (1987) Does integrated area development work? Insights from the Bicol River Basin Development Programme. *World Development* 15(2): 205–20

Kotlyakov, V.M. (1991) The Aral Sea basin: a critical environmental zone. *Environment* 33(1): 4–9, 36–8

Krutilla, J.V. and Eckstein, O. (1970) *Multiple purpose river development: studies in applied economic analysis* (2nd edn). Baltimore, University of Colombia Press

La Bounty, J.F. (1982) Assessment of the environmental effects of constructing the Three Gorges Project on the Yangtze River. In S.W. Yuan (ed.) *Energy resources and the environment*. London, Pergamon

Laconte, P. and Haimes, Y.Y. (eds) (1982) *Water resources and land-use planning: a systems approach*. The Hague, Martinus Nijhoff

Lagler, K. (1968) *Man-made lakes: planning and development*. Rome, FAO

Le Marquand, D. (1989) Developing river and lake basins for sustainable economic growth and social progress. *Natural Resources Forum* 13(2): 127–38

Levintanus, A. (1992) Saving the Aral Sea. *Environmentalist* 12(2): 85–91

Lilienthal, D.E. (1944) *TVA: democracy on the march*. New York, Harpers

Lohmann, L. (1990) Remaking the Mekong. *The Ecologist* 20(2): 61–6

Lohmann, L. (1991) Engineers move in on the Mekong. *New Scientist* 131(1777): 44–7

Long Li (1990) Major impacts of the Three Gorges Project on the Yangtze, China. *Water Resources Development* 6(1): 63–70

Lundqvist, J. (ed.) (1985) *Strategies for river basin development*. Dordrecht, Reidel

McDonald, A.T. and Kay, D. (1988) *Water resources: issues and strategies*. Harlow, Longman

Marchand, M. and Toornstra, F.H. (1986) *Ecological guidelines for river basin development* (CML Report no. 28). Leiden, Centrum voor Milieukunde Rijksuniversiteit

Martin, R. (1956) *TVA: the first twenty years*. Knoxville, Tenn., University of Tennessee Press

Middleton, N. and Thomas, D.S.G. (1992) *World atlas of desertification (UNEP)*. London, Edward Arnold

Mitchel, B. (ed.) (1989) *Integrated water management*. London, Belhaven

OAS (1978) *Environmental quality and river basin development: a model for integrated analysis and planning*. Washington, DC, Secretary-General OAS

Pearce, F. (1990) A dammed fine mess. *New Scientist* 130(1767): 36–9

Pearce, F. (1991a) Building a disaster: the monumental folly of India's Tehri Dam. *The Ecologist* 21(3): 123–8

Pearce, F. (1991b) The dam that should not be built. *New Scientist* 129(1753): 37–41

Pearce, F. (1992) Flood Plan fails to protect Bangladeshis. *New Scientist* 133(1814): 17

Pereira, H.C. (1973) *Land use and water resources in temperate and tropical climates*. Cambridge, Cambridge University Press

Pereira, H.C. (1989) *Policy and practice in the management of tropical watersheds*. London, Westview

Perera, J. (1993) A sea turns to dust. *New Scientist* 140(1896): 24–7

Petts, G. (1984) *Regulated rivers*. Chichester, Wiley

Pinto Paiva, M. (1977) The environmental impact of man-made lakes in the Amazon region of Brazil. Rio de Janeiro, ELETROBRAS (mimeo) 68 pp.

Ponting, C. (1991) *A green history of the world*. London, Sinclair-Stevenson

Quoc-Lan Nguyen (1982) The development of the Senegal River basin: an example of international cooperation. *Natural Resources Forum* 6(3): 307–19

Raphals, P. (1992) The hidden costs of Canada's cheap power. *New Scientist* 133(1808): 50–4

Rich, V. (1992) Hungary bows out of grand plan to dam the Danube. *New Scientist* 134(1822): 6

Robie, R.B. and Reynolds, R.B. (1978) The California Water Plan – past, present and future. *Progress in Water Technology* 10(3/4):69–80
Rogers, P. (1993) The value of cooperation in resolving international river basin disputes. *Natural Resources Forum* 17(2): 117–32
Rondinelli, D.A. (1980) *Spatial analysis for regional development: a case study in the Bicol Basin of the Philippines.* Tokyo, UN University
Rothé, J.P. (1968) Fill a lake, start an earthquake. *New Scientist* 39(605): 75–8
Rowntree, K. (1990) Political and administrative constraints on integrated river basin development: an evaluation of the Tana and Athi Rivers Development Authority. *Applied Geography* 10(1): 21–42
Saha, S.K. and Barrow, C.J. (eds) (1981) *River basin planning: theory and practice.* Chichester, Wiley
Sattaur, O. (1991) Greens in muddy water over Indian dam. *New Scientist* 132(1789): 16–17
Scudder, T. (1989a) Conservation versus development: river basin projects in Africa. *Environment* 31(2): 4–9; 27–32
Scudder, T. (1989b) The African experience with river basin development. *Natural Resources Forum* 13(2): 139–48
Seckler, D. (ed.) (1971) *California water: a study in resource management.* Berkeley, University of California Press
Selznick, P. (1949) *TVA and the grass roots.* Berkeley, University of California Press
Smith, N. (1972) *A history of large dams.* Secaucus, NJ., Citadel
Thomas, C. and Howlett, D. (eds) (1993) *Resource politics: freshwater and regional relations.* Milton Keynes, Open University Press
Thukral, E.G. (ed.) (1992) *Big dams, displaced people: rivers of sorrow, rivers of change.* New Delhi, Sage
UN (1970) *Integrated river basin development: report of a panel of experts.* New York, UN
Vlaar, J.C.J. (1992) Design and effectiveness of permeable infiltration dams in Burkina Faso. *Land Degradation & Rehabilitation* 3(1): 37–54
Vlaar, J.C.J. and Brasser, M.B. (1990) False expectations of labour participation in the construction of permeable infiltration dams in Burkina Faso. *Land Degradation & Rehabilitation* 2(4): 301–17
Voropaiev, G. (1984) Diversion of water resources into the Caspian Sea. *IIASA Options* 1984(2): 6–9
Ward, R. (1978) *Floods: a geographical perspective.* London, Macmillan
Widstrand, C. (1979) *The social and ecological effects of water development in developing countries.* Oxford, Pergamon
World Bank (1992) *World development report 1992: development and the environment.* Washington, DC, World Bank
Zaman, M. (ed.) (1983) *River basin development.* Dublin, Tycooly

Further reading

Adams, W.M. (1992) *Wasting the rain: rivers, people and planning in Africa.* London, Earthscan [Evaluation of water resources development in Africa]
Cummings, B. (1990) *Dam the rivers, damn the people: development and resistance in Amazonian Brazil.* Harmondsworth, Penguin [Readable text]
Goldsmith, E. and Hildyard, N. (eds) (1984) *The social and environmental effects of large dams: vol. 1: overview.* Camelford, Wadebridge Ecological Centre [With vols 2 and 3 – see references – provides a good introduction]
Pearce, F. (1992) *The dammed: rivers and the coming world water crisis.* London, Bodley Head [Readable but excessive focus on negative aspects of large projects]

CHAPTER 10 Urbanization, waste and pollution problems

The 'urban revolution'

The urbanization process is one of the most striking features of the last 200 years and the observer might be tempted to ask how sustainable today's settlements will be. Large towns or cities appeared as much as 5000 years ago and many declined or were abandoned. These 'pre-industrial cities', for example, Mohenjo-Daro (Indus Valley, South Asia), Ur (Mesopotamia), Troy (eastern Mediterranean), Athens (eastern Mediterranean), Rome (Italy), and a number of settlements in China, Japan and what is now Latin America, had relatively small populations, compared to modern cities, many of whom were employed in agriculture or trade in surrounding areas. At its height, one or the larger of these, Rome, probably had fewer than 400 000 inhabitants (Ponting 1991: 297).

The term 'urban' is a little vague and has been defined as 'the concentration of people in cities and towns'. Another definition (that of Beaujeau-Garnier and Chabot 1967: 23–31) is 'a continuous and dense agglomeration of people and dwellings'. Many would be tempted to attach a minimum population-size to a city – typically 2000 or 5000 – but even then there are considerable differences, between what constitutes a 'city' in India or Brazil or Europe. Since the eighteenth century, cities have generally become bigger and more industrialized and a greater proportion of their people work and stay within the urban boundary.

Before AD 1800 most of the world's people lived in rural settlements or towns of less than 5000 and it is unlikely that any city had reached 1 million. Somewhere between 1300 and 1800 in Europe there began a transition from cottage industries to factories which brought large numbers of workers together. Before this transition, industries located where there were fuelwood, ores and water for processing or mechanical power; in the UK such localities included the Weald, the Forest of Dean, the Cleveland Hills and parts of Derbyshire and Cornwall. After roughly AD 1600 coal-mining offered an alternative to wood (the generally accepted breakthrough being the development of iron smelting with coal in 1752). Industries began to concentrate and urban areas grew, particularly in the English Midlands, South Wales and Shropshire. Between AD 1500 and 1700 England's cities probably grew about eight-fold. Pollution had become bad enough by 1852 for Sir Humphrey Davy to be commissioned to carry out a pollution survey of non-ferrous metal smelting, arsenic and lead works in Swansea. By 1863 things were obviously little better, for an Alkali Act was passed in the UK to try and control smoke, grit and ash.

Rapid urbanization seems to reflect rapid economic growth and political causes, as well as development of industry and public transport; in the west, the private car has helped the 'urban sprawl'. Caution is necessary, for there is huge diversity between nations in respect of the causes ('push' and 'pull' factors) and character of urban change, for example LDC urban populations use far less resources than do those of DC cities.

From roughly AD 1800 an increasing number of people have come to dwell and find their livelihood, education or better healthcare within cities but since about 1945, the change has become so rapid that one can claim there has been an 'urban revolution' which around the world has reasonably common characteristics (Table 10.1; Figs 10.1 and 10.2). Governments have made little effort or have had little success in controlling urban growth; one result of this is that urban conditions are often poor and deteriorating.

Table 10.1 Percentage of world population living in urban areas (1800–2010)

Date	% of world population that was/will be urban (approx.)	
1800	2.5	
1900	10.0	(75% of UK population urbanized)
1980	50.0	
2010	>50.0	

Note: The Netherlands and UK were among first countries to urbanize; Latin America was early among LDCs to urbanize; African urbanization came late but has been rapid.
Source: Gupta 1988: 57; Lean *et al.* 1990: 21; Ponting 1991: 301; Kivell 1993: 1

Urbanization background information

An increasing percentage of the world's population lives in cities; it has been estimated that by AD 2000 three out of four Latin Americans, two out of five Asians and one out of three Africans will be urbanites (Lean *et al.* 1990: 21). By 1992 about 75 per cent of North Americans lived in cities occupying 1.5 per cent of the American land area; in Europe and Japan densities were similar or higher. In 1993 roughly half of humankind lived in cities.

Urban populations in LDCs are increasing rapidly, much faster than those of DCs (1950–85 at roughly twice the pace of DC cities). Cities of all sizes in developing countries have increasing populations: some recent UN projections seem to be a little too high and post-1980s census data are often not available.

By 1960 71 cities had reached or exceeded 1 million population; many are in LDCs. By AD 2000 there will probably be over 61 cities in LDCs with over 1 million (in 1950 only 1 LDC city exceeded 1 million – Buenos Aires) (IUCN, UNEP and WWF 1991: 104).

Most LDC governments have an 'urban bias' in their policies, i.e. they favour urban peoples.

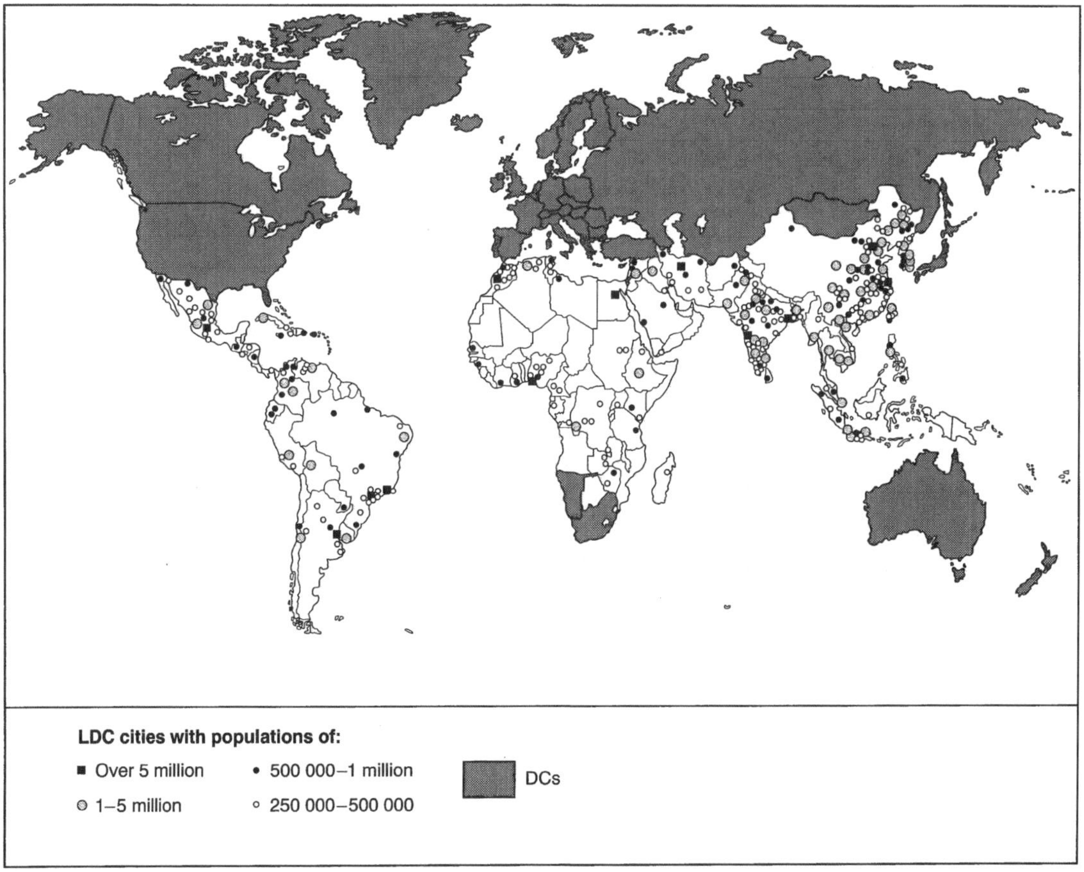

Fig. *10.1* LDC cities of over 250 000 inhabitants (1983) ⟨*Source:* Dickenson *et al.* 1983: 171 (Fig. 7.1)⟩.

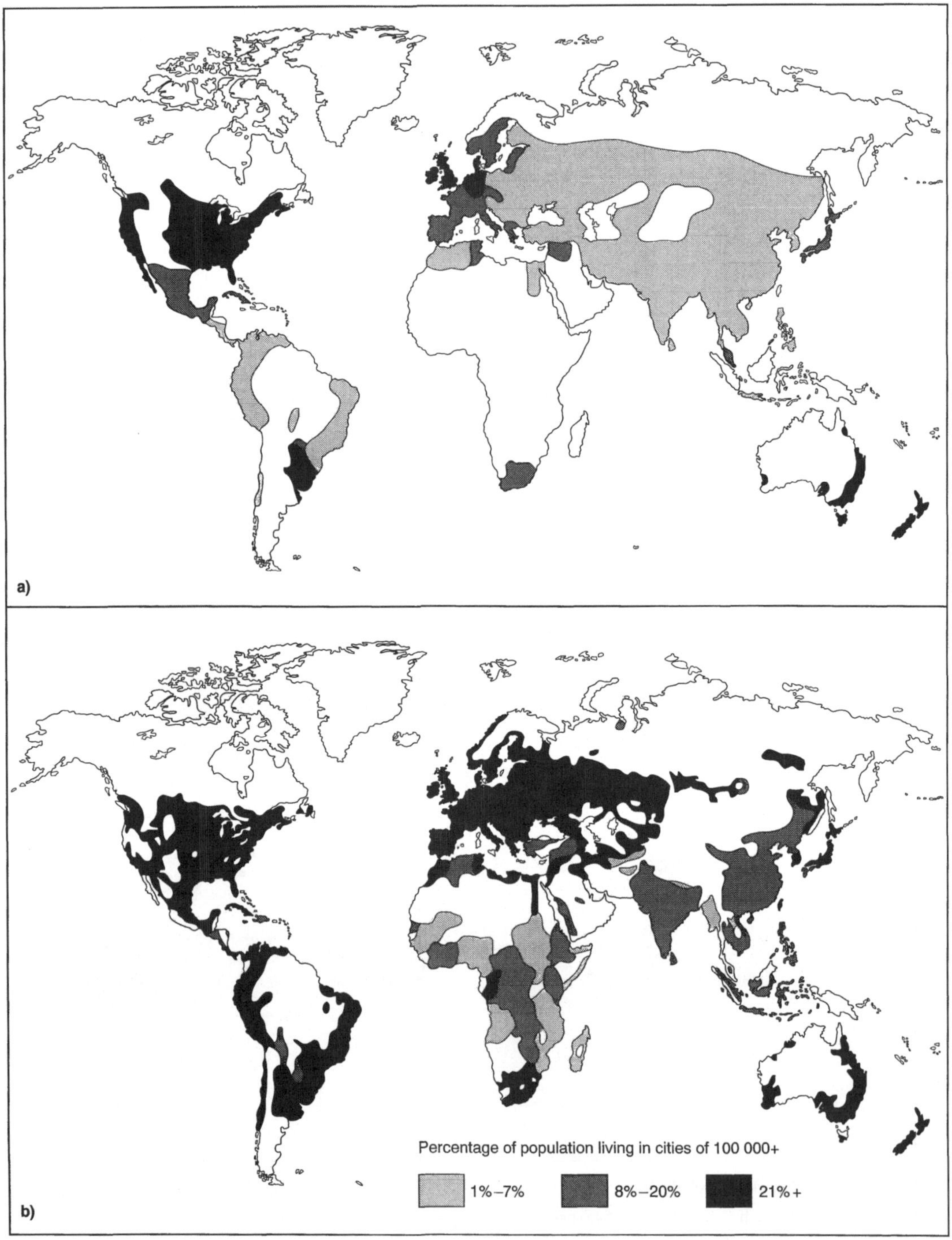

Fig. *10.2* Percentage of world population living in cities of 100 000 or more: a) in 1927, b) in 1970 ⟨*Source:* Burke and O'Hare 1984: 207–8 (Figs 9.1 and 9.2)⟩.

A number of LDCs have or will soon have mega-cities (cities of more than 5 million people). By AD 2000 there will be at least 37 LDC mega-cities, of which Seoul, Rio de Janeiro, Jakarta, Tehran, New Delhi, Bombay, Beijing and Shanghai will each have probably exceeded 13 million and several others will have exceeded 10 million. Mexico City (*c*.3.1 million in 1950, *c*.17.4 million in 1990) could by AD 2000 be over 26 million (in an urban area beset by problems of poor drainage, subsidence, earthquakes, inadequate water supply, high altitude and air conditions that lead to traffic pollution being slow to disperse). In 1992 19 of the world's 25 largest cities were in LDCs.

Many urbanites, especially in LDCs, have inadequate incomes. The distribution of people in absolute poverty is shifting from rural areas to cities and the proximity of cities at an increasing rate in DCs and LDCs. By AD 2000 over 50 per cent of the world's poor will be urban (mainly in LDCs). In 1988 an estimated 40 per cent of LDC urban dwellers lived in slums (WRI, IIED and UNEP 1988: 36). In 1989 *c*.40 per cent of poor people in the USA were urban (UNDP 1991: 31).

Cities increasingly fail to provide their people with basic needs. The gap between LDC demand for food to feed growing city populations and inadequate indigenous production surpluses has generally been filled through increasing imports of cheap DC cereals.

In the LDCs city growth is less and less from rural–urban migration and more and more from city population increase (LDC city populations tend to be young and have high birth-rates). Much of the recent and ongoing urban growth is unplanned and is stressing existing services and infrastructure which are increasingly in breakdown or decay. Overcrowded and inadequate quality housing is a widespread problem. As urban populations grow water supply, urban pollution and waste problems will increase. Already many cities are unhealthy environments with open drains, poor refuse collection and contaminated water supplies (Hardoy *et al.* 1990). A cholera epidemic began in Peru in 1990 and has since spread in Latin America, especially through urban areas with poor water and sewerage systems; Bangladesh and India also have problems with the disease.

DC city growth is slowing (some now have negative growth, but there has recently been some increase in a few European cities, for example in Germany) and there are signs of more interdependence and interchange between cities and non-city areas (some attributable to improved telecommunications). Some would go further and recognize counter-urbanization or population decentralization in DCs. Many DC cities have falling levels of sulphur emissions (air pollution); most LDC cities have increasing sulphur emissions.

The physical impacts of urbanization

Urbanized areas are among the most altered of Earth's environments and some of their problems are unique; they are also major sources of pollution. Bad housing is one of the most obvious and common physical impacts, as is overcrowding. Often cities grow in lowland areas where the soils and other conditions suit agriculture with the result that productive land is lost. Between 1958 and 1974 the USA lost an estimated 5.1 million ha to urbanization and transport landuse. The FAO estimated that world-wide between 1980 and 2000 about 1400 million ha of arable land would be lost through urban sprawl (WRI, IIED and UNEP 1988: 42). Recent estimates suggest that the UK is losing 130 square kilometres a year beneath buildings (ITE data cited in *New Scientist* 27/11/93: 8). Harrison (1992: 108) suggested that Egypt lost *c*.13 per cent of its farmland to urban sprawl between 1973 and 1985, elsewhere in LDCs and DCs loss of agricultural land is significant.

Urbanization generally results in ground surfaces becoming more impervious through compaction and installation of hard surfaced roads and buildings. Storm drains and sewers, even in those cities with inadequate provision, speed up runoff. Over-exploited and often contaminated groundwater, rapid runoff, often of silty and contaminated storm-water, and dry period surface water flows lower than those before city growth are the typical hydrological effects of urbanization. Cities are large consumers of water and may deprive surrounding areas of supplies or force relocation of people and waste of land when reservoirs are created.

Cities alter local climate, built-up land has a different albedo, heat storage characteristics and roughness than non-urban and there may also be considerable waste heat from homes and industry. In warm and cold climates city areas are likely to be warmer than their surroundings and may recycle pollution (Fig. 10.3 illustrates this urban 'heat island' effect). Urbanization pollutes air and water and generates large quantities of waste that require disposal.

City construction on soft ground or overuse of groundwater can lead to subsidence, for example

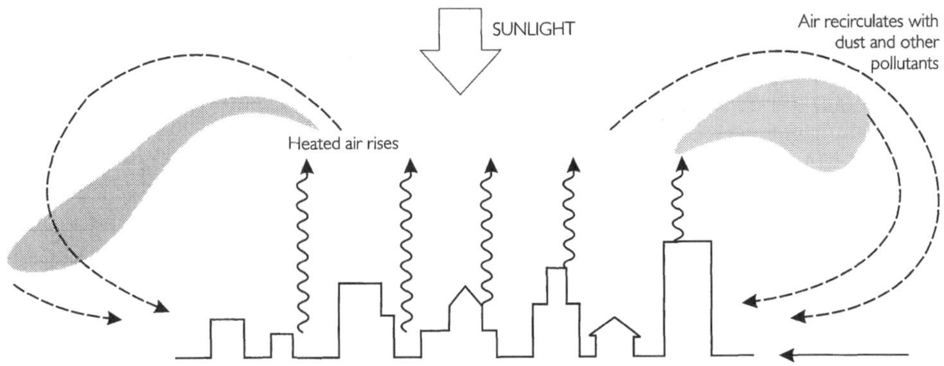

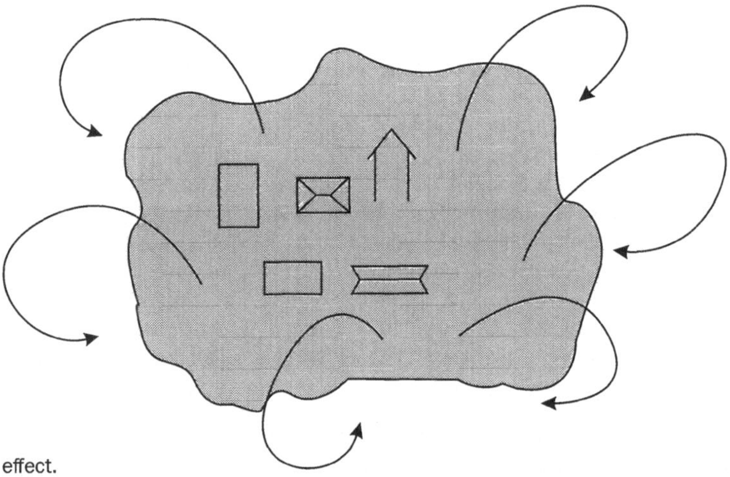

Fig. *10.3* Urban heat island effect.

Venice, Mexico City, Jakarta, Bangkok, Dhakar and parts of London are settling. Disruption of services and flooding often accompany subsidence. Squatter settlements and slums are often sited on land which is unsuitable both to the settlers and for the surrounding people. This is particularly the case where steep land is settled by the poor and becomes unstable. Rio de Janeiro, La Paz and many other cities suffer landslides and silty polluted runoff from steep areas disturbed by squatters. Some cities have slum areas on floodlands or coastal marshes where water pollution and epidemics can spread fast and where any effort to upgrade housing and provide services will be hindered by wet conditions. Industrial, power generation and energy use practices pose a threat for people through pollution hazard, explosion risk and noise.

There are LDC cities, particularly in Asia and Africa, where large numbers of people still depend on fuelwood or charcoal to supply their needs with the result that huge areas, sometimes hundreds of square kilometres, may be deforested, e.g. New Delhi receives fuelwood by rail from as much as 700 km away. Disposal of refuse and sewage from cities may result in the loss of farmland and pollution of air, groundwater and streams, even the sea. Fumes from traffic, industry and domestic fires may be carried by winds, particularly where the topography channels airflows to pollute land well beyond city-limits. Some LDC cities have perhaps the most unhealthy and unpleasant environments in the world (Hardoy *et al.* 1990). Nevertheless, much horticulture and gardening is carried on within city areas and it is often highly productive. A number of LDC cities produce large proportions of their food within urban boundaries and a few even 'export' to rural areas. Urban and peri-urban agriculture is a sector that deserves more support and research, given the number of people dependent on it and the growth of cities. Efforts to support urban agriculture have been mounted by UNICEF and the International Development Research Centre (IDRC), Canada (for a recent overview of LDC urban agriculture see *IDRC Reports* 1993; IDRC 1994).

Some wildlife has adapted well to urban environments, although not all of it is harmless and welcomed.

The socioeconomic impacts of urbanization

City populations, at least in LDCs, are growing and this is stressing infrastructure and services. Many cities in LDCs and DCs have inner suburbs in a state of stress and decay: overcrowded, poverty-stricken, crime-ridden, vandalized and polluted. There is still much to be learnt about the effects of urbanization, including overcrowding and rapid in-migration, on human attitudes and health. Many regard cities as dehumanizing and likely to cause increased mental instability and crime. Generalization is dangerous: few would hold that dense population and in-migration lead to predictable behaviour patterns and urban design, sense of community and other factors come into play and things vary from city to city.

LDCs have growing peri-urban squatter settlements with all of the problems of the inner cities and worse, little or no services and the necessity for inhabitants to travel some distance for whatever employment opportunities there are. Urban planners and managers are facing huge challenges from the inner city areas and peri-urban areas, not least of which is to provide basic housing and services. The housing problem is manifest in DCs and LDCs: in the UK in 1989 at least 400 000 were officially homeless (UNDP 1991: 31) and many probably had no fixed abode but escaped the records. Homelessness poses particular problems in cities where temperatures fall below freezing, as is the case with many Latin American, North American, Asian and European cities; in winter 1993 the authorities in Paris opened public buildings and the Metro and in London emergency shelters have been provided during particularly cold weather.

Allied to bad housing, poverty, disruption of community cohesion, and other urban conditions are public health threats. Overcrowded and poor quality housing may aid the transmission of diseases like cholera, drug-resistant TB and antibiotic-resistant pneumonia, meningitis, etc. Inner-city New York and other poor, crowded urban areas of the USA, Africa and Europe have recently seen drug-resistant TB re-emerge as a serious, growing problem (TB reached a low in New York in 1978 but has since increased three-fold) and another drain on the resources of the WHO and other healthcare bodies. Malaria, dengué (mosquito-borne haemorrhagic fever), and in South America Chagas disease (bug-borne American trypanosomiasis) are increasing in urban areas. In many, but not all, countries drug abuse and Human Immunodeficiency Virus (HIV) are growing urban (and rural) problems.

Where coal-fired power stations are sited close to or within cities or the population burn smoke-generating fuel, as is the case in eastern Europe and parts of the PRC, chronic bronchial and similar air pollution-related diseases are common. Cubatão (near São Paulo, Brazil), though now cleaned up a good deal, is infamous as one of the most polluted cities in the world with high levels of industrial chemicals and lead in the air. The long-term effects of air pollution, especially on the malnourished poor of LDC cities, are not well known and in DCs childhood respiratory complaints like asthma are increasing in some cities.

Rapid city growth generally means that little land is set aside for public space and amenity. Lack of recreational space is yet another cause of deteriorating urban quality of life in DCs and LDCs and contributes to increasing social instability. In the past child labour was a problem in some DC cities; the problem is growing in many LDC cities.

Within peri-urban areas and just beyond, it can be difficult to continue agriculture, due to pollution, theft of crops and vandalism or because urban dwellers complain of noise and smell. If peri-urban horticulture could be promoted, it might improve food supplies and provide employment for some of the poor people; the experience in the UK during the Second World War was that urban allotment gardens can be very productive. In DCs, farmland close to cities comes under pressure from demand for recreational landuse: parkland, golf courses, and surviving agriculture may adapt to include 'pick-your-own' horticulture and grazing for horses and ponies.

Some cities, like London, have tried to preserve green-belts to maintain environmental quality; in many cases (as with London) this land tends to become derelict and urbanization may simply 'jump' beyond it. Older DCs often have large areas of derelict land, sites of former heavy industry, railway marshalling yards, coal-gas works sites, etc, which have yet to be rehabilitated and landscaped or converted to new use. These areas will absorb considerable funds during the coming decades.

Derelict land may be converted to amenity areas and tree or shrub planting can help to improve city environmental quality and reduce the heat island effect a little. Landuse zoning and adequate, enforced planning regulations can help to reduce industrial hazards and the effects of some pollution.

Cities may attract labour from distant rural areas causing impoverishment, neglect of the land and other problems. In many LDCs resources may be allocated to city development even if this is not sensible or should be of lower priority than rural development. This 'urban bias' reflects the fact that decisions are made in cities, those who vote are in cities, and those who could riot and topple a government are usually urbanites. Those in rural areas are usually scattered and tend to have poor communications (Lipton 1977).

Countering urban drift could be difficult: rural people are often bombarded with media images or believe myths of idealized urban life and are encouraged to embark on urban-style consumerism that may be damaging for them physically and culturally.

Whereas cities in DCs tend to show a size gradation from large-size cities to medium-sized cities to smaller-cities, and so on, LDCs often have one 'primate city' and then a big gap between it and the next-largest. It is not unusual in LDCs for one city to hold over 25 per cent of a country's population (in the case of Montevideo, over 50 per cent of the total Uruguayan population). There are city-states, like Hong Kong or Singapore, where almost all the population are urbanized (Dickenson *et al.* 1983: 170). Primacy (primacy of 'internal colonialism') means that a LDC economy is dominated by one or a few cities so that it is difficult for there to be development elsewhere because investment, services, industry, etc, gravitates to the established centres (Barke and O'Hare 1984: 230).

Avoidance and mitigation of urbanization problems

Already in many DCs city growth has slowed; there are signs that private car use may finally be controlled more effectively and become less polluting through emission controls and new technology (electric vehicles, fuel-cell vehicles, hydrogen fuel) and improved public transport systems. Los Angeles has invested in well-tried metro technology (urban light railways) to try and reduce some of the private car traffic, a few European cities are embarking on or have similar policies (Fig. 10.4). Telecommunications advances including teleworking or telecommuting are beginning to enable people to live and work outside cities, in what Kivell (1993) described as a post-modern 'spread metropolis'. Many urban problems could be cured through more effective, more responsible local government.

Fig. *10.4* The world's first passenger railway, the Mumbles Railway (Swansea to Mumbles, c.10 km), opened 1804, converted to passenger traffic 1807. Developed into the electric tram service shown between 1929 and 1960. Deemed of no value when it was closed in 1960, there are now calls for the rails to be relaid for a new version to relieve road traffic congestion (*Source:* Swansea City Museum).

The priority for many LDC cities is the provision of basic housing, clean drinking water, sanitation and employment. Given the numbers of urban poor people, it is unlikely that governments can afford to ignore the issue for long. With limited funds, a most promising approach is the provision of cheap building materials to self-help housing groups and the installation of basic sanitation, safe drinking water and roads. Appropriate building standards will also have to be enforced: often those in LDCs are derived from DCs and are wholly unsuitable, for example roofs strong enough to withstand European snowcover in equatorial environments. There is always a risk that efforts to upgrade slums will lead to speculators buying-out the poor. Group ownership of land by members of self-help teams could give enough security of tenure to encourage improvement efforts while preventing later sale of improved dwellings by individuals.

In LDCs there must be a transition from use of fuelwood and charcoal or dung (unless it is produced on a sustainable basis) to other energy sources like hydrocarbon fuel, solar energy, electricity and biogas. In urban areas of Latin America this transition has begun and seems mainly based on bottled liquid gas. Elsewhere communications may be less good and levels of poverty can make the purchase of gas burners, regulators and bottles difficult; the transition may be to kerosene, which is easier to supply and handle.

Services like garbage collection might also be based on self-help and provide one means of employing some people. An appropriate approach to LDCs' garbage might be district or street corner skips which are collected and taken to the tip or recycling plant by a rota of local people or by cheap slum-dwellers' labour. Recycling of waste would provide employment; local authorities or aid agencies might assist by providing protective clothing, healthcare and refuse handling and processing equipment.

Pollution and waste problems

Waste products and pollution are the unpleasant price usually paid for urbanization and industrial development (and for warfare, increasingly for agricultural development and even leisure pursuits). The burden of wastes and pollution are not necessarily borne by those who benefit from development, and sometimes people far removed may be disadvantaged. When human numbers are low – especially when they are non-sedentary and have little in the way of technology – waste and pollution pose few problems. However, when people congregate, more so if they process natural materials and especially if they manufacture synthetic materials there are likely to be problems.

In the UK complaints about pollution were voiced in London as early as AD 1257 (Brimblecombe 1987); tanning, coal burning and the operation of lime-kilns caused enough nuisance for Edward I of England to legislate against air pollution in 1306 (Royal Proclamation against the use of sea coal in London's industries). John Evelyn may rate as the UK's first environmentalist author, having written a book (*Fumifugium*) on London's coal smoke pollution problem in 1661. Over the last few centuries the scale and diversity of waste and pollution has increased immensely. Pollution and environmental problems are not restricted to 'free enterprise' nations: they are world-wide. Until recently the PRC and the former USSR gave little publicity to environmental problems. Socialist humanism failed to protect nature and a few claim that environmental concern was one of the triggers for the overthrow of communism in Eurasia (Feshbach and Friendly 1992; Mnaksakanian 1992). Today there is concern and publicity but funding for remedial and avoiding action is scarce (Fisher 1992).

To summarize:

- There is an accelerating congregation of people in cities – an 'urban revolution'.
- There has been a huge increase in processing resources – an 'industrial revolution'.
- There has been a trend after about 1910, accelerating after 1945 toward the release by industry, power generation authorities and agriculturalists of artificial compounds: synthetic chemicals, radioactive materials and, since the 1980s, genetically engineered organisms.
- In general, waste and pollution has become more diverse, has increased in volume, in virulence and often persistence.

Human activity has long been causing the accumulation of various contaminants in soils, river, lake and marine sediments, for example oil spills from tankers (Table 10.2). Some of this contamination is localized (e.g. the vicinity of waste dumps), some more widely dispersed and usually less concentrated. Many dump sites have not been recorded and present a future hazard, especially if the seals deteriorate or acid deposition makes compounds more mobile (see p. 251 on chemical time bombs). Zurek *et al.* (1994) estimated that in Poland alone there were over 6000 tonnes of toxic agrochemicals buried in inadequate landfill sites. A few kilogrammes of such compounds could cause huge damage if it entered groundwater or surface waters; little trouble has occurred so far but that is no cause for complacency: the dumps must be leaking and the materials are probably slowly migrating to contaminate the environment.

Table 10.2 Some marine oil spills 1967–93

Date	Vessel/Site	Location	Material	Spill (tonnes)
1967	*Torrey Canyon*	Scillies (UK)	crude	120 000
1972	*Sea Star*	Gulf	oil	115 000
1976	*Argo Merchant*	off East USA	oil	28 000
1976	*Haven*	off Italy	oil	40 000
1976	*Showa Maru*	off Singapore	oil	70 000
1976	*Urquiola*	off Spain	oil	100 000
1978	*Andros Patria*	off Spain	oil	20 000
1978	*Amoco Cadiz*	off Brittainy	crude	220 000
1979	Ixtoc I[a]	Gulf of Mexico	crude	500 000
1979	*Burma Agate*	off Texas	oil	42 000
1979	*Atlantic Express*	off Venezuela	oil	300 000
1983	Sullom Voe[b]	Shetlands (UK)	oil	1 200
1989	*Exxon Valdez*	Alaska	oil	38 000
1989	*Kharg-5*	off Gibraltar	oil	82 000
1991	Gulf War[c]	Gulf	oil	816 000
1992	*Agean Sea*	NW Spain	oil	70 000
1993	*Maersk Navigator*	off Indonesia	oil	—
1993	*Braer*	Shetland Is.	oil	84 000

Notes:
[a] Well blow-out; [b] Spill at oil terminal: wildlife damage in coastal situation; [c] Sabotage and war damage of many onshore and offshore installations. Most of the listed spills have damaged wildlife; in some cases damage has been done by use of detergents to disperse oil. Weather conditions and type of oil determine impacts as well as size of spill and location: oil may form a 'slick' or 'mousse'. Light oil in rough seas tends to froth into floating mousse; heavy oil in cold seas tends to form a limited slick and sink. Spilt oil may be a problem for decades. Note repeated accidents in busy shipping lane areas: Shetlands, Straits of Malacca, entry to English Channel.

Pollution and waste are increasingly trans-boundary problems (i.e. involve more than one state), have in some cases upset or are upsetting global environmental balances and pose major threats to development or even survival of human-kind. In the early 1970s, Meadows *et al.* (1972) recognized pollution as one of the main 'limits to growth'.

Pollution and waste: definitions and background

Pollution

There is no precise definition:

- The addition to the environment of substances that, through either their composition or the amount released, cannot be rendered harmless by normal biological processes (Crump 1991: 201).
- The deliberate or accidental release into a shared environment and against common interest, of a problem causing by-product or waste. The by-product or waste may be injurious or a nuisance: noise, magnetic-radiation, heat, light, an excess of a harmless, non-toxic compound or something aesthetically unpleasant.
- Something in the wrong place or in the right place at the wrong time.
- Pollution is not always due to human activity: some may occur through natural processes, such as volcanic eruption, emissions from natural forest fires, weathering of rocks, etc.

Waste

It is difficult to give a precise definition and there is a lot of overlap with pollution:

- Damaged, defective or superfluous material: in some cases it may pose a hazard.
- Something the owner no longer has a current or perceived use for: the implication is often that waste is more bulky and less dangerous than pollutants.

Pollution and waste characteristics

It is usual for pollution or waste to be assigned a characteristic: hazardous, toxic, radioactive, domestic, industrial, agricultural, biohazard (organic material, e.g. live viruses or bacteria, genetically modified organisms, pests, venomous creatures, large or vicious animals), carcinogenic (cancer-inducing), mutagenic (cell mutation-inducing), tetarogenic (birth defect-inducing).

Toxic is not a very precise term; it is better to further indicate whether effect is acute (rapid effect on short exposure) or chronic (effective over a longer period of time). In LDCs workers may be malnourished and probably work longer hours with less protective measures than those in DCs; their exposure to workplace pollutants is therefore likely to be greater and their resistance lower.

Hazardous presents a threat to humans or other organisms or environment when handled or disposed of. A 'harmless' compound could be hazardous, e.g. a tanker full of milk tipped into a stream might kill fish; CFCs are chemically inert and non-toxic yet pose a serious ozone depletion hazard. Agreeing a legal definition for 'hazardous' can pose problems, for example plastic-coated copper wire may be classed as non-hazardous, but if burnt to obtain the copper it could emit toxic smoke.

Heavy metals metals which, even in minute quantities, can be harmful to living organisms, these include mercury, cadmium, nickel, copper, cobalt, lead, and so on.

Dioxin (PCDD) a hazardous, persistent pollutant that can cause liver damage, cancers and birth defects and abortion.

NO_x see Box 10.2 on p. 271.

Sewage human waste, usually mixed with storm drainage, kitchen refuse and wash-water and sometimes industrial effluent; essentially contaminated water.

Sewerage the pipework, channels, etc., built to deal with sewage.

Trash, garbage, refuse, domestic waste, household waste, municipal solid waste (MSW), landfill these are more or less synonymous ('landfill' really refers to mode of disposal, rather than type of waste); 'trash' and 'garbage' are mostly used in the USA.

Fly-tipping illicit, unauthorized disposal, often in places where it poses a nuisance or a hazard.

Concepts applied to pollution and waste

Absorptive capacity Up to a certain threshold an environment can cope with pollution and waste(s). If that threshold is exceeded or if environmental conditions change, there may be a partial or sudden complete loss of absorptive capacity; recovery may take place or could be slow or unlikely.

Coliform bacteria Presence of coliform bacteria (faecal bacteria) like *Escherichia coli* are indicative of faecal (livestock, wildlife or human) contamination. Usually expressed as a count, the higher the count the greater is the contamination.

LD-50 Tests aimed at determining 'lethal dose' under controlled conditions. One method is to expose a group or groups of animals to gradually increasing amounts of a substance, noting the point at which half die. A guide, but not an accurate indication – what kills half a group of rats may affect a group of dogs or humans differently; conditions in practice are not as constant as those in the test.

Total dissolved solids (TDS) Indicates quantity of dissolved salts and other solids: there is often a relationship between TDS and conductivity of a solution so measurement by electrical probe is possible.

Biochemical oxygen demand (BOD) A widely accepted measure of how polluting something is when contaminating water (i.e. a measure of biodegradable substances that provides an index of water pollution). The pollutant stimulates micro- and macro-organisms and these consume oxygen: if a large amount of the total available oxygen is used the pollutant is said to have a high BOD. Some pollutants have damaging effects without altering the BOD of a waterbody (BOD is usually obtained by measuring the quantity of oxygen consumed by anaerobic microbiological oxidation over five days at 20°C).

Polychlorinated biphenyls (PCBs) Toxic human-made compounds, first synthesized in 1881, used since the 1930s for manufacture of transformers, in hydraulic fluid, plastics, etc. They are very stable and toxic compounds that are difficult to destroy. Attempts to incinerate may lead to production of toxic dioxins if the combustion temperature is below *c.*1200°C. Use has been banned in EU since 1976 but are still widespread in sealed equipment and could escape during scrapping. PCBs concentrate in the tissues of higher organisms (e.g. European birds of prey and North Sea seals have high PCB levels in their bodies): there has been some recent speculation that PCBs may have lowered immunity of marine mammals leading to viral diseases.

Polluter pays principle Idea that those responsible for pollution or waste bear the cost of remedy (direct and indirect costs), rather than the polluted or some third party (often the state). Originally promoted by the OECD. Unfortunately, proving that the polluter is to blame is often difficult; when delayed or slow and insidious side-effects are concerned, even recognition that a problem exists may not be easy. For example, proving that IQ (intelligence quotient) impairment in children or increased senile dementia in elderly people is due to pollution and not other factors is difficult, even when large numbers of individuals are involved and epidemiological patterns become apparent. In practice business tends to pass on costs so 'the customer pays'.

Trans-boundary pollution Pollution which is carried across state boundaries.

Chemical time bombs Concept of a chain of events resulting in the delayed and sudden occurrence of harmful effects caused by the mobilization of chemicals (or other harmful material) stored in soils, sediments, etc, in response to a new or slow cumulative alteration of the environment or a reduction of the buffering capacity of the environment (Stigliani *et al.* 1991). There may be a long delay between accumulation and adverse effects and the manifestation is sudden. A few examples:

- Release of aluminium in soils because pH falls below a certain threshold (about 4.2 pH).
- Gradual alteration of soil by micro-organisms results in release of heavy metal(s).
- Conversion of mercury to toxic methyl mercury by bacteria (mercury accumulation may be due to gold mining, as in Amazonia or Guyana).
- Acid deposition alters grass mix, an acid-resistant but frost tender cover then develops to be suddenly totally killed-off by a severe frost the mix would have withstood.

Acceptable risk May depend on perception, etc, may influence tolerance to something.

Pollution and waste acronyms

LULU 'Locally unacceptable land use' (e.g. nuclear weapons stored in a residential area, a noisy activity near a school).

NIMBY 'Not in my back yard' (i.e. tendency for people to accept an activity as long as it is not close to them) (not exclusive to pollution and waste issues).

NIABY 'Not in anybody's back yard'.

NOPE 'Not on planet Earth'.

Forms of pollution

Pollution can be primary, i.e. have a direct effect immediately on release on the environment, or secondary, the product of interaction after release with moisture, other pollutants or sunlight (more than one of these or all three). Pollution may be local, regional or global in scale. The effects may be direct, indirect or cumulative and felt immediately or after a delay, intermittently or constantly. Until a threshold is reached, pollution may not appear to be a problem. Until a threshold is reached the environment may render the material harmless, once the threshold is exceeded the absorptive capacity may gradually or suddenly collapse. Sometime such collapse may be permanent and on collapse severe problems may arise; it is therefore important for developers to recognize and monitor thresholds.

A 'safe' background level of pollutant may become dangerous if some process causes concentration or if research exposes a new risk. Organisms may concentrate pollutants (bioaccumulation) as they metabolize (mainly by feeding or absorption through the skin), other creatures at a higher trophic level may further 'magnify' the background pollution until near the top of the food-web there are problems. Plankton in oceans and terrestrial water bodies and algae on land are often effective bioaccumulators. In tundra areas lichens accumulate radioisotopes like strontium-90 and caesium-137 from atomic weapons testing and nuclear accidents (like Chernobyl) enough to endanger caribou and reindeer and make their meat unfit for human consumption. Some substances become concentrated in the fat of higher organisms (e.g. DDT and PCBs), some in the bone (e.g. radio-isotopes like strontium-90 or radioactive iodine) and may then subject surrounding tissues to damage (Odum 1975: 103). Wastes and pollutants can also be accumulated by physical processes: tidal action, sudden rain-out by intense storms, chemical bonding to certain soil compounds, localized interception of contaminated rainfall, etc. A 'safe' background level of pollution is no guarantee that there will not be local radioactive or toxic 'hot-spots'. Some pollutants change little after release; others are unstable and may be converted into more harmful or less harmful pollutants.

Pollutants or waste initially discharged into water or the atmosphere may exchange between these two systems, for example airborne dust may settle on water and sink or polluted water may form aerosols, spray or dust. Pollutants and waste may contaminate groundwater and return to the surface at a different time and place. Pollution sources may be point, linear (e.g. a road) or extensive (e.g. a burning refuse tip or dust from a desert). Releases may be brief, more continuous, single events, random events, periodic or continuous.

The distance pollution disperses depends on many factors: on the height of the release, the temperature of the gases released, characteristics of the pollutant – including the particle size, wind (or water) speed, turbulence, favourable wind systems, inversion layers, precipitation conditions, whether any obstacle is encountered and the texture of that obstacle,

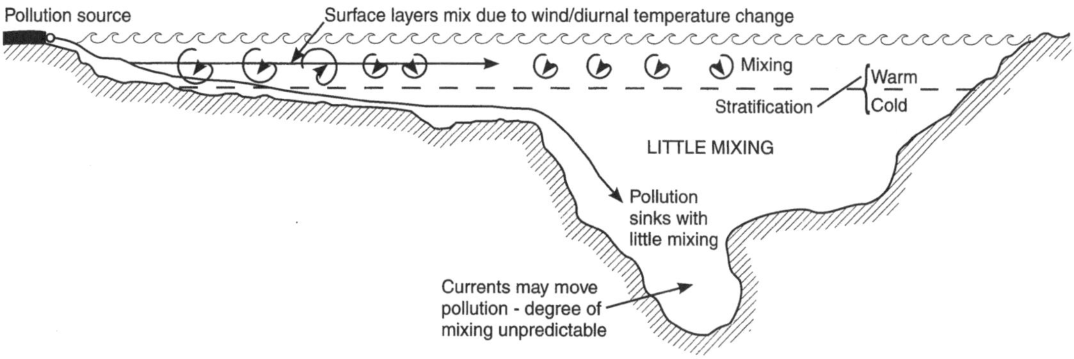

Fig. *10.5* Conditions which restrict dispersal of pollutants in the sea or lakes.

and so on. Topographical and atmospheric conditions affect dispersal in the air; similar factors within a waterbody may hinder dispersal or channel pollution to particular points. It is not uncommon for a temperature inversion in the atmosphere to effectively 'put a lid' over an area trapping pollution. Waterbodies (the sea, lakes, etc) may be stratified with only the upper few metres mixing (Fig. 10.5). Inflows of cold air into valleys or depressions, underwater currents and prevailing winds also channel pollution in certain directions. Insects and other organisms may carry pollutants, or may themselves be a nuisance or threat, and are capable of moving against the wind, through cable- or pipe-ducts into buildings, and of 'hitching' rides on vehicles, ships and aircraft.

Until recently few if any of those involved in urban or industrial development allocated funds or planned for pollution reduction and post-activity clean-up. Around the world derelict and contaminated land and pollution problems require remedial action or provoke compensation claims; those responsible may have ceased to trade, may be dead or may be untraceable. The 'polluter pays principle' has received more and more support in national and international law in recent years but in many cases the responsible parties cannot easily be traced or proven to be to blame. There is much room for improving: understanding of the risks from pollution and wastes, available standards, monitoring techniques and equipment and laws to require ongoing monitoring. Recently worries have been voiced that alpha-particle radiation can affect organisms by altering stem-cells; for decades this low-energy radiation has been seen as a very minor problem. There is increasing concern about microwave and other electromagnetic emissions: presently accepted safe levels and acceptable risks may need rethinking. With pollution, soil erosion and climatic change: it tends to be the occasional disaster event that attracts attention, not gradual but perhaps cumulative or insidious threats.

Given the diversity of wastes generated by urban and industrial development, it is difficult to give a concise overview. The strategy adopted in the following sections is to consider the problems faced in dealing with solid and liquid wastes and with airborne pollutants.

Coping with waste and pollution

In addition to the common response of ignoring the problem, there are three general strategies for coping with waste or pollution:

1 prevention
2 reclamation
3 disposal

Sewage

Before the late nineteenth century, people were often sufficiently dispersed to face relatively few sewage disposal problems: urban areas depended on cess-pits, pit latrines, street collection by nightsoil carts or sewerage systems of open channels. On the whole human waste, even if a nuisance and health hazard, was still a commodity or a resource which more often than not (if it had not gone into a stream) found its way back to the farmland (and urea sometimes to the arsenal for saltpetre manufacture).

Cities in Europe and North America began to start installing water-borne sewage disposal systems by the 1850s. In the UK Edwin Chadwick lobbied for sewerage and other public health measures following rising incidence of faecal-oral diseases and smell in London and other large cities. Joseph Bazalgette introduced an intercepting sewerage system to London in the 1850s (waste was intercepted from minor drains and held until it could be discharged at ebb-tide into the Thames); it cleaned up the city, saved lives, earned him a well-deserved knighthood and was widely copied by other cities. In the UK a number of Public Health Acts and Sanitary Acts were passed in the 1860s and 1870s and in other countries similar legislation encouraged the spread of sewerage: the infrastructure for water-transported sewage.

The 'urban revolution' has tended to shift sewage management from reclamation to disposal, to convert resource use to resource waste and pollution; only recently has a return to reclamation attracted attention. The PRC is an exception to this trend, in that much human waste is still collected and returned to the farmland.

With city growth and increase in water use, many of the world's sewerage systems are becoming overtaxed and this waste disposal consumes large amounts of water that might be used for other purposes. Many sewers in cities which grew before the 1930s are crumbling and need replacement; it is not uncommon, in LDCs especially, for systems to choke up with sand or domestic refuse. Cities like Cairo have recently spent millions of dollars upgrading existing sewerage and installing new extensions. Modern sewerage design can reduce silting-up, for example by installing stepped or ovoid cross-section pipes (see Fig. 9.7) but there are still many problems associated with water-borne sewage disposal:

- The cost of installing, extending and maintaining sewerage.
- Failure to separate storm water, sewage and industrial waste may make treatment and disposal difficult.
- Treatment of sewage before discharge is seldom satisfactory.

Alternatives to established western, water-transported sewage disposal should get more attention. Sanitary engineers have experimented with appropriate sewage systems: some advocate disposal that is not water-borne (composting latrines, etc). Some feel that existing methods can be made cheaper, for example in Brazil pilot studies show that it may be possible to link houses along a single main sewer without the problem of blockage usually feared, and at a fraction of the cost of more usual branched installations: provided people are aware of the pattern of connection they tend to police the system and report or prevent problems.

Spending money on sewers is not enough. In LDCs and DCs waste is too often simply discharged into streams, lakes and the sea: pollution and public health risks are then just shifted about in space. Tied to water-borne sewerage, authorities are faced with the need to treat the sewage to either primary, secondary or tertiary levels and this can be very expensive. EU countries will have to meet strict sewage discharge controls in AD 2000; some authorities will face a challenge to do so. Where populations are small, long sea-outfalls are uneconomic and treatment is likely to be expensive per head of population. It seems likely that chlorination of effluent from primary and secondary treatment will be less common as this treatment causes environmental damage; UV and ozone offer less damaging disinfection. Disposal of sludge will require landfilling, spreading on agricultural land or careful incineration; only the last seems likely to offer long term release from heavy metal, phosphates, etc. Fewer and fewer areas are going to be able to spread sludge on land and most or all EU countries will be prevented from sea dumping.

Levels of sewage effluent treatment

Primary sewage treatment Effluent is screened to remove large debris and is led into tanks or lagoons for aerobic decomposition (methane may be recovered to help heat and power the treatment facility); this may be speeded by air injection and heating. The sludge may be dumped on land as slurry or dewatered pellets (with caution, as it can contain pathogenic organisms, parasitic organisms, heavy metals, nitrates and phosphates), disposed of as landfill or at sea or may be composted or burnt in an incinerator (ash still requires disposal). Effluent may be released with little or no further treatment; sometimes maceration may be undertaken – the results being largely 'cosmetic' – and sometimes chlorination, ozone or UV-radiation is used to kill bacteria and some other micro-organisms before discharge. Chemicals (pesticides, petrochemicals, heavy metals, pharmaceutics, etc), excess nutrients and some harmful organisms remain in the discharged solid and liquid waste.

Secondary sewage treatment As for primary treatment but with greater efforts to remove the risk from harmful organisms, and eliminate COD and BOD demand of the discharged wastes. Commonly, liquid effluent is trickled through filter-beds or is kept in aerated tanks to assist microbiological breakdown of nutrients (reducing COD and BOD on discharge). Chlorination, ozone or UV disinfection may be used to kill most pathogens but more sophisticated treatment is needed for removal of viruses and resistant micro-organisms to meet standards like the EU 'G-standard' (chlorination is today avoided as the effluent is chemically contaminated by it, ozone or UV use avoid this). Heavy metals, phosphates, detergents and many other compounds remain in the sludge and effluent so that there may be pollution (especially eutrophication problems) on discharge. Solid waste can be composted, sent for landfill or sea disposal, spread on fields or incinerated to produce inert ash.

Tertiary sewage treatment The ultimate aim is to return harmless and aesthetically acceptable materials to the environment, possibly as a resource. Effluent is freed of all harmful substances and organisms, solids made harmless and of minimum nuisance. Some treatment plants produce effluent that is safe to drink. The cost of heavy metal, nitrate, pesticide, toxic compounds removal is so far very expensive and the technology complex and demanding of maintenance.

Alternative sewage treatment There are alternative technology options for primary treatment (and perhaps some way toward secondary) which LDCs and DCs could adopt. For example special reedbeds, water hyacinth-filled channels or other swamp communities may remove BOD, COD, some of the heavy metals and pathogenic organisms, and yield fuel, construction material or paper pulp (Fig. 10.6).

Sewage disposal systems

Many of the world's rivers, lakes and seas are so sewage polluted that they pose a health risk and are unpleasant enough to have begun to adversely affect quality of life, tourist trade and wildlife. Hopes that seawater and sunlight would rapidly render sewage harmless seem to be optimistic. There are many recorded cases of sewage pollution leading to 'red tide' conditions – planktonic algal blooms that can contaminate water, fish and shellfish enough to make them toxic to wildlife and humans. In the UK recent studies by the Department of the Environment and other researchers (e.g. in 1992 monitoring of sea-bathers at Langland Bay near Swansea) seem to link sewage-contaminated seawater with illness in people swimming well away from outfall pipes: long-held hopes that seawater and sunlight soon kill pathogenic organisms may be over-optimistic.

Fig. *10.6* Water hyacinth (*Eichhornia* spp.) A water weed which can choke irrigation channels and reservoirs but which has the potential for use in pollution control and the production of charcoal, fodder, etc.

Village or household scale biogas production can offer safe, cost-effective sewage disposal whilst also supplying gas-fuelled heating, cooking and lighting. Biogas systems are widely used in many small towns and villages, especially in the PRC and parts of India; unfortunately, even cheap systems may be too costly for very poor communities and there is an optimum mix of sewage and household waste which is not always to be found. There are a number of effective designs of waterless sewage disposal systems: earth-fill latrines, household or village composting toilets, electric incineration and chemical digestion ('Elsan'-type) toilets (from which safe waste can be periodically removed and spread on the land or safely disposed of in some other way). At the village, large farm or small town scale, composting sewage with agricultural waste like straw may prove an effective method of disposal, yielding a safe, saleable end-product.

Europe increasingly de-waters sewage and incinerates the solids (about 90 per cent of the EC total in 1984). Houston (Texas) has installed a 1500 metre-deep U-shaped borehole for wet oxidation of sewage. Continuously injected sewage is compressed by the time it reaches the base of the U-bore where oxygen is injected: enough heat is then generated to ensure that what emerges from the outlet of the U-bore is a more or less inert hygienic ash. Such systems are probably too expensive for LDC use but may have potential in DCs. Septic-tank sewerage systems are widely used and could be more widely adopted in LDC situations, provided the soil and groundwater conditions are suited and the operation and periodic removal of solids is well supervised.

Non-radioactive industrial waste and pollution

There is a tremendous diversity of industrial effluent produced during mining or processing of raw materials, transportation, manufacture of products, use of products and disposal of used products. Some is harmful, some merely a nuisance; prevention reclamation and disposal constraints and opportunities therefore vary a great deal.

Mining and processing of raw materials often contaminate streams and may easily damage coastal mangroves, saltmarshes or coral reefs. Bauxite mining, copper mining and gold mining often generate particularly toxic wastes. In Amazonia (especially Roraima and Amazonas states) and the huge Pantanal swamps of western Brazil many scattered groups of miners work shallow alluvial deposits and treat the recovered gold dust with mercury to win the metal, streams and wetlands are thus increasingly contaminated with a heavy metal that is concentrated by higher organisms. Cyanide is used by some mining companies to concentrate metal from ore and can easily poison the environment. Mine waste tips may contain sulphates and heavy metals which are leached out by rain to damage people and wildlife for decades or even centuries after mining ceases. Well-maintained mines may pollute streams if a downpour floods their pollution control systems or if a lot of dust is generated (e.g. cement production).

Effluent may be moved via city sewers, through special pipelines, by road or rail tanker or by ship. The disposal site may be a treatment plant, designed to render the material safe or safer, or it may be a dump, deep underground depository, borehole or the ocean. In addition to normal disposal there may be unintentional escapes through leaks or accidents.

Shallow terrestrial burial may allow escapes of materials to groundwater or streams, be subject to disturbance as a consequence of burrowing animals, erosion, earth movements, human interference, acid deposition or warfare. Too few disposal sites have been adequately sealed to ensure there is no escape of pollutants (in time many pollutants damage seals and escape). Record-keeping is often inadequate, even falsified, so that tracking what is disposed of where and when can be a problem. Companies hired to dispose of wastes may be irresponsible; even careful disposal can go wrong. Pollution of more than one type can leak from a disposal or spillage site and 'cocktails' of compounds can accumulate.

Wastes and pollutants may be pumped into the sea or dumped from ships into the sea, the hope being that dilution will 'treat' the effluent; unfortunately in estuarine and shallow marine environments this may not be satisfactory. Material disposed of from ships may be released relatively near or well away from the land; hazardous waste is generally sealed into drums or concrete and may be sunk in very deep water. Sealed waste containers do contain pollutants better than free dumping; however, corrosion and marine organisms may unseal the materials in time and trawlers, anchors or undersea landslides can also release materials. Once sunk into an ocean deep, there is little chance with present technology of recovery or inspection.

Old motor tyres pose a waste disposal problem: if carelessly burnt they release pollutants and the dumps are a space-wasting eyesore. The UK has a good record for reuse of tyres (remoulding about one in five car and one in two lorry tyres); even so, about 25 million tyres a year are dumped and the costs of doing so are rising. In many countries, including the UK, tyre-dumps have caught alight. Burial is no guarantee that underground combustion will not occur: vulcanized rubber does not biodegrade and can burn perhaps years after the landfill was completed; such fires can be long-lived and difficult to control and the waste products of combustion (which include heavy metals, benzene, phenols and tars) can contaminate streams and groundwater. Some cement kilns use tyres for fuel. Another promising way of tyre disposal is to use them as district heating fuel or for generating electricity (provided there is careful control of exhaust gases). Wolverhampton (UK) has a 25 megawatt waste tyre-burning electricity generating station that is expected to consume about 25 per cent of what is discarded each year; three more such installations should solve the UK's tyre problem (two car-tyres are said to provide enough electricity for an average UK home for 24 hours). These stations have scrubbing equipment to clean their flue-gases and will also produce scrap steel and zinc oxide as by-products; unfortunately, they will still emit quite large quantities of carbon.

Dumping tyres at sea may be another answer: the indications are that tyre 'reefs' attract fish, become encrusted with coral, seaweed, etc, and may offer storm protection. Studies so far indicate that there is little danger of pollutants leaching from the rubber (Mason 1993: 5), but this is not proven. There have been proposals to use scrap tyres for coastal protection, for example off the Holderness coast (northeast UK). Another possibility is to freeze with liquid nitrogen and powder tyres; the fabric, metal and rubber can then be separated and used for road construction, etc, if the costs can be kept low enough.

Industrial waste problems have arisen where land has been contaminated and then built upon, where water or food has been contaminated, where deep-well injection has led to escapes, earth tremors and groundwater contamination (e.g. in the early 1960s the Rocky Mountain Arsenal, Nevada, USA, encountered the latter problems after military waste disposal). Often waste has been left as 'harmless' or the land has been contaminated without it being noticed. When built upon such contaminated land can present serious health problems. In the UK there are over 40 000 ha of known contaminated and derelict land, such as old gasworks sites, former scrapyards, old industrial sites, etc); undoubtedly a lot of 'unknown' dumps have already been built upon in most LDCs and DCs (the most infamous being Love Canal near Niagara, USA).

Waste or pollutants that are very hazardous must either be isolated from the environment in some form of sealed containment, treated chemically or incinerated to convert them to safer materials. This sort of incineration requires more care than domestic refuse treatment and it is both costly and demands know-how, but if well managed only the lowest traces of heavy metals and other contaminants are left in the ash or escape with gases and washing fluids. Some countries have few suitable landfill sites, cannot or will not dump at sea, and are moving toward composting or incineration of sewage sludge, domestic and hazardous wastes; countries like Switzerland and Denmark incinerate over 70 per cent of their waste (compared with only 10 per cent in the UK). In 1990 the UK had only four domestic waste incineration facilities: near Pontypool, Fawley, Killamarsh and Ellesmere Port.

To avoid the emission of hazardous fumes or dust, furnaces must achieve a complete combustion at high temperature: to treat PCBs effectively requires over 1200°C for at least 60 seconds (British Medical Association 1991). Even with back-up filtration of flue gases and oxygen injection into the incinerator to aid combustion, things can go wrong; for that reason some chemical incinerators are sited in remote areas and a few are mounted onboard ships at sea (and countries like Canada are hesitating to rely on incineration). A number of incinerators near habitation have been involved in controversy or court actions over release of toxic emissions. The USA has companies that offer mobile (trailer-mounted) high-tech, high-temperature hazardous waste incinerators that can be taken by road to sites that have suffered contamination. In the future, particularly dangerous compounds may be treated in very high temperature incinerators at over 9000°C using solar power or plasma-centrifugal furnaces. Present treatments can be expensive (PCB incineration or bioremediation – treating with bacteria that attack the pollutant – is likely to cost between US$2000 and US$9000 per tonne of soil or waste treated); there will always be some waste producers who seek to avoid outlay and unscrupulous disposal companies who simply dump or partially treat material. Many LDCs cannot afford high-tech incinerators and will have to export hazardous waste to DCs for treatment. At present there is no effective way to decontaminate fissured rocks or clays that have been infiltrated by materials like PCBs, dioxins or tar compounds; more permeable soils can be ploughed up and formed into banks, treated with bacteria and nutrients, and left for bioremediation (bacterial oxidation) to reduce pollutants.

Chemical treatment of wastes ranges from simple disinfection (e.g. maceration and chlorination or ozone treatment of sewage effluent) to complex detoxification plants that can cope with nerve gas. Sometimes the end result may have enough value to compensate for treatment costs: Gourlay (1992: 189) describes an experimental treatment plant capable of cost-effective conversion of sewage into an oil substitute with a calorific value close to that of diesel fuel. Biotechnology may be applied to industrial or domestic waste treatment; bacteria and yeasts could be bred to attack various compounds (including toxic chlorinated hydrocarbons and waste oil) in bioreactors. Suitable strains of bacteria might be used to help leach pollutants from: the leacheate can be collected and treated or recycled if it is a useful material. Some plastics, intended for packaging or temporary use, have been designed to biodegrade (so far with limited success); others might be broken down by specially engineered micro-organisms.

Asbestos, which has been widely used for construction (e.g. in roofing panels or cement pipe manufacture), insulation, fire-proofing, and in vehicle brake and clutch linings, poses health problems not only during manufacture but also through dust generated when it is in use and when it is disposed of. Blue and white asbestos present the greatest threat; brown asbestos is less of a hazard. Inhalation or ingestion,

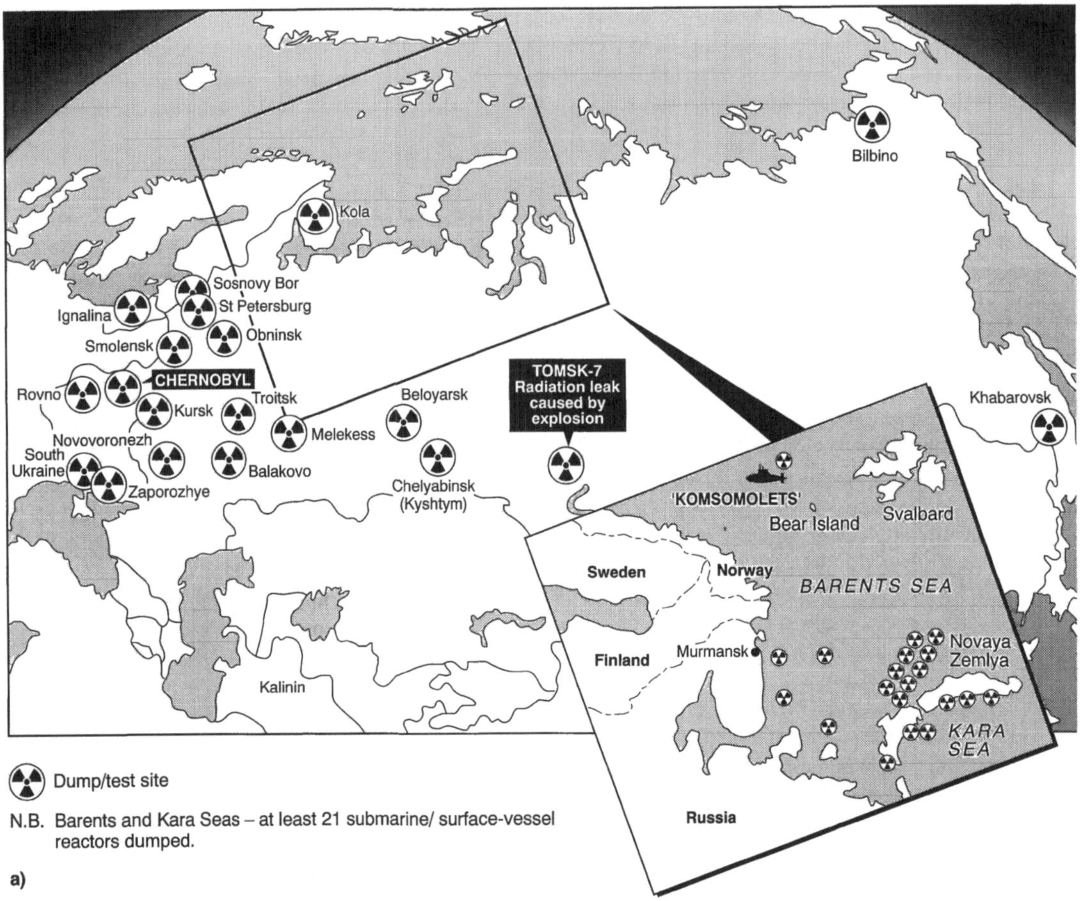

Fig. *10.7* a) Some nuclear disposal and hazard sites (*Source:* compiled from newspaper and journal articles).
Note: Komsomolets – USSR nuclear submarine sunk 1989 (reactors and nuclear warheads on board).

particularly of white or blue asbestos, causes asbestosis, a chronic, debilitating and often fatal respiratory disease like mesothelioma, sometimes decades after exposure. The dust can carry on the wind and workers using the material may contaminate their families and friends. In the USA as late as 1985 there were between 8000 and 10 000 deaths a year due to asbestos. In DCs control legislation has been tightened in recent years (especially in Sweden) but in many LDCs there are still woefully inadequate health and safety measures especially during manufacture. Vehicle components manufacturers are gradually substituting alternative materials for brake and clutch linings to reduce airborne asbestos dust.

Radioactive waste and pollution

Before the twentieth century, the only widespread threats from radioactivity were from natural emissions such as radon and a few less common gases. Production of uranium, plutonium, radium and other radioactive materials has led to contamination of miners, enrichment plant workers and the global environment as a whole through weapons use (in 1945 at Hiroshima and Nagasaki, Japan) and testing, nuclear power generation accidents and accidents with industrial and medical isotope sources. Until the 1980s the Erzgebirge region of the former East Germany was the world's third biggest uranium producer area (after the USA and Canada), mining spoil disposal was poor and today over 1000 square kilometres are badly contaminated with radioactivity. Between 1945 and 1978 there were at least 1165 nuclear test explosions world-wide – about 725 underground in Nevada (USA) alone between 1957 and 1990). The 1963 Limited (or partial) Test Ban

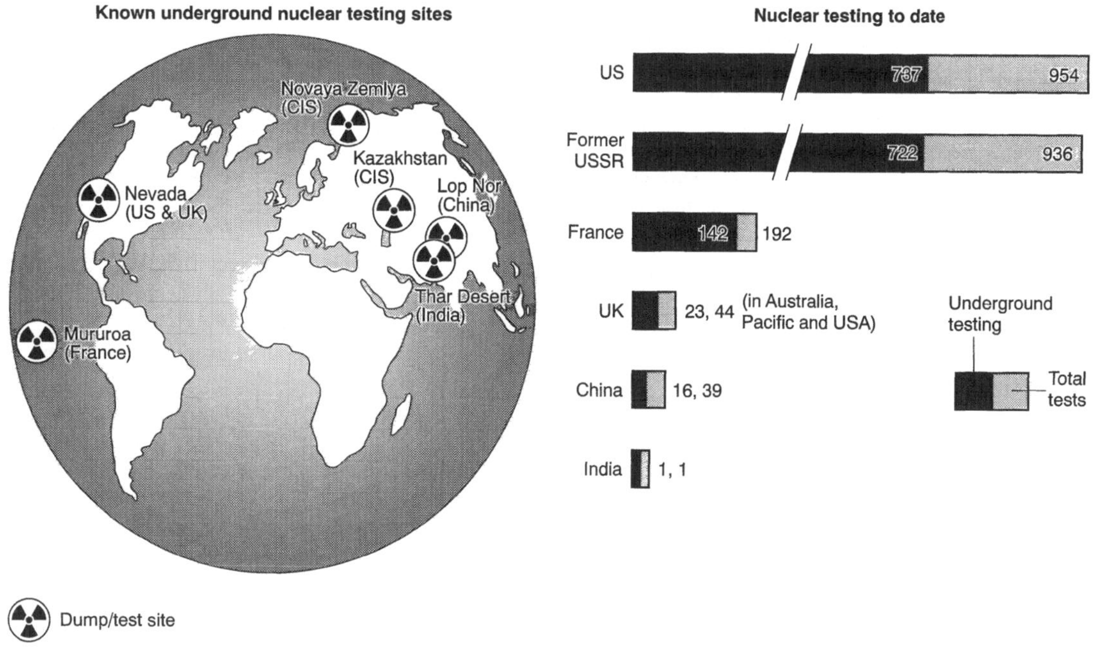

b)

Fig. *10.7* b) The World's underground test sites (*Source:* compiled from newspaper and journal articles).

Treaty ended explosions in the atmosphere, underwater and in space for its signatories. Between 1963 and 1993 tests have continued below ground. The 1967 Nuclear Weapons Test Ban ('Test Ban Treaty') virtually ended above-ground testing, although non-signatory nations may still be tempted (atmospheric tests had ceased by international agreement in 1963). Some underground test sites appear to be failing to offer complete containment of radioactivity, which may seep via fissures to groundwater and thence to rivers or the sea.

There are a number of weapons tests contaminated areas: in the deserts of southern USA, in the Soviet Arctic around Nova Zemlya (Fig. 10.7b), the Gobi Desert (PRC), near Muroroa Atoll (French Pacific), the former Soviet Central Asia, Montebello Island (Indian Ocean, Australian possession, where the UK tested three devices between 1952 and 1956) and the Maralinga test site in South Australia. At the latter site, nine UK nuclear weapons and associated tests were conducted above-ground in the 1950s and 1960s, contaminating as much as 3000 square kilometres. Some of this land has been kept clear of aboriginal peoples since 1964; further decontamination may be needed and might cost the UK£40 million (at 1993 prices). The former USSR is reported to have tested about 713 nuclear weapons at about 50 sites between 1949 and 1989; many of the pre-1963 tests were above-ground. In the region surrounding one of the sites, Semipalatinsk (Kazakhstan), there have long been high rates of cancers and birth defects among the population. By 1993 France and the CIS had declared a moratorium on all nuclear tests; the USA had restricted its (underground) testing to five a year until 1996 when a full ban is likely; the PRC were still underground testing in 1993.

The 1968 Treaty on the Non-Proliferation of Nuclear Weapons now has 140 signatories; however, many have not signed, a few have failed to abide by their agreement, and at least one country signed in the 1980s to allay suspicion that it was developing weapons. Since 1967 there has been increasing proliferation, including sub-national proliferation with the break-up of the former USSR (Ukraine and Kazakhstan now have atomic weapons). Recently there have been rumours that 'red mercury' may provide a relatively cheap and rapid route for nations or terrorists otherwise unable to acquire nuclear weapons. Monitoring the spread and possible smuggling of plutonium and red mercury is difficult.

Some of the most worrying waste disposal problems concern radioactive materials (Eisenbud 1987). These are often highly hazardous, very long-lived high-level wastes, which can generate heat enough to boil and crack or explode their containment releasing dust and gases, or they may corrode containers or prove attractive to terrorists.

Monitoring radioactivity can be difficult and there is still a lot to be learnt about 'safe' levels of exposure to gamma, beta and alpha radiation. The latter has been seen as only a minor threat and so has been poorly regulated; however, there are now fears that it could cause birth defects and cancers. There has been much disagreement in the UK over higher than expected occurrence of possibly radioactivity-related diseases, for example childhood leukaemia 'clusters' near some atomic sites (Newson 1992: 229). Radioactive, alpha-particle emitting tritium (a form of hydrogen) that has often been released from nuclear power stations and other nuclear installations as gas or liquid has recently attracted attention as a possible risk to health; it is difficult to prevent releases as it can diffuse through virtually any container, including stainless steel (generally it has been seen as little threat) (I. Fairlie 1992). Low-level activity waste containing caesium, strontium and plutonium has been pumped into the sea via a pipeline from Sellafield (Cumbria, UK) for more than four decades: there are now signs of 'hot spots' or contaminated wildlife as far away as Scotland and Eire, for example the 1990 Towyn floods (North Wales) deposited mud that proved radioactive (Gourlay 1992: 138).

Nuclear power stations, research laboratories, weapons and medical or industrial radio-isotope source manufacturers all produce waste. Nuclear waste producers face a choice: store it, reprocess it or dispose of it in the environment. Discharge to the sea, dumping in ocean deeps and reprocessing are coming under tougher restrictions (Berkhout 1991). Some waste is still sealed in drums and dumped at sea; some may be piped to the sea; some is stored in holding ponds or tanks, landfilled or incorporated into a solid matrix for burial (vitrification – incorporation into glass – being a favoured solid matrix method). For ten years there has been a moratorium on dumping nuclear waste at sea: Japan, the UK, the former USSR, France and some other nations have failed fully to observe this up to 1993.

Low-level waste is generally disposed of in shallow landfill dumps. High activity waste that cannot be reprocessed must be kept isolated from the environment for centuries or millennia. The containment facility must offer shielding to protect the surrounding environment, should cope with heat generated by the radioactivity and must withstand earthquakes, floods, terrorism, etc. The most favoured option is deep entombment and some countries now have repositories underground, hopefully in non-fissured, impermeable rocks. Such facilities are the classic 'NIMBY' activity: few people are keen to have such facilities anywhere near them (Berkhout 1991). In the UK the Nuclear Industry Radioactive Waste Executive (NIREX) is hoping to install a deep underground facility to cope with waste by AD 2010.

Vitrification, the incorporation of radioactive waste in glass, ceramic or resin, may help to reduce the risk that entombed material will get into groundwater and makes handling and storage easier.

Bulky radioactive structures such as obsolete or failed nuclear power stations, weapons production plants and submarine reactors pose problems. There has been outcry at the possibility of further sinkings of de-commissioned submarines in ocean deeps. Disassembly of nuclear facilities using robot equipment may be a future possibility; however, like the ill-fated Chernobyl nuclear power station (which suffered a meltdown of one of its reactors and release of radioactivity killing and injuring a number of people in 1986) it seems likely that many contaminated sites will simply be buried under a mound of concrete, clay or pumped-sand. Whether such containment is going to be good enough for long enough remains to be seen (the Chernobyl concrete 'sarcophagus' is already breaking up), particularly given that many sites are already very close to sea-level and future rises are likely. The UK developed civil nuclear power stations in the 1950s; in the mid-1990s state-owned Nuclear Electric and Scottish Electric run over a dozen stations of various types, some with more than one reactor. British Nuclear Fuels have constructed a nuclear waste reprocessing plant at Sellafield (the Thorpe Plant); however, this may never prove cost-effective as long as uranium prices do not rise and the construction has been delayed beyond schedule (it also raises problems over security for the weapons-grade plutonium produced). De-commissioning the UK's obsolete nuclear stations will probably cost over UK£15 billion, perhaps £30 billion (Pasqualetti 1990; *The Times* 3/5/93: 5); the UK will also have to dispose of 10 nuclear submarines by AD 2000. France is even more committed to civil nuclear power.

In late 1989 there were at least 356 nuclear power-generation reactors in 31 countries either operating or under construction. By late 1988 239 units had been shut down and 100 were being de-commissioned. In Japan, South Korea, the PRC and the USA, installations were still being built in 1993. Most of the waste has either been left *in situ* or is in storage ponds (really a temporary and vulnerable solution). While a few nations, like Sweden, plan to move away from nuclear power (no easy decision, given that in 1993 52 per cent of Swedish electricity came from nuclear generation), some depend a great deal upon it and will probably continue to. In 1989 the former USSR got roughly 14 per cent of its electricity from nuclear reactors, France 73 per cent, Japan 27 per cent, Belgium 59 per cent, the UK 23 per cent, Germany 28 per cent, Switzerland 40 per cent, Spain 40 per cent: overall roughly 17 per cent of the world's electricity is generated by nuclear reactors (Gourlay 1992: 59). While there are dangers in nuclear generation, burning natural gas is a waste of a valuable chemical feedstock and leads to greenhouse gas emissions; most if not all generation has dangers and costs and the injuries and deaths from all need to be weighed before making decisions. Energy conservation therefore makes sense as it reduces the need for more generators.

In addition to deliberate dumping of old reactors, accidents have led to the loss of an estimated 8 or more nuclear submarine reactors at sea, and there have been at least 54 re-entries of thermo-nuclear isotope-powered satellites, some of which scattered radioactive debris above or upon the Earth's surface (Gourlay 1992: 62–4). The USA and former USSR are estimated to have had about 55 000 nuclear warheads between them in 1989 (about 98 per cent of the world total). There have been a number of accidents in which nuclear weapons have been damaged or lost with the associated risk of leakage of radiation, Gourlay (1992: 62–4) suggests that at least 50 nuclear warheads have been lost at sea. Not only do warheads represent a threat if used in war, if detonated by error or if broken open, but also they require de-commissioning when obsolete and have tied up vast sums of money which might have been spent on environmental management, healthcare, etc.

A problem highlighted since the Gulf War is the storage, handling and use of depleted uranium munitions; these armour-piercing shells may burn or fragment when test-fired or used in battle, releasing radioactivity. Test ranges in Europe, the UK, the USA and the CIS may pose a risk to surrounding areas.

Electromagnetic (non-ionizing) radiation and electricity transmission-related pollution.

There are increasing electromagnetic force (EMF) emissions within houses and in the general environment from microwave ovens, radar transmitters, power cables, cellular telephones, radio and TV broadcasting, telecommunications links, computers and high-voltage wires. The risk that stray transmissions could cause difficulties with control systems in cars, aircraft, weapons, etc, is accepted and measures are taken to shield against radio-frequency pollution and to legislate to control the polluters. With more aircraft depending on fly-by-wire or computers and fibre-optic linkages rather than mechanical controls, with rail signalling depending on electronics, and more cars fitted with electronic fuel injection and ABS brakes, the risks of interference and accident are increased unless shielding is very effective and radio-frequency pollution is controlled. Radio-frequency pollution disrupts the quality of legitimate broadcasts and interferes with research by radio-telescopes; this is accepted and control is sought through legislation.

Worries that low levels of EMF may pose a health risk are not new and have recently been renewed: a family started to sue a UK electricity supply company in 1993 claiming EMF damages (*The Sunday Times* 22/8/93: 1.8). Emissions vary with voltage, type of power (DC or AC), frequency of current, whether sources are shielded, above ground or below ground. Although epidemiological studies in the USA and by the Swedish National Board for Industrial and Technological Development suggest high voltage power cables may cause childhood leukaemia, cancer and brain tumours, convincing proof is not yet available that EMF of less than 100 000 hertz, like that from high-voltage power lines, can cause health problems (Hester 1992). There may be effects on the central nervous system of organisms, on immune response, on heart rate, on DNA synthesis and on the flow of proteins across cell-membranes. Until proven completely safe, EMF should be treated seriously: it may prove necessary to shield emitting equipment much more carefully and to zone landuse to keep power lines and housing apart as far as possible.

Domestic refuse

Cities generate large quantities of refuse; in the DCs this can be 500 to 800 tonnes per day per million people. Domestic refuse may be dumped above or below ground level (Fig. 10.8), incinerated, dumped

Fig. *10.8* Refuse dumping. a) Domestic refuse. All too common a sight, an inadequately sealed tip of household waste. As is often the case the tip is on fire.

at sea, recycled or composted. UK domestic waste typically consists of (approx.) 7 per cent plastics; 8 per cent metals; 10 per cent glass; 10 per cent fines (dust); 12 per cent miscellaneous textiles; 20 per cent waste food and other easily decomposed material; 33 per cent paper products (*The Times* 14/6/93: 32).

The trend in DCs has been toward a greater proportion of packaging materials; since the 1950s much of this has been non-degradable plastics. In the UK alone, 250 000 tonnes of plastic bottles (usually made from polyethylene terephthalate – PET) were produced each year by 1992 (about 50 per cent polyethylene, 30 per cent PET and 20 per cent polyvinyl chloride – PVC). Aluminium packaging has also increased, although many communities now recover and recycle this. Associated with increased use of plastics and aluminium has been a decrease in glass, in some countries there has been a reduction of re-use of bottles. LDC refuse is likely to contain less packaging and more waste organic matter and may thus be easier to compost or generate methane from, but difficult to compress or incinerate.

Refuse may be disposed of by dumping, generally in landfill sites (Fig. 10.8a). About 90 per cent of the UK's refuse was disposed of in landfill sites in 1990 and 70–80 per cent of North America's. Ideally these are located carefully to avoid nuisance and risk of contaminating streams or groundwater: a layer of clay should be put in place before tipping and after completion used to cap and seal the tip. Many redundant tips and some of those presently in use present a serious future hazard because they are poorly sealed. Sometimes domestic refuse is mixed (co-disposal – widely used in the UK) with industrial wastes, power-station fly-ash or sewage for convenience or in the hope it will assist breakdown of harmful materials.

Provided decomposition of organic matter is vigorous, a refuse tip will generate enough heat to kill most harmful organisms. But landfill too often is not a 'last resting place': it fails to contain pollution. Recent 'tip archaeology' in the USA suggests that paper products do not decompose as fast as was hoped and that they may be a major source of

Fig. *10.8* b) Domestic and small-scale industry refuse dumped into a river.

contamination of groundwater, streams and surrounding environments because of the printing inks, waxes and sealants used for coating. A cocktail of plastics, metal waste, etc, lies beneath a usually inadequate covering, a chemical time bomb. Landfill generates methane, which means an explosion risk, perhaps for decades, after tipping. Methane, heat from spontaneous combustion, toxic compounds and subsidence limits the future landuse of rehabilitated tip sites but in some cases the gas can be collected and used for district heating, power generation or combined heat and power. Biotechnology might also improve methane production from refuse and help speed up composting. Problems with nuisance smells, variable quality of compost, difficulty selling or disposing of it and excessive levels of heavy metals are steadily being overcome.

Plastic refuse is a problem in landfill, as litter on land and adrift in the oceans. Some plastics degrade slowly and can cause considerable harm to wildlife (for example marine turtles have often been injured by floating polythene bags which they eat in mistake for jelly-fish) and equipment like pumps. Phantom fishing by lost or discarded nets is doing tremendous damage to marine wildlife; it might be reduced if biodegradability could be built into fishing equipment, but the problem is that fishermen do not want equipment to deteriorate before it is lost (possibly they could regularly treat such nets with a spray).

Considerable effort has been devoted to developing degradable plastic which rot as a result of sunlight, micro-organisms, seawater chemicals attack, etc. There has been some progress and laws in the USA, Sweden and Italy, which insist on such plastics for certain types of packaging. Germany has also started to require manufacturers to arrange for proper disposal of packaging materials. About 60 per cent of the world's shipping *should* be controlled by a 1989 treaty requiring no dumping of plastics at sea (i.e. Annex V of the International Convention for the Prevention of Pollution from Ships); a glance at any coastal strandline confirms that compliance is lax.

The main problems associated with refuse tips are as follows:

1 Nuisance and health problems (smell, noise, rodents, flies, etc), which largely depend on the speed and care of burial.
2 Risk that hazardous compounds will escape into groundwater, streams or the air: base and capping layers of clay may crack and allow leaching or gas escape.
3 Generation of methane may pose an explosion risk and is a greenhouse gas. This limits landuse close to or on abandoned pits, but may sometimes provide a source of energy if it can be tapped in a cost-effective way.
4 Birds attracted to tips may endanger flights from nearby airports.
5 Decomposition may cause 'hot-spots' which restrict re-vegetation.
6 Non-degradable materials that do not rot.

Landfill disposal becomes difficult when there are a limited number of suitable sites; the alternatives are then to compress the waste so that it can be transported more easily and takes up less room on disposal, use incineration or dump the waste at sea.

Incinerators can be used to deal with refuse, industrial waste and human bodies. In DCs municipal crematoria have had some problems with PCB and mercury emissions, the former probably caused by plastics used in coffin manufacture and the latter from mercury amalgam tooth-fillings; dioxins and lead are also released in flue gases. Refuse incinerators may provide district heating or electricity as a useful by-product. Sewage or agricultural waste (manure or straw) may sometimes be incinerated with refuse.

In the UK about 10 per cent of refuse is incinerated, costing in 1992 about twice as much as landfill disposal (Gourlay 1992: 155). In the past the UK had many more incinerators, about 338 in 1914, 295 of them with boilers to make use of the heat (in those days there were minimal controls on the smoke emissions) (Newson 1992: 155). Typically municipal incinerators operate at roughly 800°C to 1000°C, which renders most domestic and agricultural materials chemically harmless and free of organisms; it also reduces the rubbish to much less bulky, easier to handle, ash. Some incinerators pass their ash to manufacturers to produce building blocks or tiles. Hospital wastes, car tyres and plastics may require disposal at higher temperatures to reduce pollution and ensure a safe end product. If combustion, smoke and ash can be managed to ensure little or no pollution escapes to the environment, incineration offers a good means of disposal; however, some countries do not favour it (Connett and Connett 1994) and prefer composting or compression and landfill.

Recycling and reuse of waste

In the future more recycling, reuse and composting might be done at source, in the home. Some countries have national policies to encourage this; elsewhere NGOs are active, for example the US National Solid Waste Management Association and the British Land Reclamation Group (REGRO). Waste recovery and waste recycling or reuse are terms that can lead to misunderstandings: one country might recover 80 per cent of its waste paper but recycle or reuse none; another may recover only 10 per cent but recycle or reuse most of it.

Once waste is recovered there is the problem of sorting, transporting and accumulating different components: is this to be done by the state, private companies or individuals or by householders/waste generators, or a mixture of collectors? Certain types of waste can be dealt with by householders, using multiple containers or neighbourhood bottle banks, tin and waste paper collections. In some Dutch cities multiple dustbins are provided for householders to sort refuse; the system is rather clumsy but seems to work. In some US cities 'reverse vending' has been tried: a waste skip credits a company for return of cans, bottles, etc. The alternative is for authorities to sort waste at a central facility. Ferrous metals, glass and aluminium are relatively easy to sort with automatic equipment, plastics and many other materials are difficult.

In countries where authorities and households cannot afford regular door-to-door refuse collection, street corner or district dumps or skips may be cost-effective and could offer opportunities for sorting ready for recycling. In Cairo traditional rubbish collectors (*Zabbaleen*) collect refuse and charge fees to householders; as in many other LDC refuse dumps, a large informal sector 'industry' is engaged in recycling much of the wastes. The *Zabbaleen* are very efficient; some have been upgraded by the provision of tractors to pull collecting carts, they recycle about 30 per cent of city waste (cf. 2–3 per cent recycling in the UK, where refuse may be of different composition) and 80 per cent of household waste. The PRC has considerable experience of waste recovery and recycling, in Beijing roughly 13 000 people find employment in these activities (Yang and

Furedy 1993). One way of encouraging poorer members of the public to assist with waste recovery is to reward their efforts with retail discount coupons.

DCs can obviously go much further with recycling and can learn a great deal about refuse recycling and reuse from LDC informal sectors; recent studies have focused on Cairo, Calcutta and Bombay with this in mind (e.g. Bouverie 1991). Some DC recycling companies already export waste to LDCs where cheap labour can sort it by hand.

Before the 1950s the 'rag-and-bone-man' had a more simple task than today's recyclers face with numerous alloys and a huge diversity of plastics. Even if plastics or metals of the same general type can be recognized, pieces may vary in subtle ways and may be attached to some other contaminating material or have an unwanted coating that is difficult to remove (caps, labels, circuit-boards from computers, etc). Some plastics have the tendency to absorb various chemicals, which reduces their value for recycling. Unless sorting is perfect and there are few contaminants, recycled plastic has limited use and low value. Crude sorting may be sufficient if the aim is just to recover a limited range of materials like aluminium, glass or iron and to use combustible material as boiler fuel. The value of recycled plastic in DCs is not very high, making it difficult to pay for collection, sorting, washing and shredding and, if that is not enough, the recovered material is bulky for a given weight, making transport and storage costly (Engstrom 1992; Boerner and Chilton 1994). British Petroleum is constructing a pilot plant which it is hoped will convert unsorted plastic waste into an oil from which chemical companies can synthesize 'virgin' plastics (*New Scientist* 11/12/93: 22).

Recycling may not be as environmentally desirable as it first seems, a recent study (Virtanen and Nilsson 1993) suggests that waste paper processing may generate more pollution than burning it for electricity generation or district heating. Germany has strict recycling laws which have led to embarrassing 'mountains' of recovered material that is difficult to sell or store. OECD countries generated huge stockpiles of waste paper which was often difficult to reuse; in some countries it makes better sense to burn this waste for district heating (Kurth 1992). Restricting unnecessary packaging materials reduces the cost and problems of recycling (Germany has done this since 1991). Any authority involved in recycling needs to exercise common sense.

Countries vary considerably in the amount of industrial and domestic waste that they recycle; Denmark recycles over 80 per cent of its domestic waste, Switzerland about 80 per cent, Japan about 70 per cent, the UK less than 15 per cent (Johnson 1990; Gourlay 1992: 185). Estimates suggest that about 5 per cent of Japanese, 15 per cent of European and 10 per cent of US plastics are recycled; however, a recent study of UK plastics' recycling schemes found most of them to be unprofitable (Fairlie 1992). A similarly wide range of commitment to recycling is apparent with materials like glass. Glass can be recycled indefinitely and each times saves on energy compared with production of new material. Overall the EU recycles about 49 per cent of glass (the UK about 21 per cent).

Studies suggest that in countries like the UK, where there is a centralized supermarket supply system, commerce is likely to favour recycling rather than reuse. Reuse of, for example, soft drinks bottles requires a decentralized network of manufacture; supermarket-dominated retailing in the UK. USA and parts of Europe is unlikely to encourage a return to refundable bottles (S. Fairlie 1992). This is probably because non-durable products can be designed with a limited life to be recycled, and so keep sales running. If a manufacturer arranges to recycle its products, say motor cars, it might even be able to restrict sales of salvaged second-hand parts so it can control the spare-parts supply: car recycling is not new although some advertising agencies would like the pubic to think so; scrapyards have done it effectively for decades.

Steel and aluminium recovery can be worthwhile: the latter saves about 95 per cent of electricity used in making fresh aluminium. In 1992 the UK recycled about 16 per cent of aluminium cans, much lower than in the USA and one of the lowest recovery rates in Europe. Paper can be recycled up to four times before the fibres are damaged too much; in 1992 about 62 per cent of UK paper needs were met by recycling (*The Times* 14/6/93: 33). The increasing and wasteful use of disposable nappies (diapers), which may also pose hazards for those working in refuse disposal, might be countered by establishing more widely sophisticated laundry/nappy delivery/collection services: to do so consumers will need to be assured of very high, foolproof standards of hygiene.

Waste and pollution related to domestic heating

Ever since humans started fires in caves and huts there has been a health cost, a trade-off between the benefits of heat, cooking, food and thatch preservation,

Fig. *10.9* Urbanization and a haze of traffic-generated photochemical smog. Bangkok, a city which often has an air pollution problem.

vermin and larger animal deterrence and the risk of respiratory and eye diseases. Coal burning in domestic fires caused problems in Europe by the thirteenth century; by the twentieth century London, Glasgow, Manchester, New York, Milan, Belgium and many other cities and areas suffered. For about five days in December 1952 there was a London smog (in which the pH of the particles was as low as 1.6) that killed about 450 people directly and roughly 4000 over the course of the year and hospitalized over 2000; this encouraged the passing of the 1956 Clean Air Act (Elsworth 1984: 3). Further UK improvements were gradually attained through the 1968 Clean Air Act and the 1974 Control of Pollution Act, and because cheap less-polluting North Sea gas became available. Within about 30 years the UK reduced ambient smoke by about 90 per cent and ambient sulphur by 40 per cent (British Medical Association 1991: 11), but not without serious smogs in 1956, 1957 and 1962 (Newson 1992: 110). There has been progress in air pollution control in western Europe, less in eastern Europe.

The UK Clean Air Acts reduced coal burning within cities, cutting air pollution by about nine-tenths between the 1950s and 1970s. Unfortunately, energy consumption changes led to large power stations which dispersed their pollution over wide areas, contributing to trans-boundary acid deposition problems.

Waste and pollution related to transport

Before the 1950s many cities had winter smog (sulphur dioxide-rich) problems caused by domestic coal fires; in the UK and many other DCs that problem has been reduced; there has, however, been an increase in DCs and LDCs of warm weather city smog (nitrogen dioxide-rich) caused mainly by vehicle exhaust pollution (Fig. 10.9). Most petrol engine cars in Europe will be fitted with catalytic converters by late 1993; catalysers are also required in North America and other DCs and, provided these are well maintained, there should be considerable reduction of unburnt hydrocarbons and the pollution these lead to. Diesel engines emit roughly 1.5 times as much nitrous oxide as the equivalent petrol engine and slightly more sulphur dioxide. Modern diesels emit only about one-third of the unburnt hydrocarbons of an equivalent petrol engine, and a lot less (about 1 per cent) of the carbon monoxide; also the diesel is at present 15–25 per cent more fuel-efficient. The catalytic converter has reduced the advantages of the diesel; if petrol engine fuel efficiency is improved, existing advantages will disappear and the diesel will end-up 'dirtier' in terms of nitrous oxides and volatile organic carbon compounds (VOCs) emissions (already companies like VW are fitting catalysers to diesel vehicles). Lead-free petrol burnt in engines without catalytic converters releases VOCs.

In urban areas exhaust gases, especially partially burnt hydrocarbons including the dangerous VOCs (including toluene, ethylene, propylene, butanes, benzines), dust and noise generated by transport generate serious health problems. VOCs are formed mainly from diesel exhausts and are known to cause respiratory diseases and perhaps are carcinogenic (some are suspected of causing leukaemia). VOCs also play a part in tropospheric ozone production and acid deposition (diesel engines are thus not quite as 'green' as some claim, unless fitted with catalysers).

High lead levels in the air of inner city areas has been a problem for six or seven decades in some cities. About 90 per cent of the lead in the atmosphere is probably from leaded petrol; ice cores from Greenland clearly show the pattern of pollution rising after the 1750s and accelerating after 1925. Children are especially vulnerable as they accumulate the metal and may suffer reduced mental development, especially if exposed in the first few years of life. High lead levels may also reduce birth weight of children. Lead from vehicle exhaust (used since *c*.1925 to improve combustion qualities) may be compounded by drinking water rich in lead from old plumbing and by old flaking paint, especially in poor housing. These lead sources tend to be high in inner city areas and the poor are likely to suffer greater contamination. Some countries (mainly DCs) now (since about 1972 in California, USA) insist on the use of non-leaded (tetraethyl lead-free) petrol and catalytic converters for motor vehicles. There are gaps in these controls in some countries: some cars escape pollution regulations because they are too old or too small; control equipment may not be very effective or remain in peak condition for long. Nevertheless, levels in the UK have fallen from a peak in 1974. There may be a temptation for oil companies to add potentially hazardous compounds like toluene, benzene or xylene to petrol to compensate for the loss of anti-knock tetraethyl lead, a threat that needs careful monitoring.

Some countries are experimenting with biodiesel, one being oilseed rape oil processed to form rape methyl ester. This biodiesel is reputed to yield almost the same power output as regular diesel but with virtually no sulphur dioxide, fewer soots and less carbon dioxide.

Photochemical smogs are a problem in many if not most cities when there are sunny and still weather conditions. Downwind of many cities or busy road systems may be found plant damage; often the culprit is photochemical smog rich in tropospheric ozone. Damage to conifers near high alpine roads in Europe has led the Swiss authorities to force lorries on to rail carriages rather than have them climb through passes and pollute the air. The WHO consider 60 ppbv (parts per billion by volume hourly limit) of ozone to be dangerous to humans (the UN suggests 25 ppbv as a 'safe limit'); in 1992 Mexico City, where much of the ozone is generated by traffic, it has exceeded 398 ppm on more than one occasion! (Mexico City has over 4 million cars, lies at over 2000 metres altitude and has surrounding high ground which traps the pollution.) Monitoring indicated crop damage from tropospheric ozone throughout the USA by 1993; Brown *et al.* (1993: 13) suggested it might have led to a 5–10 per cent depression of harvests.

Severe winter conditions lead some DC road authorities to use salt, which contaminates the countryside, streams and groundwater. Airports in cold environments use large quantities of chemical de-icing sprays, often glycol-based, which can contaminate streams and groundwater; a few airports now have lagoons designed to intercept and hold de-icing fluid until it decomposes enough to discharge, or plan to install underground sinks that should catch and recycle the fluids or render them harmless through biotic activity.

Many would regard vehicles themselves as pollution in that they take up a great deal of room, particularly in urban areas. Cities like Singapore, Cambridge (UK), Athens and Mexico City now strictly control or seek to control entry of vehicles to inner city areas. A few enlightened, or heavily polluted, cities or countries are now investing in public transport of a sufficiently high standard to wean private motorists away from their cars. The process may be helped by severe restrictions and charges on access by private vehicles. Much of the new public transport is based on old well-proven technology: metros, trams or light railways.

Waste and pollution related to public health measures

The huge value of pesticides in reducing crop losses and spoilage, thereby ensuring more food supplies, has been discussed in chapter 8. Pesticides have another important use: the control of organisms that spread disease, such as mosquitoes, flies, lice, ticks and rats. Pollution has sometimes been the cost paid for better disease control. There are alternatives that might be used more widely to reduce pesticide applications: net-screens on windows, application of small quantities of oil or kerosene to standing water (this is a cheap means of preventing mosquito larvae from reaching maturity), stocking water bodies with mosquito- or snail-eating fish.

A health and safety problem in some countries has been the use of temporary workers or new immigrants to carry out hazardous work associated with say the manufacture of pesticides: in the USA there have been cases involving Latin American or Azorean immigrants and similar problems doubtless exist in the UK, mainland Europe and at sea on ships carrying hazardous waste. There is also some risk to people living near pesticides and other hazardous factories.

The problems of illicit dumping and export of pollution hazards

World-wide illicit disposal of waste has become more and more of a problem. Within countries 'fly-tipping' takes place and poses health threats, damages environments and wildlife, is aesthetically unpleasant and is a frequent means of side-stepping the 'polluter pays principle'. Fly-tipping may be done by a householder, a manufacturer or by a contractor paid to dispose of the waste (often the client has paid for and expects proper disposal and is being cheated).

There are two ways in which waste and pollution hazard can be transferred from a DC to a LDC: the factory, processing plant, etc, can be relocated to the LDC, or waste and pollution can be shipped from a DC to a LDC.

One difficulty with establishing hazardous industries in LDCs is that employees and local people may either not be sufficiently educated to appreciate risks, may be forced by circumstance to accept risk in return for difficult-to-get employment or may be told little about the materials they are using for fear of loss of 'trade secrets' (or the latter may be a convenient excuse) (Ives 1985: 76). A widely recognized need is for much better labelling of materials, raw materials, manufactured goods and waste, to ensure that users, carriers and bystanders know what they are, what the risks are and what the safety measures are. Safety controls can become lax in any country; however, there have been signs that some companies set up subsidiaries where pollution, health and safety controls and labour costs are more favourable (i.e. lax) than in their own – usually in LDCs.

Trans-boundary (inter-state or trans-frontier) shipment of wastes has become a huge problem (the OECD estimated in 1990 that 24 industrial nations exported waste for 'recycling' worth US$19 billion): the recipient LDC may be unaware of the true nature of the material imported, may be willing to take the risk or suffer the damage for foreign exchange or a government officer may have authorized the shipment in return for payment of bribes (Box 10.1).

Box 10.1 Export of hazardous waste: the *Khian Sea* case

In 1986 the Liberian-registered ship *Khian Sea* was chartered by a US firm to take a cargo of toxic ash that was apparently rich in pollutants like dioxin to the Bahamas. Refused permission to off-load at its destination, the vessel appears to have embarked on a world-wide hunt for a dump site. After nearly a year, waste off-loading began in Haiti but the authorities found out and forced a partial or complete reload. In 1988 the vessel returned to the USA but the cargo could not be returned to the client; the *Khian Sea* reportedly fled and was sighted near Yugoslavia in August 1988. At some point during 1988 it appears that the cargo was finally dumped at sea, possibly off Bangladesh (Gourlay 1992: 3). Many similar incidents have been recorded (e.g. the *Karin B* or *Zanoobia* cases); doubtless many more go unrecorded (some were listed in *The Times* 17/6/88: 12).

Since 1988, various countries have tried to improve controls designed to prevent export of hazardous waste. The EC introduced regulations in 1988, which, like similar legislation in the USA and UK, aimed at improving access to information so that monitoring cargoes could be easier for governments and NGOs. The Basle Convention, which came into force in 1993, was intended to regulate international trade in hazardous waste and ensure hazard is not exported to LDCs; unfortunately, although signed by 105 countries, it has many 'gaps'. Since the Convention, the trend has been for exporters to claim that waste is for 'recycling' in order to get around the agreement (a number of European and other DC countries did not ratify the Convention anyway).

Some infamous pollution disasters

Windscale

In 1957 a fire burnt for three days at the Windscale (now Sellafield) power generation, waste reprocessing and fuel production plant in Cumbria (UK). The result was the escape of considerable amounts of radioactivity which contaminated considerable areas of farmland (over 500 square kilometres). The escapes of caesium-131, -134 and -137 necessitated the destruction of quantities of milk. The health impacts are not precisely known (Crump 1991: 225) and there has been speculation that these escapes may have contributed some of the caesium-137 UK fall-out blamed on Chernobyl; however, the authorities dismiss this (Wynne 1989: 35).

Chernobyl

So far, Chernobyl is the worst nuclear accident. In 1986 a power surge blew the top off the core of reactor no. 4 and a fire burnt for 10 days with considerable escapes of radioactivity to the atmosphere (roughly 90 times as much as Hiroshima). Luckily major rivers like the Dnieper escaped serious contamination and the city of Kiev with a population of 2.5 million was not seriously affected. The escape killed about 33 people and 2000 people were contaminated enough for there to be a serious risk that they will develop cancer later in life. The clean-up, evacuation and entombment of the reactor had cost over US$10 billion by 1989; it looks as if much more would need to be spent on it in the future. An area of over 2800 square kilometres has been evacuated, affecting about 135 000 people and radioactivity rained-out from clouds has contaminated areas of land as far away as Wales and Cumbria (UK), parts of Germany and Scandinavia. Caesium-137 (with a half-life of about 30 years) has proved very persistent: some vegetation and peaty soils are holding the radioactivity and some flocks of sheep and herds of reindeer grazing these areas cannot be exploited. In the mid-1990s the caesium-137 still affects the livelihood of over 7000 Welsh sheepfarmers, placed under restrictions in 1986. With their sheep still too radioactive to market, there is a possibility that their pastures might be treated with Prussian blue (ammonium ferric hexacyanoferrite) or potassium chloride to lock the isotope in the soil and keep it from being absorbed by the vegetation.

Similar suspect RBMK-type and VVER 440/230-type nuclear reactors are in use or under construction in a number of places in the CIS, Bulgaria and India. Chernobyl has served as a relatively gentle warning of what might go wrong with civil nuclear generating plants.

Three Mile Island

In 1979 the failure of a valve in the cooling system of a civil nuclear reactor at Three Mile Island (on the Shenandoah River, Pennsylvania, USA) resulted in damaged coolant pumps, a pressure build-up explosion and near melt-down. Fortunately the core and pressure vessel remained intact. A serious escape of radioactivity was narrowly avoided but people living within 8 km were evacuated; the clean-up took 2 years and cost more than US$1 billion.

Kyshtym

Apparently in the late 1940s nuclear waste from several installations was disposed of in the Techa River (eastern Urals, former USSR). The contamination reached Lake Karachai near the town of Kyshtym, where it became concentrated in sediments and generated enough heat to lower the lake and expose the bed. These areas were covered in concrete to stop wind-blown fall-out and the worst of the waste, from nuclear weapons manufacture, was pumped into 160 tonne capacity tanks. In 1957 the cooling system for one of these tanks failed and there was a chemical explosion that scattered about 80 tonnes of high-level waste. It seems about 15 000 square kilometres were contaminated with strontium-90, 10 700 people were evacuated, 32 villages were permanently closed, thousands of convict and military forced labourers suffered radiation sickness and large areas were declared 'radio-ecological reserves' with restricted entry (Medvedev 1990).

Seveso

In 1976 ICMESA, a pharmaceutical company manufacturing trichlorophenol (a bacteriocide) in the town of Meda near Seveso (north of Milan, Italy), suffered an explosion when a reactor overheated. A cloud was emitted which contained various compounds, including an estimated 0.5 to 5.0 kg of TCDD (dioxin), a hazardous and very persistent compound (UNCTC 1985: 93). The cloud settled over about 18 square kilometres of densely settled countryside and the town of Seveso; 81 000 domestic animals were killed or subsequently had to be destroyed. An estimated 156 workers and 37 000 residents were exposed to the dioxin. This exposure led to 90 abortions and a significant rise in birth defects in the years following (*New Scientist* 29/9/83: 918); 600 people had to be evacuated and 2000 were treated for dioxin contamination (Newson 1992: 198). The authorities appear to have been told of the explosion after a considerable delay and then reacted slowly, evacuating about 1000 but then allowing 9 days to elapse before the risks were acknowledged to be serious and 14 before wider evacuation was started (UNCTC 1985: 93). The top 20 cm of soil were removed from the worst polluted areas; some was burnt locally and some stored in barrels.

Efforts to dispose of the barrels of contaminated soil and other debris cost roughly US$150 million (at late 1970s rates). In 1982 6.5 tonnes (41 barrels) of the waste were 'lost' for about eight months, having been hidden by a waste disposal contractor; they were finally found during 1983 in an abandoned abattoir in northern France. A Swiss company finally incinerated the waste in Basle in late 1984. The trial of those responsible for Seveso began 7 years after the disaster. Court actions and further leaks from the plant were still being reported in 1985 (*New Scientist* 4/7/85: 25). A consequence of the incident is widespread public mistrust of the authorities.

Tomsk-7

A chemical explosion in a military uranium waste tank (near Tomsk, CIS) in early 1993 led to quite serious contamination and highlighted the need for costly and thorough long-term care and maintenance of nuclear facilities.

Bhopal

In 1980 a chemical and fertilizer factory was opened in Bhopal (Madhya Pradesh, India), a subsidiary of the US Union Carbide Corporation (51 per cent Union Carbide stockholding, 49 per cent Indian stockholding). One of its products, a pesticide 'Sevin', required the production of methyl isocyanate (MIC), a reactive and toxic compound. In the early hours of a December morning in 1984, several 'failsafe' safety devices failed and a runaway reaction led to a gas release. A cloud of MIC blew across a nearby area of poor dwellings and contaminated about 25 square kilometres of the city. People were mainly asleep and there was confusion during the escape and in the ensuing hours as people were seeking treatment. Hospitals had no immediate information on the nature of the gas or on the best treatment; if they had had this, casualties could perhaps have been much reduced. Estimates suggest that at least 3000 people were killed more or less within minutes; a further 10 000 to 25 000 people were seriously injured; some of these subsequently died and many are permanently injured (perhaps 60 000: Shrivastrava 1992). Crump (1991: 32) put the total casualties at 500 000 injured. In 1993 officials held responsible for the disaster had still not come to trial and very few of those affected had been compensated, although Union Carbide agreed a settlement of US$470 million to the Government of India in 1989 (the victims had reportedly considered seeking US$10 billion).

Subsequent inquiries suggest that the disaster could have been reduced or avoided if there had been better safety checks on the factory, if there had been contingency plans and warning measures in force, if the medical community had been warned of the risk and if housing had been zoned away from the area. In practice rapid city growth and the need for poor people to live near their employment would probably have made effective zoning difficult (CAP 1984; Ayres and Rohatgi 1987; Weir 1988).

The post-disaster inquiries and litigation to compensate those who have suffered and prosecute any who were guilty of misconduct have been slow and difficult. Establishing whether such a case is the result of negligence, accident or sabotage is often difficult and at Bhopal raised the question of which legal system to use, that of the responsible company where the technology originated (USA) or that of the host country (India); clearly costs and compensation rates are very different. One 1993 report noted that only 2747 of 15 000 claims for compensation for death had been settled, and that on average those settlements were equivalent to only UK£2000 (well below likely USA rates of compensation). Gourlay (1992: 183) reported that Union Carbide's output of MIC had increased since the Bhopal disaster.

Love Canal

An abandoned canal in Niagara City (New York State, USA) was used for industrial chemicals landfill disposal, mainly from 1942 to 1953 with inadequate monitoring, record-keeping and site planning. The area was infilled in 1953, landscaped and a housing estate of several hundred homes and a school was built. By the 1970s problems were apparent; the 'cocktail' of contaminants, which included chlorinated hydrocarbons, were causing people on the estate a high incidence of cancers and birth defects; however, it was not until 1977 that the problem was clearly linked to pollution. Local women played a major part in exposing the problem and in the action to obtain a satisfactory solution. The area was evacuated (240 families) in 1978 and declared a Federal Disaster Area in 1980. By the early 1990s over US$14 million of lawsuits had been filed against the company blamed for the pollution (Crump 1991: 165). By 1992 clean-up and legal costs had exceeded US$250 million (Newson 1992: 164–7).

Triana

A suburb of Triana (Alabama, USA) housing mainly poor, coloured people was contaminated with a cloud of DDT from a factory in 1971. About 1200 people suffered long-term effects. The township remained contaminated for several years and only in 1979 was there clear evidence of the problem. The company concerned settled compensation in 1983.

Minimata

Between 1953 and 1983 Japanese industry polluted the seas near Minimata (synonymous with Minamata) with heavy metals. The main problem was probably dimethyl mercury from plastics manufacture, which contaminated fish; regular consumers developed 'Minimata disease' as they accumulated mercury and suffered nerve damage, anaemia and birth defects. Forty-three deaths were attributed to the pollution.

Lekkerkerk

Hazardous wastes were burnt on a town site in Lekkerkerk (The Netherlands) with inadequate monitoring and record-keeping. In the 1970s, 250 houses were built on the site. Householders became ill and following an investigation the estate was evacuated costing (at 1981 rates) UK£156 million.

Gulf War: Kuwaiti oil well and oil storage sabotage

During the conflict Iraq mined many of Kuwait's 300 to 400 oil wells and subsequently ignited them. Roughly 500 million barrels of oil were burnt and huge quantities contaminated the Persian Gulf. A cloud of smoke about 600 × 900 square kilometres formed at about 1 km altitude by February 1991. The shading caused by smoke may have added to normal chilling winter conditions and cooled the gulf seawater by 10°C or more; this, plus the toxic effects of the pollution, killed fish, plankton and other wildlife and may have affected reproduction of many creatures and the growth of important sea-grass beds (Sheppard and Price 1991). Dugong (marine animals) may be affected by the reduction in weed growth in the Gulf. The damage is most evident at the top of the food-web. Still in 1993 there were massive losses of seabird nestlings. In effect the Gulf has been the victim of the world's largest (wilful) oil spill.

Reports of soot-stained snow came from Iran, Pakistan and Kashmir (3000 km away from the conflict) and even as far away as Japan there were reported traces of pollution (Barnaby 1991). The soot was rich in sulphurous compounds and nitrous oxides and quite acidic; certainly there has been local contamination but there are few firm data on the impacts (especially longer-range effects) (Warner 1991; Bloom *et al.* 1993). Some areas of Kuwait are now darkened by soot deposits and experience increased surface temperatures in summer as a consequence of altered albedo.

Enough CO_2 and other greenhouse gases were released by the burning oil to negate a good deal of the world's global warming remedial efforts (Warner 1991).

Acid deposition

Pollutants can be deposited gradually or suddenly, close to the source of pollution or much further afield. This deposition may cause the environment to convert to more acid conditions, or accelerate a natural trend toward such a state, and is known as acidification (see Box 10.2). The expression 'acid rain' is often used; however, it is not a particularly good term as deposition can be from the air in rain, snow, mist or cloud droplets ('wet deposition'), or it may be through settling of dust, aerosols or gases ('dry deposition').

Acid deposition has been recognized within and around DC industrial and urban areas for some time; Robert Angus Smith used the term 'acid rain' in 1852 following an air pollution survey of Manchester. It is no accident that poor people generally settled on the eastern 'downwind' side of UK industrial towns which had or have greater deposition of pollutants. In the 1950s Mackereth recognized a problem of acidification of UK upland lakes and blamed it on acid rain. Diatom analysis of UK lake sediments in the 1970s showed acid conditions (remains of acid sensitive diatoms decreased as the lakes acidified) together with soot particles from coal fires after the 1800s. During the 1960s Scandinavian scientists had begun to link acidification of their waterbodies with acid deposition from Europe and the UK. Scandinavia not only lay in the path of the drifting pollution, but also had soils and waters that were vulnerable. At the 1972 UN Conference on the Human Environment

Box 10.2 Acid deposition and acidification

It is generally accepted that acid deposition is that which has a pH of less than 5.1 (Elsworth 1984: 5). Uncontaminated rainfall and other forms of precipitation are usually slightly acid with a pH typically about 5.6. A pH of 7.0 is neutral; anything below that is acidic. It should be noted that the pH-scale is not arithmetic: a slight change in pH can mean many times greater or lesser acidity, e.g. pH 4 is 10 000 times more acid than pH 8.

Critical load a concept developed in the UK and widely used as a scientific benchmark (the amount of acidity that a habitat can absorb without significant damage). As different soils vary in their response to acid deposition, a way of defining and comparing impact is vital (for a review of critical load concept see Bull *et al.* 1992). Critical loads can be used to model future acid status, given predicted reductions of SO_2 pollution.

Occult deposition sometimes applied to cloud and mist-derived pollution.

NO_x nitrous oxide (N_2O), nitric oxide (NO) and nitrogen dioxide (NO_2) are sometimes grouped as NO_x: of the three, it is NO_2 that is active in photochemical smog formation.

Some soils or waterbodies can withstand more acidification than others without damage. They are said to have some ability to buffer the pollution; the buffering may be due to alkaline material within or due to alkaline water reaching them from underlying basic (alkaline) rocks. Scandinavia has proved vulnerable because the soils and lakes tend to have little buffering capacity. Soils are generally less vulnerable than streams and ponds. In temperate and colder environments soils over slow-weathering, non-alkaline bedrocks such as granite, weathered glacial material, or sandy material are more likely to be badly affected.

in Stockholm, Sweden voiced concern over lake and stream fish kills which it attributed to acid deposition.

By the mid-1980s precipitation of pH 3.0 was not uncommon in Central Europe and some in the UK was recorded at 2.3. On average precipitation in the UK is now roughly 10 to 70 times more acid that it would be if there were no pollution. By the early 1990s western Europe, parts of North America and other badly acidified regions had suffered serious losses of freshwater fish, amphibians, and other acid-sensitive plants and animals. Acid deposition reduces the life and increases the maintenance costs of buildings, infrastructure, etc, and (as discussed on pp. 251–2) may make pollutants mobile and more of a hazard.

It is widely agreed that the main cause of acid deposition are emissions of sulphur dioxide (SO_2) and nitrogen oxides (NO_x) from combustion of fossil hydrocarbon fuels, metal smelting and biomass burning. Ammonia emitted from sewage and livestock manure and the de-nitrification of nitrogenous fertilizers also contribute. These pollutants form sulphuric and nitric acid when in contact with moisture. Volcanic eruptions, sea-spray, weathering of gypsum and gas emissions from forests, grasslands and marine plankton can lead to acid deposition (North Sea phytoplankton have been found to emit sulphur-bearing gas in large quantities in summer, enough to contribute to acid deposition, and enough sulphate particles to probably play a role in cloud formation – the latter may perhaps be a significant factor in global climatic change).

By 1988 about half of the sulphur in the Earth's atmosphere could be attributed to human activity, but the distribution was not uniform: over Europe the anthropogenic component would probably have been about 85 per cent and over the USA about 90 per cent (Rodhe and Herrera 1988: 11). Greenland ice cores show a two- to three-fold increase in sulphate and nitrate deposition during the last 100 years mainly attributable to acid deposition.

Most years the world's volcanoes vent less SO_2 than the UK's power stations did in 1987 but some eruptions release huge amounts of SO_2; for example when Mt St Helens erupted in 1980 for a while it more than doubled global levels of SO_2 (Barrow 1991: 540). Such eruptions probably have considerable effect on winter temperatures, at least for a few years after; human SO_2 emissions maintain a higher background level of acidity which has more significance in terms of acid pollution than climatic cooling. Elsworth (1984: 6) suggested that roughly 70 per cent of acid deposition was due to SO_2 pollution (much of that produced by combustion of coal), and roughly 30 per cent due to nitrogen compounds (nitrogen dioxide NO_2 and nitric oxide N_2O mainly).

The 'wet' winter smogs common in the UK and western Europe before the 1960s have been much reduced by pollution control legislation but they are still to be found in cities of eastern Europe, CIS, industrial regions of the PRC and India. Where city traffic is heavy, especially where vehicle exhaust controls are lax, emissions can undergo photochemical reactions and are a problem, particularly when conditions are still and sunny (Athens had such a smog or *nefos* in summer 1984 which hospitalized over 500 people). Peroxyacetic nitrates (PANs) are a component of most photochemical smogs; they are short-lived compounds that are toxic to organisms and damaging to construction materials, clothing, etc. PANs form when air polluted with NO and NO_2 is exposed to sunlight (tetra-ethyl lead may assist in the reaction). There is a possibility that some UK oil-fired power stations will convert to orimulsion (a sort of water and bitumen mix), a cheap alternative to oil that yields a lot of sulphur dioxide. If these stations do not have flue gas desulphurization equipment there could be a significant downwind acidification.

Whilst urban air pollution and smog are reduced in western Europe and the USA, increased electricity consumption since the 1950s has led (except where generation does not burn hydrocarbon fuel) to large power stations which together with industry release pollution from high smokestacks. The result is wide dispersal of pollutants and trans-boundary acid deposition damaging streams, lakes and shallow seas (like the Baltic), fauna and flora, soils, forests, crops, groundwater, and human-made structures. Scandinavia, western and central Europe, northeastern USA and the CIS have serious problems.

Even in regions which generate little pollution themselves, wildlife, agriculture and historic buildings now suffer and maintenance of human-made structures becomes more difficult (a classic infringement of the 'polluter pays principle'). In many regions freshwater fisheries, notably salmon and trout stocks, in Scandinavia, Europe and the USA have suffered. Other wildlife has suffered, particularly sensitive birds like the European dipper, amphibians, snails and many plant species, especially conifers. Whether acid deposition has had significant effect on oceans is uncertain, although brackish seas like the Baltic are damaged. That acid deposition is affecting the boreal forests should be a cause for concern, as they play a significant role in the global carbon cycle.

Until recently acid deposition was a problem of Europe and northeastern America; it is now becoming much more widespread, one reason being the increased combustion of coal in industrial areas of LDCs, so much so that Park (1987: xii) commented 'acid pollution is one of the most serious problems in developing countries'. Japan, North and South Korea, southern China (Rodhe *et al.* 1992), southwest India, parts of Brazil, Venezuela, Nigeria, South Africa, parts of Australia and Malaysia now have acid deposition problems. Northern polar regions have long been affected, including the high Arctic; these areas receive acid deposition in the spring, as polluted air moves in (and is visible as atmospheric 'Arctic haze' and as soot and graphite particles in the snow), in summer frontal systems tend to prevent such airmass movements. Arctic pollution has been known for over a century and includes aerosols, soot, dust, acidic material, pesticides, heavy metals and radioactivity (Heintzenberg 1989). The sources are distant – Eurasia, Europe and North America – so remedial action is unlikely to attract much support in the short term. A possibility is that the haze will trap more solar radiation and warm the Arctic environment enough to raise problems. Slow-growing lichen and mosses may accumulate decades of pollution and grazing animals that crop these plants then get heavy doses of pollutants (Soroos 1992). Tundra vegetation appears to be very vulnerable to acid deposition damage.

It is possible to map areas of vulnerable soil and vegetation and to superimpose forecasts of future acid deposition (Barrow 1991: 65): large areas of tropical Latin America, especially Amazonia, have soils already acid and with high concentrations of aluminium which could become toxic to plants if subject to acidification. Tropical upland forests which intercept mist and rain are probably vulnerable to acidification; conservation managers should monitor rainfall so as to be warned of acidification.

Fig. 10.10 gives the global picture (early 1980s) (NO_x deposition patterns were similar). Clearly many regions well away from industrial development or large urban areas have at least occasional acid deposition.

Acid deposition degrades the environment and its organisms in a number of ways; it may:

1 Damage plants and animals directly.
2 Alter soil chemistry or structure.
3 Alter plant metabolism.
4 Alter metabolism or species diversity of soil micro-organisms leading to change in fertility or soil chemistry.
5 Damage artificial and natural structures.
6 Mobilize compounds in soils, waste dumps and water (notably phosphates, heavy metals, aluminium).

The impact of acid deposition is variable: some localities may be exposed to prevailing winds or may get a localized intense storm of acid rain or may suddenly be overwhelmed by snow-melt in which

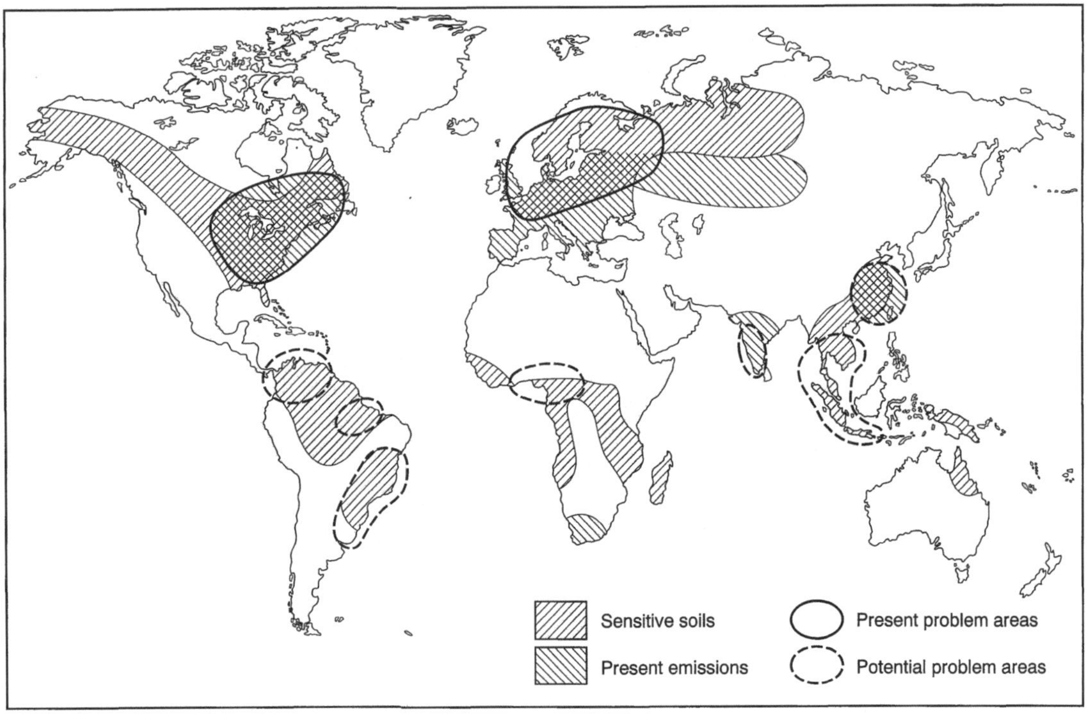

Fig. *10.10* Schematic map showing areas that currently have acidification problems and areas, where, given their sensitive soils and expected acid emission patterns, acidification is likely to become a problem ⟨*Source:* based on Rodhe and Herrera 1988: 27 (Fig. 1.8)⟩.

there is the accumulated deposition from a whole winter. Soils vary in buffering capability and those which receive a dressing of ammonium-rich fertilizer, which tends to decrease soil pH, are also at more risk; indeed in some cases it may be fertilizer application that is the primary cause of acidification (Kennedy 1992). Alkaline soils may sometimes become more fertile if there is acid deposition. It is possible to map areas at risk from acid deposition. Soils tending toward acidity and those rich in aluminium or heavy metals post particular problems. Many tropical soils are quite acid and contain a lot of aluminium (for example large areas of northwest Amazonia and South East Asia); only slight acid deposition may raise the pH above 6.0, mobilize aluminium leading to plant growth problems and possibly soil crust or pan formation (Moran 1987; Rodhe and Herrera 1988: 24; Rodhe 1989). The vegetation of tropical uplands, like those of South East Asia, seem particularly vulnerable (Bhatti *et al.* 1992). In 1993 the World Bank started funding an Asian Acid Rain Monitoring Project.

Maps of acid damage to soils in the UK have been prepared by the Institute of Terrestrial Ecology and show more serious damage in the south. Acid deposition may increase the mobility of aluminium, lead, and other compounds in soil (and possibly raise content in crops) groundwater and streams posing a public health risk. The UK already has a problem with old lead domestic water pipes; further rise in acidity of domestic water supplies will increase the risk to consumers, especially young children. (Water companies are now removing lead piping; the cost is passed on to the consumer.)

Often more than one process is operating, for example, soil may have H_2SO_4 forming within it as 'dry' acid deposition products are broken down; at the same time there may be an input of 'wet' acid rain. Pollutants like tropospheric ozone may also be present so that attributing blame to acidity may not be straightforward.

Acid deposition can have direct, secondary or cumulative effects. By the time there are obvious signs of acid deposition, much damage to sensitive components of the ecosystem will already have been done. There are likely to be 'thresholds' or levels of acid deposition that, once exceeded, result in severe effects, although there was little apparent problem before that level was reached; in order to establish reliable scientific criteria, the concept of critical load has been developed. Those seeking to counter the problem should be aware of these thresholds.

Acid deposition damages vegetation but the process can be difficult to unravel and may vary even from plant to plant of the same species. Injury may be in the form of visible damage to plant structures (leaves, buds, etc), or there may be a decline in productivity or ability to reproduce, or increased vulnerability to some other difficulty: frost, insect damage, drought, other pollution, etc. Plants may not be damaged directly; it may be that symbiotic bacteria or fungi are affected and a plant then has less support in its quest for nutrients. If it is winter and broad-leaved trees have no leaves, they intercept little acidity; if it is spring and there are delicate shoots and leaves, damage may be great; the annual loss of broad-leaved tree foliage may cause episodic increases in acidity as leaves rot and release a year's accumulation of pollutants. The efficiency of interception by conifers and broad-leaved trees also differs because leaf or needle shape differs. Broad-leaved trees in temperate environments lose leaves annually and rid themselves of leaf pollution more effectively than conifers; also broad-leaved trees may sometimes shed leaves and regrow new ones after a severe pollution episode.

If deposition is moderate to high, the point will probably come where leaf-shedding makes little difference because acid accumulates in the soil. But if deposition is slight to moderate and episodic, different soil micro-organisms in broad-leaved and coniferous forests will probably vary in ability to counter acidity build-up. Possibly conifers have suffered more because they grow in exposed positions where they collect pollution especially well, where thin soil and extremes of temperature place them under some stress, which pollution then renders excessive.

Lichens are good indicators of air quality, because they rely on dissolved solids in rainwater for nutrients: they are thus directly affected by any increase in acidity (plants growing in the soil are likely to enjoy more 'buffering' from chemicals in the soil which may counter the acid deposition. In general lichen species diversity decreases as SO_2 levels increase. Even in the non-industrial areas of the UK acid-tolerant lichens are increasing and acid-intolerant ones are decreasing (Park 1987: 97). The effect of acid deposition on tropical epiphytic (parasites like lichens, ferns, bromelliads, orchids, etc) or acid sensitive plants will probably be considerable (Fig. 10.11).

Fig. *10.11* Epiphytic plants, and other plants growing in already acid and nutrient poor soil and those which require alkaline soil are vulnerable to acid deposition. Pitcher plant growing in acid soil in an area likely to suffer acid deposition as a result of industrial development (near Ipoh, Malaysia).

Tree die-back syndrome, or *waldeschäden* or *waldesterben* ('forest death') was noted first in the 1960s in the Black Forest (Germany), and in the 1970s in mountain regions of western Germany and the Alps. Since about 1981 it has been widely held that acid deposition is to blame and the problem seems to be spreading. German tree-ring studies suggest that the problem has increased between 1960 and 1985.

Initially, die-back is usually patchy; as conifers become 'sick', it is quite common to have increased growth of lichen on their leaves and branches. The vulnerability of trees (and possibly other types of vegetation) seems to depend on many factors which include: position, altitude, soil, moisture availability, i.e. the effects of die-back are synergistic (Barrow

1991: 63). Die-back takes place on very different soils and under quite different climatic conditions; also it seems to correlate in Europe with recent increase in the corrosion of glass in churches so acid pollution does appear a likely trigger for die-back. However, Park (1987: 110) observed that in some areas trees apparently suffering die-back had SO_2-sensitive lichens flourishing upon them, apparently contradicting assumed causes.

Some conifers are more susceptible than others: in Europe silver fir (*Abies alba*) seems to be the first affected, then Norway spruce (*Picea abies*); a ranking of sensitivity has been established. Broad-leaved trees appear to be less susceptible, but in Europe and North America they are increasingly showing signs of die-back. In Europe beech (*Fraxinus* spp.) and oaks (*Quercus* spp.) are generally the first broad-leaved trees to show signs of damage.

It is difficult to assess the costs of forest die-back: the process can be slow (taking up to 40 years) and is underway in many areas without having yet manifest itself; nevertheless the problem is very serious and is increasing. In 1984 about half of the EEC's agricultural land was under forest. Forests in Europe (including those in Germany, The Netherlands, Switzerland, Austria, Czechoslovakia and the UK), the USA (especially Vermont) and the CIS have already been widely affected and in Germany, Finland, Sweden, France, the costs in terms of lost timber and need for replanting are already considerable. In 1988 the cost of acid deposition to Scottish foresters was estimated to be roughly UK£25 million (Milne 1988: 56). Elsworth (1984: 18) estimated over half of Germany's coniferous forests were showing signs of die-back, and about 560 000 ha were 'devastated'. By 1985 the damage was probably costing the former West Germany roughly US$300 million a year (McCormick 1988: 27).

Natural acid deposition can damage vegetation, for example in Nicaragua in the 1950s volcanic emissions caused quite widespread forest damage (Rodhe and Herrera 1988: 126). Non-ferrous metals smelters have a particularly bad acid deposition record: the Trail Copper Smelter (British Columbia), Anaconda Aluminium Smelter (Montana, USA) and Sudbury Copper Smelter (Ontario) have caused widespread vegetation damage. Recently a number of aluminium and other metal smelters have been established in Amazonia, some using charcoal; the impact on the environment is yet to be established but could be considerable.

Fears are growing that if acid precipitation is not checked, vast areas of soil, particularly those with a weak buffering capacity, may suffer degradation that could affect yield potential and wildlife. Soil acidification begins when rate of acid input exceeds the rate of neutralization resulting from weathering. Acid pollution of soil tends to have the following effects:

1 Increasing acidity slows down breakdown of litter (in the soil or on the surface) by soil organisms.
2 Increasing acidity interferes with plant roots' absorption of soluble material (moisture and nutrients).
3 Cations (K^+, Ca^{++}, Mg^{++} and Na^+) get leached out of the soil to be replaced by hydrogen and aluminium ions and toxic heavy metals are likely to be mobilized.

The effect of acid deposition on peatlands is still uncertain. The South Pennines (UK) blanket peats have suffered acid deposition since the 1750s. One effect has been the decrease of the plants vital for peat formation, notably *Sphagnum* spp. The number of *Sphagnum* spp. has generally declined in the Pennines, and as a consequence of the decline there has been widespread erosion (in part also due to trampling by humans and livestock and occasional fires). The increased acidity of these UK uplands has led to severe reduction in species diversity of higher plants as well as bryophytes and consequently the domination of the vegetation by two species: *Eriophorum vaginatum* and *Vaccinium myrtilius* (Usher and Thompson 1988).

Separating crop damage due to acidification of soil from that directly caused to crop plants by acid deposition, low altitude ozone attack and a multiplicity of plant diseases, pest damage, various pollutants and environmental factors (such as drought) is very difficult. It may well be that acid deposition has made a crop more vulnerable to other things. There are nitrophile plants that like acid soils; it is also possible that high acidity and atmospheric sulphur could, as many urban UK rose growers are aware, discourage crop pests like aphids and fungal diseases. Wellburn (1988: 52) reported that cereals and grasses might benefit from *slight* acid deposition (but once levels rose above 60 ppmv SO_2, species like the European grass *Lollium perenne* L. began to suffer).

Some crops are particularly vulnerable to acid deposition, including rye, salad vegetables, barley, oats, wheat, tomatoes, apples and pears. Elsworth (1984: 48) and McCormick (1985: 27) have made crude estimates of the value of crop losses due to acid deposition; the latter suggested that it was costing Europe US$500 million a year (in 1985).

Some soil scientists believe that natural soil processes and agricultural management practices play a much more important role than atmospheric pollution in acidification. Even if that is the case, acid deposition is one more unwelcome pressure on the environment. In 1993 attention seems to have swung away from acidification to global warming: this diversion of interest and resources is probably mistaken.

Pollution avoidance, controls and mitigation

Pollution can be countered in a number of ways:

1 *Risk assessment studies* a process of seeking to ascertain in advance what logical consequences of a problem or accident would be, and an effort to judge the probability of malfunctions or accidents; it may permit the preparation of contingency plans, installation of warning devices, safety equipment or redesign of potential polluting equipment to minimize risk.
2 *Health and safety at work measures* the workforce, employers and possibly unions monitor and act to avoid and remedy problems.
3 *State inspection and legislation* governments or commissioned specialists monitor industry, services, transport, etc, and enforce avoidance or remedial work.
4 *Consumer protection* government and often NGOs monitor and act (or lobby governments and industry to act) to avoid or remedy pollution and waste problems.
5 *Individual action* individuals and, where law allows, groups of individuals act through media statements and courts to restrain polluters.
6 *Market forces* it may pay polluters, in terms of marketing, to be seen to be 'clean'; it can be profitable to recover wastes, and it could cost less to avoid pollution and waste than pay to dispose of it or meet fines (unfortunately fines today are often set too low). The USA appears to be taking this route: US power stations can now sell or buy sulphur dioxide rights ('SOxcreds'), which allows stations with cheap sulphur rich coal to buy credits from a station that finds it cheaper to fit pollution control equipment or which has clean energy sources. It could be a problem if pollution ends up moving to areas where it can do harm: the market forces make sense to polluters and allows national targets to be met but could fail to regulate pollution at regional level.

Pollution control requires agreements: there must also be 'teeth' (means of enforcement and adequate penalties), standards (to judge each situation by a common yardstick) and monitoring; without these, efforts are likely to be in vain. It is also necessary to have alternatives to the pollution or waste generating activity or cost-effective technology to make it possible to counter the problem. In the UK, legislation since the 1950s has largely cured local acid deposition and some other waste and pollution problems. To cure regional and international acid deposition and other waste and pollution problems demands cooperation between nations; progress toward controls has tended to be slow, even when a threat is apparent and there is relatively little sacrifice required to counter it. However, there have been a number of promising agreements to control pollution and waste made between 1972 and 1990:

- 1954 International Convention for the Prevention of Pollution of the Seas by Oil.
- 1972 Convention on the Prevention of Marine Pollution by Dumping Wastes and Other Matter (London Dumping Convention).
- 1973 International Convention for the Prevention of Pollution from Ships (MARPOL Convention).
- 1975 Initiative on Acid Pollution (President Carter) USA.
- 1979 International Convention on Long Range Transboundary Air Pollution ('Geneva Convention') (West Germany not very supportive).
- 1979–80 Convention on Long-Range Transboundary Air Pollution, Geneva (Norway and Sweden active in organization: 32 countries sign, including USA and the EC). In effect the first anti-acid deposition agreement.
- 1980 Canada and USA sign Memorandum of Intent on Transboundary Air Pollution (a commitment to joint research).
- 1980 OECD gives independent confirmation of Scandinavian charges.
- 1982 Multilateral Conference on Acidification of the Environment (Stockholm) – recognized pollution and noted that there were solutions available – proposed '30% Club' (supported by Canada, Austria, Switzerland, West Germany, Sweden, Denmark, Norway and France).
- 1983 West Germany tries to get the EC to insist on catalytic converters on all cars built after 1986 and introduction of unleaded-petrol.

- 1983 Two-year moratorium on dumping of low-level radioactive waste into the sea, made permanent in 1988 (the London Convention; – 27 nations signed but many, including the UK, USA, CIS and France, did not). This became a permanent ban on nuclear dumping at sea in 1994.
- 1983 Cooperative Program for Monitoring and Evaluation of the Long Range Transmission of Air Pollution in Europe (EMEP). Signatories undertake to develop, without delay, policies and strategies to reduce pollution.
- 1984 Ottowa Conference: the '30% Club' formed (10 members pledged to reduce SO_2 emissions to at least 30 per cent of their 1980 levels by 1993).
- 1985 Helsinki Protocol – the '30% Club' has 21 signatories (in 1982 32 had expressed support); UK, Greece, Eire, USA, Poland and France refuse to join '30% Club' (for a range of reasons). Strengthened the 1979–80 Convention on Long-Range Transboundary Air Pollution.
- 1988 EC *proposal* for two-stage conversion to lead-free petrol by 1989 for all new and 1991 for all vehicles and to adopt US Emission Control standards in 1995.
- 1988 Sophia Protocol – seeks to freeze nitrogen oxide emissions at 1973 levels.
- 1989 Basle Convention on the Control of Transboundary Movements of Hazardous Wastes and their Disposal – intended to reduce shipments of hazardous waste to LDCs. The Convention was not binding and a little later the EC pledged not to export waste to 68 LDCs.
- 1989 International Convention for the Prevention of Pollution from Ships. By 1991 accepted by 41 nations, including most DCs, but representing only about 60 per cent of world's shipping. Annex V of this Convention seeks to control releases of waste garbage from ships, in particular plastics debris. Unfortunately this Convention does not cover lost or discarded fishing nets.
- 1990 Hague agreement to reduce water-borne and airborne waste dumping in North Sea.

Despite these agreements, some problems will prove difficult to counter, for example lost fishing nets do a great deal of damage to wildlife but it would be difficult to make them degradable as users would be unable to afford frequent replacement.

Pollution control need not cost huge sums of money: often recovery of by-products gives a lucrative return. Building and servicing pollution avoidance or reduction facilities provides employment, and control is generally much cheaper than the consequences of pollution and clean-up efforts. Already pollution control and avoidance have become big business in DCs.

There is a need for standards that inform of safe levels and practices and which allow objective comparison. Hazard warning markings on materials could be improved. The UK in 1977 adopted a system of 'hazchem' markings that indicate the nature of the material and the safety measures that it requires. This has proved very useful but it still has shortcomings. Better liaison between multinational and governmental bodies and NGOs should improve pollution avoidance and control. An example of a liaison aid for NGOs is the International Organization of Consumer Organizations. Trade unions can play an important role in educating workers in health and safety: 'empowerment' informing workers so they can act. The 85 signatories of GATT undertook in 1985 to try to restrict exports of hazardous materials; such an undertaking is important, for GATT agreements have the potential to do much to hinder pollution controls.

The UNEP (Geneva) maintains an International Register of Potentially Toxic Chemicals (IRPTC) which holds details of characteristics, and the laws and standards of various countries. The IRPTC incorporated data from the OECD Chemicals Programme and shares data with the EC's Environmental Chemicals Data and Information Network (ECDIN); which cooperates with the WHO. Also based at Geneva and managed by the WHO is the International Programme on Chemical Safety (IPCS), which runs a database and issues guidelines on the transport, use and disposal of chemicals. In 1992 UNEP published an Earthwatch booklet to try and increase governments' awareness of the risks of chemical pollution (UNEP 1992). In 1986 the US Government set up an advanced pollution releases reporting system, the Toxic Release Inventory (TRI), which collects information on toxic chemical releases/escapes and assists in the identification of pollutants.

Once wastes, pollutants or precursors of pollution have been released, there can be little control, only mitigation (with compounds like heavy metals mitigation can be difficult; see next section and Fig. 10.12).

Fig. 10.12 Cast-netting for fish, Tocantins River, Brazil. Already affected by the Tucuruí Dam, fish stocks are likely to suffer further as agriculture and industry develop in the region. The greatest threats are mercury pollution from gold mining and careless use of pesticides.

Case study: gold-diggers and mercury pollution in Brazil

Garimpeiros are independent gold-diggers who work in small groups and total as many as 650 000 to 850 000 in Brazilian Amazonia (Greer 1992; *South* March 1989: 12), perhaps a million in Brazil as a whole. Since the mid-1970s there has been a gold 'rush' in Brazil and *garimpeiros*, excavating mainly alongside the streams or sorting shallow alluvial deposits from floodplains or gathered from rivers using barges or divers, have caused widespread mercury pollution which threatens aquatic and floodland environments and human health in Amazonia, the Pantanal wetlands (southwest Brazil) and in the Serra Pelada region of Pará (northeastern Brazil).

The mercury is used to trap and to separate gold from the debris. As over 80 per cent of the gold is sold to unofficial dealers, accurate estimates of production and thus a good idea of pollution levels is impossible; Malm *et al.* (1990) felt that it is likely that between 1979 and 1985 over 100 tonnes of mercury got into the Madeira River alone, and then on into the rest of Amazonia (De Lacerda *et al.* 1989; Smith *et al.* 1991). Greer (1993) suggested that 90 to 120 tonnes a year were discharged into Amazonian streams (especially the Madeira and Tapajós), which seems an underestimate in view of the previous assessments.

Mercury accumulates in organisms and threatens the health and long-term survival of fish, wildlife and humans. Creatures feeding well up the food chain are especially at risk. Greer (1993) called it 'the paupers' poison' because the miners are exposed to the fumes and because fish accumulate the poison and the poor cannot afford to switch to other sources of protein. In some states (e.g. Rondônia) people are now avoiding some of the fish that they traditionally consumed. Samples of river water, hair from people living alongside rivers and fish tissue have shown frighteningly high levels of mercury. Toxic methyl mercury will be released from sediment (some of which have 1500 times the normal background levels) by microbial action for decades even if mercury use could be immediately controlled (Martinelli *et al.* 1988; Greer 1993). Should the environment change, perhaps through acid deposition, there may be increased mercury release from sediments – a 'chemical time bomb' situation.

The drug trade probably 'launders' some of its profits by buying gold; the miners are driven to mine by poverty and hope, so controlling gold mining may prove difficult. There may be suitable alternatives to mercury: Smith *et al.* (1991: 314) noted that in Colombia in colonial times gold-miners used a foaming plant sap to capture gold particles; perhaps it is worth promoting such a compound or a synthetic as an alternative to mercury? Provided it is effective, and proves cheap enough, the reduced health risk to miners should ensure that it spreads fast. Recent studies by researchers from University College, London, aim to provide ways of recovering mercury from the flumes and when the gold–mercury amalgam is heated, so reducing escapes to the environment (Coghlan 1994). Initially miners may need to be educated and possibly supported to change equipment and techniques.

Pollution mitigation

Pollution control involves costs and perhaps the forgoing of presently enjoyed benefits, at least in the short term. However, there may be by-products or benefits resulting from control which, in addition to reduced pollution, help to compensate (and may sometimes even more than pay for pollution control or waste reduction).

Pollution mitigation may be achieved by siting an activity, for example a factory, away from housing or crops. This is no solution to pollution but it can lessen the impacts. Successful siting depends on having an effective planning authority and enforcement of landuse zoning.

Much pollution and waste control has been and remains a *laissez-faire* process which depends on the polluter being willing to conform. In the USA a major move away from that approach was the passing of the 1970 National Environmental Policy Act (NEPA) and with it the setting up of an Environmental Protection Agency. A number of countries have subsequently made similar provisions.

There does seem to have been some progress in control of acid deposition. In western Europe emissions of SO_2 peaked between 1975 and 1978 and are at present declining as a consequence of vehicle and industrial emission controls and economic recession, which has cut industrial fuel use. In eastern Europe the situation is less encouraging, with ageing industrial plant, heavy use of sulphur-rich coal and little money to introduce pollution control improvements.

Countries like the UK have abundant opportunities for further emission controls which could reduce atmospheric pollution considerably if funds are made available to support them (see next section). Levels of building insulation in the UK and some other DCs are very poor: it would be quite easy to improve this through better building standards, provision of grants for uprating, tax incentives or penalties. In the case of the UK, upgrading all domestic housing to the latest building standards would save vast amounts of energy, would make long-term sense and should be affordable.

In most DCs a lot more could be done to encourage the use of public transport and discourage use of private cars: this too could be easy to achieve. In LDCs there has often been a trend away from traditional building methods which maintained comfortable temperatures to less suitable buildings and energy-wasteful air-conditioning.

Ways of reducing sulphur and other emissions

There are four main methods discussed below; the second, third and fourth options will not reduce carbon dioxide emissions:

1 Burn less hydrocarbon fuel.
2 Treat hydrocarbon fuel.
3 Burn fuel in such a way that emissions are reduced.
4 Remove pollutants after burning from exhaust gases.

Burn less hydrocarbon fuel Fuel consumption can be reduced by improving efficiency of energy use through cutting energy losses (insulating homes and other buildings, recovering waste heat, improving furnaces, boilers and motors, through the use of alternative sources of heat and power) or cutting energy demands (people adapt demands in response to change in attitudes, taxation or legislation). In the UK huge improvements could be made to energy conservation, particularly through better domestic insulation and low energy appliances (Fig. 10.13). Alternative energy sources include hydropower, nuclear power, tidal power, wind power, hydrogen, alcohol, geothermal energy, exploitation of differences in ocean temperatures, solar energy. Use of gas turbines and magnetohydrodynamic (MHD)-generation may prove more efficient and less polluting than present boiler–steam-turbines–generator installations.

Fig. *10.13* Low energy lightbulbs,. UK retail price (1993) approx. £15 for 60 W size, cf. £0.60 for conventional 60 W bulb. Expected savings of 75 per cent energy and a life for the bulb of eight times the conventional type. Adoption unlikely to be fast unless price is much reduced ⟨*Source:* Department of the Environment (UK) Energy Efficiency Office (March 1993) and Miss Alessandra Muto⟩.

Nuclear generation might be made much safer if thorium 233 were used as a fuel. The proposed thorium nuclear reactor (under study at CERN) would require a beam of protons to sustain fission which means instant control and virtually no chance of a Chernobyl-type accident, nor the dangerous by-product plutonium with its pollution and arms proliferation risks (*The Times* 5/12/93: 5).

Treat hydrocarbon fuel Remove pollutants before burning: coal and oil can be treated to remove some of their sulphur content (coal may have as much as 5 per cent sulphur and oils up to 3 per cent). Sometimes simply crushing and washing coal with water considerably reduces pollution and, because there may be less ash produced and a better burn, this may be quite cost-effective. Sulphur can be removed from oil using solvents (this is done in Japan) or by biotechnology-enhanced leaching (see chapter 8). 'Smokeless' coal-derivatives like coke have long been available (however, care should be taken to ensure that there is no pollution during treatment to produce the cleaner fuel).

Burn fuel in such a way that emissions are reduced New furnaces are designed to ensure better combustion of coal or oil; the fluidized-bed combustion furnace is one form. Car engines may be of the lean-burn type, in which a sparse fuel mixture is more fully burnt and better control of fuel–air mixture can markedly reduce pollution. With furnaces and engines many pollutants result from *incomplete* combustion of hydrocarbons.

Remove pollutants after burning from exhaust gases Air can be injected into the hot vehicle exhaust gases to ensure fuller combustion and the gas stream can be passed through a catalytic converter. To use a catalytic converter, fuel must be free of lead tetra-ethyl 'anti-knock' additive (i.e. requires lead-free petrol). From 1993 all new cars sold in the EU must have a catalytic converter fitted; as long as these are well maintained they effectively cut nitrogen oxide emissions and some sulphur. Catalytic converters can also be fitted to factories or power stations although it may not be cost-effective to retro-fit to old installations.

Furnace injection consists of adding a compound before or during combustion. One method uses urea solution, another ammonia. The result is reduced emission of NO_x and SO_2. Furnaces (power stations, heating plants, smelting plants, waste incinerators, etc) may have their flue gases 'wet scrubbed'; that is, a spray of water, lime or chemicals removes pollution which is collected as a slurry for sale or disposal, preferably where there will be no land degradation. 'Dry scrubbing' is the addition of crushed lime to fuel or the exhaust or the use of ammonia sprays to remove pollution.

Flue gas desulphurization (FGD) by injecting crushed lime with the fuel or into the flue gases (limestone-gypsum process) effectively removes sulphur dioxide. The processes require supplies of limestone and disposal of gypsum, its by-product; some can be sold to the construction industry, but the UK Drax Power Station (4000 MW) alone produces about 2000 tonnes a day (in addition to huge amounts of pulverized fuel ash): disposal from all generation therefore presents a challenge. Some methods of dry scrubbing produce a by-product that can be used as fertilizer. FGD is widely and effectively used in Germany and Japan. Flue gases may also have pollutants removed using electrostatic precipitators or fine filters.

Some pollution control technology has long been available: Battersea Power Station (London) had flue gas desulphurization facilities in 1929 that were up to 90 per cent efficient.

Costing pollution control

There may be situations where it is cost-effective to apply alkaline material (e.g. agricultural lime or magnesium salts) to the land, ponds or streams to counter acidification (locally this may damage lime-sensitive species). It may be possible to adapt landuse to acidification: select crops, fish and tree species that resist acid soils and waters, high aluminium levels, etc. It is, however, remedial action, even at best not a permanent or complete solution. Ultimately, a combination of national/international pollution control and local remedial action and monitoring is required (this approach is needed for many other aspects of environmental management).

The main problem in controlling acid deposition has been that of cost; other pollution problems may be complicated by inadequate knowledge and lack of control or treatment techniques. When technology is available there are often limited resources to spend, and when pollution and waste presents a transboundary or even global threat there is some incentive for the costs to be borne by more than one nation.

International agreement on sulphur emissions has made some progress: 34 countries signed the 1979 (Geneva) Convention on the Control of Transboundary Transport of Polluted Air. Further signatures in 1982 brought it into force. A statement of principle, rather than a set of rules, the Convention issued proposals in 1983 for its signatories to each limit pollution to 30 per cent of their 1980 levels. In 1984 the signatories met and some made the '30% Club' a binding international agreement on reduction of sulphur emissions; 21 countries signed that agreement at Helsinki in 1985 (14 hesitated, including the USA, the UK and Poland, mainly for economic reasons). There are currently (1994) negotiations to add a sulphur protocol to the UN–EC Convention on Long-Range Transboundary Air Pollution and to set targets for reducing SO_2 emissions in Europe and North America by AD 2000 to 50 per cent of 1980 levels for each country; in some cases targets may need to be relaxed as they receive sulphur emissions from neighbouring nations. Nevertheless, there has been progress, for example the UK has had falling SO_2 emissions for some years.

In 1993 the US Environmental Protection Agency sold rights to emit sulphur dioxide to power companies. This raised money for pollution control and gave purchasers 'allowances'; if they exceed those allowances, fines of up to US$2000 per tonne of SO_2 can be imposed or the company can try to purchase extra allowances at the market price. These measures are felt to be an aid in encouraging the fitting of pollution control measures.

References

Ayres, R. and Rohatgi, P. (1987) *Bhopal: lessons for technical decision makers.* Laxenberg, Austria, International Institute for Applied Systems Analysis

Barke, M. and O'Hare, G. (1984) *The third world: diversity, change and interdependence.* Edinburgh, Oliver and Boyd

Barnaby, F. (1991) Environmental impact of the Gulf War. *The Ecologist* 21(4): 166–72

Barrow, C.J. (1991) *Land degradation: development and breakdown of terrestrial environments.* Cambridge, Cambridge University Press

Beaujeu-Garnier, J. and Chabot, G. (1967) *Urban geography.* London, Longman

Berkhout, F. (1991) *Radioactive waste: politics and technology.* London, Routledge

Bhatti, N., Streets, D.G. and Foell, W.K. (1992) Acid rain in Asia. *Environmental Management* 16(4): 541–62

Bloom, S., Miller, J.M. and Winkler, P. (eds) (1993) *Hidden casualties: environmental health and political consequences of the Gulf War.* London, Earthscan

Boerner, C. and Chilton, K. (1994) False economy: the folly of demand-side recycling. *Environment* 36(1): 6–15

Bouverie, J. (1991) Recycling in Cairo: a tale of rags to riches. *New Scientist* 130(1775): 52–5

Brimblecombe, P. (1987) *The big smoke: a history of pollution in London since medieval times.* London, Methuen

British Medical Association (1991) *Hazardous waste and human health.* Oxford, Oxford University Press

Brown, L.R. *et al.* (1993) *State of the world 1993.* New York, W.W. Norton

Bull, K., Hall, J., Steenson, D., Smith, C. and Cresser, M. (1992) Critical loads of acid deposition for soils: the U.K. approach. *Endeavour* 16(3): 132–8

CAP (1984) *The lessons of Bhopal: a community action resource manual.* Penang, Consumer Association of Penang (PO Box 1045 Penang, Malaysia)

Coghlan, A. (1994) Midas touch could end Amazon's pollution. *New Scientist* 141(1916): 18

Connett, P. and Connett, E. (1994) Municipal waste incineration: wrong question, wrong answer. *The Ecologist* 24(1): 14–20

Crump, A. (1991) *Dictionary of environment and development: people, places, ideas and organizations.* London, Earthscan

De Lacerda, L.D., Pfeiffer, W.C., Ott, A.T. and Da Silveira, E.G. (1989) Mercury contamination in the Madeira River, Amazon: Hg inputs to the environment. *Biotropica* 21: 91–3

Dickenson, J.P., Clarke, C.G., Gould, W.T.S., Prothero, R.M., Siddle, D.J., Smith, C.T., Thomas-Hope, E.M. and Hodgkiss, A.G. (1983) *A geography of the third world*. London, Methuen
Eisenbud, M. (1987) *Environmental radioactivity, from natural, industrial and military sources*. London, Academic Press
Elsworth, S. (1984) *Acid rain*. London, Earthscan
Engstrom, K. (1992) The recycling of plastic bottles collected from public waste. *Endeavour* 16(3): 117–21
Fairlie, I. (1992) Tritium: the overlooked nuclear hazard. *The Ecologist* 22(5): 228–32
Fairlie, S. (1992) Long distance, short life: why big business favours recycling. *The Ecologist* 22(6): 276–83
Feshbach, M. and Friendly, A. (1992) *Ecocide in the USSR*. London, Arum
Fisher, D. (1992) *Paradise deferred: environmental policy-making in central and eastern Europe*. London, Royal Institute of International Affairs
Gourlay, K.A. (1992) *World of waste: dilemmas of industrial development*. London, Zed
Greer, J. (1993) The price of gold: environmental costs of the new gold rush. *The Ecologist* 23(3): 91–6
Gupta, A. (1988) *Ecology and development in the third world*. London, Routledge
Hardoy, J.E., Mitlin, D. and Satterthwaite, D. (eds) (1990) *The poor die young: housing and health in third world cities*. London, Earthscan
Harrison, P. (1992) *The third revolution: population, environ-ment and a sustainable world*. Harmondsworth, Penguin
Heintzenberg, J. (1989) Arctic haze: air pollution in polar regions. *Ambio* XVIII(1): 50–5
Hester, G.L. (1992) Electric and magnetic fields: managing an uncertain risk. *Environment* 34(1): 6–11; 25–32
IDRC (1994) *Cities feeding people: an examination of urban agriculture in East Africa*. Ottawa, International Development Research Centre
IDRC Reports (1993) Farming in the city: the rise of urban agriculture. *IDRC Reports* 21(3)
IUCN, UNEP and WWF (1991) *Caring for the earth: a strategy for sustainable living*. London, Earthscan
Ives, J.H. (1985) *The export of hazards*. London, Routledge and Kegan Paul
Johnson, J. (1990) Waste that no one wants. *New Scientist* 127(1733): 50–5
Kennedy, I. (1992) *Acid soil and acid rain* (2nd edn). Chichester, Wiley
Kivell, P. (1993) *Land and the city: patterns and processes of urban change*. London, Routledge
Kurth, W. (1992) The mounting pile of waste paper. *The OECD Observer* no. 174: 27–30
Lean, G., Hinrichsen, D. and Markham, A. (1990) *WWF atlas of the environment*. London, Arrow
Leonard, H.J. *et al.* (1989) *Environment and the poor: development and strategies for a common agenda*. New Brunswick, NJ., Transaction
Lipton, M. (1977) *Why poor people stay poor: a study of urban bias in world development*. London, Temple Smith
Mackereth, F.J.H. (1966) Some chemical observations on post-glacial lake sediments. *Philosophical Transactions of the Royal Society, London*. series B, 250: 165–213
Malm, O., Pfeiffer, W., Souza, C.M.M. and Reuther, R. (1990) Mercury pollution due to gold mining in the Madeira River basin, Brazil. *Ambio* XIX(1): 11–15
Mannion, A.M. (1991) *Global environmental change*. Harlow, Longman
Martinelli, L.A., Ferreira, J.R., Fosberg, B.R. and Victoria, R.L. (1988) Mercury contamination in the Amazon: a gold rush consequence. *Ambio* XVII(4): 252–4
Mason, P. (1993) Spare tyres drop into uncharted waters. *The Times* 5/2/93
Mather, A.S. (1986) *Land Use*. Harlow, Longman
McCormick, J. (1988) *Acid earth: the global threat of acid pollution*. London, Earthscan
Meadows, D.H., Meadows, D.L., Randers, J. and Behrens, W.W. III (1972) *The limits to growth: a report of the Club of Rome's project on the predicament of mankind*. New York, Universe Books
Medvedev, Z.A. (1990) The environmental destruction of the Soviet Union. *The Ecologist* 20(1): 24–9
Milne, R. (1988) Conservation clouds choke Britain's forests. *New Scientist* 117(1604): 27
Mnaksakanian, R.A. (1992) *Environmental legacy of the former Soviet Union*. Edinburgh, Centre for Human Ecology
Moran, E.F. (1987) Monitoring fertility degradation of agricultural lands in the lowland tropics. In P.D. Little and M.M. Horowitz (eds) *Lands at risk in the third world: local-level perspectives*. Boulder, Colo., Westview
Newson, M. (ed.) (1992) *Managing the human impact on the natural environment: patterns and processes*. London, Belhaven
Odum, E.P. (1975) *Ecology: the link between the natural and the social sciences* 2nd edn). London, Holt, Rinehart and Winston
Park, C.C. (1987) *Acid rain: rhetoric and reality*. London, Methuen
Pasqualetti, M.J. (ed.) (1990) *Nuclear decommissioning and society: public links to a new technology*. London, Routledge
Ponting, C. (1991) *A green history of the world*. London, Sinclair-Stevenson
Rodhe, H. (1989) Acidification in a global perspective. *Ambio* XVIII(3): 155–60
Rodhe, H. and Herrera, R. (eds) (1988) *Acidification in tropical countries* (SCOPE 36). Chichester, Wiley
Rodhe, H., Callaway, J. and Dianwu, Z. (1992) Acidification in Southeast Asia: prospects for coming decades. *Ambio* XXI(2): 148–50
Sheppard, C. and Price, A. (1991) Will marine life survive the Gulf War? *New Scientist* 129(1759): 36–40
Shrivastrava, P. (1992) *Bhopal: anatomy of a crisis* (2nd edn). London, Paul Chapman
Small, R.D. (1991) Environmental impact of fires in Kuwait. *Nature* 350(6313): 11–12
Smith, R.A. (1852) On the air and rain of Manchester. *Memoirs and proceedings of the Manchester Literary and Philosophical Society*, 2: 207–17
Smith, N.J.H., Alvirn, P., Homma, A., Folesi, I. and Serrao, A. (1991) Environmental impacts of resource exploitation in Amazonia. *Global Environmental Change* 1(4): 313–20
Soroos, M.S. (1992) The odyssey of Arctic haze: toward a global atmospheric regime. *Environment* 34(1): 6–12; 25–7

Stigliani, W.M., Doelman, P., Salomons, W., Schulin, R., Smidt, G.R.B. and Van Der Zee, S.E.A.T.M. (1991) Chemical time bombs: predicting the unpredictable. *Environment* 33(4): 4–9; 26–30

UNCTC (1985) *Environmental aspects of activities of transnational corporations: a survey*. New York, UNCTC

UNDP (1991) *Human development report 1991* (for UN Development Programme). Oxford, Oxford University Press

UNEP (1992) *Chemical pollution: a global review*. Geneva (Paláis de Nations), United Nations Environment Programme

Usher, M.B. and Thompson, D.B. (eds) (1988) *Ecological change in the uplands* (British Ecological Society Special Publication no. 7). Oxford, Blackwell Scientific

Virtanen, Y. and Nilsson, S. (1993) *Environmental impacts of waste paper recycling*. London, Earthscan in association with IIASA (International Institute for Applied Systems Analysis)

Warner, F. (1991) The environmental consequences of the Gulf War. *Environment* 33(5): 6–9; 25–6

Weir, D. (1988) *The Bhopal syndrome*. London, Earthscan

Weir, D. and Schapiro, M. (1981) *Circle of poison: pesticides and people in a hungry world.* San Francisco, Calif., Institute for Food and development Policy

Wellburn, A. (1988) *Air pollution and acid rain: the global threat of acid pollution*. London, Earthscan

WRI, IIED and UNEP (1988) *World resources 1988–9*. New York, Basic Books

Wynne, B. (1989) Sheep farming after Chernobyl: a case study in communicating scientific information. *Environment* 31(2): 10–15; 33–8

Yang, Shi and Furedy, C. (1993) Recovery of wastes for recycling in Beijing. *Environmental Conservation* 20(1): 79–82}

Zurek, J., Samsonowicz, A. and Gosk, E. (1994) Chemical dump sites in Poland. *Land Degradation & Rehabilitation* 4(4): 415–20

Further reading

Anderson, V. (1993) *Energy efficient policies*. London, Routledge [Examines policies for promoting energy efficiency to counter acidification and global warming]

Girardet, H. (1993) *The Gaia atlas of cities: new directions for sustainable urban living*. London, Gaia Books [Readable introduction to the ecology of settlement, urban pollution and what can be done to make city living more pleasant and less damaging]

Gunnerson, C.G. (1978) *Environmental impacts of civil engineering*. Washington, DC, American Society of Civil Engineers [Good overview, a little dated]

Hardoy, J.E., Mitlin, D. and Satterthwaite, D. (1993) *Environmental problems in third world cities*. London, Earthscan [Looks at the problems, their causes and practical solutions]

Holdgate, M.W. (1979) *A perspective on environmental pollution*. London, Cambridge University Press [Slightly dated but comprehensive introduction]

McCormick, J. (1988) *Acid earth: the global threat of acid pollution*. London, Earthscan [Good introduction]

Pearce, F. (1986) *Acid rain*. Harmondsworth, Penguin [Basic introduction]

Wellburn, A. (1988) *Air pollution and acid rain: the biological impact*. Harlow, Longman [Introduction to biological impacts]

CHAPTER 11 The greening of development

The green movement

There was environmentalist activity long before the use of the terms 'green' (meaning environmentally benign), 'greening' (meaning environmental improvement) and 'green movement' (a social or cultural movement, which in spite of considerable diversity contains a common element of environmental consciousness or concern) (e.g. Leopold 1949). There is little about green philosophy that is new; it has even been compared to that of the 'Diggers' of the 1640s (Hill 1972; Weston 1986), and with that of many Victorian radicals from the 1820s onwards, for example Robert Owen. What is new is the vision and the drive, and the extent to which interest has spread.

The use of green terminology increased after the mid-1980s, at first being used in politics, and then spreading as a popular, friendly alternative to environment or environmentalist, particularly in media discussion of development issues. Part of the phenomenon of greening has been increased popular interest and, more recently, marketing and advertising exploitation. In North America, Europe and many LDCs, 'green NGOs' (an NGO is a local, regional or international non-governmental organization) have massively expanded their activities since the 1970s. One problem has been the spread of 'green science' (which is sometimes rhetoric rather than based on proper investigation): development decision-makers must learn to weed out unreliable green advocacy and care must be exercised to ensure that rational enquiry is not discouraged. The green lobby can sometimes react in an emotive way to questions which require careful investigation and possibly discourage such research.

Because of the mistrust that many greens have for mass politics and centralized bureaucracy (which comprise what many of them see as a 'factory mentality of the nineteenth century' pervading western and communist states in the twentieth century) they might be considered postmodern and post-industrial. Although green often implies politicized environmentalism – Yearley (1991: 3) suggested that greens are a 'political manifestation of a cultural shift' – many are not politically active, indeed, eschew politics and some, like members of Friends of the Earth (a NGO pressure group), profess to be non-political.

The green movement has affected philosophy and religion. Spiritual ecologists include those who focus on established western religion, for example, Matthew Fox (OP) (an American Dominican priest accused by some of neo-paganism) has made proposals since the early 1980s for extensive greening of Judaeo-Christian religion to form a more anthropocentric and cosmological 'creation spirituality' (M. Fox 1983; 1989; Merchant 1992; 124). Other spiritual ecologists look to pre-Christian religions of Europe and America or the Orient for inspiration to transform human consciousness so as to have reverence for creation.

The green movement may be seen as a social movement, spawned from environmentalism (Redclift 1984: 6), which goes deeper than politics to embrace ethics and fundamental attitudes to life. Many would claim that greening is essentially a 'cultural' attack on the ills of modern society and economics to be contrasted with the 'economic' attack by socialists (W.M. Adams 1990: 71). What would probably have been called Ghandian in the mid-1970s is now likely to be called green. A frequently stated view among greens is that non-green political approaches are bankrupt; an East German utopian Marxist 'turned green', who was jailed for publishing a book criticizing orthodox

Marxism, Rudolf Bahro (1982; 1984) noted in the 1980s that 'The Greens are to Marx and Marxism what Einstein was to Newton and Newtonian physics ... a qualitative transformation of a system whose time was up'.

A huge range of people operate as greens: socialists, conservatives, intellectuals, poor people, Bhuddists, Christians, Muslims, etc. *The Sunday Times* (UK) noted that 'While socialist students may be swarming the rigging of Greenpeace, their Tory [i.e. of right-wing political persuasion] parents are rejecting E-additives ... and fighting new housing on greenfield sites' (4/9/88: B1). At the risk of overgeneralizing, most greens would share a worry that present industrial nations are pursuing an unsustainable, probably dangerous development path (Porritt 1984: 15).

Some common green characteristics

Listed below are characteristics common to the various green movements (based on Porritt 1984: 10, 15; Spretnak and Capra 1985: xx; Merchant 1992: 159).

The 'four pillars of greens'

1 Ecology.
2 Social responsibility.
3 Grass roots democracy.
4 Non-violence.

The 'six values of greens'

1 Decentralization 'think globally, act locally' (a dictum attributed to Hazel Henderson) (Henderson 1981).
2 Community-based economics.
3 Post-patriarchal principles.
4 Respect for diversity.
5 Global responsibility.
6 Future focus.

Other green characteristics

1 Holistic approach.
2 Disillusionment with modern unsustainable development paths.
3 A shift in emphasis away from philosophy of means to ends.
4 A shift away from growth economics.
5 A shift toward human development goals.
6 A shift from quantitative to qualitative values and goods.
7 A shift from impersonal and organizational to interpersonal and personal.
8 Commonly a feminist interest.

Seeds of the green movement

Clearly, a concise, accurate definition of green is difficult: the term is used loosely and greens vary a great deal in 'shade' (romantic, anarchistic, utopian, etc). Green thought has evolved and is evolving fast; politicians and marketing bodies are trying to make use of it. Perhaps the most useful definition of the green movement is 'the political voice of citizen movements' (Spretnak and Capra 1985: 3). Greens may be said to have grown from partially American seeds (Devall and Sessions 1985; Spretnak and Capra 1985: xvii), for example the writings of Henry Thoreau, Theodor Roszak (Roszak 1979), Ivan Illich, Aldo Leopold, Martin Luther King and Charles Reich. Reich (1970), writing about the possibilities for a new development ethic after the demise of the corporate state, was probably the first to use the expression 'greening'.

The seeds of the green movement germinated first in New Zealand, where the Values Party appeared in the late 1960s, but soon died out. There was also early interest in Tasmania, and in 1979 Switzerland elected what was probably the world's first green member of a national government. In Germany greens appeared a little later but survived. The UK Ecology Party, founded in 1973 (as the People's Party, later changing its name to Ecology Party), may be the oldest in Europe (Yearley 1991: 88); however, more influence has come from the former West Germany (the UK Ecology Party changed its name again in 1985 to the Green Party, and in the 1989 European elections won 15 per cent of total UK votes).

Somehow out of an anti-establishment, 'grass-roots'-orientated, Ghandian (even 'hippy') counter-culture, the green movement developed and became a force in German politics. The founding activists of the German green movement included Petra Kelly (the 'Joan of Arc of the Green movement'), Rudolph Bahro and Gert Baskian (some of the activists have been called 'red-greens'). Kelly worked in the USA with the Civil Rights Movement, returning to West Germany in the 1970s (the sunflower symbol of the German greens may have been derived via Kelly from the US Civil Rights Movement).

By the late 1970s the Federal Republic (West Germany) was seen by many of its people to be in a nuclear 'front-line' position, and there were signs of pollution damage to its forests, enough to stir action (Porritt 1984: 12). Herbert Gruhl formed the Green Action Future in the Federal Republic in 1978 using the slogan: 'We are neither left nor right, we are in front'. An attempt by West German greens (Die Grünen) to get seats in the West German lower house of parliament (the Bundestag) in 1979 failed, but in 1983 the movement managed to place 27 members – with 5.6 per cent of total votes (Spretnak and Capra 1985). Some of this success may reflect exploitation of a favourable, complex political system, rather than indicate a much higher popularity of green politics in Germany than the rest of Europe (the West German greens have been examined by Hülsberg 1988; Kolinsky 1989). In the UK the lack of proportional representation prevented actual election of green Members of Parliament (MPs); in the USA the cumbersome and costly political system dominated by Republicans and Democrats is probably what has prevented the election of greens.

Green and greening are increasingly applied to development thinking. In a modern world dominated by capitalism, the state, technology and science, there are those who support a 'red' (socialist) route to change and others who argue for a 'green' approach (Redclift 1984; Friberg and Hettne 1985). There have been efforts to develop a green socialism, particularly by those who see environmental difficulties as largely caused by social problems, but with limited success (Weston 1986: 4; Ryle 1988).

W.M. Adams (1990: 3) wondered if there really has been a greening of development or whether it is just a convenient rhetorical flag under which ships of very different kinds could sail. As a consequence of courtship from the West German left, it was often noted in the 1970s that Die Grünen (the greens) were 'like tomatoes, they start green and gradually turn red'. However, although left, right and centre political parties, plus advertising and marketing interests have tried in various countries, they have found it difficult to control the green movement.

The spread of the green movement

In the former West Germany the voice of the greens has been their newspaper *Die Grünen*. Kelly and Bahro presented green ideas to the USA from 1983 by means of lectures and media interviews, making rather slow progress (indeed they have remained a movement rather than a political party in the USA). By 1984 Canada, Austria, Switzerland, Belgium, The Netherlands, Sweden and Italy and a number of other countries had active green movements (Clark 1985; Merchant 1992: 157). By the late 1970s there was green activity in Australia and in Malaysia (e.g. NGOs like the Consumer Association, Penang and *Alam Sekitar*). A green party (*Pani Panchayat*) was established in India by 1985. By 1991 countries like Brazil and Mongolia had green parties and there was an active green movement in eastern Europe and the former USSR (for a guide to green activity, see Parkin 1989).

A sign that the green movement had sufficient power to disturb established governments came in 1985, when French intelligence officers sabotaged the Greenpeace vessel *Rainbow Warrior* in Auckland Harbour (New Zealand) to try and prevent it from engaging in non-violent protest at nuclear testing on Mururoa Atoll in the Pacific. One of *Rainbow Warrior*'s crew members was killed by the mine used in the sabotage (Young 1991: 61). Greenpeace had originally been established in Canada to send an anti-nuclear weapons protest ship into the Aleutian Island, where the USA proposed to carry out atomic tests. Greenpeace and Friends of the Earth have now become well-established NGOs that operate worldwide; both emphasize political neutrality and do not support political parties but they strongly support and propagandize the green cause.

In late 1988 UK Prime Minister Margaret Thatcher made a 'green speech' on the environment (the Royal Society Speech to the Conservative Party Conference of 1988). Public interest in the UK was such in the late 1980s that *The Green Consumer Guide* (Elkington and Haines 1988) reached the top of the bestsellers list and by late 1988 most UK newspapers had appointed environmental or green correspondents or even editors. In June 1989 the Green Party polled 14.5 per cent of votes in the UK elections to the European Parliament. This did not go unobserved: by the next British elections all major parties had greened, probably taking support away from the UK Green Party so that the success it had enjoyed in 1989 fell away by 1990, until in late 1993 UK support for the Green Party was probably around 1 per cent of the electorate. At that time in the UK, Europe and USA advertising agencies started actively to exploit green issues.

In 1993, in spite of a reduced support from the electorate, the green movement was well established in Europe, the USA, many LDC countries and the former Soviet bloc, but not very strongly in Japan. While authors like Atkinson (1991a; 1991b) have tried to clarify issues, the route that the green movement will take in the future is not clear: some of the initial popularity seems to have worn off but may well return.

Broad groupings of greens

Although there is some overlap between greens and environmentalists, it is possible to sketch broad groups of greens (based on Porritt 1984: 4–5; Weston 1986: 20; Taylor 1991).

Conservationists/traditionalists

These are heirs to the nineteenth-century romantic liberal rejection of industry and materialism; they are less interested in drastic change of attitudes and lifestyle than some greens. The group includes traditional conservationists like members of the UK Royal Society for the Protection of Birds or the Council for the Protection of Rural England and in the USA of the National Audubon Society or Sierra Club.

Reformists

They have no particular tradition and are midway between the previous and following groupings. They tend to be single issue groups with problem-orientated aims, for example a group opposed to construction of a new airport or road or rail route.

Radical environmentalists

These draw ideas from sources like Kropotkin, Thoreau, Aldo Leopold, Godwin, etc. They recognize the need for considerable change of attitudes and lifestyles because environmental problems arise. They seek alteration of political choices, economic systems, social inequalities, etc, often with a holistic, multi-issue approach. There is a considerable range, from 'moderates' like Friends of the Earth to extremists like Earth First! who espouse militant tactics like 'ecotage' (sabotage of things and people they see as a threat to the environment), 'ecovangelists' (who profess reverence for environment, not just stewardship) and even shamanists (a schism has opened between 'practical' and 'spiritual' factions of Earth First!).

Ecofeminists

These believe that women need to organize to achieve sustainable development.

Shades of green

Although environmentalists and greens all share some sort of feeling of fighting for survival and perceive a need to protect the Earth's environment and organisms, they span a very wide spectrum (see chapter 1, pp. 12–13). In the USA the green movement seems to have been particularly prone to splitting into factions. One group of factions (with a high profile in but not confined to the USA) are the 'deep greens', who hold that human populations are too high and present levels of consumption in DCs cannot be sustainable and so seek various routes for retreat from industrial consumerism. One point of view is that environmental awareness has evolved from the 'shallow', human-centred ideas of the 1960s and 1970s to progressively 'deeper' views of the 1990s. Many deep greens were initially inspired by the writings of Arne Naess (1972; 1989) (for an outline of deep ecology see Devall and Sessions 1985; Dobson 1990). One deep green faction (the 'deep'/'shallow' distinction being first made by Arne Naess in 1972) are the (often-anti-humanist) eco-anarchists, notably the Earth First! movement who are sufficiently ecocentric to be prepared to use violence to further their aims (Lewis 1992). An Earth First! tactic which has attracted attention is 'monkey-wrenching' – the sabotage of things they perceive to threaten the Earth (probably inspired by the 1975 writings of Abbey in the USA). In general, shallow greens view humans as separate from environment, while deep greens do not recognize such separation (W. Fox 1984).

Lewis (1992) felt that many strident greens were 'mired in ideology' and could end up wreaking environmental damage (his critique of radical environmentalism and eco-extremism should be read by all who are interested in the green movement).

Supporters of green consumerism argue that it is possible for green growth. However, many greens, including supporters of Deep Ecology, doubt that green growth is possible and seek radical change of consumption habits 'for the good of the planet'; for trying to impose such views they are sometimes accused of green fascism.

Social ecologists hold that environmental crisis is rooted in social crisis; an eminent exponent of this green philosophy is Murray Bookchin (Lowe and Goydev 1983; Atkinson 1991a; 1991b: 223 provides a bibliography). At another extreme is Lewis (1992), who suggests 'promethian environmentalism', which embraces the harmonization of technology with a new environmental vision, even disengagement from the natural world to try and protect nature through high-tech development. This is an approach that deserves more attention.

Green law

Since the 1960s, and especially after the USA's 1970 National Environmental Policy Act, there have been growing numbers of lawyers specializing in environmental issues; journals and professional groups have also appeared, mainly in DCs. In general, law has been slow to adapt to the demands of development and environmental management (the situation with respect to shared international river basins has been discussed in chapter 9). Sands (1993) has recently reviewed the international environmental law situation.

Corporate greens

Some of the world's large companies each control more funds than some DC governments and enjoy sufficient management stability to be in a better position to embark on long-term planning than most LDCs. Recent developments with GATT are likely to strengthen MNCs and TNCs, which are often seen as incapable of anything other than profit-driven activities, yet a number are founded by highly principled people and profess a sense of public duty. Even the less altruistic realize the value of good public relations, and this may prompt them to encourage green activities. For example, McDonalds phased out ozone-damaging blown plastic packaging from 1987 and companies like AT & T and Siemens have made ozone protection a policy and insist that their component suppliers comply (Porter and Brown 1991).

Business certainly has to develop a greener corporate culture on both the production and the marketing sides (Elkington and Burke 1987; Winter 1988; Davis 1991; Smith 1992). For a review of how European business has reacted to environmental issues see Vaughn and Mickle (1993); for a more American coverage see Allenby and Richards (1994). Schmidheiny (1992) presents numerous case studies of corporate green activity (and is a text promoted by the Business Council for Sustainable Development – a body active at UNCED, the 'Rio Earth Summit').

Many large businesses now have environmental advisors on their boards or full-time environmental officers (environmental awareness at board level should speed up adoption of green measures). Already there are promising signs of many companies shifting from regarding environmental protection as a necessary evil to viewing it as an essential aspect which can also generate business (S.R. Taylor 1992). There are also indications that managers are becoming more proactive with respect to environment; S.R. Taylor (1992) noted that firms who are slow to add pollution controls to their products may find that countries with strict rules are no longer a market.

Undoubtedly some companies can make profits from green awareness and environmental problems, by exploiting opportunities without really improving the environment. For example, worries about ozone thinning means increased opportunity for selling sun-protection creams and sunglasses; concern about the quality of domestic water supplies can support sales of household water filtration devices. The question is, will such companies feed any of their profits to environmental research or rehabilitation, given that it might mean reduction of threat and reduced sales?

Controls on green marketing and 'eco-labelling' need to be tightened to prevent abuse; in recent years the NGO Friends of the Earth have made 'Green Con' awards to companies or bodies which make misleading green claims in their advertising. In the USA bodies like the Coalition for Environmentally Responsible Economies are promoting voluntary mechanisms for corporate self-governance to improve environmental protection and pursue sustainable development.

In Europe and the USA environmental auditing has become lucrative business, e.g. an estimated 10 per cent of UK capital investment in 1993 underwent 'green screening' to ensure that it reached BS7750 (*The Times* 14/6/93: 32). The Environmental Management Systems Standard BS7750 (the British Standard of 'environmental respectability'), developed by the British Standards Institute and published in draft in 1992, is a voluntary standard that confirms an organization has an environmental policy and complies with it. BS7750 consists of three elements: an environmental policy, a documented environmental management system, and a register of effects on the environment. A company adopting BS7750 is committed to cycles of self-improvement by internal environmental auditing. Advantages to the company include possible savings in insurance premiums, perhaps reduced waste, lower energy costs and more customer confidence. BS7750 was launched in spring 1994 and should help companies reduce waste, consumption and pollution.

The EU is introducing an Eco-Management and Audit Regulation in 1995 (again voluntary adoption) which will require companies to implement an environmental management system and submit to at least three-yearly environmental audits, revealing the results. These developments may not be as beneficial as they at first glance appear: there is a risk that smaller companies will find it more difficult to afford audits and controls than large ones, a deterioration in terms of competition.

It is increasingly common for companies to undergo a pre-acquisition environmental audit when changing ownership or buying land. Banks and insurance companies are also using audits to try and foresee liability and risks.

Some insurance and investment companies and even credit card companies offer environmentally supportive investment and banking schemes; for example loans for energy conservation buildings, loans for construction with traditional materials or restoration of historic buildings, investment in activities that are assumed to do no environmental damage, donation of some investment profit to conservation activities (some companies active in environmentally aware investment are listed in Beaumont 1992: 193).

The problem is to encourage and coordinate MNCs and TNCs to play a greater part in environmental care, also to weed out those that profess to be green, but which in reality are just using this as a marketing ploy or adopting an environmental etiquette. A promising development was the creation of a Business Council for Sustainable Development at the 1992 Rio 'Earth Summit' by 48 executives from large corporations (for details see Holmberg *et al.* 1993: 15). Bodies like Greenpeace have business wings and publish propaganda aimed at business; there are already link-ups between corporations and environmental bodies, e.g. in the USA McDonalds in 1991 entered into an agreement with the Environmental Defense Fund, giving the Fund access to information to assess how environmentally damaging waste might be reduced. The result was a reduction of McDonald's packaging waste by 80 per cent.

Aid and the environment

There are a wide range of forms and practices of aid: recipients and donors may be governments, bodies, groups of people or individuals; international agencies, international NGOs and groupings of governments may act as donors. Sometimes donors contribute aid directly to recipients, sometimes via an intermediary such as an NGO or a UN body. When aid is government to government it is termed bilateral aid; when several governments or an international organization have contributed it is multilateral aid.

Frequently aid is 'tied' or is conditional, i.e. requirements are attached to the funding: a recipient may have to behave in a particular way or a percentage of the aid is to be used to buy donor nation goods and services. The latter arrangement is known as 'aid for trade provision' and it is not unknown for obsolete, overpriced or unsuitable goods or services to be traded (Hayter 1989: 21; 92). (At the time of writing, the Pergau Dam scandal had just broken in the UK.) Aid may be in the form of funding, foodstuffs or other supplies, sometimes training or secondment of skilled personnel rather than donation of goods or funds. There are situations where conditionality makes sense (environmental care or improvement may be a condition), others where it is perceived as neo-protectionism or neo-colonialism, as an extra cost and as a sign there is a risk that support could be diverted.

There has been criticism of FAO aid policy for its impact on the environment (Cross 1991; Dinham 1991; Hildyard 1991). Japan has increased overseas aid in recent years and now provides more than the USA, spread over more countries. At the UNCED in 1992 Japan offered more aid for environment than any other nation; sceptics have hinted that this might be for reasons of self-interest, not just altruism. Much of Japan's aid is tied to its export or resource import policy and, according to Forrest (1991), there has been a tendency to support large superprojects which have sometimes caused serious environmental impacts. The aid may have been well intended but even providing roads can cause problems. The World Bank has recognized these risks and recently pledged not to finance commercial logging where forests were at risk. In practice the World Bank has run into criticism for failure to put such policies into practice (*The Observer* 26/1/92: 4): environmentally benign aid is not easy to achieve. What to a donor seems like sensible safeguards to avoid unwanted environmental (and socioeconomic) impacts may appear to a recipient to be 'excuses' for more conditionality, delay and perhaps loss of a portion of funding to pay for appraisals, safeguards and remedial measures and intrusion into sovereignty.

To combat global environmental problems will require considerable aid from DCs to LDCs. At the 1992 UNCED 'Earth Summit' some richer nations were clearly reluctant to commit themselves to this either for fear of what it would do to their economic growth or because they wished to ensure tight control over how the aid was spent. The 'democratization' of the CIS and its allies has meant less wasteful spending on arms and propaganda in both the east and west but it may also result in DCs diverting aid there that would in the past have gone to LDCs. Since the 1992 Earth Summit there have been signs of a shift from a world divided east and west to one divided north and south.

Wildlife poaching, deforestation and soil erosion often follow road or dam construction and resettlement which may be funded by aid (Hayter 1989: 39) and negative impacts may offset, even overshadow benefits. Proposals to fund the eradication of tsetse flies in Africa should be weighed against the possibility that people and livestock will spread into areas where there is presently a risk of trypanosomiasis (sleeping sickness) spread by the flies. Such aid would remove a constraint preventing people from degrading the land, and could lead to widespread pesticide poisoning of wildlife, people and livestock; the results could be far worse than the benefits obtained (Linear 1982; 1985).

Academics and aid agencies have examined environmental (and socioeconomic) aspects of aid, although surprisingly little has been published, given the huge sums and considerable impacts involved (P. Adams and Solomon 1985; Hayter 1989). Hayter (1989: 83) lists a number of reasons why this might be. A number of agencies have published or use in-house guidelines and there have increasingly been efforts to assess impacts prior to the granting of assistance. The way has been led by the World Bank, which established an Office of Environmental Affairs in 1970 (Warford and Partow 1989) and in the UK by the Overseas Development Administration (ODA), which has established environmental appraisal procedures (ODA 1984; 1989a; 1989b).

Debt and the environment

Efforts by LDCs to repay foreign debts acquired in the early 1980s (see Fig. 3.4) could have encouraged short-sighted resource exploitation, and may have diverted funds away from environmental management. The LDC debt crisis broke in 1982; soon claims were made that there was a link between debt and environmental degradation; however, there seems to be little clear proof for or against this assertion (Reed 1992: 143; UN 1992). Two main impacts of debt on LDC environments have been recognized: first money diverted to servicing debt is unavailable for environmental management; and second, resources and land may be put under pressure to earn foreign exchange to pay interest or pay off debt. Structural readjustment policies have been blamed for triggering or exacerbating the proposed link between debt and environmental degradation. Following OPEC oil price increases in 1974, recession had by 1975 driven down export prices for LDC commodities; heavy borrowing in the 1970s to finance large projects also helped lead to debt. To try and stimulate growth and ensure debt repayment, bodies like the World Bank started structural readjustment lending in the 1980s – conditional loans which seldom considered environmental impacts (Reed 1992). Some recognize a debt crisis between 1982 and 1990 (George 1992). By 1988 the flow of funds to the 15 most indebted LDCs had *reversed*, i.e. there was a net flow of money out of the poor LDCs to DCs (Ekins 1992: 28); by the 1990s Latin American and subSaharan countries were spending roughly 25 per cent of their total foreign exchange each year servicing debt (Davidson *et al.* 1992: 161). In spite of paying US$6500 million a month interest between 1982 and 1990, debtor countries were still 61 per cent more indebted in 1991. Often the lender banks are enjoying tax relief on the debts in their own countries.

In 1991 under the 'Trinidad Terms' the 'Paris Club' of creditors agreed to cancel some debts (Davidson *et al.* 1992: 174). However, LDC debt may also have an ultimate impact on DCs: if the struggle to earn foreign exchange to service debts damages biodiversity and upsets global environmental conditions (Kahn and McDonald 1990; George 1992).

Linkages between economics and environment are often complex and caution should be exercised when simple debt–damage relationships are recognized. Debt-servicing is not the only reason why an LDC may exploit rural resources: it might be to support urban facilities or industrialization.

Debt-for-nature swaps

Debt-for-nature swaps (DNS) are forms of aid that can help improve the environment, but which are regarded by some, especially in the recipient countries and NGOs, with suspicion or hostility (Hayter 1989: 258; Patterson 1990: 31; Fidler 1991; for a critique see Mahony 1992: 102; Sarkar and Ebbs 1992). Thomas Lovejoy of the US Smithsonian Institution has been credited with the idea of the DNS in 1984.

There are a range of related measures:

- debt–equity swaps
- debt–debt swaps
- debt–rescue swaps
- repayment in national currency of a debt to an environment fund
- debt write-off in return for environmental activities.
- debt–for–conservation swaps
- debt–for–exports swaps
- debt–for–drugs swaps.

All these involve conversion of hard currency debt to local currency debt and investment in the debtor country. Debt–debt swaps are the interchange of foreign loans between creditors. Debt–rescue swaps ('buy-back') are the re-purchase of a country's debt in secondary markets.

DNS were set up in Ecuador and Bolivia in 1987, and subsequently in Bolivia, Costa Rica, the Philippines, the Dominican Republic, Mexico, Malagasy, Jamaica, Cameroon (e.g. Mt Korup Conservation Project) and many other countries (Patterson 1990; Mahony 1992: 98). There are various forms of DNS: some are bond-based (a central bank pays interest on a bond created, usually for an NGO, over a period typically of five to seven years); some are government policy programmes (under which the recipient government commits itself to implement a policy or initiatives aimed at improving the environment or conservation).

Broadly, a (bond-based) DNS works as follows: a bank or lending agency realizes that it will probably never recoup a loan debt; it therefore seeks to sell the debt, at a discount, to another (donor), which then releases cash to the debtor country, in local currency so the donation goes further, to support environmental or conservation activity (McNeely 1989; Pearce 1989). In one recent DNS each US$1.0 bought US$3.0 of debt and over three to five years the donor made about 20 per cent return on the aid, so the foreign bank got some debt money back (if a US bank is involved it should be able to write-off some of the expense against taxes). The debtor country avoids defaulting and still retains control over conservation or environmental activities (Simons 1988).

Criticisms have been levelled at DNSs:

1. They offer limited potential to pay off existing LDC debts (i.e. are tiny compared with typical indebtedness).
2. In practice DNS may be used to 'smear' indigenous environmental groups efforts, i.e. opponents of environmental protection spread rumours of foreign interference to divert attention from real issues (see Cleary 1991: 133).
3. There may be difficulties in adopting DNSs in some countries because of their different accounting and regulatory systems.
4. There is no guarantee of ongoing protection or care.
5. Some see it as an erosion of LDC sovereignty.
6. If operated through NGOs, DNSs may not assist or train local agencies.
7. They do little to change commercial forces that may damage the environment.
8. They have so far been applied to a limited range of activities, mainly park and reserve establishment and maintenance.
9. Mahony (1992) argues the main beneficiaries are the debt-seller banks.

Whatever the faults, DNS are one source of funds for environmental action.

Trade and the environment

Trade affects environment, in many cases environmental damage is 'driven' by it, for example:

- trade affects rates of deforestation;
- trade changes affect demand for animal and plant products and may be a major reason why a species is endangered;
- trade affects global carbon dioxide levels;
- trade affects extraction of mineral resources, production of food and commodity crops;
- trade affects levels of pollution in developed countries;
- trade affects pollution controls in developing countries.

Some forms of trade can be less damaging than others: export of forest products is less damaging than logging and may also discourage deforestation if it is carefully controlled and local people benefit. A UK store chain, the Body Shop, has recently tried to encourage environmentally benign forest product trade, minimizing middlemen profits. However, such products may have limited markets which places a limit on what can be achieved by this route.

Falling commodity prices on the world market often lie at the root of land degradation, as farmers getting poor returns on crops, yet committed to purchasing inputs, are forced to expand the area farmed or to intensify production leading to environmental degradation and marginalization. The problem cannot be easily solved by going back to a pre-cashcrop economy.

In 1974 the Group of 77 (G77) – a coalition of about 100 LDCs, plus a number of non-aligned countries – put forward demands for a New International Economic Order (NIEO) to the UN General Assembly. The NIEO included demands for new commodity agreements, alteration of what were seen as 'unfair' patents laws and general North–South economic reform. The main vehicle for reform is the General Agreement on Tariffs and Trade (GATT), founded in 1947 to liberalize world trading.

Free trade and GATT

Free trade means that countries can obtain materials and continue to over-expand production; it can also mean production impacts (pollution, goods, demand for goods) are felt over a wider area. In 1993 over 25 per cent of world trade was in primary products: timber, food, fish, minerals, etc.

Awareness that free trade can lead to environmental damage is not new. Free trade within the Roman Empire forced down grain prices and led to large landowners with many slaves practising careless commercial farming while smaller careful agriculturalists were forced out of business. Richard Cobden was aware of the environmental implications for the UK of the Repeal of the Corn Laws (1846), legislation which had protected farmers from falling wheat prices, with free trade landowners drained and cleared more land and intensified landuse. Yet many now hope that GATT agreement might reduce the flood of cheap grain which depresses produce prices and acts as a brake on efficient agriculture in LDCs.

Free trade could pose serious environmental problems or might ease existing difficulties (Bown 1990; Morris 1990; Shrybman 1990; Porter and Brown 1991; 134–41; Davidson *et al.* 1992: 174). Considerable efforts have been made to remove trade barriers and DCs and LDCs have been actively negotiating to redraw the rules of the international economic system and trade. GATT is the primary body of international trade law first drafted in the USA in 1947–8 to establish rules for the conduct of international trade – the hope being to lower tariff barriers erected in the 1930s that are held to be a hindrance to world development (in effect GATT is a forum for an ongoing attempt to renegotiate world trade rules). GATT rules will soon cover about 90 per cent of the world's trade in goods. There were eight meetings to discuss GATT prior to 1985–6, when the latest round (the 'Uruguay Round') was launched; subsequent negotiations were inconclusive, and the Uruguay Round should have been concluded in 1990–1 (Anon 1992). Meetings seeking to rewrite trade rules in the hope that it would increase LDC exports stalled over various disagreements, in particular on cutting subsidies to agriculture, especially in the EU where the French farming lobby opposed agreements reached in Washington in 1992 (Blair House Agreements) to lower farm subsidies (World Bank 1992: 175). In 1993 in Tokyo the 'Quad Group' of GATT (Japan, USA, Canada and the EU) agreed to abolish tariffs on pharmaceutics, construction equipment, farming equipment, furniture, beer and spirits and steel (the latter remained under control of other agreements – quotas, etc), and to reduce tariffs on several other tradesgoods. In December 1993, all 117 participating countries and organizations agreed to open markets (to come into force in July 1995). A new Multilateral Trade Organization will then replace GATT and will 'police' trade to see that it is fair and free.

While the GATT agreement may open markets and introduce new intellectual property rights, there could also be problems. GATT proposals could lead to the situation where countries could no longer limit the value of agricultural or other goods imported, that is it would cut import quotas and controls; for example, for two decades the Multi-fibre Arrangement (part of GATT agreements established in 1974 and designed to allow DC industry time to adjust to low-cost LDC competition in clothing and textiles) has restricted textile and textile raw material imports into DCs (Tussie 1987). There are worries that GATT could favour DCs and hinder LDCs ability to develop biotechnology and pursue development free of dependency (Raghava 1990).

Anything deemed a barrier to trade would probably be illegal under GATT and this could mean that GATT might upset environmental management because:

1 Much environmental legislation could be interpreted as a non-tariff trade barrier.

2 There may be reduced export controls to discourage logging, trade in endangered species, etc.
3 Reduced import restrictions might remove opportunities to counter trade in hardwoods, endangered species, etc.
4 Increased opportunity to sell commodities like beef, sugar, etc, around the world might well encourage increased forest clearance and poor land management in certain countries to boost production.
5 Countries may think twice about spending money on pollution control if another country does not and they compete to sell the similar goods on equal terms.
6 It may be less easy without the 'threat' of trade restrictions to encourage countries to reduce carbon dioxide emissions or other pollution.
7 LDCs may reduce domestic food prices and raise more export crops (as has happened in Brazil).
8 Marginalization of smaller farmers could result: they could then damage the land as do the larger farmers practising industrial (agrochemical-using) agriculture.
9 There is a risk foreign inputs and MNC controls will increase leading to more dependency.
10 If free trade leads to reduced home production, there is a risk of problems or conflict if overseas supplies fail.
11 It could be difficult to pass and enforce environment and resource management or health protection laws, for example some of the Convention on International Trade in Endangered Species (CITES) restrictions. Denmark has banned sale of beverages in non-reusable containers; in 1993 importers who felt this was an unfair barrier to trade tried to revoke the law through the European Court (largely unsuccessfully – Denmark's approach was deemed good for the environment and was supported by the Court). The USA placed an embargo on import of Mexican canned yellow-fin tuna, which it felt was caught in ways that endangered marine mammals, subsequently there has been considerable dispute over this restriction especially since NAFTA (see following discussion). Free trade in beef seems to have led to rangeland and game animals damage in countries like Botswana. Breakdown of CITES restrictions would be a particular blow, given that trade in wildlife in 1993 amounted to as much as US$8 billion (about 30 per cent of which was illegal).

GATT might help environmental management if:

1 It ended tariff barriers keeping up produce prices that cause farmers to overstress land for profit.
2 It reduced dumping of cheap US and European food surpluses that make it difficult for LDC producers to get a fair price and so are unable to leave land fallow or invest in land improvement, erosion control, etc.
3 It led to free trade in timber and freedom for LDCs to produce and sell finished wood products to DCs, this might reduce logging (Lang and Yu 1992).

There is now a debate on 'free trade versus sustainable agriculture' focusing on the likelihood that farmers will reduce short-term costs to compete in a free market (Ritchie 1992). Another argument is for a 'new protectionism', i.e. a reduction in the volume of trade as an alternative to free trade to cure present market problems (Lang and Hines 1993).

GATT has set up a group on Environmental Means and International Trade, and bodies like the OECD are keen to harmonize free trade and environment (De Miraman and Stevens 1992). A study by Zarsky (1994) examined the implications of GATT agreements and the need for environmentally sound rules to control GATT. It would certainly be wise to seek greater integration of GATT and bodies like the UN Commission on Trade and Development (UNCTAD). The North American Free Trade Agreement (NAFTA) agreed between the USA, Canada and Mexico in early 1994 does seem to be likely to cause marginalization of small farmers and will generate problems over anti-pollution and questionable animal hormone use controls and possibly discourage those considering organic farming (Ritchie 1992). Another organization seeking more free trade and economic growth is the Asia Pacific Economic Cooperation (APEC) group, founded in 1989 as a loose grouping of 15 nations. It seems highly likely that the world trade in the twenty-first century will be shaped by GATT, NAFTA, EU and APEC. There are already indications that considerable economic expansion will occur in the Asia–Pacific region.

Clearly there is a need for environmentally sound trade rules applied to all and with none of the exceptions and loopholes that allow and encourage production and export of produce that leads to land degradation or other negative ecological impacts.

Women and environment

One could argue that Rachael Carson was the first to popularize environmental concern in 1962. Women (initially voicing objections to use of feathers from rare birds in the fashion trade) played a major role in founding one of the first conservation bodies, the Royal Society for the Protection of Birds (RSPB) in 1889. The green movement has since its outset had a strong element of feminism and many of its activists in Europe are women. A woman, Gro Harlem Brundtland (Prime Minister of Norway), chaired the World Commission on Environment and Development and deserves much of the credit for spreading the idea of 'sustainable development' (a term first used by Barbara Ward).

Publications on women and environment and by women on environment have multiplied in recent years (Shiva 1988; Rodda 1991; Sontheimer 1991; *The Economist* January/February 1992: special issue on 'feminism, nature and development' Braidotti *et al.* 1993; Momsen 1991).

In India forest conservation and the spread of social forestry has been assisted by women's movements like 'Chipko' (*Chipko Andolan*) in Uttar Pradesh. The Chipko Movement originated spontaneously in the Reni Forest Area of Uttar Pradesh in 1973 when women used passive resistance tactics to stop commercial logging (Dankelman and Davidson 1988: 48). In Kenya the women's Green Belt Movement plant trees to rehabilitate degraded lands, in Brazil women's environmental groups have opposed Amazonian deforestation (e.g. the Açao Democrática Feminina Gaúcha) and have been active in seeking to improve conditions in the *favelas* (squatter settlements around major cities like Rio). In Bolivia (and other countries) the YWCA (Young Women's Christian Association) have been active in tree planting and other environmental work.

It is vital that efforts to protect, improve or rehabilitate the environment involve women because:

- Roughly 50 per cent of the world's people are women who do about 75 per cent of the work, get 10 per cent of income and own 1 per cent of property (Ekins 1992: 73).
- Women are often the main producers of food (therefore users of land) (Boserüp 1989); Davidson *et al.* (1992: 73) suggested that 60–90 per cent of African food production was by women.
- Women often collect fuelwood, dung, forest resources and sometimes fish, thus they are in closer contact with the environment than men.
- Women are the first educators world-wide and are thus in an excellent position to spread environmental education.
- Increasingly men may move out of an area to engage in migrant labour; women are left permanently or seasonally in control of resource use.
- The percentage of women in poverty compared with men is greater so they are more likely to be compelled to degrade the environment to survive.
- Education of women in many parts of the world appears to have led to a reduction in family size and improvement in health: population control could be combined with environmental care.

In spite of the aforementioned reasons for involving women in development and environmental protection, governments and agencies still tend to overlook them. Social forestry will more often than not have to deal with women, yet there are very few trained women foresters in the world. Within the green movement there are those who hold that feminism can counter paternalistic attitudes that result in exploitation rather than stewardship of the environment and so offer a route to sustainable development (environmental feminism or ecofeminism). Ecofeminists claim that exploitation of women and nature have much in common (Seager 1993), one line of argument being that women are closer to nature and less aggressive and destructive than men. Originating around 1970, ecofeminism has already diversified into liberal, Marxist, socialist and cultural factions.

References

Abbey, E. (1975) *The monkeywrench gang*. New York, Avon

Adams, P. and Solomon, L. (1985) *In the name of progress: the underside of development aid*. London, Earthscan

Adams, W.M. (1990) *Green development: environment and sustainability in the third world*. London, Routledge

Allenby, B.R. and Richards, D.J. (eds) (1994) *The greening of industrial ecosystems*. Washington, DC, National Academy Press

Anon (1992) Where should GATT go next? *Nature* 360(6401): 195–6

Atkinson, A. (1991a) *Green utopias: the future of modern environmentalism*. London, Zed

Atkinson, A. (1991b) *Principles of political ecology*. London, Belhaven

Bahro, R. (1982) *Socialism and survival.* London, Heretic
Bahro, R. (1984) *From red to green: interviews with* New Left Review. London, Verso
Beaumont, J. (1992) Managing the environment: business opportunities and responsibilities. *Futures* 24: 187–205
Bookchin, M. (1989) *Remaking society.* New York, Rose
Boserüp, E. (1989) *Women's role in economic development.* London, Earthscan
Bown, W. (1990) Trade deals a blow to the environment. *New Scientist* 128(1742): 20–1
Braidotti, R., Charkiewicz, E., Hèausler, S. and Wieringa, S. (1993) *Women, the environment and sustainable development.* London, Zed
Clark, R. (1985) The grey and the green (review). *New Scientist* 105(1438): 35
Cleary, D. (1991) The greening of the Amazon. In D. Goodman and M. Redclift (eds) *Environment and development in Latin America: the politics of sustainability.* Manchester, Manchester University Press
Conroy, C. and Litvinoff, M. (1988) *The greening of aid: sustainable livelihoods in practice.* London, Earthscan
Cross, D. (1991) FAO and aquaculture: ponds and politics in Africa. *The Ecologist* 21(2): 73–6
Dankelman, I. and Davidson, J. (1988) *Women and environment in the third world.* London, Earthscan
Davidson, J., Myers, D. and Chakraborty, M. (1992) *No time to waste: poverty and the global environment.* Oxford, OXFAM
Davis, J. (1991) *Greening Business.* Oxford, Basil Blackwell
De Miraman, J. and Stevens, C. (1992) The trade/environmental policy balance. *The OECD Observer* 176(January/July 1992): 25–7
Devall, B. and Sessions, G. (1985) *Deep ecology: living as if nature mattered.* Salt Lake City, Utah, Gibbs M. Smith; London, Peregrine
Dinham, B. (1991) FAO and pesticides: promotion or proscription? *The Ecologist* 21(2): 61–5
Dobson, A. (1990) *Green political thought.* London, Unwin Hyman
Ekins, P. (1992) *A new world order: grass roots movements for global change.* London, Routledge
Elkington, J. and Burke, T. (1987) *The green capitalists – industry's search for environmental excellence.* London, Gollancz
Elkington, J. and Haines, J. (1988) *The green consumer guide: from shampoo to champagne.* London, Victor Gollancz
Fidler, S. (1991) Trade-off of a heavy burden. *The Financial Times* 22/5/91: 12
Forrest, R.A. (1991) Japanese aid and the environment. *The Ecologist* 21(1): 24–32
Fox, M. (1983) *Original blessing.* Santa Fe, NM, Bear
Fox, M. (1989) A call for a spiritual renaissance. *Green Letter* 5(1): 4; 16–17
Fox, W. (1984) Deep ecology: a new philosophy of our time? *The Ecologist* 14(5–6): 194–200
Friberg, M. and Hettne, B. (1985) The greening of the world: towards a non-deterministic model of global processes. In H. Addo, S. Amin, G. Aseniero, A.G. Frank, M. Friberg, F. Frobel, J. Heinrichs, B. Hettne, O. Kreye and H. Seki (eds) *Development as social transformation: reflections on the global problematique.* Sevenoaks, Hodder and Stoughton
George, S. (1992) *The debt boomerang: how third world debt harms us all.* London, Pluto
Hayter, T. (1989) *Exploited earth: Britain's aid and the environment.* London, Earthscan
Henderson, H. (1981) Thinking globally, acting locally: ethics of the dawning solar age. In H. Henderson (ed.) *The politics of the solar age: alternatives to economics.* New York, Anchor Books, pp. 354–405.
Hildyard, N. (1991) An open letter to Eduoard Saouma, Director-General of FAO. *The Ecologist* 21(2): 43–6
Hill, C. (1972) *The world turned upside down: radical ideas during the English Revolution.* London, Temple Smith (Harmondsworth, Penguin edn 1975)
Holmberg, J., Thomson, K. and Timberlake, L. (1993) *Facing the future: beyond the Earth Summit.* London, Earthscan
Hülsberg, W. (1988) *The German greens: a social and political profile* (English trans. by G. Fagan). London, Verso
Kahn, J.R. and McDonald, J.A. (1990) *Third world debt and tropical deforestation.* Binghamption, NY, New York State University at Binghamption
Kolinsky, E. (eds.) (1989) *The greens in West Germany: organization and policy making.* Oxford and Munich, Berg
Kropotkin, P. (1974) *Fields, factories and workshops tomorrow* (ed. by C. Ward). London, Unwin (original published 1899 as *Fields, factories and workshops*)
Lang, T. and Hines, C. (1993) *The new protectionism: protecting the future against free trade.* London, Earthscan
Lang, T. and Yu, D. (1992) Free trade versus the environment: a debate. *Our Planet* (UNEP) 4(2): 12–13
Leopold, A. (1949) *A sand country almanac.* New York, Oxford University Press
Lewis, M. (1992) *Green delusions: an environmentalist critique of radical environmentalism.* Durham, N.C., Duke University Press
Linear, M. (1982) Gift of poison: the unacceptable face of development aid. *Ambio* XI(1): 2–8
Linear, M. (1985) *Zapping the third world: the disaster of development aid.* London, Zed
Lowe, P. and Goydev, J. (1983) *Environmental groups in politics.* London, Allen and Unwin
Mahony, R. (1992) Debt-for-nature swaps: who really benefits? *The Ecologist* 21(3): 97–103
McNeely, J.A. (1989) How to pay for conserving biological diversity. *Ambio* XVIII(6): 303–18
Merchant, C. (1992) *Radical ecology: the search for a livable world.* London, Routledge
Momsen, J.H. (1991) *Women and development in the third world.* London, Routledge.
Morris, D. (1990) Free trade the great destroyer. *The Ecologist* 20(5): 190–5
Naess, A. (1972) The shallow and the deep, long-range ecology movement. *Inquiry* 16: 95–100
Naess, A. (1989) *Ecology, community and lifestyle.* Cambridge, Cambridge University Press
ODA (1984) Checklist for screening environmental aspects in aid activities (mimeo, amended 1984). London, Overseas Development Administration
ODA (1989a) *The environment and the British aid programme.* London, Overseas Development Administration

ODA (1989b) *Manual of environmental appraisal.* London, Overseas Development Administration
O'Riordan, T. (1976) *Environmentalism.* London, Pion
Parkin, S. (1989) *Green parties: an international guide.* London, Heretic
Patterson, A. (1990) Debt-for-nature swaps and the need for alternatives. *Environment* 32(10): 4–13, 31–4
Pearce, F. (1989) Kill or cure? Remedy for rainforests. *New Scientist* 123(1682): 37–41
Porritt, J. (1984) *Seeing green: the politics of ecology explained.* Oxford, Basil Blackwell
Porter, G. and Brown, J.W. (1991) *Global environmental politics.* Boulder, Colo., Westview
Raghavan, C. (1990) *Recolonization: GATT, the Uruguay Round and a new global economy.* London, Zed
Redclift, M. (1984) *Development and the environmental crisis: red or green alternatives?* London, Methuen
Reed, D. (ed.) (1992) *Structural readjustment and the environment.* London, Earthscan
Reich, C. (1970) *The greening of America.* New York, Random House
Ritchie, M. (1990) GATT, agriculture and environment: the US double zero plan. *The Ecologist* 20(6): 214–20
Ritchie, M. (1992) Free trade versus sustainable agriculture: the implications of NAFTA. *The Ecologist* 21(5): 221–7
Rodda, A. (1991) *Women and environment.* London, Zed
Roszak, T. (1979) *Person/planet: the creative disintegration of industrial society.* London, Victor Gollancz
Ryle, M. (1988) *Ecology and socialism.* London, Radius
Sands, P. (ed.) (1993) *Greening international law.* London, Earthscan
Sarkar, A.U. and Ebbs, K.L. (1992) A possible solution to tropical troubles? Debt-for-nature swaps. *Futures* 24: 653–68
Schmidheiny, S. (1992) *Changing courses: a global business perspective on development and the environment.* Cambridge, Mass., MIT Press
Schumacher, E.G. (1974) *Small is beautiful: a study of economics as if people really mattered.* London, Abacus (US edn 1974, Harper and Row)
Seager, J. (1993) *Earth follies: feminism, politics and environment.* London, Earthscan
Shiva, V. (1988) *Staying alive: women, ecology and environment in India.* London, Zed
Shrybman, S. (1990) Free trade vs. the environment: the implications of GATT. *The Ecologist* 20(1): 30–4
Simons, P. (1988) Costa Rica's forests are reborn. *New Scientist* 120(1635): 43–7
Smith, D. (ed.) (1992) *Business and the environment: implications for the new environmentalism.* London, Paul Chapman
Sontheimer, S. (ed.) (1991) *Women and the environment: a reader.* London, Earthscan
Spretnak, C. and Capra, F. (1985) *Green politics: the global promise.* London, Paladin (US edn 1984)
Taylor, B. (1991) The religion and politics of Earth First! *The Ecologist* 21(6): 258–66
Taylor, S.R. (1992) Green management: the next competitive weapon. *Futures* 24: 669–80
Tussie, D. (1987) *The less developed countries and the world trading system: a challenge to the GATT.* London, Frances Pinter
UN (1992) *Debt and the environment: converging crisis.* Washington, DC, United Nations
UN (1993) *Agenda 21: program of action for sustainable development.* Washington, DC, United Nations
Vaughn, D. and Mickle, C. (1993) *Environmental profiles of European business.* London, Earthscan
Warford, J. and Partow, Z. (1989) Evolution of the World Bank's environmental policy. *Finance and Development* December: 5–8
Weston, J. (ed.) (1986) *Red and green: the new politics of the environment.* London, Pluto
Winter, G. (1988) *Business and environment: a handbook of industrial ecology with 22 checklists for procedural use.* New York, McGraw-Hill
World Bank (1992) *World development report 1992: development and the environment.* Oxford, Oxford University Press
Yearlly, S. (1991) *The green case: a sociology of environmental issues, arguments and politics.* London, Harper Collins
Zarsky, L. (1994) *Borders and the biosphere: environment, development and world trade rules.* London, Earthscan

Further reading

Dankelman, I. and Davidson, J. (1988) *Women and environment in the third world.* London, Earthscan [Introduction to women/environment interrelationship in LDCs]
Deziron, M. and Bailey, L. (1991) *A dictionary of European environmental organisations.* Oxford, Basil Blackwell [Lists European agencies, NGOs, etc, concerned with the environment]
Hayter, T. (1989) *Exploited earth: Britain's aid and the environment.* London, Earthscan [Controversial critique which includes some consideration of aid and environment relationship]
Lang, T. and Hines, C. (1993) *The new protectionism: protecting the future.* London, Earthscan [Warns of GATT impacts]
Porritt, J. (1984) *Seeing green: the politics of ecology explained.* Oxford, Basil Blackwell [Popular introduction to the green movement]
Spretnak, C. and Capra, F. (1985) *Green politics: the global promise.* London, Paladin (US edn 1984) [Introduction to the green movement]

CHAPTER 12 Future prospects

To reduce future environmental problems, the following will be required:

- new development ethics
- global cooperation
- improved environmental monitoring
- better environmental management.

New development ethics

There has already been a shift away from the west's *laissez-faire* and the communist bloc's equally environmentally callous approaches to development. Humanity does seem to have made progress: most of those who before the 1950s would have hunted big game with a rifle now use a camera; international agreement has led most whaling nations to stop the practice (Fig. 12.1); the 'super-powers' have made progress towards atomic weapons control; major world conferences are being held on environmental matters; environmental issues are on the political agenda in many countries. Today there are both awareness and hopes but a long way to go to develop practical working approaches and policies that will be accepted by all or most DCs and LDCs. It is difficult to separate environmental management from development ethics and issues of global cooperation. Already there is sufficient interest to have generated journals like *Environmental Ethics*.

Global cooperation

Despite a number of UN 'development decades' and declarations of intent to counter problems like desertification, environmental progress has been limited. Monitoring and international exchange of views and data, however, and negotiations on environmental problems have progressed. One can also recognize what Pirages (1978: 11) called an 'ecopolitical paradigm' evolving, the recognition of the finite nature of the planet and the inextricable interdependence of its component states. The shift caused if this paradigm is accepted will be painful for some states and individuals if resource-intensive growth and pursuit of self-interest are discouraged. Failure to make the shift, in effect 'cutting the LDCs adrift' in the manner described in Hardin's 'lifeboat ethics', could be even more painful: stopping hungry and desperate people from affecting the DCs would be difficult (an issue discussed with some pessimism in 1993 by the UK historian Paul Kennedy).

Improved environmental monitoring

Monitoring is vital:

- for assembling and updating baseline data
- for recognizing emergence of environmental problems
- for keeping track of the development of environmental problems and whether control, mitigation or avoidance measures are having results.

UNEP ideas on global environmental monitoring were supported at the 1972 Stockholm Conference; subsequently in 1975 UNEP established the Global Environmental Monitoring System (GEMS). Based in Nairobi, Kenya GEMS seeks to acquire and hold data for better environmental management, particularly of air, food and water resources but also, since

Fig. *12.1* Derelict whale-catcher ships (*Dias, Albatross* and *Petrel*) moored at Grytviken whaling station, South Georgia (subAntarctic). Grytviken was one of several stations on South Georgia (a number of other subAntarctic islands also had stations and there were ship-based factories as well) and functioned between 1904 and 1965. During those six decades the annual catch at Grytviken was seldom less than 3000 and frequently exceeded 6000 whales. Mercifully, these stations have remained closed.

the early 1980s, long-range transport of pollutants, health-related issues, ocean monitoring and terrestrial renewable resources. GEMS is now one of the foremost environmental monitoring agencies, disseminating its information via the International Environmental Information Network (INFOTERRA). UNEP also coordinates on Earthscan Programme.

From 1985 GEMS has fed satellite remote sensing data, collected by platforms like SPOT-1 (launched 1986), to the Global Resource Information Database (GRID). GRID is a sophisticated geographical information system that collects and makes available monitoring data on environmental issues ranging from desertification to endangered large mammals. An independent international research unit founded in 1975 to assist international organizations, and funded by the UNEP, WHO and GEMS, is the Monitoring and Assessment Research Centre (MARC) based at the University of London. MARC concentrates on biological and ecological monitoring, particularly pollution (some data-gathering is for retrospective monitoring – sampling of ancient lake sediments or ice deposits to provide a database reaching back in time). UNEP publish from their monitored data an annual *Environmental Data Report*, the IUCN from time to time release *Red Data Books* on endangered plant or animal species, and have maintained a Conservation Monitoring Centre. In 1980 the latter Centre was upgraded and became the World Conservation Monitoring Centre (based at Cambridge, UK) charged with collecting and databasing information on biodiversity, protected areas, habitat change and endangered species. The International Whaling Commission (established 1946) monitors and maintains databases on whale stocks.

A number of NGOs monitor biological and environmental resources, for example Greenpeace, Friends of the Earth and the World Wide Fund for Nature (formerly the World Wildlife Fund). With scarce resources there will often have to be selection of monitoring 'priorities', for example, measurement of an aquatic environment should tell quite a lot about the terrestrial and atmospheric environments surrounding it.

Getting better environmental monitoring presents less of a challenge than the improvement of environmental management. Before the 1970s, resource management decision-makers tended to ask:

- Is it technically feasible?
- Is it financially viable?
- Is it legally permissible?

It is now widely (but not widely enough) accepted that an additional question should be asked:

- Is it environmentally sound?

An answer to the last question can be arrived at with the support of a combination of environmental and social impact assessment, hazard assessment, and an awareness of the development context, but the art is far from perfect.

Reviewing progress with impact assessment Smith (1993: 1) noted that, despite two decades of evolution of a myriad of techniques, present practices often appeared unable to help prevent environmental problems. He argued that what is needed is a process for resource management and environmental planning that provides for the achievement of the goal of sustainability. To do this impact assessment should be designed as a bridge linking environmental assessment and the policies of resource management (Smith 1992: 12, 328), in effect integrating resource assessment, impact assessment and development studies.

Recent global exchange of views and data and negotiations on environmental problems

In June 1992 the UN held its largest ever conference in Rio de Janeiro – the UN Conference on Environment and Development (UNCED). Attended by representatives of 178 governments (some sources claim 185), 120 heads of state and about 1400 attending on behalf of NGOs and other bodies, the Rio 'Earth Summit' was probably the greatest gathering of those influencing development ever to have taken place. Originally proposed by Gro Harlem Brundtland and supported by Maurice Strong, UNCED spent 12 days discussing the world's environmental problems and related development issues (for reviews see Haas *et al.* 1992; Holmberg *et al.* 1993; Keating, 1993).

Was UNCED a success?

An answer to this question depends on one's standpoint; there were fewer concrete commitments than had been hoped for, nevertheless UNCED and the Global Forum (a parallel meeting in Rio of NGOs) did reach consensus on many issues and is probably a start to many more negotiations. As Holmberg *et al.* (1993) noted in their review of the 'Earth Summit' there is little scientific, let alone political, consensus on many of the issues debated. It is also rather unfair to charge that the results fell short of *stated* goals: most meetings do. An optimistic view of the achievements would be that UNCED set in motion a process that may pressure governments, business and perhaps ultimately individuals to make enough changes to reduce environmental and developmental problems. It was a time for taking stock, rather than a lost opportunity: as one UK journalist commented; 'it pressed people's noses against problems'.

UNCED achieved much more than formal documents; it brought together large numbers of NGOs, representatives of marginalized peoples, etc and there have doubtless been important but difficult to measure 'downstream effects' (Grubb *et al.* 1993). Within many of the agreements is the implicit expectation, and in many cases acceptance by DCs, that solving global problems will mean funding from DCs to LDCs. Much of the 'hidden agenda' at Rio concerned either issues of sovereignty, DC to LDC aid and issues relating to profit from biotechnology or the costs of pollution control. When it came to DCs actually agreeing aid to LDCs to solve environmental and poverty problems much less than had been hoped for was forthcoming.

However, little was done at Rio to address problems caused by increasing human population, indeed, the expression 'population increase' was generally avoided. Hopes that a climate convention would be agreed and lead to a significant greenhouse gas emissions cut (of say 60 per cent) were not fully realized.

The Summit generated five formal documents:

- Convention on Biodiversity
- Convention on Climatic Change

- Statement on Forest Principles
- Rio Declaration on Environment and Development
- An action programme: 'Agenda 21'.

Convention on Biodiversity

Essentially a treaty to preserve flora and fauna (which came into force in late 1994). The draft was sponsored by the UNEP and at Rio 153 nations, excluding the USA, signed. Signatories indicated that they would seek to protect biodiversity and to use it sustainably (Holmberg *et al.* 1993: 20–2). This combines protection with opportunities for exploitation and has been hailed as 'the greatest triumph' of UNCED. The USA was reluctant to sign for fear that it would conflict with patent rights protecting its commercial development of biodiversity for biotechnology.

Not part of the Convention on Biodiversity, but announced at Rio and related, is the Darwin Initiative for the Survival of the Species. The Darwin Initiative is intended to provide UK and associated researchers with funds for research on conservation and sustainable use of the world's biological resources.

Convention on Climatic Change

Rio produced a Framework Convention on Climatic Change signed by over 150 nations (some call it the Climate Change Convention) – what is essentially a tentative first step toward solving global warming by agreeing controls on greenhouse gas emissions (Holmberg *et al.* 1993: 25–8). The EU failed to agree a joint ratification; instead a number of European countries made their own agreements, probably undertaking lower cuts than joint action would have required of them and allowing poorer EU countries to delay ratification (*The Times* 4/12/93: 10). The USA was not especially supportive of this agreement and CO_2 emission controls were not as tough as they might have been, largely due to US objections. It has been argued that emphasis on global climate change is diverting attention and resources from other pressing problems: soil degradation, pollution, poverty alleviation, etc (Westcoat 1993).

Statement on Forest Principles

This non-legally binding statement of principles seeks global consensus on conservation and sustainable development of all types of forest (Holmberg *et al.* 1993: 28–31). It was not the hoped-for forest protection treaty and ensures that LDCs have sovereign right to develop their forests if they so wish. In effect logging was not adequately addressed.

Rio Declaration on Environment and Development

The Rio declaration started life as the Earth Charter of basic principles for the conduct of nations and peoples with respect to the environment and development. The draft Charter got bogged down and the Declaration evolved, hopefully an inspirational expression of a new global ethic (for details see Holmberg *et al.* 1993: 30–2 and 43–4).

An action programme: 'Agenda 21'

'Agenda 21' is really a 'yardstick' or action plan and sets out a list of tasks, targets, costs, priorities and suggested responsibilities for the twenty-first century covering a wide range of aspects of the environment. It contains useful ideas (Holmberg *et al.* 1993: 30–3) but many are not operationally workable and it has been criticized as being too 'bland' and is not, as was originally intended (during preparations for Rio), a programme of action to implement an 'Environmental Charter' which had been drafted at Geneva and New York in 1991. It does provide a reference against which to measure future progress. The Earth Charter was to have been a declaration of basic principles of sustainable development but it was not signed at Rio. 'Agenda 21' is the only UNCED document really to embrace environment *and* development.

Other Earth Summit achievements

A (UN) Commission on Sustainable Development was set up by the UN General Assembly to police the programme agreed at Rio. It is charged with ensuring bodies involved in promoting development seek sustainable strategies.

The Earth Summit initiated three funding mechanisms (vague funding pledges make it difficult to judge how much there will be to allocate):

1. *The Earth Increment* a fund to be administered by the World Bank to help LDCs implement 'Agenda 21'.
2. *The UN Global Environmental Facility* initially set up by the World Bank in 1990, this is a channel for 'green aid' which will fund the main achievements of Rio (treaties on biodiversity and climate change). The Facility is jointly administered by the World Bank, the UNDP and UNEP. However, in the early 1990s the funds available were not large (only about US$1.5 billion to last three years).

3 *Bilateral Aid from DCs to LDCs to Help with Agenda 21* some DCs have made donations or have stated an intent to do so.

The G77 nations' idea of establishing a Green Fund had little support from donor countries

UNCED helped to establish a Business Council for Sustainable Development, a group of executives keen to stimulate commercial involvement in UNCED and to improve environmental responsibility of commerce.

It was agreed at Rio to form an International Negotiating Committee for a Convention to Combat Desertification. The plan was to hold an international conference on desertification in 1994 (in late 1993 it seemed unlikely that this and any anti-desertification treaty would be agreed before 1995, due to disagreements about whether Africa should be given priority).

Many other targets and meetings were agreed at Rio; these include the designation of the 1990s as the International Decade for Natural Disaster Reduction; an International Conference on Population and Development (1994); a World Conference on Women (1995); development of an Earth Charter (for a full list see Holmberg *et al.* 1993: 40).

In 1994 a Global Forum '94 was held at Manchester (UK) and later that year in the same city there will be a Partnership for Change meeting, both aimed at coordinating NGO responses to UNCED. The focus of Global Forum '94 will be the world's cities.

Better environmental management

A crucial step toward improved environmental management is to assess why problems have not so far been solved or why solutions have taken so long. Trudgill (1990) looked at barriers to solving environmental problems, providing conceptual structures for policy-makers, educators, environmental managers, etc. Somehow the international community must decide which problems merit priority and implement solutions.

Environmental managers can adopt three main approaches: advisory, economic and regulatory.

Advisory approaches

- advice, leaflets
- media information (may be 'semi-covert', i.e. carried in entertainment programmes and articles)
- education
- demonstration (e.g. model farms)

Economic approaches

- taxes
- grants or loans
- subsidies
- quotas

Regulatory

- standards controlling things like agrochemicals or activities
- restriction of damaging practices
- licensing of potentially damaging activities

Responses to environmental problems can be antidotal or corrective, the former address symptoms, the latter seek to correct root causes (Boyden 1987). More corrective responses are desirable.

An obvious step forward to better environmental management is for a state to establish a sufficiently powerful environmental ministry. In the last few decades most states have established ministries, although these often lack strength. Another promising development is to improve integration between environmental ministries and others responsible for agriculture, housing, industry, and so on. In the USA the Environmental Protection Agency (EPA) has had some success at promoting environmental management and enforcing protective measures. The European Commission has established a European Environmental Agency (with weaker powers of enforcement than the US EPA) to collect, analyse and distribute information; ensure that monitoring data are comparable from country to country in Europe; promote forecasting; improve assessment, counter-measures against and prevention of environmental problems.

In some situations agroecological zoning could be a useful aid. This depends upon satellite imagery and storage of the collected data in geographical information systems (GIS). the GIS are used to try and match landuse to ecology of local area units and these can then be monitored to check progress, and if need be to penalize those who misuse resources.

There are signs that some NGOs have already evolved into globally influential bodies, for example Greenpeace. Another possibility is that groups of NGOs will network to address global issues.

References

Boyden, S. (1987) *Western civilisation in biological perspective*. Oxford, Clarendon Press

Grubb, M., Koch, M., Thomson, K., Munson, A. and Sullivan, F. (1993) *The Earth Summit agreements: a guide and assessment*. London, Earthscan

Haas, P.M., Levy, M.A. and Parson, E.A. (1992) Appraising the Earth Summit: how should we judge success? *Environment* 34(8): 13–15; 34–6

Holmberg, J., Thomson, K. and Timberlake, L. (1993) *Facing the future: beyond the Earth Summit*. London, Earthscan

Keating, M. (1993) *The Earth Summit's agenda for change: a plain language version of Agenda 21*. Geneva, Centre for Our Common Future, Palais Wilson, 52 Rue des Pâquis, CH-1201

Kennedy, P.M. (1993) *Preparing for the twenty-first century*. London, Harper Collins (Fontana edn UK 1994)

Pirages, D. (1978) *Global ecopolitics: the new context for international relations*. North Scituate, Mass., Duxbury

Smith, L.G. (1993) *Impact assessment and sustainable resource management*. Harlow, Longman

Trudgill, S. (1990) *Barriers to a better environment: what stops us solving environmental problems?* London, Belhaven

Westcoat, J.L. jnr (1993) Resource management: UNCED, GATT, and global change. *Progress in Human Geography*. 17(2): 232–40

Further reading

Environment 34(8) October 1992 issue [coverage of UNCED Rio]

Harrison, P. (1993) *The third revolution: population, environment and a sustainable world.* Harmondsworth, Penguin [Readable, constructive treatment of global problems which asks what can realistically be done]

The Ecologist (1993) *Whose common future? Reclaiming the commons*. London, Earthscan [Reviews what was achieved at UNCED and discusses what is seen as a crucial issue in environment and development – world-wide enclosure of commons]

Index